Advances in Temporal Logic

APPLIED LOGIC SERIES

VOLUME 16

Managing Editor

Dov M. Gabbay, *Department of Computer Science, King's College, Londen, U.K.*

Co-Editor

John Barwise, *Department of Philosophy, Indiana University, Bloomington, IN, U.S.A.*

Editorial Assistant

Jane Spurr, *Department of Computer Science, King's College, London, U.K.*

SCOPE OF THE SERIES

Logic is applied in an increasingly wide variety of disciplines, from the traditional subjects of philosophy and mathematics to the more recent disciplines of cognitive science, computer science, artificial intelligence, and linguistics, leading to new vigor in this ancient subject. Kluwer, through its Applied Logic Series, seeks to provide a home for outstanding books and research monographs in applied logic, and in doing so demonstrates the underlying unity and applicability of logic.

Advances in Temporal Logic

edited by

HOWARD BARRINGER
University of Manchester, United Kingdom

MICHAEL FISHER
Manchester Metropolitan University, United Kingdom

DOV GABBAY
King's College London, United Kingdom

and

GRAHAM GOUGH
University of Manchester, United Kingdom

KLUWER ACADEMIC PUBLISHERS
DORDRECHT / BOSTON / LONDON

A C.I.P. Catalogue record for this book is available from the Library of Congress.

ISBN 978-90-481-5389-3

Published by Kluwer Academic Publishers,
P.O. Box 17, 3300 AA Dordrecht, The Netherlands.

Sold and distributed in North, Central and South America
by Kluwer Academic Publishers,
101 Philip Drive, Norwell, MA 02061, U.S.A.

In all other countries, sold and distributed
by Kluwer Academic Publishers,
P.O. Box 322, 3300 AH Dordrecht, The Netherlands.

Printed on acid-free paper

Contents

Preface

Time is a fascinating subject and has long since captured mankind's imagination, from the ancients to modern man, both adult and child alike. It has been studied across a wide range of disciplines, from the natural sciences to philosophy and logic. Today, thirty plus years since Prior's work in laying out foundations for temporal logic, and two decades on from Pnueli's seminal work applying of temporal logic in specification and verification of computer programs, temporal logic has a strong and thriving international research community within the broad disciplines of computer science and artificial intelligence. Areas of activity include, but are certainly not restricted to:

Pure Temporal Logic, e.g. temporal systems, proof theory, model theory, expressiveness and complexity issues, algebraic properties, application of game theory;

Specification and Verification, e.g. of reactive systems, of real-time components, of user interaction, of hardware systems, techniques and tools for verification, execution and prototyping methods;

Temporal Databases, e.g. temporal representation, temporal querying, granularity of time, update mechanisms, active temporal databases, hypothetical reasoning;

Temporal Aspects in AI, e.g. modelling temporal phenomena, interval temporal calculi, temporal nonmonotonicity, interaction of temporal reasoning with action/knowledge/belief logics, temporal planning;

Tense and Aspect in Natural Language, e.g. models, ontologies, temporal quantifiers, connectives, prepositions, processing temporal statements;

Temporal Theorem Proving, e.g. translation methods, clausal and non-clausal resolution, tableaux, automata-theoretic approaches, tools and practical systems.

This volume presents a selection of papers originally presented at an international conference on temporal logic held in Manchester (ICTL97). It was the second such international conference dedicated to temporal logic (the first being in Bonn in 1994), both of which aimed to create bridges between the various communities working in temporal logic.[1] The meeting started with a day of superb tutorial presentations (not included in this volume):

[1] A third conference is being planned for 2000.

Mark Steedman: Temporality in Natural Language

Peter Ladkin: Temporal logic of Actions

Facheim Bacchus: Temporal logic and Planning

Jan Chomicki & David Toman: Temporal Databases

Then, over and above the technical sessions detailed in this volume, the conference attendees were delighted to thought-provoking keynote addresses from two of the community's leading researchers, **Amir Pnueli** and **Allen Emerson**. And then, for lighter relief before the conference dinner, the audience was fascinated by a special lecture from Chris Burton on "The Rebuild of the Baby Machine (Manchester Mark 1)", the remarkable story about the construction of an authentic (fully operational) replica of the world's first stored-program computer, originally operational on June 21st, 1948.

The editors are most grateful to the programme committee members for their efforts and thought in organising the scientific programme, to the keynote speakers and tutorial presenters. Thanks are also due to Prentice-Hall, John Wiley, Kluwer and Research Studies Press for supporting the conference. Finally, special thanks must be given to our local colleagues and secretaries, from Hulme Hall and the Department of Computer Science at the University of Manchester, especially Bill Mitchell, Lynn Howarth and Pat Rarity who ensured the whole meeting ran without a hitch.

Howard Barringer, Michael Fisher, Dov Gabbay and Graham Gough

Programme Committee

Howard Barringer	(University of Manchester, UK)
Gerd Brewka	(GMD Bonn, Germany)
Jan Chomicki	(Monmouth University, USA)
Allen Emerson	(Austin, Texas)
Michael Fisher	(MMU, UK)
Nissim Francez	(Technion, Israel)
Dov Gabbay	(Imperial College, London)
Joe Halpern	(Cornell University, USA)
Hans Kamp	(IMS, Stuttgart, Germany)
Peter Ladkin	(Bielefeld, Germany)
Angelo Montanari	(Udine, Italy)
Istvan Nemeti	(Math Institute, Hungary)
Hans Jürgen Ohlbach	(Max-Planck-Institut, Saarbruecken, Germany)
Amir Pnueli	(Weizmann Institute, Israel)
Wojciech Penczek	(University of Warsaw, Poland)
Antonio Porto	(Univ Nova de Lisboa, Portugal)
Mark Reynolds	(King's College London, UK)
Willem Paul de Roever	(Kiel University, Germany)
Eric Sandewall	(Linkoeping University, Sweden)
Andrzej Szalas	(University of Warsaw, Poland)
Yde Venema	(Amsterdam, The Netherlands)

Additional Reviewers

Claudio Bettini	Mirna Bognar	Anthony Bonner
Carlo Combi	Clare Dixon	Shmuel Katz
Beata Konikowska	Marta Kwiatkowska	Yassine Lakhnech
Peter McBrien	Szabolcs Mikulas	Bill Mitchell
Ian Pratt	Michael Siegel	Jerzy Skurczynski
David Toman		

Proceedings Editors

Howard Barringer, Michael Fisher, Dov Gabbay and Graham Gough.

A Hierarchy of Modal Event Calculi: Expressiveness and Complexity

Iliano Cervesato (iliano@cs.cmu.edu)
Department of Computer Science
Carnegie Mellon University
5000 Forbes Avenue - Pittsburgh, PA 15213-3891
Phone: (412)-268-3069 - Fax: (412)-268-5576

Massimo Franceschet (tfranceschet@uniud.it) and
Angelo Montanari (montana@dimi.uniud.it)
Dipartimento di Matematica e Informatica
Università di Udine
Via delle Scienze, 206 - 33100 Udine, Italy
Phone: +39-432-558477 - Fax: +39-432-558499

Abstract. We consider a hierarchy of modal event calculi to represent and reason about partially ordered events. These calculi are based on the model of time and change of Kowalski and Sergot's Event Calculus (*EC*): given a set of event occurrences, *EC* allows the derivation of the maximal validity intervals (MVIs) over which properties initiated or terminated by those events hold. The formalisms we analyze extend *EC* with operators from modal logic. They range from the basic Modal Event Calculus (*MEC*), that computes the set of all current MVIs (MVIs computed by *EC*) as well as the sets of MVIs that are true in some/every refinement of the current partial ordering of events ($\Diamond$-/$\Box$-MVIs), to the Generalized Modal Event Calculus (*GMEC*), that extends *MEC* by allowing a free mix of boolean connectives and modal operators. We analyze and compare the expressive power and the complexity of the proposed calculi, focusing on intermediate systems between *MEC* and *GMEC*. We motivate the discussion by using a fault diagnosis problem as a case study.

1. Introduction

The *Event Calculus*, abbreviated *EC* (Kowalski and Sergot, 1986), is a simple temporal formalism designed to model situations characterized by a set of *events*, whose occurrences have the effect of starting or terminating the validity of determined *properties*. Given a possibly incomplete description of when these events take place and of the properties they affect, *EC* is able to determine the *maximal validity intervals*, or *MVIs*, over which a property holds uninterruptedly. The algorithm *EC* relies on for the verification or calculation of MVIs is polynomial (Cervesato et al., 1995b). It can advantageously be implemented as a logic program. Indeed, the primitive operations of logic programming languages can be exploited to express boolean combinations of MVI computations and limited forms of quantification.

H. Barringer et al. (eds.), Advances in Temporal Logic, 1–20.

The range of the queries that can be expressed in *EC* is, however, too limited for modeling realistic situations, even when permitting boolean connectives. Expressiveness can be improved either by extending the representation capabilities of *EC* to encompass a wider spectrum of situations (e.g. by permitting precondition-triggered events), or by enriching the query language of the formalism. The first alternative is discussed in (Cervesato et al., 1997); in this paper, we explore extensions to the query language relatively to a specific subclass of *EC* problems.

We limit our investigation to problems consisting of a fixed set of events that are known to have happened, but with incomplete information about the relative order of their occurrences (Cervesato et al., 1995a; Cervesato et al., 1996; Cervesato et al., 1993; Dean and Boddy, 1988; Franceschet, 1996; D.C. Moffat, 1990). In these situations, the MVIs derived by *EC* bear little relevance since the acquisition of additional knowledge about the actual event ordering might both dismiss current MVIs and validate new MVIs (Cervesato et al., 1993). It is instead critical to compute precise variability bounds for the MVIs of the (currently underspecified) actual ordering of events. Optimal bounds have been identified in the set of *necessary MVIs*, or *□-MVIs*, and the set of *possible MVIs*, or *◇-MVIs*. They are the subset of the current MVIs that are not invalidated by the acquisition of new ordering information and the set of intervals that are MVIs in at least one completion of the current ordering of events, respectively.

The *Modal Event Calculus*, *MEC* (Cervesato et al., 1995a), extends *EC* with the possibility of inquiring about □-MVIs and ◇-MVIs. The enhanced capabilities of *MEC* do not raise the polynomial complexity of *EC*, but they are still insufficient to model effectively significant situations. This limitation has been overcome with the *Generalized Modal Event Calculus*, *GMEC* (Cervesato et al., 1996). This formalism reduces the computation of □-MVIs and ◇-MVIs to the derivation of basic MVIs, mediated by the resolution of the operators □ and ◇ from the modal logic *K1.1*, a refinement of *S4* (Segerberg, 1971). The query language of *GMEC* permits a free mix of boolean connectives and modal operators, recovering the possibility of expressing a large number of common situations, but at the price of making the evaluation of *GMEC* queries an NP-hard problem (as we will show in Section 5). In this paper, we refine the taxonomy of the modal event calculi by considering two intermediate formalisms between *MEC* and *GMEC*, as shown in Figure 1. The queries of the *Modal Event Calculus with External Connectives*, *ECMEC*, allow combining computations of MVIs, □-MVIs and ◇-MVIs by means of boolean connectives. The approach followed in *ICMEC*, the *Modal Event Calculus with Internal Connectives*, is dual: boolean combinations of MVI computations can be pre-

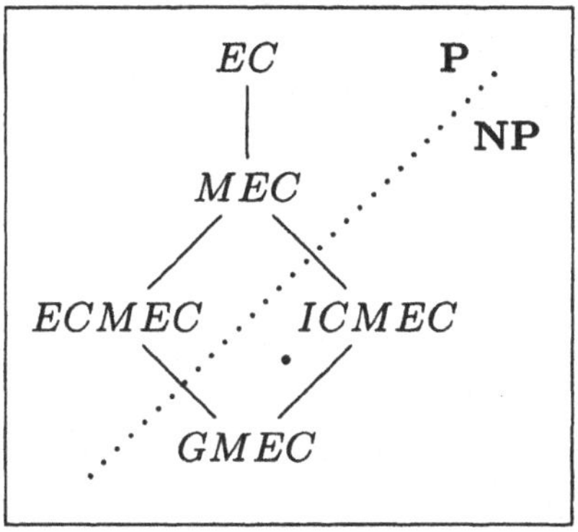

Figure 1. A Hierarchy of Event Calculi

fixed by either □ or ◇. Both *ECMEC* and *ICMEC* retain enough of the expressive power of *GMEC* to allow a faithful and usable representation of numerous common situations. However, while the problem of evaluating an *ICMEC* query is still NP-hard, *ECMEC* admits polynomial implementations, making this formalism an ideal candidate for the formalization of a number of applicative domains.

The main contributions of this paper lie in the investigation of intermediate modal event calculi between *MEC* and *GMEC*, and in the individuation of *ECMEC* as an expressive but tractable sublanguage of the latter. The paper provides also a formal analysis of the complexity of various modal event calculi, notably *MEC* and *GMEC*. We invite the interested reader to consult (Franceschet, 1996) for a more detailed discussion of the topics treated in this paper, and for the proofs of the statements we mention. In order to make the discussion more concrete, we interleave the presentation of our various modal event calculi with the formalization of an applicative example. The paper is organized as follows. We give an informal description of our case study in Section 2. Section 3 defines the modal extensions of *EC* we examine and discusses their main properties. In Section 4, we come back to our case study and provide a formalization. Section 5 gives a complexity analysis for the calculi considered in this paper. Finally, Section 6 summarizes the main contributions of the paper and outlines directions of future work.

2. The Application Domain: an Informal Description

In this section, we introduce a real-world case study taken from the domain of fault diagnosis. In Section 4, we will compare the formalizations obtained by encoding it in *GMEC*, *ICMEC* and *ECMEC*.

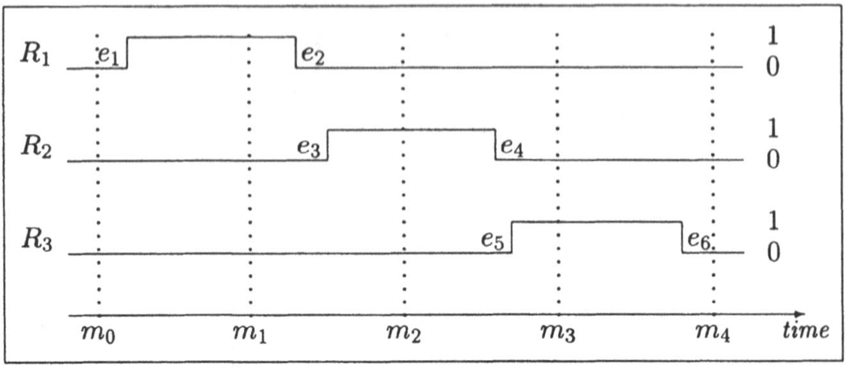

Figure 2. Expected Register Behavior and Measurements

We focus our attention on the representation and information processing of fault symptoms that are spread over periods of time and for which current expert system technology is particularly deficient (Nökel, 1991). Consider the following example, which diagnoses a fault in a computerized numerical control center for a production chain.

> *A possible cause for an undefined position of the tool magazine is a faulty limit switch S. This cause can however be ruled out if the status registers R_1, R_2 and R_3 show the following behavior: from a situation in which all three registers contain the value* 0*, they all assume the value* 1 *in successive and disjoint time intervals (first R_1, then R_2, and last R_3), and then return to* 0.

Figure 2 describes a possible sequence of transitions for R_1, R_2 and R_3, that excludes the eventuality of S being faulty. In order to verify this behavior, the contents of the registers must be monitored over time. Typically, each value (0 or 1) of a register persists for at least t time units. Measurements are made at fixed intervals (sampling periods), asynchronously with the change of value of the status registers. In order to avoid losing register transitions, measurements must be made frequently enough, that is, the sampling period must be less than t. However, it is not possible to avoid in general that transitions of *different* registers take place between two consecutive measurements, making it impossible to recover their relative order.

This situation is depicted, in the case of our example, in Figure 2, where dotted lines indicate measurements. Moreover, we have given names to the individual transitions of state of the different registers. In this specific situation, the values found at measurements m_0 and m_1 allow us to determine that R_1 has been set during this interval (transition e_1). The contents of the registers at measurement m_2 let us

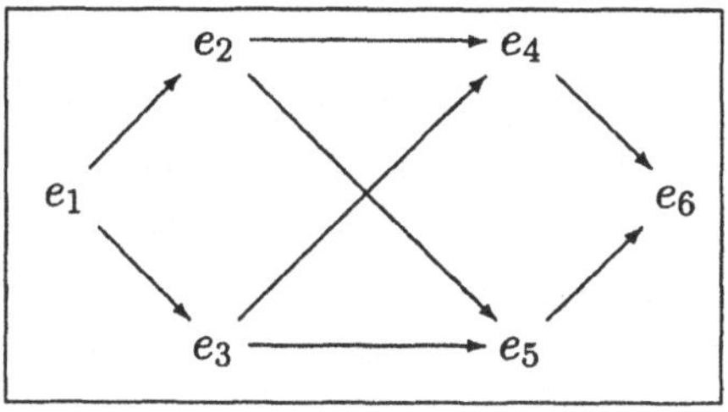

Figure 3. Ordering of the Events

infer that R_1 has been reset (transition e_2) and that the value of R_2 has changed to 1 (transition e_3). We know that both e_2 and e_3 have taken place after e_1, but we have no information about the relative order of these transitions. Similarly, m_3 allows us to infer that, successively, R_2 has been reset (e_4) and R_3 has been set (e_5), but cannot decide which among e_4 and e_5 has taken place first. Finally, m_4 acknowledges that R_3 has successively been reset to 0 (e_6). The available information about the ordering of transitions is summarized in Figure 3.

As we will see in Section 4, this example can be easily formalized in *EC*. We will however need the expressive power of a modal version of this formalism to draw conclusions about the possibility that the switch S is faulty.

3. A Hierarchy of Modal Event Calculi

In this section, we recall the syntax and semantics of the modal event calculi *MEC* (Cervesato et al., 1995a) and *GMEC* (Cervesato et al., 1996) and adapt it to define the intermediate calculi *ECMEC* and *ICMEC*. We present some relevant properties of these formalisms. These systems form the linguistic hierarchy shown in Figure 1. Implementations in the language of hereditary Harrop formulas have been given in (Franceschet, 1996), together with formal proofs of soundness and completeness with respect to the specifications below. Space limitations do not allow us to further discuss this aspect.

The *Event Calculus* (*EC*) (Kowalski and Sergot, 1986) and the modal variants we consider aim at modeling situations that consist of a set of events, whose occurrences over time have the effect of initiating or terminating the validity of properties, some of which may be mutually exclusive. We formalize the time-independent aspects of a situation by means of an *EC-structure*, defined as follows.

DEFINITION 3.1. (*EC-structure*)

A structure *for the* Event Calculus *(abbreviated* EC-structure*) is a quintuple* $\mathcal{H} = (E,\ P,\ [\cdot\rangle,\ \langle\cdot],\ ]\cdot,\cdot[)$ *such that:*

- $E = \{e_1, \ldots, e_n\}$ *and* $P = \{p_1, \ldots, p_m\}$ *are finite sets of* events *and* properties, *respectively.*
- $[\cdot\rangle : P \to \mathbf{2}^E$ *and* $\langle\cdot] : P \to \mathbf{2}^E$ *are respectively the* initiating *and* terminating map *of* $\mathcal{H}$. *For every property* $p \in P$, $[p\rangle$ *and* $\langle p]$ *represent the set of events that initiate and terminate p, respectively.*
- $]\cdot,\cdot[\subseteq P \times P$ *is an irreflexive and symmetric relation, called the* exclusivity relation, *that models exclusivity among properties.* □

Unlike the original presentation of *EC* (Kowalski and Sergot, 1986), we focus our attention on situations where the occurrence time of events is unknown. Indeed, we only assume the availability of incomplete information about the relative order in which these events have happened. Therefore, we formalize the time-dependent aspects of an *EC* problem by providing a (strict) *partial order* for the involved event occurrences. We write $W_{\mathcal{H}}$ for the set of all partial orders on the set of events E of an *EC*-structure $\mathcal{H}$ and use the letter w to denote individual orderings, or *knowledge states*, in $W_{\mathcal{H}}$ (that is, a knowledge state w is an irreflexive and transitive subset of $E \times E$). Given $w \in W_{\mathcal{H}}$, we will sometimes call a pair of events $(e_1, e_2) \in w$ an *interval*. For reasons of efficiency, implementations generally represent the current situation w as a binary *acyclic* relation o, from which w can be recovered as the transitive closure o^+ of o (Cervesato et al., 1996; Cervesato et al., 1995b). In the following, we will often work with *extensions* of an ordering w, defined as any element of $W_{\mathcal{H}}$ that contains w as a subset. We define a *completion* of w as any extension of w that is a total order. We denote with $\mathrm{Ext}_{\mathcal{H}}(w)$ and $\mathrm{Comp}_{\mathcal{H}}(w)$ the set of all extensions and the set of all completions of the ordering w in $W_{\mathcal{H}}$, respectively. We will drop the subscript $_{\mathcal{H}}$ when clear from the context.

Given a structure $\mathcal{H}$ and a knowledge state w, *EC* offers a means to infer the *maximal validity intervals*, or *MVIs*, over which a property p holds uninterruptedly. We represent an MVI for p as $p(e_i, e_t)$, where e_i and e_t are the events that initiate and terminate the interval over which p maximally holds, respectively. Consequently, we adopt as the *query language* of *EC* the set $\mathcal{A}_{\mathcal{H}} = \{p(e_1, e_2) : p \in P \text{ and } e_1, e_2 \in E\}$ of all such property-labeled intervals over $\mathcal{H}$. We interpret the elements of $\mathcal{A}_{\mathcal{H}}$ as propositional letters and the task performed by *EC* reduces to deciding which of these formulas are MVIs and which are not, with respect to the current partial order of events. This is a problem of model checking.

In order for $p(e_1, e_2)$ to be an MVI relatively to the knowledge state w, (e_1, e_2) must be an interval in w, i.e. $(e_1, e_2) \in w$. Moreover, e_1 and e_2 must witness the validity of the property p at the ends of this interval by initiating and terminating p, respectively. These requirements are enforced by conditions (i), (ii) and (iii), respectively, in the definition of valuation given below. The maximality requirement is caught by the meta-predicate $nb(p, e_1, e_2, w)$ in condition (iv), which expresses the fact that the validity of an MVI must not be *broken* by any interrupting event. Any event e which is known to have happened between e_1 and e_2 in w and that initiates or terminates a property that is either p itself or a property exclusive with p interrupts the validity of $p(e_1, e_2)$. These observations are formalized as follows.

DEFINITION 3.2. (*EC intended model*)

Let $\mathcal{H} = (E,\ P,\ [\cdot\rangle,\ \langle\cdot],\ ]\cdot,\cdot[)$ *be a EC-structure. The* intended EC-model *of* $\mathcal{H}$ *is the propositional valuation* $\upsilon_{\mathcal{H}} : W_{\mathcal{H}} \rightarrow 2^{\mathcal{A}_{\mathcal{H}}}$, *where* $\upsilon_{\mathcal{H}}$ *is defined in such a way that* $p(e_1, e_2) \in \upsilon_{\mathcal{H}}(w)$ *if and only if*

i. $(e_1, e_2) \in w$;

ii. $e_1 \in [p\rangle$;

iii. $e_2 \in \langle p]$;

iv. $nb(p, e_1, e_2, w)$, *where* $nb(p, e_1, e_2, w)$ *iff*

$$\begin{array}{rl} \neg\exists e \in E. & (e_1, e) \in w \wedge (e, e_2) \in w \\ \wedge & \exists q \in P. ((e \in [q\rangle \vee e \in \langle q]) \wedge (]p, q[\vee p = q))). \end{array} \qquad \Box$$

The set of MVIs of an *EC* problem $(\mathcal{H}, w)$ is not stable with respect to the acquisition of new ordering information. Indeed, as we move to an extension of w, current MVIs might become invalid and new MVIs can emerge (Cervesato et al., 1993). The *Modal Event Calculus*, or *MEC* (Cervesato et al., 1995a), extends the language of *EC* with the possibility of enquiring about which MVIs will remain valid in every extension of the current knowledge state, and about which intervals might become MVIs in some extension of it. We call intervals of these two types *necessary MVIs* and *possible MVIs*, respectively, using $\Box$-*MVIs* and $\Diamond$-*MVIs* as abbreviations. Formally, the query language $\mathcal{B}_{\mathcal{H}}$ of *MEC* consists of formulas of the form $p(e_1, e_2)$, $\Box p(e_1, e_2)$ and $\Diamond p(e_1, e_2)$, for every property p and events e_1 and e_2 defined in $\mathcal{H}$. We intend representing in this way the property-labeled interval $p(e_1, e_2)$ as an MVI, a $\Box$-MVI and a $\Diamond$-MVI, respectively. Clearly, $\mathcal{A}_{\mathcal{H}} \subseteq \mathcal{B}_{\mathcal{H}}$.

In order to provide *MEC* with a semantics, we must shift the focus from the current knowledge state w to all knowledge states that are reachable from w, i.e. $\text{Ext}_{\mathcal{H}}(w)$, and more generally to $W_{\mathcal{H}}$. Now, by definition, w' is an extension of w if $w \subseteq w'$. Since $\subseteq$ is a reflexive partial order, $(W_{\mathcal{H}}, \subseteq)$ can be naturally viewed as a finite, reflexive, transitive and antisymmetric modal frame. If we consider this frame together with the straightforward modal extension of the valuation $\upsilon_{\mathcal{H}}$ to an arbitrary knowledge state, we obtain a modal model for *MEC*.

DEFINITION 3.3. (*MEC intended model*)

Let $\mathcal{H} = (E,\ P,\ [\cdot\rangle,\ \langle\cdot],\ ]\cdot,\cdot[)$ *be an EC-structure. The* MEC-frame $\mathcal{F}_{\mathcal{H}}$ *of* $\mathcal{H}$ *is the frame* $(W_{\mathcal{H}}, \subseteq)$. *The* intended MEC-model *of* $\mathcal{H}$ *is the modal model* $\mathcal{I}_{\mathcal{H}} = (W_{\mathcal{H}}, \subseteq, \upsilon_{\mathcal{H}})$, *where the propositional valuation* $\upsilon_{\mathcal{H}} : W_{\mathcal{H}} \rightarrow 2^{\mathcal{A}_{\mathcal{H}}}$ *is defined as in Definition 3.2. Given* $w \in W_{\mathcal{H}}$ *and* $\varphi \in \mathcal{B}_{\mathcal{H}}$, *the truth of* φ *at* w *with respect to* $\mathcal{I}_{\mathcal{H}}$, *denoted by* $\mathcal{I}_{\mathcal{H}}; w \models \varphi$, *is defined as follows:*

$$\begin{array}{lll} \mathcal{I}_{\mathcal{H}}; w \models p(e_1,e_2) & \textit{iff} & p(e_1,e_2) \in \upsilon_{\mathcal{H}}(w) \ \textit{ as from Definition 3.2;} \\ \mathcal{I}_{\mathcal{H}}; w \models \Box p(e_1,e_2) & \textit{iff} & \forall w' \in W_{\mathcal{H}} \textit{ s.t. } w \subseteq w',\ \mathcal{I}_{\mathcal{H}}; w' \models p(e_1,e_2); \\ \mathcal{I}_{\mathcal{H}}; w \models \Diamond p(e_1,e_2) & \textit{iff} & \exists w' \in W_{\mathcal{H}} \textit{ s.t. } w \subseteq w' \textit{ and } \mathcal{I}_{\mathcal{H}}; w' \models p(e_1,e_2). \end{array}$$

A MEC-formula φ *is* valid *in* $\mathcal{I}_{\mathcal{H}}$, *written* $\mathcal{I}_{\mathcal{H}} \models \varphi$, *if* $\mathcal{I}_{\mathcal{H}}; w \models \varphi$ *for all* $w \in W_{\mathcal{H}}$. □

We will drop the subscripts $_{\mathcal{H}}$ whenever this does not lead to ambiguities. Moreover, given a knowledge state w in $W_{\mathcal{H}}$ and a *MEC*-formula φ over $\mathcal{H}$, we write $w \models \varphi$ for $\mathcal{I}_{\mathcal{H}}; w \models \varphi$. Similarly, we abbreviate $\mathcal{I}_{\mathcal{H}} \models \varphi$ as $\models \varphi$.

It is interesting to notice that it would have been equivalent to consider completions instead of extensions in the previous definition. Instead, the two notions are not interchangeable in general in the refinements of *MEC* discussed in the remainder of this section, as shown in (Cervesato et al., 1997; Franceschet, 1996).

To determine the sets of □- and ◇-MVIs, it is possible to exploit necessary and sufficient *local conditions* over the given partial order, thus avoiding a complete (and expensive) search of all the consistent extensions of the given order. More precisely (Cervesato et al., 1996), a property p necessarily holds between two events e_1 and e_2 if and only if the interval (e_1, e_2) belongs to the current order, e_1 initiates p, e_2 terminates p, and no event that either initiates or terminates p (or a property incompatible with p) will ever be consistently located between e_1 and e_2. Similarly, a property p may possibly hold between e_1 and e_2 if and only if the interval (e_1, e_2) is consistent with the current ordering, e_1 initiates p, e_2 terminates p, and there are no already known interrupting events between e_1 and e_2. This is precisely expressed by the following lemma (Cervesato et al., 1996).

LEMMA 3.4. (*Local conditions*)

Let $\mathcal{H} = (E,\ P,\ [\cdot\rangle,\ \langle\cdot],\ ]\cdot,\cdot[)$ *be a EC-structure. For any pair of events* $e_1, e_2 \in E$*, any property* $p \in P$*, and any* $w \in W_{\mathcal{H}}$*,*

– $\mathcal{I}_{\mathcal{H}}; w \models \Box p(e_1, e_2)$ *if and only if*

i. $(e_1, e_2) \in w$;

ii. $e_1 \in [p\rangle$;

iii. $e_2 \in \langle p]$;

iv. $nsb(p, e_1, e_2, w)$, *where* $nsb(p, e_1, e_2, w)$ *iff*

$$\begin{array}{lll} \neg\exists e \in E. & (e, e_1) \notin w & \wedge \quad e \neq e_1 \\ \wedge & (e_2, e) \notin w & \wedge \quad e \neq e_2 \\ \wedge & \exists q \in P. ((e \in [q\rangle \vee e \in \langle q]) \wedge (]p, q[\vee p = q))). & \end{array}$$

– $\mathcal{I}_{\mathcal{H}}; w \models \Diamond p(e_1, e_2)$ *if and only if*

i. $(e_2, e_1) \notin w$;

ii. $e_1 \in [p\rangle$;

iii. $e_2 \in \langle p]$;

iv. $nb(p, e_1, e_2, w)$. □

The *Generalized Modal Event Calculus, GMEC* (Cervesato et al., 1996), broadens the scope of *MEC* by interpreting a necessary MVI $\Box p(e_1, e_2)$ and a possible MVI $\Diamond p(e_1, e_2)$ as the application of the operators □ and ◇, respectively, from an appropriate modal logic to the MVI $p(e_1, e_2)$. On the basis of this observation, it extends the language of *MEC* by allowing the combination of property-labeled intervals by means of propositional connectives and modal operators. The query language of *GMEC* is defined as follows.

DEFINITION 3.5. (*GMEC-language*)

Let $\mathcal{H} = (E,\ P,\ [\cdot\rangle,\ \langle\cdot],\ ]\cdot,\cdot[)$ *be an EC-structure. Given the EC-language* $\mathcal{A}_{\mathcal{H}}$*, the* GMEC-language *of* $\mathcal{H}$*, denoted* $\mathcal{L}_{\mathcal{H}}$*, is the modal language with propositional letters in* $\mathcal{A}_{\mathcal{H}}$ *and logical operators in* $\{\neg, \wedge, \vee, \Box, \Diamond\}$. □

Clearly, $\mathcal{B}_{\mathcal{H}} \subseteq \mathcal{L}_{\mathcal{H}}$. Definition 3.3 can be easily generalized to GMEC as follows.

DEFINITION 3.6. (*GMEC intended model*)

Let $\mathcal{H} = (E,\ P,\ [\cdot\rangle,\ \langle\cdot],\ ]\cdot,\cdot[)$ *be an EC-structure. The* GMEC-frame $\mathcal{F}_{\mathcal{H}}$ *and the* intended GMEC-model $\mathcal{I}_{\mathcal{H}}$ *of* $\mathcal{H}$ *are defined as in*

Definition 3.3. Given $w \in W_{\mathcal{H}}$ *and* $\varphi \in \mathcal{L}_{\mathcal{H}}$*, the truth of* φ *at* w *with respect to* $\mathcal{I}_{\mathcal{H}}$*, denoted by* $\mathcal{I}_{\mathcal{H}}; w \models \varphi$*, is defined as follows:*

$$
\begin{array}{lll}
\mathcal{I}_{\mathcal{H}}; w \models p(e_1, e_2) & \textit{iff} & p(e_1, e_2) \in \upsilon_{\mathcal{H}}(w) \ \textit{as in Definition 3.2;} \\
\mathcal{I}_{\mathcal{H}}; w \models \neg\varphi & \textit{iff} & \mathcal{I}_{\mathcal{H}}; w \not\models \varphi; \\
\mathcal{I}_{\mathcal{H}}; w \models \varphi_1 \wedge \varphi_2 & \textit{iff} & \mathcal{I}_{\mathcal{H}}; w \models \varphi_1 \ \textit{and} \ \mathcal{I}_{\mathcal{H}}; w \models \varphi_2; \\
\mathcal{I}_{\mathcal{H}}; w \models \varphi_1 \vee \varphi_2 & \textit{iff} & \mathcal{I}_{\mathcal{H}}; w \models \varphi_1 \ \textit{or} \ \mathcal{I}_{\mathcal{H}}; w \models \varphi_2; \\
\mathcal{I}_{\mathcal{H}}; w \models \Box\varphi & \textit{iff} & \forall w' \in W_{\mathcal{H}} \ \textit{s.t.} \ w \subseteq w', \mathcal{I}_{\mathcal{H}}; w' \models \varphi; \\
\mathcal{I}_{\mathcal{H}}; w \models \Diamond\varphi & \textit{iff} & \exists w' \in W_{\mathcal{H}} \ \textit{s.t.} \ w \subseteq w' \ \textit{and} \ \mathcal{I}_{\mathcal{H}}; w' \models \varphi. \ \Box
\end{array}
$$

The attempt of characterizing *GMEC* within the rich taxonomy of modal logics reveals *Sobocinski logic*, also known as *system K1.1* (Segerberg, 1971), as its closest relative. Syntactically, this logic extends *S4* for the validity of the formula $\Box(\Box(p \rightarrow \Box p) \rightarrow p) \rightarrow p$, added as a further axiom to the traditional formulation of that system. Semantically, it is characterized by the class of the finite, reflexive, transitive and antisymmetric frames, i.e. by the class of all finite partial orders. The relationship between *GMEC* and *K1.1* is captured by the following theorem, where the validity relation of Sobocinski logic has been indicated as $\models_{K1.1}$.

THEOREM 3.7. (*GMEC and K1.1*)

Each thesis of K1.1 is a valid formula of GMEC, i.e., for each GMEC-formula φ*, if* $\models_{K1.1} \varphi$*, then* $\models \varphi$*.* ■

Since the intended *GMEC*-model $\mathcal{I}_{\mathcal{H}}$ is based on a finite, reflexive, transitive and antisymmetric frame, Theorem 3.7 immediately follows from the (soundness and) completeness of *K1.1* with respect to the class of all finite partial orders (Segerberg, 1971). From the above syntactic characterization of Sobocinski logic, every formula valid in *S4* is valid in *K1.1*. Therefore, Theorem 3.7 permits lifting to *GMEC* the following well-known equivalences of *S4*.

COROLLARY 3.8. (*Some equivalent GMEC-formulas*)

Let φ*,* φ_1 *and* φ_2 *be GMEC-formulas. Then, for every knowledge state* $w \in W$*,*

- $w \models \Box\neg\varphi$ *iff* $w \models \neg\Diamond\varphi$
- $w \models \Diamond\neg\varphi$ *iff* $w \models \neg\Box\varphi$
- $w \models \Box(\varphi_1 \wedge \varphi_2)$ *iff* $w \models \Box\varphi_1 \wedge \Box\varphi_2$
- $w \models \Diamond(\varphi_1 \vee \varphi_2)$ *iff* $w \models \Diamond\varphi_1 \vee \Diamond\varphi_2$
- $w \models \Box\Box\varphi$ *iff* $w \models \Box\varphi$
- $w \models \Diamond\Diamond\varphi$ *iff* $w \models \Diamond\varphi$
- $w \models \Box\Diamond\Box\Diamond\varphi$ *iff* $w \models \Box\Diamond\varphi$
- $w \models \Diamond\Box\Diamond\Box\varphi$ *iff* $w \models \Diamond\Box\varphi$ ■

Also specific properties of *K1.1* turn out to be useful in order to implement *GMEC*. The following equivalences can be obtained by exploiting the *McKinsey formula*, $\Box\Diamond\varphi \rightarrow \Diamond\Box\varphi$, valid in *K1.1* (but not in *S4*).

COROLLARY 3.9. (*Further equivalent GMEC-formulas*)

Let φ be a GMEC-formula. Then, for every knowledge state $w \in W$,

- $w \models \Box\Diamond\Box\varphi$ *iff* $w \models \Box\Diamond\varphi$
- $w \models \Diamond\Box\Diamond\varphi$ *iff* $w \models \Diamond\Box\varphi$ ■

An interesting consequence of Corollaries 3.8 and 3.9 is that each *GMEC*-formula φ is logically equivalent to a formula of one of the following forms: ψ, $\Box\psi$, $\Diamond\psi$, $\Box\Diamond\psi$, $\Diamond\Box\psi$, where the main operator of ψ is non-modal. In (Cervesato et al., 1996), we provided *GMEC* with a sound and complete axiomatization in the language of hereditary Harrop formulas that heavily exploits the above reductions. Unfortunately, there is no way, in general, of reducing formulas of the form $\Box(\varphi_1 \vee \varphi_2)$ and $\Diamond(\varphi_1 \wedge \varphi_2)$ (such a reduction would be quite significant from a computational point of view): we only have that $(\Box\varphi_1 \vee \Box\varphi_2) \rightarrow \Box(\varphi_1 \vee \varphi_2)$ and $\Diamond(\varphi_1 \wedge \varphi_2) \rightarrow (\Diamond\varphi_1 \wedge \Diamond\varphi_2)$. Furthermore, the attempt at overcoming these difficulties by adding to *K.1.1* the axiom $\Box(p \vee q) \rightarrow (\Box p \vee \Box q)$, distributing the $\Box$ operator over $\vee$, or, equivalently, the axiom $(\Diamond p \wedge \Diamond q) \rightarrow \Diamond(p \wedge q)$, has the effect of collapsing *K.1.1* onto the Propositional Calculus as stated by the following (general) theorem (Franceschet, 1996).

THEOREM 3.10. (*Collapse of Modal Logics onto the Propositional Calculus*)

The addition of the axiom $\Box(p \vee q) \rightarrow (\Box p \vee \Box q)$ to the axiom system of any modal logic over T causes its collapse onto the Propositional Calculus. ■

In the following, we propose two *intermediate* modal event calculi, that lie linguistically between *MEC* and *GMEC*. They are aimed at extending the expressive power of *MEC*, while preserving as much as possible its computational efficiency. Unlike *GMEC*, the proposed calculi only allow a restricted mix of boolean and modal operators. The first calculus, called *ICMEC* (*Modal Event Calculus with Internal Connectives*), can be obtained from *MEC* by replacing atomic formulas (property-labeled intervals) with propositional formulas (boolean combinations of property-labeled intervals), that is, *ICMEC*-formulas are propositional formulas, possibly prefixed by at most one modal operator. The query language of *ICMEC* is defined as follows.

DEFINITION 3.11. (*ICMEC-language*)
Let $\mathcal{H} = (E,\ P,\ [\cdot\rangle,\ \langle\cdot],\ ]\cdot,\cdot[)$ *be an EC-structure. The* ICMEC-language *of* $\mathcal{H}$ *is the class of formulas* $\mathcal{C}_{\mathcal{H}} = \{\varphi, \Diamond\varphi, \Box\varphi : \varphi$ is a boolean combination of formulas over $\mathcal{A}_{\mathcal{H}}\}$. *Any element of* $\mathcal{C}_{\mathcal{H}}$ *is called an ICMEC-formula.* □

Clearly, we have $\mathcal{B}_{\mathcal{H}} \subseteq \mathcal{C}_{\mathcal{H}} \subseteq \mathcal{L}_{\mathcal{H}}$. The semantics of *ICMEC* is given as for *GMEC*. The second calculus, called *ECMEC* (*Modal Event Calculus with External Connectives*), extends *MEC* by supporting boolean combinations of *MEC*-formulas. The query language of *ECMEC* is defined as follows.

DEFINITION 3.12. (*ECMEC-language*)
Let $\mathcal{H} = (E,\ P,\ [\cdot\rangle,\ \langle\cdot],\ ]\cdot,\cdot[)$ *be an EC-structure. The* ECMEC-language *of* $\mathcal{H}$ *is the class of formulas* $\mathcal{D}_{\mathcal{H}} = \{\varphi : \varphi$ is a boolean combination of formulas over $\mathcal{B}_{\mathcal{H}}\}$. *Any element of* $\mathcal{D}_{\mathcal{H}}$ *is called an ECMEC-formula.* □

Clearly, we have $\mathcal{B}_{\mathcal{H}} \subseteq \mathcal{D}_{\mathcal{H}} \subseteq \mathcal{L}_{\mathcal{H}}$, completing in this way the diagram in Figure 1. Again, the semantics of this modal event calculus is given as in the case of *GMEC*.

We will now consider the expressiveness of the intermediate calculi we just defined by showing how they can be used to encode our case study from Section 2. In Section 5, we will instead analyze the complexity of these calculi.

4. A Formalization of the Application Domain

In this section, we give a formalization of the example presented in Section 2, and use various modal event calculi to draw conclusions about it. The situation depicted in Figure 2 can be represented by the *EC*-structure $\mathcal{H} = (E,\ P,\ [\cdot\rangle,\ \langle\cdot],\ ]\cdot,\cdot[)$, whose components are defined as follows:

- $E = \{e_1, e_2, e_3, e_4, e_5, e_6\}$;
- $P = \{R_1, R_2, R_3, R_{12}, R_{13}, R_{21}, R_{23}, R_{31}, R_{32}\}$;
- $[R_1\rangle = \{e_1\}$, $[R_2\rangle = \{e_3\}$, $[R_3\rangle = \{e_5\}$, $[R_{12}\rangle = [R_{13}\rangle = \{e_2\}$, $[R_{21}\rangle = [R_{23}\rangle = \{e_4\}$, $[R_{31}\rangle = [R_{32}\rangle = \{e_6\}$;
- $\langle R_1] = \{e_2\}$, $\langle R_2] = \{e_4\}$, $\langle R_3] = \{e_6\}$, $\langle R_{21}] = \langle R_{31}] = \{e_1\}$, $\langle R_{12}] = \langle R_{32}] = \{e_3\}$, $\langle R_{13}] = \langle R_{23}] = \{e_5\}$;
- $]\cdot,\cdot[= \emptyset$.

We have represented transitions as events with the same name, and denoted by R_i the property that register R_i has value 1, for $i = 1, 2, 3$. Furthermore, we have denoted by R_{ij} the fact that the end point of the interval over which R_i assumes value 1 precedes the starting point of the interval over which R_j takes value 1, for $i, j \in \{1, 2, 3\}$ and $i \neq j$. Properties R_{ij} can be exploited to order the time intervals over which the status registers are set to 1.

The partial order of transitions, described in Figure 3, is captured by the following (current) knowledge state:

$$o = \{(e_1, e_2), (e_1, e_3), (e_2, e_4), (e_2, e_5), (e_3, e_4), (e_3, e_5), (e_4, e_6), (e_5, e_6)\}.$$

Consider the formulas

$$\begin{aligned}
\varphi &= R_1(e_1, e_2) \wedge R_{12}(e_2, e_3) \wedge R_2(e_3, e_4) \wedge R_{23}(e_4, e_5) \wedge R_3(e_5, e_6);\\
\psi_1 &= R_1(e_1, e_2) \wedge R_{12}(e_2, e_3) \wedge R_2(e_3, e_4);\\
\psi_2 &= R_2(e_3, e_4) \wedge R_{23}(e_4, e_5) \wedge R_3(e_5, e_6).
\end{aligned}$$

which are in the query language of *ECMEC*, *ICMEC* and *GMEC*. In order to verify that the switch S is not faulty, we must ensure that the registers R_1, R_2 and R_3 display the expected behavior in all refinements of the current knowledge state o. This amounts to proving that the *GMEC*-formula $\Box\varphi$ is true in o. If this is the case, the fault is to be excluded. If we want to determine the existence of at least one extension of o where the registers behave as displayed in Figure 2, we must verify the truth of $\Diamond\varphi$ in o. If this *GMEC*-formula is true, we cannot be sure whether S is faulty or not. Finally, formulas ψ_1 and ψ_2 (which are subformulas of φ), prefixed by the modal operators $\Box$ or $\Diamond$, can be exploited to locally verify the behavior of status registers.

Since we have that $o^+ \models \Diamond\varphi$ and $o^+ \not\models \Box\varphi$, the knowledge contained in o entitles us to assert that the fault is possible but not certain. Moreover, we are not able to localize the fault. In fact, we have that both $o^+ \not\models \Box\psi_1$ (the fault may involve the relative transition of registers R_1 and R_2) and $o^+ \not\models \Box\psi_2$ (the fault may involve the relative transition of R_2 and R_3 as well). Let us denote with o_1 the state of knowledge obtained by adding the pair (e_2, e_3) to o. As in the previous state, we have that $o_1^+ \models \Diamond\varphi$ and $o_1^+ \not\models \Box\varphi$. In this case, however, we are able to localize the possible fault. Since $o_1^+ \models \Box\psi_1$, we can conclude that the fault does not involve the relative transition of registers R_1 and R_2. On the contrary, $o^+ \not\models \Box\psi_2$, and hence the relative transition of registers R_2 and R_3 may be incorrect. Assume now that, unlike the actual situation depicted in Figure 2, we extend o with the pair (e_3, e_2). Let us denote the resulting state with o_2. We have that $o_2^+ \not\models \Diamond\varphi$, that is, the evolution of the values in the registers hints at a fault in switch S. Finally, let us refine o_1 by adding the pair (e_4, e_5), and call o_3 the resulting knowledge state. In this case, we can infer $o_3^+ \models \Box\varphi$, and hence we can conclude that the switch S is certainly not faulty.

The formulas we have used so far belong to the language of both *GMEC* and *ICMEC*. As we will see in Section 5, this is unfortunate since model checking in these languages is intractable. However, the results presented in Section 3 postulate the existence of approximations of these formulas, within the language of *ECMEC*, that have a polynomial validity test. We will use these formulas to analyze the example at hand.

By Corollary 3.8, $\Box\varphi$ is equiprovable with the *ECMEC*-formula:

$$\begin{aligned}\varphi' \;=\;& \Box R_1(e_1, e_2) \wedge \Box R_{12}(e_2, e_3) \wedge \Box R_2(e_3, e_4) \wedge \Box R_{23}(e_2, e_3) \wedge \\ & \Box R_3(e_5, e_6).\end{aligned}$$

Therefore, we can use *ECMEC* and φ' to establish whether the switch S is fault-free or is possibly defective. For example, since $o_3^+ \models \varphi'$ entails $o_3^+ \models \Box\varphi$, we can exclude the possibility of a misbehavior of S in situation o_3.

The best *ECMEC*-approximation of $\Diamond\varphi$ we can achieve is the formula

$$\begin{aligned}\varphi'' \;=\;& \Diamond R_1(e_1, e_2) \wedge \Diamond R_{12}(e_2, e_3) \wedge \Diamond R_2(e_3, e_4) \wedge \Diamond R_{23}(e_2, e_3) \wedge \\ & \Diamond R_3(e_5, e_6)\end{aligned}$$

which is *not* equivalent to $\Diamond\varphi$. However, we know that for every knowledge state $w \in W_H$, if $w \models \Diamond\varphi$, then $w \models \varphi''$. We can use this fact to draw negative consequences about our example. In particular, we can use φ'' to determine that S is faulty assuming the trace o_2. Indeed, we have that $o_2^+ \not\models \varphi''$, from which it must be the case that $o_2^+ \not\models \Diamond\varphi$. This allows us to conclude that the behavior of S is certainly incorrect.

Finally, in the knowledge state o, both $o^+ \not\models \varphi'$ and $o^+ \models \varphi''$ hold. The former implies $o^+ \not\models \Box\varphi$, and thus a faulty behavior of S cannot be excluded in the current state. Instead, $o^+ \models \varphi''$ neither allows us to conclude that $o^+ \models \Diamond\varphi$ nor that $o^+ \not\models \Diamond\varphi$. In this case, using *ECMEC* we are not able to establish whether the system is certainly faulty or not. The same holds for the knowledge state o_1.

5. Complexity Analysis

This section is dedicated to studying the complexity of the various modal event calculi presented in Section 3. We model our analysis around the satisfiability relation $\models$ given in Definitions 3.3 and 3.6, but we also take into account the numerous results that permit improving its computational behavior (in particular, the locality conditions for the computation of $\Box$-MVIs and $\Diamond$-MVIs — Lemma 3.4). This approach is sensible since these specifications can be directly turned into the clauses of logic programs implementing these calculi (Franceschet, 1996).

The notion of cost we adopt is as follows: we assume that verifying the truth of the propositions $e \in [p\rangle$, $e \in \langle p]$ and $]p, p'[$ has constant cost $O(1)$, for given event e and properties p and p'. Although possible in principle, it is disadvantageous in practice to implement knowledge states so that the test $(e_1, e_2) \in w$ has constant cost. We instead maintain an acyclic binary relation o on events whose transitive closure o^+ is w (cf. Section 3). Verifying whether $(e_1, e_2) \in w$ holds becomes a reachability problem in o and it can be solved in quadratic time $O(n^2)$ in the number n of events (Cervesato et al., 1995b).

Given an *EC*-structure $\mathcal{H}$, a knowledge state $w \in W_{\mathcal{H}}$ and a formula φ relatively to any of the modal event calculi presented in Section 3, we want to characterize the complexity of the problem of establishing whether $\mathcal{I}_{\mathcal{H}}; w \models \varphi$ is true (which is an instance of the general problem of model checking). We call the triple $(\mathcal{H}, w, \varphi)$ an *instance* and generally prefix this term with the name of the calculus we are considering. In the following, we will show that, given an instance $(\mathcal{H}, w, \varphi)$, model checking for φ in the intended model $\mathcal{I}_{\mathcal{H}}$ is polynomial in *EC*, *MEC* and *ECMEC*, while it is NP-hard in *ICMEC* and *GMEC*. The reason for such a different computational behavior for the various modal event calculi is that *MEC* and *ECMEC* can exploit local conditions for testing (boolean combinations of) atomic formulas, possibly prefixed by a modal operator, while the latter two cannot avoid of explicitly searching the whole set of extensions of the given partial ordering, whose number is, in general, exponential with respect to the number of events.

Given an *EC*-instance $(\mathcal{H}, w, \varphi)$, the cost of the test $w \models \varphi$ can be derived to be $O(n^3)$ directly from the relevant parts of Definition 3.2, as proved in (Cervesato et al., 1995b). Exploiting the local conditions yields an identical bound in the case of *MEC*:

THEOREM 5.1. (*Complexity of model checking in MEC*)
Given a MEC-instance $(\mathcal{H}, w, \varphi)$, *the test* $w \models \varphi$ *has cost* $O(n^3)$. ■

Unlike *EC* and *MEC*, where the input formula φ is an atomic formula, possibly prefixed by a modal operator in *MEC*, φ can be arbitrarily large in the case of the remaining calculi in the hierarchy in Figure 1. As a consequence, the dimension of the input formula, that does not come into play in the complexity analysis of *EC* and *MEC*, becomes a relevant parameter for the analysis of the cost of the remaining calculi. Thus, their complexity will be measured in terms of both the dimension k of the input formula (the number of occurrences of atomic formulas it includes) and the size n of the input structure (the number of events in E).

An *ECMEC*-formula φ is the boolean combination of a number of *MEC*-formulas. If φ contains k atomic formulas, this number is precisely k. Therefore, by virtue of Definition 3.6, the test for φ can be reduced to the resolution of k *MEC* problems. Thus, *ECMEC* has polynomial complexity.

THEOREM 5.2. (*Complexity of model checking in ECMEC*)
Given an ECMEC-instance $(\mathcal{H}, w, \varphi)$, *the test* $w \models \varphi$ *has cost* $O(kn^3)$. ■

The placement of the modal operators in *ICMEC* prevents us, in general, from being able to use local conditions in tests. An exhaustive exploration of the extensions of the current knowledge state is unavoidable. This raises the complexity of the problem beyond tractability, as expressed by the following theorem.

THEOREM 5.3. (*Complexity of model checking in ICMEC*)
Given an ICMEC-instance $(\mathcal{H}, w, \varphi)$, *the test* $w \models \varphi$ *is NP-hard.*

Proof.
If φ is a propositional formula, then model checking reduces to verifying whether it evaluates to true with respect to the current state of knowledge. It has a polynomial cost. The remaining cases are $\varphi = \Diamond\psi$ and $\varphi = \Box\psi$.

We first prove that if $\varphi = \Diamond\psi$, then model checking in *ICMEC* is NP-complete. It is easy to see that this problem belongs to NP. Indeed, in order to establish whether $w \models \Diamond\psi$ holds, we non-deterministically generate extensions w' of w, and then test the truth of ψ in w' until an extension where ψ holds is found. There are exponentially many such extensions. Since the formula ψ does not include any modal operator, the test in each extension is polynomial. In order to prove that the considered problem is NP-hard, we define a (polynomial) reduction of 3SAT (Garey and Johnson, 1979) into *ICMEC*.

Let q be a boolean formula in 3CNF, $p_1, p_2, .., p_n$ be the propositional variables that occur in q, and $q = c_1 \wedge c_2 \wedge \ldots \wedge c_m$, where $c_i = l_{i,1} \vee l_{i,2} \vee l_{i,3}$ and for each i, j either $l_{i,j} = p_k$ or $l_{i,j} = \neg p_k$ for some k.

Let us define an *EC*-structure $\mathcal{H} = (E,\ P,\ [\cdot\rangle,\ \langle\cdot],\ ]\cdot,\cdot[)$ such that:

$E = \{e(p_i), e(\neg p_i) : 1 \leq i \leq n\}$;
$P = \{p_i : 1 \leq i \leq n\}$;
$[p_i\rangle : \{e(p_i)\}$ and $\langle p_i] : \{e(\neg p_i)\}$, for $1 \leq i \leq n$;
$]\cdot,\cdot[= \emptyset$.

Moreover, let $w = \emptyset$ and $\psi = c'_1 \wedge c'_2 \wedge \ldots \wedge c'_m$, where $c'_i = l'_{i,1} \vee l'_{i,2} \vee l'_{i,3}$, and for each i, j, if $l_{i,j} = p_k$, then $l'_{i,j} = p_k(e(p_k), e(\neg p_k))$, otherwise ($l_{i,j} = \neg p_k$) $l'_{i,j} = \neg p_k(e(p_k), e(\neg p_k))$. It is not difficult to see that $w \models \Diamond\psi$ if and only if q is satisfiable.

Let us show now that if $\varphi = \Box\psi$, then model checking in *ICMEC* is NP-hard. We prove this result by defining a (polynomial) reduction of the problem of propositional validity into *ICMEC*.

Let q be a boolean formula in 3DNF, $p_1, p_2, .., p_n$ be the propositional variables that occur in q, and $q = d_1 \vee d_2 \vee \ldots \vee d_m$, where $d_i = l_{i,1} \wedge l_{i,2} \wedge l_{i,3}$ and for each i, j, either $l_{i,j} = p_k$ or $l_{i,j} = \neg p_k$. We define the *EC*-structure $\mathcal{H} = (E,\ P,\ [\cdot\rangle,\ \langle\cdot],\ ]\cdot,\cdot[)$ as in the previous subcase. Let $w = \emptyset$ and $\psi = d'_1 \vee d'_2 \vee \ldots \vee d'_m$, where $d'_i = l'_{i,1} \wedge l'_{i,2} \wedge l'_{i,3}$, and for each i, j, if $l_{i,j} = p_k$, then $l'_{i,j} = p_k(e(p_k), e(\neg p_k))$, otherwise ($l_{i,j} = \neg p_k$) $l'_{i,j} = \neg p_k(e(p_k), e(\neg p_k))$. It is straightforward to see that $w \models \Box\psi$ if and only if q is valid in propositional logic. ∎

Since *ICMEC* is a linguistic fragment of *GMEC*, Theorem 5.3 allows us to conclude that model checking for *GMEC* is NP-hard too.

COROLLARY 5.4. (*Complexity of model checking in GMEC*)

Given a GMEC-instance $(\mathcal{H}, w, \varphi)$, *the test* $w \models \varphi$ *is NP-hard.* ∎

We summarize the results obtained in this section in the following table.

Calculus	*EC*	*MEC*	*ECMEC*	*ICMEC*	*GMEC*
Parameters	n events	n events	n events k atomic formulas	n events k atomic formulas	n events k atomic formulas
Model checking	$O(n^3)$	$O(n^3)$	$O(kn^3)$	NP-hard	NP-hard

6. Conclusions and Further Developments

In this paper, we have established a hierarchy of modal event calculi by investigating *ECMEC* and *ICMEC* as intermediate languages between the modal event calculus *MEC* (Cervesato et al., 1995a) and the generalized modal event calculus *GMEC* (Cervesato et al., 1996). In particular, we showed that *ECMEC* retains enough of the expressive power of *GMEC* while admitting an efficient polynomial implementation in the style of *MEC*. We supported our claims by showing the formalization of an example from the applicative domain of fault diagnosis. Moreover, we gave a rigorous analysis of the complexity of the modal event calculi we considered.

We are developing the proposed framework in several directions. First, the complexity results given in Section 5 can actually be improved. In the proof of Theorem 5.3, we showed that checking *ICMEC*-formulas of the form $\Diamond\psi$ is NP-complete. Since $\Box = \neg\Diamond\neg$, it easily follows that checking $\Box\psi$ formulas is co-NP(-complete). Thus, the whole problem of testing $w \models \varphi$ (for *ICMEC*) involves either a polynomial check, or an NP-check, or a co-NP check. This means that it can be computed by a Turing machine which can acces an NP-oracle and runs in deterministic polynomial time, and hence the problem is in P^{NP} (since only one call to the oracle is needed, it is actually in $P^{NP[1]}$) (Stockmeyer, 1987). We are currently working at the characterization of the exact complexity of model checking in both *ICMEC* and *GMEC*.

Another issue of interest when working with *EC* and in its modal refinements is the *generation of MVIs*, which can be solved using the same logic programs that implement model checking. In this problem, we replace some, possibly all, events in a formula φ by logical variables and ask which instantiations of these variables make φ true. The problem of MVI generation can still be viewed as a problem of model checking in modal event calculi extended with limited forms of existential quantification. Let *QEC*, *QMEC*, *ECQMEC*, *ICQMEC*, and *GQMEC* respectively be the quantified counterparts of *EC*, *MEC*, *ECMEC*, *ICMEC*, and *GMEC*. It is possible to show that model checking for all quantified modal event calculi essentially lies in the same complexity class of model checking for their unquantified counterparts, except for *ECMEC*, for which the addition of quantification makes the problem NP-hard.

Finally, we are considering the effects of the addition of preconditions to our framework. This step would enlarge the range of applicability of the modal event calculi. However, as proved in (Dean and Boddy, 1988), an indiscriminated use of preconditions immediately makes the problem NP-hard. Nevertheless, we believe that a formal study of various modal event calculi with preconditions can shed some light on the dynamics of preconditions, and possibly lead to polynomial approximations of the computation of MVIs. Preliminary results in this direction can be found in (Cervesato et al., 1997).

ACKNOWLEDGMENTS

We would like to thank the reviewers for their useful comments. The first author was partially supported by NFS grant CCR-9303383 and by a scholarship for specialization overseas from the University of Udine. The work of the third author was partially supported by the CNR project *Ambienti e strumenti per la gestione di informazioni temporali.*

References

Cervesato, I., L. Chittaro, and A. Montanari: 1995a, 'A Modal Calculus of Partially Ordered Events in a Logic Programming Framework'. In: L. Sterling (ed.): *Proceedings of the Twelfth International Conference on Logic Programming — ICLP'95*. Kanagawa, Japan, pp. 299–313.

Cervesato, I., L. Chittaro, and A. Montanari: 1995b, 'Speeding up temporal reasoning by exploiting the notion of kernel of an ordering relation'. In: S. Goodwin and H. Hamilton (eds.): *Proceedings of the Second International Workshop on Temporal Representation and Reasoning — TIME'95*. Melbourne Beach, FL, pp. 73–80.

Cervesato, I., L. Chittaro, and A. Montanari: 1996, 'A General Modal Framework for the Event Calculus and its Skeptical and Credulous Variants'. In: W. Wahlster (ed.): *Proceedings of the Twelfth European Conference on Artificial Intelligence — ECAI'96*. Budapest, Hungary, pp. 33–37. Extended and revised version submitted for publication, July 1996.

Cervesato, I., M. Franceschet, and A. Montanari: 1997, 'Modal Event Calculi with Preconditions'. In: R. Morris and L. Khatib (eds.): *Fourth International Workshop on Temporal Representation and Reasoning — TIME'97*. Daytona Beach, FL, pp. 38–45.

Cervesato, I., A. Montanari, and A. Provetti: 1993, 'On the Non-Monotonic Behavior of the Event Calculus for Deriving Maximal Time Intervals'. *International Journal on Interval Computations* **2**, 83–119.

D.C. Moffat, G. R.: 1990, 'Modal Queries about Partially-ordered Plans'. *Journal of Expt. Theor. Artificial Intelligence* **2**, 341–368.

Dean, T. and M. Boddy: 1988, 'Reasoning about Partially Ordered Events'. *Artificial Intelligence* **36**, 375–399.

Franceschet, M.: 1996, 'Una Gerarchia di Calcoli Modali degli Eventi: Espressività e Complessità (in Italian)'. Tesi di Laurea in Informatica, Università di Udine, Italy. To appear as a Research Report in English.

Garey, M. and D. Johnson: 1979, *Computing and Intractability: A Guide to the Theory of NP-Completeness.* Freeman & Cie.

Kowalski, R. and M. Sergot: 1986, 'A Logic-Based Calculus of Events'. *New Generation Computing* **4**, 67–95.

Nökel, K.: 1991, *Temporarily Distributed Symptoms in Technical Diagnosis.* Springer-Verlag.

Segerberg, K.: 1971, 'An Essay in Classical Modal Logic'. *Uppsala Filosofiska Studier.*

Stockmeyer, L.: 1987, 'Classifying the computational complexity of problems'. *Journal of Symbolic Logic* **52**(1), 1–43.

Release Logics for Temporalizing Dynamic Logic

Orthogonalizing modal logics

Jeroen Krabbendam* and John-Jules Meyer
Utrecht University, Department of Mathematics and Computer Science, P.O. Box 80089, NL-3508 TB Utrecht, The Netherlands ({krabb,jj}@cs.ruu.nl)

Abstract.

Time and action are the two modalities which are combined in one logical system in this paper, in order to allow the partial specification of systems. In any composition of dynamic and temporal logic there is mutual interference. The dynamical component takes time and the temporal component should have an effect on the state space. The plain union of both logics cannot handle the mutual interference properly. After a discussion on the interference of action and temporal components, a multi-modal **S5** logic, called Release Logic, is introduced to act as an orthogonalizing intermediate logic. With the now three modal components, two different temporal logics will be constructed. One logic allows instantaneous actions only and the other logic allows actions which take some time to complete.

Release Logic can be viewed as the logic of controlled ignorance. It is a modal logic which is particularly well-fitted for partial specifications. Several specifications will be presented, most of them as one-liners.

Keywords: linear temporal logic, dynamic logic, partial system specification

1. Introduction

The undesired, but inevitable, effect of computations is that they take time to execute. The effects of actions can be described as changes of the state in a state space together with an advance of time. In general, however, models of logic for describing program and system behaviour focus on one of these aspects only, or identify both of them. The changing state model is the approach of dynamic logic. It links the changes of the variables, propositions and formulae directly to named actions, but it lacks the concept of time. The aspect of time can adequately be described using a temporal logic, but — usually — there is no direct link between the action and its effects on the state space since the actions are anonymous, hence their effects tucked away in the models.

This paper brings together both approaches, thereby following the accepted interpretations and intuition of both dynamic and temporal logic as close as possible. First, it is shown that an unconditioned join of the two modal logics of action and time raises the question about the influence of both modalities on each others terrain. It will be shown that the standard interpretation and intuition of dynamic and linear dynamic logic are in a way incompatible if the dynamic component spends time and the temporal component effects the state space.

The purpose of this paper is to introduce a multi-dimensional modal **S5** - logics for constructing a logic which integrates standard dynamic and linear

* Supported by NWO under grant 612-23-419 (DEIRDRE) and ESPRIT BRWC project 8319 (ModelAge)

H. Barringer et al. (eds.), Advances in Temporal Logic, 21–45.

temporal logics. The combined logic can serve as a specification tool for system behaviour. It is particularly well suited for focussing on parts of the system, while still taking a global notice of the possible effects of the environment in which this part operates.

The founding fathers are, not surprisingly, a (propositional) dynamic logic à la Harel (Harel, 1979) and a linear temporal logic à la Manna and Pnueli ((Manna and Pnueli, 1992) and (Manna and Pnueli, 1995)). These two modal logics are well-known and well-understood, but only simple sublogics will be used here. What follows is, apart from introducing the components action and time, a brief sketch of the problems involving the combination of these two modalities.

This is followed by an introduction on a new modal logic, a simple **S5** -logic, which serves as a glue between the (seemingly) incompatible modal operators of time and action. This logic is called Release Logic since it 'releases' some assumptions on the current state. If the release logic is properly chosen, it creates enough room for the other modal operators to change the state space *and* time, thus solving the vexed question.

2. The mutual interference of time and action

The structure of this section is as follows: First the two modal logics, dynamic and temporal, are defined, because, on the one hand they reflect the basic structure of computer programs and systems, namely action and time and on the other hand the definitions and theorems will be reused thereafter. Axioms will be listed as well as standard Kripke models and soundness and completeness theorems. It should be noted that the dynamic and temporal logic are basic ones, to illustrate the idea. A further expansion and enrichment of the logics is possible.

In the sequel, it will be shown that within the accepted settings of both frameworks, it is hard to describe their combined effects since several paradoxes can be derived. In the section thereafter a simple solution to these paradoxes will be proposed, still allowing the standard interpretations and intuition of both dynamic and temporal logic.

Familiarity with modal logics is assumed. Many books on this subject give an indepth description of its mathematical structure, for instance (Goldblatt, 1992), (Hughes and Cresswell, 1984), (Hughes and Cresswell, 1996) and (Meyer and van der Hoek, 1996). Algebraic approaches can be found in (Goldblatt, 1993), (Henkin et al., 1971) and (Henkin et al., 1985).

2.1. Unstructured Propositional Dynamic Logic

The action logic in this paper, is a dynamic logic **PDL** , allowing for unstructured components only. In fact, it is a simplified version of Harel's dynamic logic (Harel, 1979). In the sequel, the set $\mathcal{ACT}$ is the set of unstructured actions.

Definition 2.1. (Formulae of the language $\mathcal{LAN}_{PDL}$)

The set $\mathcal{FOR}_{PDL}$ of formulae of the language $\mathcal{LAN}_{PDL}$ has the following inductive definition:

- The set $\mathcal{PROP}$ of proposition variables is in the set of formulae
- If φ is a formula then $\neg\varphi$ is a formula
- If φ and ψ are formulae then $(\varphi \rightarrow \psi)$ is a formula
- If φ is a formula and $\alpha \in \mathcal{ACT}$, then $[\alpha]\varphi$ is a formula

The formula $[\alpha]\varphi$ should be read as: all terminating executions of program α establish the truth of φ. The modal operator $\langle\alpha\rangle$ is defined as $\neg[\alpha]\neg$, and the accepted reading of $\langle\alpha\rangle\varphi$ is: there exist a terminating execution of program α that ends with φ being true.

A Kripke model with standard interpretations is used for the dynamic logic.

Definition 2.2. (Kripke model for **PDL**)

A Kripke model $\mathfrak{M}_{PDL} = \langle \mathbb{S}, \pi, \mathcal{R}_{[.]} \rangle$ ($\mathfrak{M}$ for short) consists of:

- A non empty set $\mathbb{S}$ of states.
- A mapping $\pi : \mathbb{S} \rightarrow \mathcal{PROP} \rightarrow \mathbb{B}$.
- A mapping $\mathcal{R}_{[.]} : \mathcal{ACT} \rightarrow 2^{(\mathbb{S}\times\mathbb{S})}$ from the set $\mathcal{ACT}$ to relations. A relation $\mathcal{R}_\alpha$ specifies the transitions from states to states induced by the execution of action α.

Propositional dynamic logic in this variant is a **K** modal logic, hence it has the following axioms.

Definition 2.3. (Axioms of **PDL**)

Suppose $\alpha \in \mathcal{ACT}$. The axiom schemes of the logic **PDL** are:

TAUT	All instances of propositional tautologies
$\mathbf{K}_\alpha$	$[\alpha](\varphi \rightarrow \psi) \rightarrow ([\alpha]\varphi \rightarrow [\alpha]\psi)$

The derivation rules are standard.

Definition 2.4. (Rules of **PDL**) Let $\alpha \in \mathcal{ACT}$:

MP	$\varphi, \varphi \rightarrow \psi \models \psi$	(Modus Ponens)
$\mathbf{Nec}_\alpha$	$\varphi \models [\alpha]\varphi$	(Necessitation)

Suppose $\mathcal{PDL}$ is the class of **PDL** Kripke models. The **PDL** is a multi modal variant of the logic **K** , its soundness with respect to the class $\mathcal{PDL}$ is immediate, and its completeness can easily be proved with a Henkin's proof, using a canonical model cf. (Chellas, 1993), (Meyer and van der Hoek, 1996). Note that given a the set of formulae the canonical model is unique.

Definition 2.5. (Canonical model for modal logic **PDL**)

Let $\mathfrak{M}^{PDL} = \langle \mathbb{S}^{PDL}, \pi^{PDL}, \mathcal{R}^{PDL}_{[.]} \rangle$ be the canonical Kripke model for the modal logic **PDL** .

Theorem 2.6. (Soundness and completeness axiom systems satisfying axioms **PDL**)

The axiom system **PDL** is sound and complete with respect to the class of Kripke models $\mathcal{PDL}$.

2.2. Linear Temporal Logic

As in the case of dynamic logic, a plain propositional linear temporal logic (**PTL**) is used. It allows the future next operator $\bigcirc$ and no more modalities[1]. Excellent introductions to linear temporal logic can be found in (Manna and Pnueli, 1992) and (Kröger, 1987).

Definition 2.7. (Formulae of the language $\mathcal{LAN}_{PTL}$)

The set $\mathcal{FOR}_{PTL}$ of formulae of the language $\mathcal{LAN}_{PTL}$ has the following inductive definition:

- The set $\mathcal{PROP}$ of proposition variables is in the set of formulae
- If φ is a formula then $\neg\varphi$ is a formula
- If φ and ψ are formulae then $(\varphi \to \psi)$ is a formula
- If φ is a formula then $\bigcirc\varphi$ is a formula

The formulae $\bigcirc\varphi$ should be read as: next φ, i.e. at the next point of time, φ is true. A Kripke model is used for the semantics of this temporal logic.

Definition 2.8. (Kripke model for **PTL**)

A Kripke model $\mathfrak{M}_{PTL} = \langle \mathbb{S}, \pi, \mathcal{R}_{\bigcirc} \rangle$ ($\mathfrak{M}$ for short) consists of:

- The set $\mathbb{S}$ of states.
- A mapping $\pi : \mathbb{S} \to \mathcal{PROP} \to \mathbb{B}$.
- A relation $\mathcal{R}_{\bigcirc} : 2^{(\mathbb{S}\times\mathbb{S})}$ specifying the transitions to a next time point.

The proposed linear temporal logic satisfies the **TAUT**, **K** and, because it is linear, **D** as well. Moreover, since the logic is not branching time, the axiom **D** can be reversed, thus strengthened to **D!**.

Definition 2.9. (Axioms of **PTL**)

The axiom schemes of the logic **PTL** are:

TAUT	All instances of propositional tautologies
$\mathbf{K}_{\bigcirc}$	$\bigcirc(\varphi \to \psi) \to (\bigcirc\varphi \to \bigcirc\psi)$
$\mathbf{D!}_{\bigcirc}$	$\bigcirc\varphi \leftrightarrow \neg\bigcirc\neg\varphi$

The derivation rules are standard, cf. definition 2.4, with $\mathbf{Nec}_\alpha$ replaced by $\mathbf{Nec}_{\bigcirc}$ which reads $\varphi \models \bigcirc\varphi$.

Definition 2.10. (Functional relation) A relation $\mathcal{R}$ is functional iff $\forall_s \exists!_t : s\mathcal{R}t$.

In the sequel, $\mathcal{PTL}$ is the class of functional **PTL** Kripke models. The **PTL** is a functional logic **KD!**, its soundness with respect to the class $\mathcal{PTL}$ is immediately, and its completeness can be proved with a Henkin's proof, using a canonical model again. A similar proof can be found in (Andréka et al., 1995).

Definition 2.11. (Canonical model for modal logic **PTL**)

Let $\mathfrak{M}^{PTL} = \langle \mathbb{S}^{PTL}, \pi^{PTL}, \mathcal{R}_{\bigcirc}^{PTL} \rangle$ be the canonical Kripke model for the modal logic **PTL**.

[1] Clearly, a temporal logic should incorporate the always $\Box$-operator. The main issue of this paper is to show how to combine temporal and dynamic logic. The $\Box$-operator is not essential for the technique presented in this paper. Although the logics with the $\Box$-operator are sound and complete as well, completeness proofs would be more complicated. In an extended paper, the temporal logic does allow the always operator in its formulae.

If a system **S** satisfies axiom **D!** then the canonical model is functional.

Lemma 2.12. $\mathbf{D!} \in \mathbf{S} \Rightarrow (\forall t_0 : t_0 \in \mathbb{S} : (\exists! t_1 : t_1 \in \mathbb{S} : t_0(\mathcal{R}_{\bigcirc}^{PTL})t_1))$

Corollary 2.13. $\mathfrak{M}^{PTL} \in \mathcal{PTL}$

Theorem 2.14. (Soundness and completeness axiom systems satisfying axioms **PTL**)

The axiom system **PTL** is complete with respect to the class of Kripke models $\mathcal{PTL}$.

Proof. The completeness of **PTL** follows from the functionality of the relation $\mathcal{R}_{\bigcirc}^{PTL}$ in the canonical model $\mathfrak{M}^{PTL}$.

2.3. THE FRICTION IN THE MARRIAGE OF TIME AND ACTION

The most simple approach to a program's behaviour in a (time) dynamic environment is a combination of the two modalities $[\alpha]$ and $\bigcirc$, thus writing down formulae like $[\alpha]\bigcirc\varphi$, which should be read as: one timeframe after calculation α is completed, φ holds. Note that program α does not necessarily consumes time.

As the program α effects the state space, the question arises whether the temporal operator $\bigcirc$ has effect on the state space as well. Two extreme viewpoints will be discussed.

First, assume that the next operator does not have any effect on the state space, thus, the next operator *freezes* the state space. An example of this model is like the so-called step-semantics. In this model the actions can be instantaneous and the time is modelled as a step in another direction, when all other activities are suspended. This relates to work done by Gabbay (Gabbay, 1976) and Jánossy et al (Jánossy et al., 1996). The fact that the actions are instantaneous is unimportant in what follows. In a modal logic setting, the effect φ of a concatenation of two actions α and β, each followed by a pause of one timeframe each, would read: $[\alpha]\bigcirc[\beta]\bigcirc\varphi$.

In evaluating formula $[\alpha]\bigcirc[\beta]\bigcirc\varphi$ only the *last* state is the state in which φ is evaluated. In the not so rare case that actions behave independently from the clock, the freeze model has some undesired properties. In the sequence of transitions, it does not matter when the temporal transitions are being done. As a result, the formulae $[\alpha]\bigcirc[\beta]\bigcirc\varphi$, $[\alpha][\beta]\bigcirc\bigcirc\varphi$, $\bigcirc[\alpha]\bigcirc[\beta]\varphi$, and $\bigcirc\bigcirc[\alpha][\beta]\varphi$ (and more) are logically equivalent (see also figure 1).

As a consequence, the temporal component can be shifted out of the formula. Moreover, in the situation that φ is not time dependent, the formula $\bigcirc\varphi \leftrightarrow \varphi$ is valid. The basic assumption, however, is to put a statement about the temporal aspect and its effect on the state space. The combination of the two assumptions — the actions behave independent of the clock and formula φ is independent of the time — creates an even stranger situation. In this case the four formulae above collaps into $[\alpha][\beta]\varphi$.

Hence, the *freeze* approach leads to two pseudo paradoxes, $[\alpha]\bigcirc\varphi \leftrightarrow \bigcirc[\alpha]\varphi$ and, in special cases, leads to the more problematic validity $\bigcirc\varphi \leftrightarrow \varphi$.

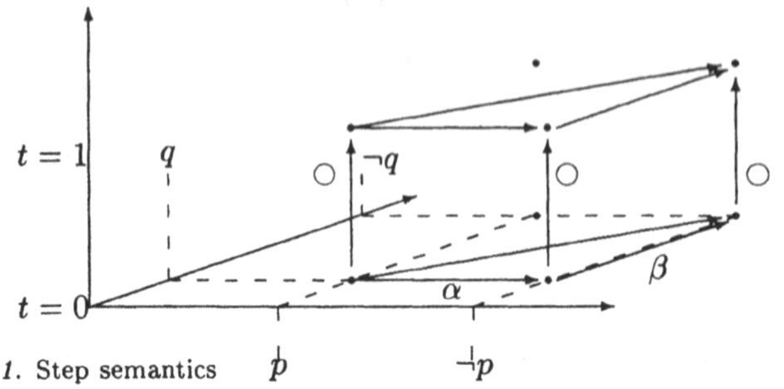

Figure 1. Step semantics

In another approach, the next operator models the effect of an unknown, and sometimes even unpredictable, environment. In this setting, during a time-frame *anything* can happen. In particular, an effect of action α can be reversed. This is the ultimate *chaos* paradox as every effect of an action is immediately lost after pausing for a even the slightest moment, hence $[\alpha]\varphi \to [\alpha]\bigcirc\neg\varphi$, if φ is deniable. If action α consumes one timeframe, what can be achieved by $\bigcirc$ can also be achieved by α, thus $\bigcirc\varphi \to [\alpha]\varphi$. If α is instantaneous the paradox reads $\bigcirc\varphi \not\to [\alpha]\bigcirc\varphi$. The point is that the temporal modality does not specify one single path through the time space, but all paths, see figure 2.

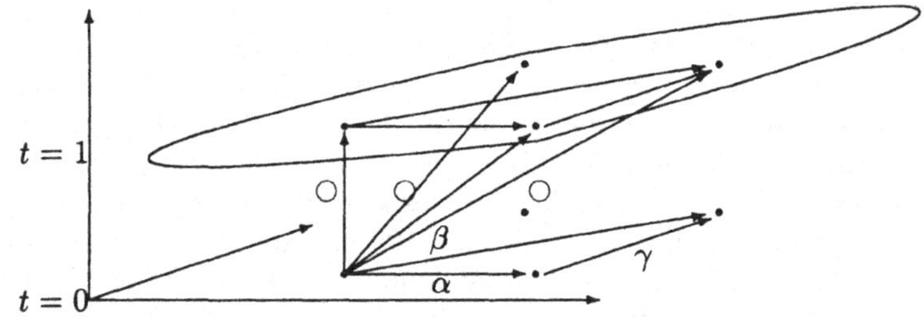

Figure 2. Chaos model

2.4. Analyzing the paradoxes: orthogonalization

Clearly, if there is no interference between a program α and the temporal modal operator $\bigcirc$, the combined logic is meaningless. The freeze pseudo paradox is an indication of exactly that.

If α takes some time to perform its action, and $\bigcirc$ changes the time only, the operational interpretation of a formula $[\alpha]\bigcirc\varphi$ is that after the execution of α nothing can happen, except the advance to the next point of time. This is a situation which is not very realistic when specifying programs and systems which operate in parallel or in a dynamic, but more or less anonymous, environment. It would be preferable if the specifications allow some influence of the environment, but in such a way that (part of) what α achieved is maintained.

In the ultimate chaos model, other system parts should be able to perform their actions during a certain period of time, but there is no way to say anything significant about the effects, nor the effects of α. Obviously there is a way in between. The temporal component should be allowed to have a limited effect on the state space. In general terms, however, the effect is unspecified, hence a general idea of the effect of other systems parts suffice.

In the next section, a modality $\Delta_{\mathcal{P}}$ will be introduced which allows to specify effect $\mathcal{P}$ which is limited but unspecific at the same time.

3. Specifying irrelevancy with Release Logic

In this section an indexed **S5** modality will be proposed, in fact a simple extension to the **S5** -logic. The semantics of the logic concentrate on (non-empty) classes of states, not on single states as most propositional modal logics do. The logic supports reasoning about classes of states, without mentioning these classes specifically. In fact, the logic has properties like φ holds in all states modulo a certain issue. The irrelevancy, or deviation, is to be found in a index $\mathcal{P}$, it specifies what to ignore. The logic builder can, by choosing the appropriate type of index $\mathcal{P}$, create any logic of (equivalence) classes. This approach relates to work done by Tarski et al (Henkin et al., 1971) and Nemeti (Nemeti, $\star$).

In isolation, the logic is very simple; it is a(nother) multi modal **S5** . However, it allows a subtle reasoning about slightly different states, with several applications. More on multi modal logics can be found in (Venema, 1992).

3.1. The syntax of PRL

The language $\mathcal{LAN}_{PRL_{\mathcal{A}}}$ is the language of a propositional release logic $\mathbf{PRL}_{\mathcal{A}}$. The language allows propositional formulae and, additionally, the deviation operators.

The deviation operator Δ is indexed by an arbitrary index $\mathcal{P} \in \mathcal{A}$. Although the language $\mathcal{LAN}_{PRL_{\mathcal{A}}}$ and all other structures are relative with respect to the anonymous set $\mathcal{A}$, this index will be dropped.

Definition 3.1. (Formulae of $\mathcal{LAN}_{PRL}$)

Let $\mathcal{A}$ be a set. The set $\mathcal{FOR}_{PRL}$ of formulae of the language $\mathcal{LAN}_{PRL}$ has the following inductive definition:

- The set $\mathcal{PROP}$ of proposition variables is in the set of formulae
- If φ is a formula then $\neg\varphi$ is a formula
- If φ and ψ are formulae then $(\varphi \rightarrow \psi)$ is a formula
- If φ is a formula and $\mathcal{P} \in \mathcal{A}$, then $\Delta_{\mathcal{P}}\,\varphi$ is a formula

In general, the formula $\Delta_{\mathcal{P}}\,\varphi$ should be read as: φ holds for all $\mathcal{P}$ deviations, or, φ holds modulo $\mathcal{P}$ deviations. Another interpretation would be that any constraint $\mathcal{P}$ on φ is released, hence the name release logic.

The ∇-operator, the possible deviation operator, is the dual of the Δ-operator. An informal release logical interpretation is: φ holds for some $\mathcal{P}$ deviation, or, φ within $\mathcal{P}$-release range.

Definition 3.2. (Abbreviation)

The possible deviation operator $\nabla_{\mathcal{P}}$ is defined as: $\nabla_{\mathcal{P}} \stackrel{\text{def}}{=} \neg\Delta_{\mathcal{P}}\neg$

Abbreviations and conventions used in the literature on **S5** -logics apply here as well, as far as they are meaningful and do not violate the rules of the language of PRL. For instance, the deviation modalities have the same binding power as the modal **S5** -operator.

3.2. A SEMANTIC MODEL FOR **PRL**

In seeking a close link to the aforementioned modal **S5** logic, the meta level reasoning about the interpretation of release logic is by means of a slightly adapted Kripke model. The release operator $\Delta_{\mathcal{P}}$ is linked to a semantic equivalence relation $\mathcal{R}_{\Delta_{\mathcal{P}}}$, this relation clusters the states. The type of the argument $\mathcal{P}$ will not be specified here — each release logic should specify its own. An example would be that the states are the same, modulo $\mathcal{P}$, or that they agree on everything but $\mathcal{P}$. However, there is one additional property of the $\Delta_{\mathcal{P}}$ modal operator which is shared with all release logics and that is a partial order structure on the indices. Note no detailed assumption on the property of the transition relations $\mathcal{R}_{\Delta}$ is necessary in order to define a Kripke model for release logics.

Definition 3.3. (Kripke model for **PRL**)

Let $\mathcal{A}$ be a set. A Kripke model $\mathfrak{M}_{PRL} = \langle \mathbb{S}, \pi, \mathcal{R}_{\Delta}\rangle$ ($\mathfrak{M}$ for short) consists of:

- A non empty set $\mathbb{S}$ of states.
- A mapping $\pi : \mathbb{S} \to \mathcal{PROP} \to \mathbb{B}$.
- A mapping $\mathcal{R}_{\Delta} : \mathcal{A} \to 2^{(\mathbb{S}\times\mathbb{S})}$ from the set $\mathcal{A}$ to relations. The relation itself specifies the transitions from states to states induced by an element of $\mathcal{A}$.

All appropriate definitions and conventions about Kripke models, commonly used in the **S5** -case, are assumed to be redefined in a natural manner.

Definition 3.4. (Semantics **PRL**)

Let $\mathcal{A}$ be a set. The satisfaction relation $\models_{PRL}$ follows the standard inductive definition but includes the next rule. If orderly, the index $_{PRL}$ will be omitted.

$$\mathfrak{M}, s \models \Delta_{\mathcal{P}}\varphi \quad \Leftrightarrow \quad (\forall t : s(\mathcal{R}_{\Delta_{\mathcal{P}}})t : \mathfrak{M}, t \models \varphi)$$

The ∇-operator is the possible deviation operator, and it is the dual of the Δ-operator. On the semantical level, the formula $\nabla_{\mathcal{P}}\varphi$ is true if φ holds at at least one $\mathcal{P}$-deviated state. With abbreviation 3.2 the semantics of the ∇-operator can easily be derived.

Lemma 3.5. (Semantics derived operator)

$$\mathfrak{M}, t \models \nabla_{\mathcal{P}}\varphi \quad \Leftrightarrow \quad (\exists s : t(\mathcal{R}_{\Delta_{\mathcal{P}}})s : \mathfrak{M}, s \models \varphi)$$

The way the $\mathcal{R}_{\Delta}$-relations are defined influences the semantics of the logic and hence the properties of the operator itself.

Release logic is the logic of controlled ignorance and, since it is **S5**, might be viewed as a kind of epistemic logic. As said before, the operational interpretation of modal operator $\Delta_{\mathcal{P}}$ is modulo $\mathcal{P}$. Relation $\mathcal{R}_{\Delta_{\mathcal{P}}}$ allows *any* transition to a state has that is precisely the same, modulo $\mathcal{P}$.

In general, the set $\mathcal{A}$ will be chosen in such a way that, when being semantically translated, the subformula $\Delta_{\mathcal{P}}$ identifies some states — typically more than one. Now if a property is true in a set of states it is definitely true in all subsets of this collection of states. It suffices to define a partial order on the indices of the modality Δ in order to implement this property into the modal release logic.

Let $\sqsubseteq$ be a partial order on set $\mathcal{A}$. Then, the *only* assumption for a relation $\mathcal{R}_{\Delta_{\mathcal{P}}}$ to be a relation corresponding to the modal release operator $\Delta_{\mathcal{P}}$ is that if $\mathcal{P} \sqsupseteq \mathcal{Q}$ than via $\mathcal{R}_{\Delta_{\mathcal{P}}}$ *more* states are reachable then via $\mathcal{R}_{\Delta_{\mathcal{Q}}}$. Hence, release models are monotonic in this sense.

Definition 3.6. (**PRL** Kripke model)

Let $(\mathcal{A}, \sqsubseteq)$ be a partial order. A Kripke model $\mathfrak{M} = \langle \mathbb{S}, \pi, \mathcal{R}_{\Delta} \rangle$, where $\pi : \mathbb{S} \to \mathcal{PROP} \to \mathbb{B}$ is a *release* Kripke model if, for all $\mathcal{P}, \mathcal{Q} \in \mathcal{A}$:

- $\mathcal{R}_{\Delta_{\mathcal{P}}}$ is an equivalence relation
- $\mathcal{P} \sqsubseteq \mathcal{Q} \Rightarrow \mathcal{R}_{\Delta_{\mathcal{P}}} \subseteq \mathcal{R}_{\Delta_{\mathcal{Q}}}$

Hence, in a release model, the partial order on set $\mathcal{A}$ is monotonic with respect to the relation $\mathcal{R}_{\Delta}$. In fact, this property represents the only (additional) axiom of release logic.

In the sequel, the set $\mathcal{A}$ will be the set that generates the modal operators $\Delta_{\mathcal{P}}$, for all $\mathcal{P} \in \mathcal{A}$. The logic **PRL** is therefore identified with $\mathbf{PRL}_{\mathcal{A}}$.

3.3. Axiomatization of PRL

The logic **PRL** is a multi-modal **S5** logic, each component satisfies axiom system **S5** and an additional axiom, reflected by the partial order property of the modal Δ-operator. The logic **PRL**, therefore, needs the axiom scheme S5 and one additional axiom. Axiom **PO** (partial order) axiom expresses the partial order property between two release modalities.

Definition 3.7. (Axioms of **PRL**)

Suppose $\mathcal{P}, \mathcal{Q} \in \mathcal{A}$. The axioms of the logic **PRL** are:

TAUT	All instances of propositional tautologies	
$\mathbf{K}_{\Delta}$	$\Delta_{\mathcal{P}} (\varphi \to \psi) \to (\Delta_{\mathcal{P}} \varphi \to \Delta_{\mathcal{P}} \psi)$	
$\mathbf{T}_{\Delta}$	$\Delta_{\mathcal{P}} \varphi \to \varphi$	
$\mathbf{4}_{\Delta}$	$\Delta_{\mathcal{P}} \varphi \to \Delta_{\mathcal{P}} \Delta_{\mathcal{P}} \varphi$	
$\mathbf{5}_{\Delta}$	$\nabla_{\mathcal{P}} \varphi \to \Delta_{\mathcal{P}} \nabla_{\mathcal{P}} \varphi$	
PO	$\Delta_{\mathcal{P}} \varphi \to \Delta_{\mathcal{Q}} \varphi$	, if $\mathcal{P} \sqsupseteq \mathcal{Q}$

Definition 3.8. (Rules of **PRL**)

The derivation rules of the logic **PRL** are standard, let $\mathcal{P} \in \mathcal{A}$:

MP	$\varphi, \varphi \to \psi \models \psi$	(Modus Ponens)
$\mathbf{Nec}_{\Delta}$	$\varphi \models \Delta_{\mathcal{P}} \varphi$	(Necessitation)

The shorthand for a logic system satisfying the above axioms and rules is therefore **S5 PO** .

The partial ordering on the indices of the release operator are *anti* monotonic with respect to the material implication.

Lemma 3.9. Let $(\mathcal{A}, \sqsubseteq)$ be a partial order for **PRL** , $\mathcal{P}, \mathcal{Q} \in \mathcal{A}$, then $\Delta_.$ is anti monotonic with respect to the logical connective $\rightarrow$,

3.4. Soundness and completeness of PRL

Since **PRL** is a derivate of the logic **S5** , its soundness and completeness rely on the same results of **S5** . Moreover, as the proofs used for **S5** are compositional, the proofs can be re-used. However, the class of models and the canonical model have to be rebuilt, but the guidelines are identical, see (Chellas, 1993), (Hughes and Cresswell, 1984), (Hughes and Cresswell, 1996) and (Meyer and van der Hoek, 1996).

Definition 3.10. (Class of Kripke models for a logic **PRL**) Let $\mathcal{A}$ be a set. $\mathcal{PRL}$ is the class of **PRL** Kripke models.

Lemma 3.11. (Axiom **PO** is sound)

Let $\mathcal{A}$ be set, $\mathcal{P}, \mathcal{Q} \in \mathcal{A}$ and $\mathcal{P} \sqsupseteq \mathcal{Q}$. For all $\mathfrak{M} = \langle \mathbb{S}, \pi, \mathcal{R}_\Delta \rangle \in \mathcal{PRL}$, where $s \in \mathbb{S}$ and $\varphi \in \mathcal{FOR}_{PRL}$:

$\mathfrak{M}, s \models \Delta_\mathcal{P}\, \varphi \rightarrow \Delta_\mathcal{Q}\, \varphi$

Theorem 3.12. (Soundness of system **PRL**)

The axiom system **PRL** (= **S5 PO**) is sound with respect to the class of Kripke models $\mathcal{PRL}$.

Proof. Referring to literature ((Chellas, 1993), (Meyer and van der Hoek, 1996) etc.), the only obligation left is the soundness of axiom **PO** , which is sound by lemma 3.11.

Heading for completeness, the standard construction of a canonical model for multi modal logics provides a canonical model for **PRL** .

Definition 3.13. (Canonical model for modal logic **PRL**)

Let $\mathfrak{M}^{PRL} = \langle \mathbb{S}^{PRL}, \pi^{PRL}, \mathcal{R}_\Delta^{PRL} \rangle$ be the canonical Kripke model for the modal logic **PRL** .

Since the logic **PRL** satisfies axiom **K** , the truth and consistency lemmas immediately hold for **PRL** .

Lemma 3.14. (Truth lemma for **PRL**)

For any maximal consistent set Φ: $(\forall \varphi :: \mathfrak{M}^{PRL}, \Phi \models \varphi \Leftrightarrow \varphi \in \Phi)$

Lemma 3.15. (Consistency lemma for logic **PRL**)

$(\exists \mathfrak{M} : \mathfrak{M} \in \mathcal{PRL} : \mathfrak{M}, s \models \Phi)$

If a system **S** satisfies axioms **PO** then there is an inclusion relation between two ordered transition relations of the canonical model.

Lemma 3.16. Let $\mathcal{P}, \mathcal{Q} \in \mathcal{A}$, and, moreover, $\mathcal{P} \sqsubseteq \mathcal{Q}$ then:

$\mathbf{PO} \in \mathbf{S} \Rightarrow \mathcal{R}_{\Delta_\mathcal{P}}^{PRL} \subseteq \mathcal{R}_{\Delta_\mathcal{Q}}^{PRL}$

Corollary 3.17. $\mathfrak{M}^{PRL} \in \mathcal{PRL}$

Theorem 3.18. (Completeness axiom systems satisfying axioms **PRL**)

The axiom system **PRL** (= **S5 PO**) is complete with respect to the class of Kripke models $\mathcal{PRL}$.

Proof. For re-applying the Henkins proof, it suffices to point out that the canonical model $\mathfrak{M}^{PRL}$ reflects the inclusion structure between its relations, cf. corollary 3.17. □

3.5. Poset and lattice properties of Release Logics

The poset structure of a release logic is reflected by axiom **PO** , on the semantical level by the partial order on its relations, see definition 3.6. Several **PRL** validities can be derived, using combinations of the axioms **T** , **4** and **PO** .

The axioms **T** and **4** are used in the following lemma and **PO** is applied in its corollary.

Lemma 3.19. For all $\mathcal{P} \in \mathcal{A}$:

$\mathcal{R}_{\Delta_\mathcal{P}} \supseteq \mathbf{Id}$

$\mathcal{R}_{\Delta_\mathcal{P}} \supseteq \mathcal{R}_{\Delta_\mathcal{P}} \circ \mathcal{R}_{\Delta_\mathcal{P}}$

A corollary can be proved by multiple application of lemma 3.19 and definition 3.6.

Corollary 3.20. For all $\mathcal{P}, \mathcal{Q} \in \mathcal{A}$:

$\mathcal{R}_{\Delta_\mathcal{P}} = \mathcal{R}_{\Delta_\mathcal{P}} \circ \mathcal{R}_{\Delta_\mathcal{P}}$

$\mathcal{R}_{\Delta_\mathcal{P}} \cup \mathcal{R}_{\Delta_\mathcal{Q}} \subseteq \mathcal{R}_{\Delta_\mathcal{P}} \circ \mathcal{R}_{\Delta_\mathcal{Q}}$

Corollary 3.21. For all $\mathcal{P}, \mathcal{Q} \in \mathcal{A}$:

$\models \Delta_\mathcal{P}\, \varphi \leftrightarrow \Delta_\mathcal{P}\, \Delta_\mathcal{P}\, \varphi$

$\models \Delta_\mathcal{P}\, \varphi \wedge \Delta_\mathcal{Q}\, \varphi \leftarrow \Delta_\mathcal{P}\, \Delta_\mathcal{Q}\, \varphi$

Many partial orders $(\mathcal{A}, \sqsubseteq)$ have an additional lattice structure $(\mathcal{A}, \sqcup, \sqcap, \sqsubseteq)$.

Definition 3.22. (Modalities originating from a poset)

Let $(\mathcal{A}, \sqcup, \sqcap, \sqsubseteq)$ be a lattice. The set $\ _\mathcal{A}$ (shortened to) is the set of modalities generated by $\mathcal{A}$:

$\ _\mathcal{A} = \{\Delta_\mathcal{P} \mid \mathcal{P} \in \mathcal{A}\}$

If $\mathcal{A}$ has a lattice structure, this structure can be found in the set of modalities.

Lemma 3.23. Let $(\mathcal{A}, \sqcup, \sqcap, \sqsubseteq)$ be a lattice and $\mathcal{P}, \mathcal{Q} \in \mathcal{A}$. Then:

$\Delta_{\mathcal{P} \sqcup \mathcal{Q}} \in$ and $\Delta_{\mathcal{P} \sqcap \mathcal{Q}} \in$

Corollary 3.24. Let $(\mathcal{A}, \sqcup, \sqcap, \sqsubseteq, \perp, \top)$ be a *complete* lattice, then:

$\Delta_\perp, \Delta_\top \in$

The lattice structure has, obviously, effect on the number of modalities in the logic, but *not* necessarily on the (number of) axiom schemes. Yet the complete lattice structure initiates more constraints on the relations, and more validities. The latter are depicted in the leftmost picture of figure 3.

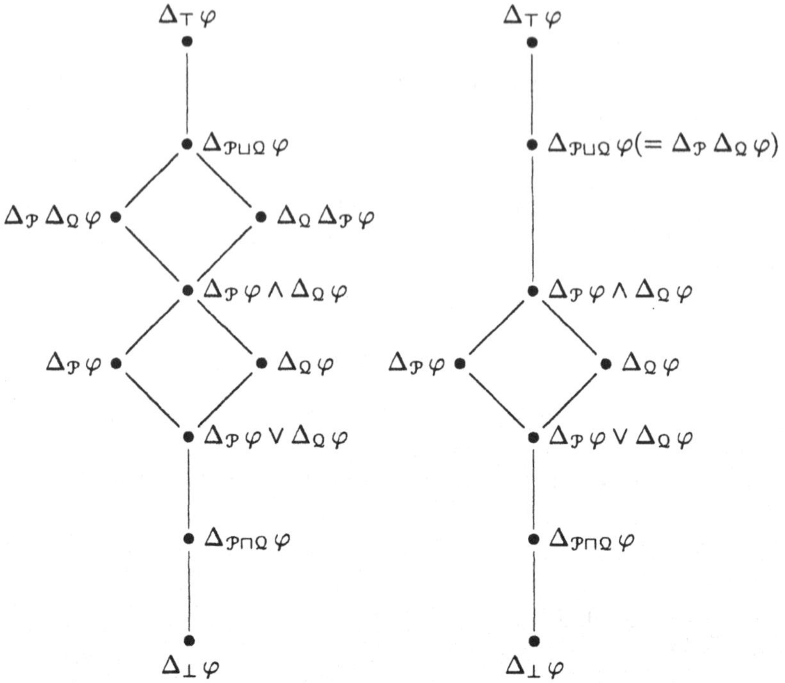

Figure 3. Lattice properties: a) General case. b) Full state space

Lemma 3.25. For all $\mathcal{P}, \mathcal{Q} \in \mathcal{A}$:
$\mathcal{R}_{\Delta_{\mathcal{P}\sqcup\mathcal{Q}}} \supseteq \mathcal{R}_{\Delta_\mathcal{Q}} \circ \mathcal{R}_{\Delta_\mathcal{P}}$
$\mathcal{R}_{\Delta_{\mathcal{P}\sqcap\mathcal{Q}}} \subseteq \mathcal{R}_{\Delta_\mathcal{P}} \cap \mathcal{R}_{\Delta_\mathcal{Q}}$

Corollary 3.26. For all $\mathcal{P}, \mathcal{Q} \in \mathcal{A}$:
$\models \Delta_{\mathcal{P}\sqcup\mathcal{Q}}\,\varphi \rightarrow \Delta_\mathcal{P}\,\Delta_\mathcal{Q}\,\varphi$
$\models \Delta_{\mathcal{P}\sqcap\mathcal{Q}}\,\varphi \leftarrow \Delta_\mathcal{P}\,\varphi \vee \Delta_\mathcal{Q}\,\varphi$

Models of special interest are those in which, for every valation, there is a state. Collecting the valuation function results in all valuation functions. Note that a state could still have many identical copies.

Definition 3.27. Let $\mathfrak{M} = \langle \mathbb{S}, \pi, \mathcal{R} \rangle$ be a model.
$\mathfrak{M}$ has a full state space iff $\pi.\mathbb{S} = \mathcal{PROP} \rightarrow \mathbb{B}$

To obtain equality in the first line of lemma 3.25, a full state space has to be assumed, see alse the rightmost picture in figure 3.

Lemma 3.28. Let $\mathfrak{M} = \langle \mathbb{S}, \pi, \mathcal{R} \rangle \in \mathcal{PRL}$ be a model with a full state space. Then:
$\mathcal{R}_{\Delta_{\mathcal{P}\sqcup\mathcal{Q}}} = \mathcal{R}_{\Delta_\mathcal{Q}} \circ \mathcal{R}_{\Delta_\mathcal{P}}$
$\mathcal{R}_{\Delta_{\mathcal{P}\sqcap\mathcal{Q}}} = \mathcal{R}_{\Delta_\mathcal{P}} \cap \mathcal{R}_{\Delta_\mathcal{Q}}$

The effect of the lemma on the Hasse diagrams is in figure 3, note the implosion of the upper diamond.

Corollary 3.29. Let $\mathfrak{M} = \langle \mathbb{S}, \pi, \mathcal{R} \rangle \in \mathcal{PRL}$ be a model with a full state space, φ a formula. Then:

$$\mathfrak{M} \models \Delta_{\mathcal{P} \sqcup \mathcal{Q}}\, \varphi \quad \leftrightarrow \quad \Delta_{\mathcal{P}}\, \Delta_{\mathcal{Q}}\, \varphi$$
$$\mathfrak{M} \models \Delta_{\mathcal{P}}\, \Delta_{\mathcal{Q}}\, \varphi \quad \leftrightarrow \quad \Delta_{\mathcal{Q}}\, \Delta_{\mathcal{P}}\, \varphi$$

3.6. An example of Release Logic

Release logic is a tool to cluster states into equivalent classes. In one single modal logic, many different equivalence classes can exist, by chosing the appropriate set $\mathcal{A}$ and its release operators $\Delta_{\mathcal{P}}$, $\mathcal{P} \in \mathcal{A}$.

An example of release logic is the logic $\mathbf{PRL}_{2^{\mathcal{PROP}}}$, thus instantiate $\mathcal{A}$ with $2^{\mathcal{PROP}}$. Hence, $\mathcal{P} \subseteq \mathcal{PROP}$ and the release operator $\Delta_{\{p,q\}}$, for example, creates equivalence classes of states modulo the valuation of p and q. More examples of release logic can be found in (Krabbendam and Meyer, 1997).

4. Combining Release Logic with other logics

Release logic can be integrated with many other modal logics. In particular, it can be combined with dynamic logic as well as temporal logic. The next sections describe briefly the combinations, approximation and orthogonal time logics respectively.

4.1. Approximation Logic: Release Logic × Dynamic Logic

For adding release logic to dynamic logic, an obvious choice for an instantiation of the set $\mathcal{A}$ in the release logic (definition 3.1) is the set of entities on which the actions take effect. If the valuation function π is in terms of propositions, then $\mathcal{A} = 2^{\mathcal{PROP}}$, see section 3.6. If the valuation function handles variables (e.g. $\pi : \mathbb{S} \to \mathcal{VAR} \to \mathbb{N}$), then set $\mathcal{A} = 2^{\mathcal{VAR}}$. In the sequel, it is assumed that $\pi.s$ is defined on propositions, see section 3.6, and the lattice $(\mathcal{A}, \sqcup, \sqcap, \sqsubseteq, \bot, \top) = (2^{\mathcal{PROP}}, \cup, \cap, \subseteq, \varnothing, \mathcal{PROP})$. To relate a specific release operator to a specific action, the modality Δ_α will be introduced.

Definition 4.1. (Effected propositions)

Let $\alpha \in \mathcal{ACT}$. The function $\mathcal{EP}$ collects the effects and interests of actions, thus $\mathcal{EP} : \mathcal{ACT} \to 2^{\mathcal{PROP}}$.

Typically, $\mathcal{EP}$ is the set of all effects, i.e. $\mathcal{EP}(\alpha) = \{p \in \mathcal{PROP} \mid (\exists s, t : s\mathfrak{R}_\alpha t : \pi.s.p \neq \pi.t.p)\}$, but this is not necessarily so.

For all $\alpha \in \mathcal{ACT}$ the modal operator Δ_α is a shorthand for $\Delta_{\mathcal{EP}(\alpha)}$. It specifies an equivalence class modulo the effects and interests of action α.

The language $\mathcal{LAN}_{PAL}$ is the language of a propositional approximation logic **PAL** . The language incorporates dynamic and release operators.

Definition 4.2. (Formulae of $\mathcal{LAN}_{PAL}$)

The set $\mathcal{FOR}_{PAL}$ of formulae of the language $\mathcal{LAN}_{PAL}$ has the following inductive definition:

- The set $\mathcal{PROP}$ of proposition variables is in the set of formulae

- If φ is a formula then $\neg\varphi$ is a formula
- If φ and ψ are formulae then $(\varphi \to \psi)$ is a formula
- If φ is a formula and $\alpha \in \mathcal{ACT}$, then $[\alpha]\varphi$ and $\Delta_\alpha \varphi$ are formulae

A natural join of the Kripke models of definitions 2.2 and 3.3 suffices. Note however, due to the assumption above, that in spite of the notation Δ_α, $\mathcal{R}_\Delta$ maps *sets of propositions*, not actions, onto the booleans.

Definition 4.3. (Kripke model for **PAL**)

A Kripke model $\mathfrak{M}_{PAL} = \langle \mathbb{S}, \pi, \mathcal{R}_{[.]}, \mathcal{R}_\Delta \rangle$ ($\mathfrak{M}$ for short) consists of:

- A non empty set $\mathbb{S}$ of states.
- A mapping $\pi : \mathbb{S} \to \mathcal{PROP} \to \mathbb{B}$.
- A mapping $\mathcal{R}_{[.]} : \mathcal{ACT} \to 2^{(\mathbb{S}\times\mathbb{S})}$ models the actions.
- A mapping $\mathcal{R}_\Delta : 2^{\mathcal{PROP}} \to 2^{(\mathbb{S}\times\mathbb{S})}$ models the release operators.

The semantic interpretations are exactly the same as in the cases **PAL** and **PRL** .

Definition 4.4. (**PAL** Kripke model)

A Kripke model $\mathfrak{M} = \langle \mathbb{S}, \pi, \mathcal{R}_{[.]}, \mathcal{R}_\Delta \rangle$, is an *approximation* Kripke model if $\langle \mathbb{S}, \pi, \mathcal{R}_\Delta \rangle \in \mathcal{PRL}$, so is a *release* Kripke model, $\langle \mathbb{S}, \pi, \mathcal{R}_{[.]} \rangle \in \mathcal{PDL}$ and, moreover, for all $\alpha \in \mathcal{ACT}, \mathcal{P} \subseteq \mathcal{PROP}$:

$$\mathcal{R}_{\Delta_\mathcal{P}} \circ \mathcal{R}_\alpha = \mathcal{R}_\alpha \circ \mathcal{R}_{\Delta_\mathcal{P}} \quad , \text{if } \mathcal{P} \cap \mathcal{EP}(\alpha) = \varnothing$$
$$\mathcal{R}_{\Delta_\alpha} \subseteq \mathcal{R}_{\Delta_\alpha} \circ \mathcal{R}_\alpha$$
$$\mathcal{R}_\alpha \subseteq \mathcal{R}_{\Delta_\alpha}$$

All axioms of the logics **PDL** and **PRL** are present in the axiom system for **PAL** . There are three axioms that specify the interaction between the dynamic and the release operator. Obviously, if the modalities $[\alpha]$ and $\Delta_\mathcal{P}$ share no common effect or interest on the state space, they can be swapped. Hence, if $\mathcal{P} \cap \mathcal{EP}(\alpha) = \varnothing$ then $[\alpha]\Delta_\mathcal{P}\varphi \leftrightarrow \Delta_\mathcal{P}[\alpha]\varphi$.

If action α operates entirely in the class $\mathcal{EP}(\alpha)$, its effects are absorbed by the release operator Δ_α and, moreover, it can freely operate within this class. This leads to $[\alpha]\Delta_\alpha\varphi \to \Delta_\alpha\varphi$ (**ABS**$_\alpha$, absorb) and $\Delta_\alpha\varphi \to [\alpha]\varphi$ (**ACT**$_\alpha$, act freely) respectively.

Definition 4.5. (Axioms of **PAL**)

Suppose $\mathcal{P}, \mathcal{Q} \subseteq \mathcal{PROP}$. The axioms of the logic **PAL** are:

TAUT	All instances of propositional tautologies	
$\mathbf{K}_\alpha$	$[\alpha](\varphi \to \psi) \to ([\alpha]\varphi \to [\alpha]\psi)$	
$\mathbf{K}_\Delta$	$\Delta_\mathcal{P}(\varphi \to \psi) \to (\Delta_\mathcal{P}\varphi \to \Delta_\mathcal{P}\psi)$	
$\mathbf{T}_\Delta$	$\Delta_\mathcal{P}\varphi \to \varphi$	
$\mathbf{4}_\Delta$	$\Delta_\mathcal{P}\varphi \to \Delta_\mathcal{P}\Delta_\mathcal{P}\varphi$	
$\mathbf{5}_\Delta$	$\nabla_\mathcal{P}\varphi \to \Delta_\mathcal{P}\nabla_\mathcal{P}\varphi$	
PO	$\Delta_\mathcal{P}\varphi \to \Delta_\mathcal{Q}\varphi$	, if $\mathcal{P} \supseteq \mathcal{Q}$
$\mathbf{SWAP}_{\Delta\alpha}$	$[\alpha]\Delta_\mathcal{P}\varphi \leftrightarrow \Delta_\mathcal{P}[\alpha]\varphi$	, if $\mathcal{P} \cap \mathcal{EP}(\alpha) = \varnothing$
$\mathbf{ABS}_\alpha$	$[\alpha]\Delta_\alpha\varphi \to \Delta_\alpha\varphi$	
$\mathbf{ACT}_\alpha$	$\Delta_\alpha\varphi \to [\alpha]\varphi$	

The rules are copied from the logics **PDL** and **PRL** .

Definition 4.6. (Class of Kripke models for a logic **PAL**)

$\mathcal{PAL}$ is the class of **PAL** Kripke models.

Lemma 4.7. For all $\mathfrak{M} = \langle \mathbb{S}, \pi, \mathcal{R}_{[.]}, \mathcal{R}_{\Delta} \rangle \in \mathcal{PAL}$, where $s \in \mathbb{S}$ and $\varphi \in \mathcal{FOR}_{PRL}$, the axioms $\mathbf{SWAP}_{\Delta\alpha}$, $\mathbf{ABS}_{\alpha}$ and $\mathbf{ACT}_{\alpha}$ are sound.

Theorem 4.8. (Soundness of system **PAL**)

The axiom system **PAL** is sound with respect to the class of Kripke models $\mathcal{PAL}$.

Proof. The proof is a composition of the proofs of theorems 2.6 and 3.12 and lemma 4.7. □

The logic **PAL** can be proved complete again, by adding the results from its ancestor logics.

Definition 4.9. (Canonical model for modal logic **PAL**)

Let $\mathfrak{M}^{PAL} = \langle \mathbb{S}^{PAL}, \pi^{PAL}, \mathcal{R}_{[.]}^{PAL}, \mathcal{R}_{\Delta}^{PAL} \rangle$ be the canonical Kripke model for the modal logic **PAL** .

Lemma 4.10. Let system **S** contain system **S5** .

$$\begin{array}{lll} \mathbf{SWAP}_{\Delta\alpha} \in \mathbf{S} & \Rightarrow & \mathcal{R}^S_{\alpha} \circ \mathcal{R}^S_{\Delta_\alpha} = \mathcal{R}^S_{\Delta_\alpha} \circ \mathcal{R}^S_{\alpha} \\ \mathbf{ABS}_{\alpha} \in \mathbf{S} & \Rightarrow & \mathcal{R}^S_{\Delta_\alpha} \subseteq \mathcal{R}^S_{\Delta_\alpha} \circ \mathcal{R}^S_{\alpha} \\ \mathbf{ACT}_{\alpha} \in \mathbf{S} & \Rightarrow & \mathcal{R}^S_{\alpha} \subseteq \mathcal{R}^S_{\Delta_\alpha} \end{array}$$

Corollary 4.11. $\mathfrak{M}^{PAL} \in \mathcal{PAL}$

Theorem 4.12. (Completeness axiom systems satisfying axioms **PAL**)

The axiom system **PAL** is complete with respect to the class of Kripke models $\mathcal{PAL}$.

Proof. By re-using the proofs of theorems 2.6 and 3.18 and application of lemma 4.11. □

An elementary computer science application of approximation logic is mutual recovery of crashes in system components. Let φ be a safety criterion. The specification for component α, able to recover from a fault in component β, would read in **PAL** : $\Delta_\beta\, [\alpha]\varphi$.

The analogue specification for component β can directly be combined with this one, thus $\Delta_\alpha\, [\beta]\varphi \wedge \Delta_\beta\, [\alpha]\varphi$.

Another, closely related, example would be the pre-emptive fault tolerancy. This is, an action creates a situation in which formula φ, safety for example, is immune for a crash in other component of the system. This would read: $[\alpha]\Delta_\beta\, \varphi \wedge [\beta]\Delta_\alpha\, \varphi$. Thus, action α establishes φ and β is by no means to undo it and vice verse.

4.2. Orthogonal Time Logic: Release logic × Temporal Logic

Linear temporal logic specifies one path through time. With release logic one can build a cluster of states, a bundle of paths of time. The idea is to use a release logic which has *no interference at all* with the temporal modal operator, in other words, the modalities release and time are orthogonal.

Assuming this orthogonality, the release logic that is part of the approximation logic **PAL** can directly be plugged in the linear temporal logic **PTL** .

The language $\mathcal{LAN}_{POTL}$ is the language of a orthogonal time logic **POTL** . The language is composed of temporal and release operators.

Definition 4.13. (Formulae of $\mathcal{LAN}_{POTL}$)

The set $\mathcal{FOR}_{POTL}$ of formulae of the language $\mathcal{LAN}_{POTL}$ has the following inductive definition:

- The set $\mathcal{PROP}$ of proposition variables is in the set of formulae
- If φ is a formula then $\neg\varphi$ is a formula
- If φ and ψ are formulae then $(\varphi \rightarrow \psi)$ is a formula
- If φ is a formula and $\mathcal{P} \subseteq \mathcal{PROP}$, then $\bigcirc\varphi$ and $\Delta_{\mathcal{P}}\varphi$ are formulae

Again a join of two Kripke models suffices, in this case the Kripke models of **PTL** and **PRL** , see definitions 2.8 and 3.3.

Definition 4.14. (Kripke model for **POTL**)

A Kripke model $\mathfrak{M}_{POTL} = \langle \mathbb{S}, \pi, \mathcal{R}_{\bigcirc}, \mathcal{R}_{\Delta} \rangle$ ($\mathfrak{M}$ for short) consists of:

- A non empty set $\mathbb{S}$ of states.
- A mapping $\pi : \mathbb{S} \rightarrow \mathcal{PROP} \rightarrow \mathbb{B}$.
- A relation $\mathcal{R}_{\bigcirc} \subseteq \mathbb{S} \times \mathbb{S}$ models the next operator.
- A mapping $\mathcal{R}_{\Delta} : 2^{\mathcal{PROP}} \rightarrow 2^{(\mathbb{S}\times\mathbb{S})}$ models the release operators.

The semantic interpretations are exactly the same as in the cases **POTL** and **PRL** . This is accomplished through the equivalence properties of the release operator in combination with the property that the order in applying $\mathcal{R}_{\bigcirc}$ and $\mathcal{R}_{\Delta}$ is irrelevant.

Definition 4.15. (**POTL** Kripke model)

A Kripke model $\mathfrak{M} = \langle \mathbb{S}, \pi, \mathcal{R}_{\bigcirc}, \mathcal{R}_{\Delta} \rangle$, is a *orthogonal time* Kripke model if:

$\langle \mathbb{S}, \pi, \mathcal{R}_{\Delta} \rangle \in \mathcal{PRL}$, hence is a *release* Kripke model

$\langle \mathbb{S}, \pi, \mathcal{R}_{\bigcirc} \rangle \in \mathcal{PTL}$, hence is a functional Kripke model

$\mathcal{R}_{\Delta_{\mathcal{P}}} \circ \mathcal{R}_{\bigcirc} = \mathcal{R}_{\bigcirc} \circ \mathcal{R}_{\Delta_{\mathcal{P}}}$, where $\mathcal{P} \in \mathcal{A}$

All axioms of the logics **PTL** and **PRL** are present in the axiom system for **POTL** . Since the logics are completely orthogonal, there is no interference, hence the modalities can be swapped.

Definition 4.16. (Axioms of **POTL**)

Suppose $\mathcal{P}, \mathcal{Q} \subseteq \mathcal{PROP}$. The axioms of the logic **POTL** are:

TAUT	All instances of propositional tautologies	
$\mathbf{K}_{\Delta}$	$\Delta_{\mathcal{P}}(\varphi \rightarrow \psi) \rightarrow (\Delta_{\mathcal{P}}\varphi \rightarrow \Delta_{\mathcal{P}}\psi)$	
$\mathbf{K}_{\bigcirc}$	$\bigcirc(\varphi \rightarrow \psi) \rightarrow (\bigcirc\varphi \rightarrow \bigcirc\psi)$	
$\mathbf{D!}_{\bigcirc}$	$\bigcirc\varphi \leftrightarrow \neg\bigcirc\neg\varphi$	
$\mathbf{T}_{\Delta}$	$\Delta_{\mathcal{P}}\varphi \rightarrow \varphi$	
$\mathbf{4}_{\Delta}$	$\Delta_{\mathcal{P}}\varphi \rightarrow \Delta_{\mathcal{P}}\Delta_{\mathcal{P}}\varphi$	
$\mathbf{5}_{\Delta}$	$\nabla_{\mathcal{P}}\varphi \rightarrow \Delta_{\mathcal{P}}\nabla_{\mathcal{P}}\varphi$	
PO	$\Delta_{\mathcal{P}}\varphi \rightarrow \Delta_{\mathcal{Q}}\varphi$	, if $\mathcal{P} \supseteq \mathcal{Q}$
$\mathbf{SWAP}_{\Delta\bigcirc}$	$\bigcirc\Delta_{\mathcal{P}}\varphi \leftrightarrow \Delta_{\mathcal{P}}\bigcirc\varphi$	

The rules are copied from the logics **PTL** and **PRL** .

Lemma 4.17. For all models $\mathfrak{M} = \langle \mathbb{S}, \pi, \mathcal{R}_{\bigcirc}, \mathcal{R}_{\Delta} \rangle \in \mathcal{POTL}$, the swap axiom $\mathbf{SWAP}_{\Delta\bigcirc}$ is sound.

Theorem 4.18. (Soundness of system **POTL**)
The axiom system **POTL** is sound with respect to the class of Kripke models $\mathcal{POTL}$.

Proof. The proof is a composition of the proofs of theorems 2.14 and 3.12 and lemma 4.17. □

The logic is complete.

Definition 4.19. (Canonical model for modal logic **POTL**)
Let $\mathfrak{M}^{POTL} = \langle \mathbb{S}^{POTL}, \pi^{POTL}, \mathcal{R}_{\bigcirc}^{POTL}, \mathcal{R}_{\Delta}^{POTL} \rangle$ be the canonical Kripke model for the modal logic **POTL** .

Lemma 4.20. Let system **S** contain system $S5_{\Delta}$. Then $\mathbf{SWAP}_{\Delta\bigcirc} \in \mathbf{S} \Rightarrow \mathcal{R}_{\bigcirc}^{S} \circ \mathcal{R}_{\Delta_{\mathcal{P}}}^{S} = \mathcal{R}_{\Delta_{\mathcal{P}}}^{S} \circ \mathcal{R}_{\bigcirc}^{S}$

Corollary 4.21. $\mathfrak{M}^{POTL} \in \mathcal{POTL}$

Theorem 4.22. (Completeness axiom systems satisfying axioms **POTL**)
The axiom system **POTL** is complete with respect to the class of Kripke models $\mathcal{POTL}$.

Proof. By re-using the proofs of theorems 2.14 and 3.18 and applying lemma 4.21. □

The logic **POTL** on its own is not very exciting. It has an application in the next section, where it will be used to accept actions in the framework. One remark is important though. The axiom $\mathbf{SWAP}_{\Delta\bigcirc}$ indicates that time has no influence on the cardinality of the valuation function, i.e. as time passes, no opportunity, that is, valuation function on a state, will be lost, nor gained. It is easy to drop this axiom and construct a logic **POTL′** which is sound and complete again. Anyway, in doing so, the orthogonality between the temporal and release modalities is lost.

5. Temporalizing Propositional Dynamic Logic

This section briefly describes the threefold combination of release, dynamic and temporal logic. In fact, this threefold combination is an orthogonal twofold union of approximation logic and orthogonal time logic, whose unchanged axioms and properties are identifiable in the temporalized propositional dynamic logic PTDL, see figure 4.

The time constraints on the actions finally prescribe the interconnecting axioms and properties between the dynamic and release modalities and the temporal end release modalities, without the aforementioned friction between time and action.

5.1. The orthogonal approach

The temporalized dynamic logic comes in two flavours, one with instantaneous actions and one allowing an action to take some time to complete. Many

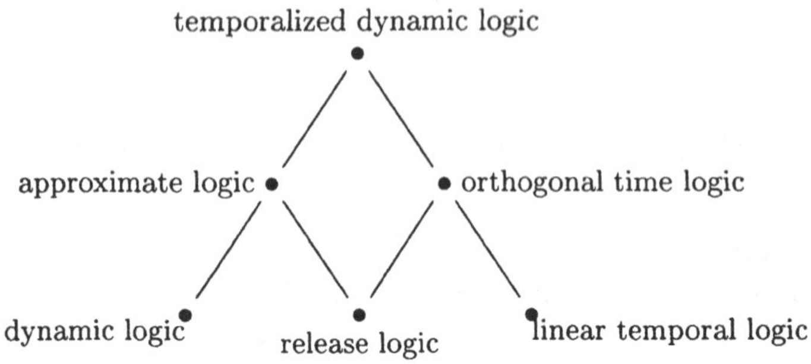

Figure 4. The (genea-)logical tree

definitions apply for both variants and there are shared properties since the instantaneous logic is a sublogic of the time spending logic. The common parts are the subject of this section. The instantaneous sublogic of **PTDL** will gain an index $_i$, when it is necessary to make a distinction.

The language $\mathcal{LAN}_{PTDL}$ is the language of a temporalized dynamic logic **PTDL** . The modalities in the language are dynamic, temporal and release operators.

Definition 5.1. (Formulae of $\mathcal{LAN}_{PTDL}$)

The set $\mathcal{FOR}_{PTDL}$ of formulae of the language $\mathcal{LAN}_{PTDL}$ has the following inductive definition:

- The set $\mathcal{PROP}$ of proposition variables is in the set of formulae
- If φ is a formula then $\neg\varphi$ is a formula
- If φ and ψ are formulae then $(\varphi \rightarrow \psi)$ is a formula
- If φ is a formula, $\alpha \in \mathcal{ACT}$ and $\mathcal{P} \subseteq \mathcal{PROP}$, then $\bigcirc\varphi$, $[\alpha]\varphi$ and $\Delta_{\mathcal{P}}\,\varphi$ are formulae

The release operator used in the approximation and orthogonal time logics is the same, and the Kripke models of definitions 4.3 and 4.14 can be merged.

Definition 5.2. (Kripke model for **PTDL**)

A Kripke model $\mathfrak{M}_{PTDL} = \langle \mathbb{S}, \pi, \mathcal{R}_{[.]}, \mathcal{R}_{\bigcirc}, \mathcal{R}_{\Delta} \rangle$ ($\mathfrak{M}$ for short) consists of:

- A non empty set $\mathbb{S}$ of states.
- A mapping $\pi : \mathbb{S} \rightarrow \mathcal{PROP} \rightarrow \mathbb{B}$.
- A mapping $\mathcal{R}_{[.]} : \mathcal{ACT} \rightarrow 2^{(\mathbb{S}\times\mathbb{S})}$ models the actions.
- A mapping $\mathcal{R}_{\bigcirc} \subseteq \mathbb{S} \times \mathbb{S}$ models the next future operator.
- A mapping $\mathcal{R}_{\Delta} : 2^{\mathcal{PROP}} \rightarrow 2^{(\mathbb{S}\times\mathbb{S})}$ models the release operators.

The semantic interpretations are exactly the same as in the cases **PAL** and **POTL** .

The axioms, which the instantaneous and time consuming logics have in common, are plain copies from the the logics **PAL** and **POTL** .

Definition 5.3. (Shared axioms of **PTDL**)

Suppose $\mathcal{P}, \mathcal{Q} \subseteq \mathcal{PROP}$. Among the axioms of the logics **PTDL** are:

TAUT	All instances of propositional tautologies	
$\mathbf{K}_\alpha$	$[\alpha](\varphi \to \psi) \to ([\alpha]\varphi \to [\alpha]\psi)$	
$\mathbf{K}_\bigcirc$	$\bigcirc(\varphi \to \psi) \to (\bigcirc\varphi \to \bigcirc\psi)$	
$\mathbf{K}_\Delta$	$\Delta_\mathcal{P}(\varphi \to \psi) \to (\Delta_\mathcal{P}\varphi \to \Delta_\mathcal{P}\psi)$	
$\mathbf{T}_\Delta$	$\Delta_\mathcal{P}\varphi \to \varphi$	
$\mathbf{4}_\Delta$	$\Delta_\mathcal{P}\varphi \to \Delta_\mathcal{P}\Delta_\mathcal{P}\varphi$	
$\mathbf{5}_\Delta$	$\nabla_\mathcal{P}\varphi \to \Delta_\mathcal{P}\nabla_\mathcal{P}\varphi$	
PO	$\Delta_\mathcal{P}\varphi \to \Delta_\mathcal{Q}\varphi$	, if $\mathcal{P} \supseteq \mathcal{Q}$
$\mathbf{SWAP}_{\Delta\alpha}$	$[\alpha]\Delta_\mathcal{P}\varphi \leftrightarrow \Delta_\mathcal{P}[\alpha]\varphi$	, if $\mathcal{P} \cap \mathcal{EP}(\alpha) = \varnothing$
$\mathbf{SWAP}_{\Delta\bigcirc}$	$\bigcirc\Delta_\mathcal{P}\varphi \leftrightarrow \Delta_\mathcal{P}\bigcirc\varphi$	

The rules are copied from the logics **PDL** and **PRL** .

A general temporalized dynamic logic model has the following structure.

Definition 5.4. (**PTDL** Kripke model)

A Kripke model $\mathfrak{M} = \langle \mathbb{S}, \pi, \mathcal{R}_{[.]}, \mathcal{R}_\bigcirc, \mathcal{R}_\Delta \rangle$, is a temporalized dynamic Kripke model if

$\langle \mathbb{S}, \pi, \mathcal{R}_{[.]}, \mathcal{R}_\Delta \rangle \in \mathcal{PAL}$

$\langle \mathbb{S}, \pi, \mathcal{R}_\bigcirc, \mathcal{R}_\Delta \rangle \in \mathcal{POTL}$

and, moreover, for all $\alpha \in \mathcal{ACT}, \mathcal{P} \subseteq \mathcal{PROP}$: $\mathcal{R}_{\Delta_\mathcal{P}} \circ \mathcal{R}_\alpha = \mathcal{R}_\alpha \circ \mathcal{R}_{\Delta_\mathcal{P}}$, if $\mathcal{P} \cap \mathcal{EP}(\alpha) = \varnothing$

5.2. Instantaneous actions

If a temporalized dynamic logic has instantaneous actions only, the release operator effectively separates the temporal and action components, it serves as a transition modality. The $\mathbf{ACT}_\alpha$ and $\mathbf{ABS}_\alpha$ axioms are therefore the same as in the logic **PAL** .

Definition 5.5. (Axioms of instantaneous $\mathbf{PTDL}_i$)

Suppose $\mathcal{P}, \mathcal{Q} \subseteq \mathcal{PROP}$. The axioms of the logic $\mathbf{PTDL}_i$ are the axioms from definition 5.3 and:

$\mathbf{ABS}_\alpha$ $[\alpha]\Delta_\alpha\varphi \to \Delta_\alpha\varphi$

$\mathbf{ACT}_\alpha$ $\Delta_\alpha\varphi \to [\alpha]\varphi$

Definition 5.6. ($\mathbf{PTDL}_i$ Kripke model)

A Kripke model $\mathfrak{M} = \langle \mathbb{S}, \pi, \mathcal{R}_{[.]}, \mathcal{R}_\bigcirc, \mathcal{R}_\Delta \rangle$, is an instantaneous temporalized dynamic Kripke model ($\mathbf{PTDL}_i$ model) if it is a **PTDL** model (definition 5.4), and, for all $\alpha \in \mathcal{ACT}$:

$\mathcal{R}_{\Delta_\alpha} \subseteq \mathcal{R}_{\Delta_\alpha} \circ \mathcal{R}_\alpha$

$\mathcal{R}_\alpha \subseteq \mathcal{R}_{\Delta_\alpha}$

Definition 5.7. (Class of Kripke models for a logic $\mathbf{PTDL}_i$)

$\mathcal{PTDL}_i$ is the class of $\mathbf{PTDL}_i$ Kripke models.

Theorem 5.8. (Soundness of system $\mathbf{PTDL}_i$)

The axiom system $\mathbf{PTDL}_i$ is sound with respect to the class of Kripke models $\mathcal{PTDL}_i$.

Proof. The proof is a composition of the proofs of theorems 4.8 and 4.18. □

The completeness of the logic $\mathbf{PTDL}_i$ is a result from the properties of its building blocks.

Definition 5.9. (Canonical model for modal logic $\mathbf{PTDL}_i$)
Let $\mathfrak{M}^{PTDL_i} = \langle \mathbb{S}^{PTDL_i}, \pi^{PTDL_i}, \mathcal{R}_{[.]}^{PTDL_i}, \mathcal{R}_{\bigcirc}^{PTDL_i}, \mathcal{R}_{\Delta}^{PTDL_i} \rangle$ be the canonical Kripke model for the modal logic $\mathbf{PTDL}_i$.

Lemma 5.10. $\mathfrak{M}^{PTDL_i} \in \mathcal{PTDL}_i$

Proof. By lemmas 4.10 and 4.20. □

Theorem 5.11. (Completeness axiom systems satisfying axioms $\mathbf{PTDL}_i$)
The axiom system $\mathbf{PTDL}_i$ is complete with respect to the class of Kripke models $\mathcal{PTDL}_i$.

Proof. Lemma 5.10 and a re-application of the proofs of theorems 4.12 and 4.22. □

5.3. Actions that take time

The interesting case is if the actions of the temporalized dynamic logic consume time, the logic is named $\mathbf{PTDL}_t$. All definitions in section 5.1 apply and are shared with $\mathbf{PTDL}_i$.

Referring to the axioms of the instantaneous logic $\mathbf{PTDL}_i$, the number of temporal modal operators on the left hand side and the right hand side of the arrows are the same. This is not the case with actions, the axioms $\mathbf{ABS}_\alpha$ and $\mathbf{ACT}_\alpha$ (definition 5.5) are out of balance.

Definition 5.12. (Duration)
The mapping $\|.\| : \mathcal{ACT} \to \mathbb{N}$ determines the time spent by action $\alpha \in \mathcal{ACT}$.

The time spent on action α is assumed to be constant, see the co-domain of mapping $\|.\|$. The temporal transition of an action α can now be expressed as $\bigcirc^{\|\alpha\|}$ and this will be used to re-balance axioms $\mathbf{ABS}_\alpha$ and $\mathbf{ACT}_\alpha$.

Definition 5.13. (Axioms of instantaneous $\mathbf{PTDL}_t$)
Suppose $\mathcal{P}, \mathcal{Q} \subseteq \mathcal{PROP}$. The axioms of the logic $\mathbf{PTDL}_t$ are the axioms from definition 5.3 and:

$\mathbf{ABS}_\alpha$	$[\alpha]\Delta_\alpha \varphi \to \bigcirc^n \Delta_\alpha \varphi$	, if $n = \|\alpha\|$
$\mathbf{ACT}_\alpha$	$\bigcirc^n \Delta_\alpha \varphi \to [\alpha]\varphi$	, if $n = \|\alpha\|$

Definition 5.14. ($\mathbf{PTDL}_t$ Kripke model)
A Kripke model $\mathfrak{M} = \langle \mathbb{S}, \pi, \mathcal{R}_{[.]}, \mathcal{R}_{\bigcirc}, \mathcal{R}_{\Delta} \rangle$, is a time spending temporalized dynamic Kripke model ($\mathbf{PTDL}_t$ model) if it is a $\mathbf{PTDL}$ model (definition 5.4), and, for all $\alpha \in \mathcal{ACT}$, where $n = \|\alpha\|$:

$\mathcal{R}_{\Delta_\alpha} \circ \mathcal{R}_{\bigcirc}^n \subseteq \mathcal{R}_{\Delta_\alpha} \circ \mathcal{R}_\alpha$

$\mathcal{R}_\alpha \subseteq \mathcal{R}_{\Delta_\alpha} \circ \mathcal{R}_{\bigcirc}^n$

Definition 5.15. (Class of Kripke models for a logic $\mathbf{PTDL}_t$)
$\mathcal{PTDL}_t$ is the class of $\mathbf{PTDL}_i$ Kripke models.

Taking $\|\alpha\| = 0$, for all $\alpha \in \mathcal{ACT}$, it is obvious that $\mathbf{PTDL}_i \sqsubseteq \mathbf{PTDL}_t$, thus $\mathcal{PTDL}_i \subseteq \mathcal{PTDL}_t$ etc.. The index $_t$ will be omitted in the rest of the paper.

Lemma 5.16. For all $\mathfrak{M} = \langle \mathbb{S}, \pi, \mathcal{R}_{[.]}, \mathcal{R}_{\bigcirc}, \mathcal{R}_{\Delta} \rangle \in \mathcal{PTDL}$, where $s \in \mathbb{S}$ and $\varphi \in \mathcal{FOR}_{PTDL}$, the axioms, $\mathbf{ABS}_\alpha$ and $\mathbf{ACT}_\alpha$ are sound.

Theorem 5.17. (Soundness of system **PTDL**)

The axiom system **PTDL** is sound with respect to the class of Kripke models $\mathcal{PTDL}$.

Proof. Add lemma 5.16 to the proofs of theorems 4.8 and 4.18. □

The completeness proof of the logic **PTDL** is a re-run of the completeness proof for its instantaneous variant.

Definition 5.18. (Canonical model for modal logic **PTDL**)

Let $\mathfrak{M}^{PTDL} = \langle \mathbb{S}^{PTDL}, \pi^{PTDL}, \mathcal{R}_{[.]}^{PTDL}, \mathcal{R}_{\bigcirc}^{PTDL}, \mathcal{R}_{\Delta}^{PTDL} \rangle$ be the canonical Kripke model for the modal logic **PTDL** .

Lemma 5.19. $\mathfrak{M}^{PTDL} \in \mathcal{PTDL}$

Proof. By lemmas 4.10 and 4.20. □

Theorem 5.20. (Completeness axiom systems satisfying axioms **PTDL**)

The axiom system **PTDL** is complete with respect to the class of Kripke models $\mathcal{PTDL}$.

Proof. Lemma 5.19 in combination with the proofs of theorem 4.12 and theorem 4.22. □

5.4. Oneline specification examples and the Yale Shooting Problem

The temporalized dynamic logics described in this paper can be used to specify systems and programs. Three oneliners examplify the temporalized dynamic logic **PTDL** as a tool for partial specification.

- The first specification example is an advertisement about machines. It reads $[\alpha]\varphi \rightarrow \bigcirc^{(\|\alpha\| \div 2)} \nabla_\beta \varphi$ and must be a promotion campaign for machine β, since β can perform the same actions *at least* twice as fast as machine α. Note however, that if φ is a safety criterium, β probably fails. In that case, the conclusion should be strenghtened with $[\beta]\varphi$.
- The second specification is about a parameter, which can only measured k seconds after completion of system part α. During this period of time, a part β is active and not allowed to interfere with the measurement: $[\alpha](\varphi \rightarrow \Delta_\beta \bigcirc^k \varphi)$. Note that β acts as it likes, *except* for any inteference with respect to φ.
- The last example is about an action α, which establishes the truth of φ, no matter any interference by β. A program β has (exactly) k seconds to react on this event and undo φ. The specification is: $[\alpha]\Delta_\beta \varphi \rightarrow [\alpha]\Delta_\beta \bigcirc^k [\beta]\neg\varphi$. It is even possible for program β to falsify φ no matter what α is doing during the pause and the run of β: $[\alpha]\Delta_\beta \varphi \rightarrow$

$[\alpha]\Delta_\beta \bigcirc^k [\beta]\Delta_\alpha \neg\varphi$. The reader is invited to check that there is a lock free solution to the specification due to the temporal aspects of the specification.

Finally, the release logic solves the Yale Shooting problem, a standard specification problem in Artificial Intelligence. In short, many models allow the satisfiability of $[\text{load}]\bigcirc[\text{fire}]\text{alive}$, stating that if the gun is loaded and a second later fired at a person, it is possible he will *not* be hurt, since the gun could be unloaded in during the pause. This is in fact the *chaos* interpretation. Adding a release operator ensures the reliability of the gun by prohibiting to unload the gun: $[\text{load}]\bigcirc\Delta_{\overline{\text{load}}}[\text{fire}]\text{dead}$. Note that the temporal operator $\bigcirc$ is now the *freeze* next operator.

5.5. The passing objects example

In this section, a specification example is treated in some more detail.

Suppose there are identical three robots $\mathcal{A}$, $\mathcal{B}$ and $\mathcal{C}$. The first robot puts objects on a table $\mathcal{X}$. Robot $\mathcal{B}$ lifts them again, moves its arm together with the object towards table $\mathcal{Y}$, where the object will be released. Robot $\mathcal{C}$ is to remove it from table $\mathcal{Y}$ again etc. The layout is depicted in figure 5.

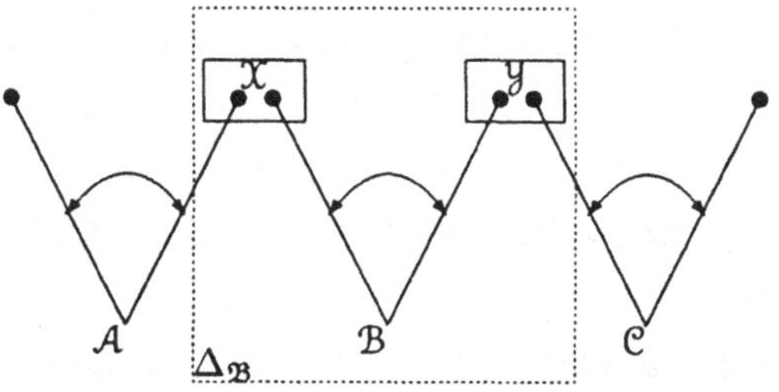

Figure 5. Passing objects

The designer's task is to specify robot $\mathcal{B}$. Hence, whether robot $\mathcal{A}$ has any interference with another robot or production system other than robot $\mathcal{B}$ is irrelevant, since the only concern of the design is the specification of robot $\mathcal{B}$. What *is* interesting is that the arms of $\mathcal{A}$ and $\mathcal{B}$ should not collide. Of course, collisions between $\mathcal{B}$ and $\mathcal{C}$ should be avoided as well, but that should not bother robot $\mathcal{A}$. Unnecessary arm movements and unnecessary delays should be avoided as well. Moreover, the designer should minimize his meddlesomeness with the neighbouring robots $\mathcal{A}$ and $\mathcal{C}$. However, in order to avoid unwanted sitations, such as collisions, there should be *some* restrictions on $\mathcal{A}$ and $\mathcal{C}$.

The programs of all robots can be decribed as $(\text{pick}; \text{rotR}; \text{drop}; \text{rotL})^*$, the effects of the robots $\mathcal{A}$, $\mathcal{B}$ and $\mathcal{C}$ are collected in sets $\mathfrak{A}$, $\mathfrak{B}$ and $\mathfrak{C}$ respectively. Hence, if robot $\mathcal{A}$ is picking up an object, it resides above table $\mathcal{X}$, a position identified with proposition $\mathcal{A}_{\text{at}\mathcal{X}}$. Clearly, $\mathcal{A}_{\text{at}\mathcal{X}} \in \mathfrak{A}$.

Avoiding a collision with robots $\mathcal{A}$ and $\mathcal{B}$ involved can be stated as:

$$\neg(\mathcal{A}_{\text{at}}\mathcal{X} \wedge \mathcal{B}_{\text{at}}\mathcal{X}) \tag{5.1}$$

The problem is of course, that mechanical apparatus tend to behave with a certain delay. Now if $\bigcirc^{\leq 1}\varphi \equiv \varphi \wedge \bigcirc\varphi$, the constraint can be:

$$\neg(\bigcirc^{\leq 1}\mathcal{A}_{\text{at}}\mathcal{X} \wedge \bigcirc^{\leq 1}\mathcal{B}_{\text{at}}\mathcal{X}) \tag{5.2}$$

But, if ones interpretates the $\bigcirc$-operator as a *freeze*-operator, the whole system is frozen and unacceptable delays are slipped into the specification. The solution is to be aware of the sphere of influence of *all* robots in the system and add release operators to ensure enough room for manoeuvre for all robots to do what they like, *except* for robots $\mathcal{A}$ and $\mathcal{B}$, which are restricted only if they are in a position where a collision can occur. So, both propositions in equation 5.2 should be preceded by $\Delta_{\mathfrak{C}}$ to ensure robot $\mathfrak{C}$ is free to act, *and* they are preceded by $\Delta_{\mathcal{B}\backslash\mathcal{B}_{\text{at}\mathcal{X}}}$ and $\Delta_{\mathcal{A}\backslash\mathcal{A}_{\text{at}\mathcal{X}}}$ respectively. So, the tolerant formula, where no unnecessary blocking occurs reads:

$$\neg(\Delta_{\mathfrak{C}}\,\Delta_{\mathfrak{B}\backslash\mathcal{B}_{\text{at}\mathcal{X}}}\,\bigcirc^{\leq 1}\mathcal{A}_{\text{at}}\mathcal{X} \wedge \Delta_{\mathfrak{C}}\,\Delta_{\mathfrak{A}\backslash\mathcal{A}_{\text{at}\mathcal{X}}}\,\bigcirc^{\leq 1}\mathcal{B}_{\text{at}}\mathcal{X}) \tag{5.3}$$

If the state space is full, by lemma 3.28, the concatenating release operators can be combined, thus reducing complexity issues and will then be:

$$\neg(\Delta_{\mathfrak{C}\sqcup(\mathfrak{B}\backslash\mathcal{B}_{\text{at}\mathcal{X}})}\,\bigcirc^{\leq 1}\mathcal{A}_{\text{at}}\mathcal{X} \wedge \Delta_{\mathfrak{C}\sqcup(\mathfrak{A}\backslash\mathcal{A}_{\text{at}\mathcal{X}})}\,\bigcirc^{\leq 1}\mathcal{B}_{\text{at}}\mathcal{X}) \tag{5.4}$$

The second example is a part of the behavioural specification of robot $\mathcal{B}$. Obviously, the robot should pick up an object if it arrives at table $\mathcal{X}$: $\mathcal{B}_{\mathcal{B}_{\text{at}\mathcal{X}}} \rightarrow [\text{pick}]\mathcal{B}_{\text{has-object}}$. Now this action should again be tolerant with respect to robot $\mathfrak{C}$, as well as, to some respect, to robot $\mathcal{A}$. But if there is danger of collisions, it should be possible to withdraw from table $\mathcal{X}$, *without* having lifted an object. So the specification needs two release operators again:

$$\mathcal{B}_{\mathcal{B}_{\text{at}\mathcal{X}}} \rightarrow [\text{pick}]\Delta_{\mathfrak{C}}\,\Delta_{\mathfrak{A}\backslash\mathcal{A}_{\text{at}\mathcal{X}}}\,\mathcal{B}_{\text{has-object}} \tag{5.5}$$

Of course, if the state space is full, one could write:

$$\mathcal{B}_{\mathcal{B}_{\text{at}\mathcal{X}}} \rightarrow [\text{pick}]\Delta_{\mathfrak{C}\sqcup(\mathfrak{A}\backslash\mathcal{A}_{\text{at}\mathcal{X}})}\,\mathcal{B}_{\text{has-object}} \tag{5.6}$$

Finally, equation 5.5 has no explicit temporal component. The time is hided in the now time consuming modal operator [pick], and, this equation is therefore an example of the language in the logic **PAL** .

5.6. Conclusion

As shown, the additional modal release operator gives a framework for orthogonalizing temporal and dynamic modalities. The effect of the $\bigcirc$-operator is precisely the intended effect of the step semantics — freezing the state space and advancing one timeframe. Possible effects on the state space can be expressed by release operators. As a consequence, the freeze paradox is valid in the cases where it is most natural, namely if the temporal aspect is irrelevant. On the other hand, if time is relevant, the paradoxes are avoided. The orthogonal component of the combined logics is the glue between the interfering time and action modalities.

The actions in the logic can be instantaneous and this is a subcase of the actions spending a certain amount of time. Both approaches are supported with **PTDL**.

System specifications, with special focus on parts of the system only, can now be expressed in a logic composed of action and temporal operators. The release logic provides the tools to separate the system and its environment, while still taking into account the influence of the environment.

The logics which use release operators to combine time and action do not suffer from the chaos paradox $\bigcirc\varphi \to [\alpha]\varphi$, neither from the freeze paradox $\bigcirc[\alpha]\varphi \leftrightarrow [\alpha]\bigcirc\varphi$ (section 2.3). It is, however, possible to add the freeze 'paradox' as an axiom. The implication is then that the actions' behaviours are independent of time. Such logic can be useful if the part of the system to be specified is instantaneous, but its environment is not. In such a case, a definition $\bigcirc_{\mathcal{P}} = \bigcirc\Delta_{\mathcal{P}}$ describes the uncertainty with respect to $\mathcal{P}$ because of advancing time.

5.7. Future research

The first step will be the construction of an interval logic with modal operators $\bigcirc_{\mathcal{T}}$, where $\mathcal{T} \subseteq \mathbb{N}$. This expansion allows an action α to utilize a nondeterministic amount of time and bring in the future always operator $\Box$-operator.

Release logic in isolation has some theoretical significance in modelling the semantic structure with syntactical constraints. For example, with a single additional axiom it is possible to enforce the existence of all valuations in the state space. With another axiom all states differ in at least n valuations (Krabbendam and Meyer, 1997).

In combination with other logics, release logic provides a modal logic tool to orthogonalize the modal components.

Future research focuses on enriching the components of temporalized dynamic logic, such as concatenation of actions, (non-)deterministic choice and iteration, thus implementing the standard dynamic logic. Moreover, release logic can also provide means to construct logics for specifying true concurrency (Krabbendam and Meyer, 1997).

References

Andréka, H., V. Goranko, S. Mikulás, I. Németi, and I. Sain: 1995, 'Effective Temporal Logics of Programs'. In: L. Bolc and A. Szałas (eds.): *Time and Logic - A computational Approach*. London, UK: University College London Press Ltd., pp. 51–129.

Chellas, B.: 1993, *Modal Logic, An introduction*. New York, USA: Cambridge University Press.

Gabbay, D.: 1976, 'Two-dimensional Propositional Tense Logic'. In: A. Kasher (ed.): *Language in Focus*. Dordrecht, NL: Reidel, pp. 569–583.

Goldblatt, R.: 1992, *Logics of Time and Computation*, Vol. 7 of *CSLI Lecture Notes*. Stanford, USA: Center for the Study of Language and Information, second edition.

Goldblatt, R.: 1993, *Mathematics of Modality*, Vol. 43 of *CSLI Lecture Notes*. Stanford, USA: Center for the Study of Language and Information.

Harel, D.: 1979, *First-Order Dynamic Logic*, Vol. 64 of *Lecture Notes in Computer Science*. Berlin, Germany: Springer Verlag.

Henkin, L., J. Monk, and A. Tarski: 1971, *Cylindric Algebras - Part 1*, Vol. 64 of *Studies in Logic and the Foundations of Mathematics*. Amsterdam, NL: North-Holland.

Henkin, L., J. Monk, and A. Tarski: 1985, *Cylindric Algebras - Part 2*, Vol. 115 of *Studies in Logic and the Foundations of Mathematics*. Amsterdam, NL: North-Holland.

Hughes, G. and M. Cresswell: 1984, *A Companion to Modal Logic*. London, UK: Methuen.

Hughes, G. and M. Cresswell: 1996, *A New Introduction to Modal Logic*. London, UK: Routledge.

Jánossy, A., Á. Kurucz, and Á. Eiben: 1996, 'Combining Algebraizable Logics'. *Notre Dame Journal of Formal Logic* **37**(2), 366–380.

Krabbendam, J. and J.-J. C. Meyer: 1997, 'Release logic as a specification tool for true concurrency'. Technical report, Utrecht University, Utrecht, NL. In preparation.

Kröger, F.: 1987, *Temporal Logic of Programs*, Vol. 8 of *EATCS Monographs on Theoretical Computer Science*. Berlin, Germany: Springer-Verlag.

Manna, Z. and A. Pnueli: 1992, *The Temporal Logic of Reactive and Concurrent Systems — Specification*. New York, USA: Springer Verlag.

Manna, Z. and A. Pnueli: 1995, *Temporal Verification of Reactive Systems — Safety*. New York, USA: Springer Verlag.

Meyer, J.-J. and W. van der Hoek: 1996, *Epistemic Logic for AI and Computer Science*. Cambridge, UK: Cambridge University Press.

Nemeti, I.: $\star$, 'Combining Algebraizable Logics'. regularly updated version available via ftp:/circle.math-inst.hu/pub/algebraic-logic.

Venema, Y.: 1992, 'Many-dimensional modal logic'. Ph.D. thesis, University of Amsterdam, Amsterdam, NL.

Compositional Verification of Timed Statecharts

Francesca Levi (levifran@di.unipi.it)
Dipartimento di Informatica, Università di Pisa
Corso Italia 40
56100 Pisa, Italy

Abstract. We propose a compositional proof system for checking real-time properties of a dicrete timed process language $\mathcal{TSP}$ with minimal and maximal delays associated to actions. In order to capture the quantitative aspect of time we consider a discrete extension of μ-calculus with freeze quantification over clocks and clock constraints. The language $\mathcal{TSP}$ is parametric in the set of basic actions and it is characterized by an operator of process refinement, which permits to suitably model a discrete timed version of statecharts. The proof system is proved to be sound in general and complete for the class of regular finite state processes. It is indeed complete for processes corresponding to statecharts.

1. Introduction

Nowadays many applications are time-critical. Typical examples are transportation systems, railways control systems and signal-processing systems. The languages for programming real-time reactive systems are required to express timing contraints on the interaction of the system with the environment and their correctness depends also on the quantitative aspect of time. Hence, there is an increasing interest in the study of both the practical and the theoretical aspects of real-time logics (Alur and Henzinger, 1992; Alur and Henzinger, 1990), that are capable of expressing quantitative temporal properties. In this framework compositional verification is very important for developing in a modular way complex specifications. As opposed to verification over the whole system, it makes it possible to leave undefined parts of the system and still to reason about it. Moreover, it has the advantage that, when a part is modified, only the verification concerning this particular component must be redone.

In this paper we propose a compositional verification method for a discrete timed process language called $\mathcal{TSP}$. It is a CCS-like language where actions can be constrained between a minimal and maximal delay and it is parametric in the set of basic actions and in two operations over actions of parallel composition and conflict. Moreover, it is characterized by an operator of process refinement and by explicit locations.

The design of $\mathcal{TSP}$ is motivated by the aim to have a quite general and simple process language that is adequate for modelling a discrete timed version of statecharts with minimal and maximal delays

H. Barringer et al. (eds.), Advances in Temporal Logic, 47–70.

associated to transitions. Statecharts (Harel, 1987) is a synchronous formalism for the specification of reactive and real-time systems that overcome the limitations of state-transition diagrams by introducing notions of *parallelism*, *hierarchy* and *broadcast communication*. The language is based on the synchrony hypothesis (Berry and Gonthier, 1992):"the reaction of the system to inputs is instantaneous". Due to well-known causality problems the formal definition of the instantaneous reaction of the system (*step*) to inputs is quite controversial (Harel et al., 1987; Pnueli and Shalev, 1991; Maggiolo-Schettini et al., 1996). In the approach of (Pnueli and Shalev, 1991) a step is given by a maximal set of parallel transitions that are enabled with respect to the set of current inputs and that satisfy some additional requirements of consistency and causality. (Levi, 1996b; Levi, 1997) show a translation of classical untimed statecharts into an untimed version of $\mathcal{TSP}$ which agrees with the approach of Pnueli and Shalev. The basic idea is to build in a compositional way a process with locations corresponding to the basic states of the statechart and actions corresponding to transitions and representing the associated communication *trigger/action*, where *trigger* expresses an enabling condition on input events and *action* gives the set of instantaneuosly generated events. The main problem is that due to the non-modularity of the semantics the input-output behaviour is not sufficient for compositionality. The proposed solution is to take into account in basic actions also the causal relation between inputs and outputs. The operator of refinement allows us to naturally represent the statecharts hierarchy. The translation between statecharts and the process language can be suitably extended to the timed discrete case (Levi, 1997). However, the interest of the language is not limited to statecharts, since it can be instantiated to other process calculi with different communication mechanisms, such as Synchronous CCS (Milner, 1983) and the Calculus of Broadcasting Systems (Prasad, 1995), by properly choosing the set of basic actions and the related operations.

In order to express quantitatively bounded temporal properties we consider a discrete real-time extension of propositional μ-calculus (Kozen, 1983; Pratt, 1981) (RTL_μ) with freeze quantification over clocks and clock constraints. The logic is powerful enough to express the typical real-time properties of bounded response and bounded invariance. We define a compositional proof system for checking RTL_μ properties of $\mathcal{TSP}$ processes. In the syle of (Andersen et al., 1994) local and compositional reasoning are combined toghether. There are two types of proof rules:

1. logical rules that work on the structure of the formula. Fixpoint formulas are treated by a generalization of the method of tagging fixpoints (Winskel, 1991).

2. rules for the next modalities that work on the structure of the process and reduce a satisfaction problem for a compound process into satisfaction problems for its components.

In this approach the construction of a proof is driven by the syntax of the formula and of the process without reference to the underlying transition system which is typically infinite. The proof system is proved to be sound in general and complete for the class of regular finite state processes. It is indeed complete for processes corresponding to timed statecharts. A similar proof system for an untimed version version of $\mathcal{TSP}$ has been developed in (Levi, 1997; Levi, 1996a).

The paper is organized as follows. Section 2 presents the language $\mathcal{TSP}$ and its labelled transition system semantics. The logic RTL_μ is introduced in section 3. Finally, section 4 presents the compositional proof system.

2. The language $\mathcal{TSP}$

We introduce the process language $\mathcal{TSP}$. We assume a set *Act* of basic actions, a set *Loc* of locations and a set *Var* of variables. Two operations are assumed on basic actions, i.e. $\times : Act \times Act \rightarrow 2^{Act}$ (*parallel composition*) and $\# : Act \cup \{*\} \rightarrow 2^{Act}$ (*conflict*). We require $\times$ and $\#$ to satisfy the following properties:

1. for each $a_1, a_2, a_3 \in Act$, $a_1 \times (a_2 \times a_3) = (a_1 \times a_2) \times a_3$ (associativity), where $\times$ is the straightforward extension to sets;

2. for each $a_1, a_2, a_3, a_4 \in Act$, such that $a_2 \in \#(a_5)$ and $a_4 \in \#(a_6)$ for some $a_5, a_6 \in Act$, if $a \in a_1 \times a_2$ and $a \in a_3 \times a_4$, then $a_2 = a_4$.

The operation of parallel composition is used to encode the communication among parallel components. The operation of conflict relies on the assumption that the enabling of actions depends on the environment inputs. Intuitively, if $\#(a) = \emptyset$ then a is autonomous and it requires no enabling from the environment. In contrast, an action $a' \in \#(a)$ espresses conditions on the environment which prevent the performance of a. For instance, if a requires an input i, then an action representing " i is not produced" is in conflict. In this case we require that a does not depend on timing constraints. The special action $*$ is used for obtaining a generic behaviour of the environment.

DEFINITION 1. *Processes p of $\mathcal{TSP}$ are defined as follows:*

$$b ::= nil \mid b_1 + b_2 \mid [m, M]a.t$$
$$t ::= \ell :: b \mid rec\, x.t \mid t \bigtriangledown p \mid x$$
$$p ::= \ell :: t \mid \ell :: p_1 \times p_2$$

where $a \in Act$, $x \in Var$, $\ell \in Loc$, $m \in N$, $M \in N \cup \infty$ and $m \leq M$, $M \neq 0$. For $a \in Act$ with $\#(a) \neq \emptyset$, we require $m = 0$ and $M = \infty$.

The constructors $+$, *rec*, $\times$ stand as usual for *choice*, *recursion* and *parallel composition*. Timed prefixing is denoted by $[m, M]a.$. Roughly speaking, the action a can be performed only after its minimal delay m and cannot be delayed after its maximal delay M. The constructor of process refinement $\bigtriangledown$ is typical of our language and permits very compact and hierarchical specifications. It is inspired by the statecharts hierarchy. Locations are introduced as the counterparts of basic states in statecharts.

For $p = \ell :: t$ and $p = \ell :: p_1 \times p_2$, $main(p) = \{\ell\}$ and $loc(p) = \{\ell\} \cup loc(t)$, $loc(p) = \{\ell\} \cup loc(p_1) \times loc(p_2)$ respectively, where $loc(\ell :: b) = \{\ell\}$, $loc(rec\, x.t) = loc(t[rec\, x.t/x])$, $loc(t \bigtriangledown p) = loc(t) \cup loc(p)$ and $loc(x) = \emptyset$. Moreover, $rel(p) = loc(p) \setminus main(p)$. In other words $loc(p)$ represents the current configuration, namely the set of entered locations (basic states). In the following we consider well-formed and closed processes, where the locations of all subprocesses are distinct. We denote by $\mathcal{CTSP}$ the set of closed and well-formed processes.

2.1. The semantics of $\mathcal{TSP}$ processes

We define a labelled transition system semantics of closed $\mathcal{TSP}$ processes. Such a semantics is based on the idea of considering clocks corresponding to locations. The value of a clock represents the time that has been elapsed by the process in the corresponding location. States are given by a process together with an assignment of values to its clocks and transitions represent actions of the process together with an increment of one time unit in the values of clocks.

We need some preliminary concepts. For a set C of clocks, $\pi : C \to N$ is a *clock assignment* over C. For $\delta \in N$, we denote by $\pi + \delta$ the clock assignment, such that for each $x \in C$, $\pi(x) + \delta$ (and analogously for $\pi - \delta$). Moreover, for $C_1 \subseteq C$, $\pi[C_1 := 0]$ denotes the assignment such that, $\pi(y) = \pi[C_1 := 0](y)$, for $y \notin C_1$, and $\pi[C_1 := 0](x) = 0$ for $x \in C_1$.

DEFINITION 2. *Let C be a set of clocks. A* clock constraint *is inductively defined by the grammar $\omega ::= x\mathcal{R}n \mid x + n\mathcal{R}y + m \mid \omega_1 \wedge \omega_2 \mid \neg\omega$ with $x, y \in C$ and $n, m \in N$, $\mathcal{R} \in \{\geq, \leq\}$.*

We denote by $\Omega(C)$ the set of clock constraints over clocks C. Clock constraints of $\Omega(C)$ are evaluated with respect to clock assignments over C ($\pi \models \omega$).

We define a function $delay : \mathcal{TSP} \rightarrow \Omega(rel(p))$ as follows:

$$\begin{array}{l}
delay(\ell :: t) \equiv delay(t) \\
delay(\ell :: p_1 \times p_2) \equiv delay(p_1) \wedge delay(p_2) \\
delay(rec\, x.t) \equiv delay(t[rec\, x.t/x]) \\
delay(t \bigtriangledown p) \equiv delay(t) \wedge delay(p) \\
delay(\ell :: b) \equiv \left\{ \begin{array}{ll} \ell \leq delay(b) & delay(b) \neq \infty \\ true & \text{otherwise} \end{array} \right\} \\
delay(nil) \equiv \infty \\
delay(a.t) \equiv \infty \\
delay([m, M]a.t) \equiv M \\
delay(b_1 + b_2) \equiv min(delay(b_1), delay(b_2));
\end{array}$$

Intuitively, $delay(p)$ gives the conditions that an assignment to clocks of p must satisfy. These conditions are used to force a timed action to be performed whenever its maximal delay is reached.

Let C be a set of clocks, $\pi : C \rightarrow N$ over C and $C_1 \subseteq C$. We define $proc_{C_1}(\pi) : C_1 \rightarrow N$, as $\pi(x) = proc_{C_1}(\pi)(x)$, for each $x \in C_1$. For C_1 and C_2 with $C_1 \cap C_2 = \emptyset$, and for $\pi_1 : C_1 \rightarrow N$ and $\pi_2 : C_2 \rightarrow N$, we define $\pi_1 \cup \pi_2 : C_1 \cup C_2 \rightarrow N$, such that $\pi_1 \cup \pi_2(x) = \pi_i(x)$, for $x \in C_i$ and $i \in \{1, 2\}$. Moreover, for $C_1 \subseteq C$ and C_2, such that $C_1 \cap C_2 = \emptyset$, and for $\pi : C \rightarrow N$, we denote by $reset(\pi, C_1, C_2) : C_1 \cup C_2 \rightarrow N$ the clock assignment $proc_{C_1}(\pi) \cup \{C_2 := 0\}$.

Transitions are labelled by actions obtained as follows.

DEFINITION 3. *Let Act be a set of basic actions. We define $\mathcal{S}(Act)$ as the set of* semantic actions *such that, for each $a \in Act$, $\tau \in \{\epsilon, \bar{\epsilon}\}$, $a : \tau \in \mathcal{S}(Act)$.*

There are two types of actions: *non-idle* (ϵ) and idle ($\bar{\epsilon}$). A non-idle action represents a real action of the process, while an idle action represents the inability to do anything. This distinction is essential for expressing in the $\mathcal{TSP}$ framework the maximality of reaction as required by the synchrony of statecharts. The idea is that idle actions represent an environment behaviour, which does not enable any non-idle action of the process.

The operation $\times$ can be trivially extended to semantic actions $\mathcal{S}(Act)$. For $\alpha_i = a_i : \tau_i \in \mathcal{S}(Act)$, for $i \in \{1, 2\}$, we define $\alpha_1 \times \alpha_2$ as the set $\{a : \tau \in \mathcal{S}(Act)\}$, such that $a \in a_1 \times a_2$ and $\tau = \epsilon$ if $\tau_i = \epsilon$, for some $i \in \{1, 2\}$, while $\tau = \bar{\epsilon}$ otherwise. For $\alpha = a : \tau$, we use the notation $type(\alpha) = \tau$.

Assume $p \in \mathcal{TSP}$ to be a closed process over actions Act and locations Loc. We define the labelled transition system semantics of p as $LTS(p) = (S, s_0, \mathcal{S}(Act), \mapsto)$ where

1. S is the set of timed states (p', π), such that p' is a closed process and π is an assignment over $rel(p)$, such that $\pi \models delay(p)$;
2. $s_0 = (p, \pi_0)$ with $\pi_0(\ell) = 0$, for each $\ell \in rel(p)$;
3. the transition relation $s_1 \overset{\alpha}{\mapsto} s_2$ is obtained by the rules in table I.

Let us briefly explain the semantics. The main feature is the ability to express maximal reactions, because of the synchrony of statecharts. By rule $non-idle$ action a can be performed nondeterministically whithin its minimal and maximal delay. The resulting assignment for $\ell_1 :: t$ is obtained by setting to one the value of the clocks of t, since the corresponding locations have been entered and time has advanced of one unit. If $\#(a) = \emptyset$, then a does not depend on the environment and can be delayed until M. This is expressed by rule $idle$ by considering a generic behaviour of the environment ($b \in \#(*)$). On the other hand, if $\#(a) \neq \emptyset$, then action a does not depend on temporal constraints by definition and the process can be idle only if the environment has a behaviour which does not enable a. This is obtained by rule $idle$ by taking $a' \in \#(a)$, which is in conflict with a. Note that it must be the case that $\pi_1 + 1 \models delay(\ell_1 :: \ell_2 :: [m, M]a.t)$ so that when the maximal delay is reached action a is forced to happen. Rule $choice_1$ is standard. On the other hand, by rule $choice_2$ an idle action must be an idle action of both processes so that the reaction is maximal. The constructor $\triangledown$ is similar to the one of choice. However, the choice between the two processes is not symmetric as expressed by rules $\triangledown_1$ and $\triangledown_2$. The side condition of $\triangledown_2$ guarantees that no timed action of $\ell :: t$ has reached its maximal delay. Rule $\times$ is the classical rule of parallel synchronous composition and is based on the operation $\times$ over actions, that intuitively realizes the communication between parallel components. We refer the reader to (Levi, 1997) for a more careful presentation of the connection of the semantics of $\mathcal{TSP}$ with that of timed statecharts.

3. The logic RTL_μ

We introduce a discrete version of real-time μ-calculus which is negation free and has freeze quantification over a set of specification clocks. We adopt a generalization of the tagged fixpoints method for local model checking (Winskel, 1991; Stirling and Walker, 1991).

Table I. The semantics of $\mathcal{TSP}$

$$\frac{\pi_1 \models m \leq \ell_2 \leq M}{(\ell_1 :: \ell_2 :: [m,M]a.t, \pi_1) \overset{a:\epsilon}{\mapsto} (\ell_1 :: t, \{loc(t) := 1\})}\ non-id$$

$$\frac{\pi_1 + 1 \models delay(\ell_1 :: \ell_2 :: [m,M]a.t) \quad (b \in \#(a) \vee (b \in \#(*) \wedge \#(a) = \emptyset)}{(\ell_1 :: \ell_2 :: [m,M]a.t, \pi_1) \overset{b:\bar{\epsilon}}{\mapsto} (\ell_1 :: \ell_2 :: [m,M]a.t, \pi_1 + 1)}\ id$$

$$\frac{b \in \#(*)}{(\ell_1 :: \ell_2 :: nil, \pi_1) \overset{b:\bar{\epsilon}}{\mapsto} (\ell_1 :: \ell_2 :: nil, \pi_1 + 1)}\ nil$$

$$\frac{\begin{array}{l}(\ell_1 :: \ell_2 :: b_i, \pi) \overset{\alpha}{\mapsto} (p, \pi') \\ type(\alpha) = \epsilon\end{array}}{(\ell_1 :: \ell_2 :: b_1 + b_2, \pi \overset{\alpha}{\mapsto} (p, \pi')}\ choice_1$$

$$\frac{\begin{array}{l}\{(\ell_1 :: \ell_2 :: b_i, \pi) \overset{\alpha}{\mapsto} (\ell_1 :: \ell_2 :: b_i, \pi'_i)\}_{i \in \{1,2\}} \\ type(\alpha) = \bar{\epsilon}\end{array}}{(\ell_1 :: \ell_2 :: b_1 + b_2, \pi) \overset{\alpha}{\mapsto} (\ell_1 :: \ell_2 :: b_1, \pi'_1 \cup \pi'_2)}\ choice_2$$

$$\frac{\begin{array}{l}(\ell :: t, \pi_1) \overset{\alpha}{\mapsto} (p', \pi'_1) \\ type(\alpha) = \epsilon\end{array}}{(\ell :: t \triangledown p, \pi_1 \cup \pi_2) \overset{\alpha}{\mapsto} (p', \pi'_1)}\ \triangledown_1$$

$$\frac{\begin{array}{l}(p, \pi_1) \overset{\alpha}{\mapsto} (p', \pi'_1) \\ type(\alpha) = \epsilon \quad \pi_2 + 1 \models delay(\ell :: t)\end{array}}{(t \triangledown p, \pi_1 \cup \pi_2) \overset{\alpha}{\mapsto} (t \triangledown p', \pi'_1 \cup (\pi_2 + 1))}\ \triangledown_2$$

$$\frac{\begin{array}{l}(\ell :: t, \pi_1) \overset{\alpha}{\mapsto} (\ell :: t, \pi'_1) \\ (p, \pi_2) \overset{\alpha}{\mapsto} (p, \pi'_2) \\ type(\alpha) = \bar{\epsilon}\end{array}}{(\ell :: t \triangledown p, \pi_1 \cup \pi_2) \overset{\alpha}{\mapsto} (\ell :: t \triangledown p, \pi'_1 \cup \pi'_2)}\ \triangledown_3$$

$$\frac{(\ell :: t[rec\ x.t/x], \pi) \overset{\alpha}{\mapsto} (p, \pi')}{(\ell :: rec\ x.t, \pi) \overset{\alpha}{\mapsto} (p, \pi')}\ rec$$

$$\frac{\{(p_i, \pi_i) \overset{\alpha_i}{\mapsto} (p'_i, \pi'_i)\}_{i \in \{1,2\}} \quad \alpha \in \alpha_1 \times \alpha_2}{(\ell :: p_1 \times p_2, \pi_2 \cup \pi_2) \overset{\alpha}{\mapsto} (\ell :: p'_1 \times p'_2, \pi'_2 \cup \pi'_2)}\ \times$$

Assume a set of specification clocks C, a set of logical variables VAR, a set of propositional symbols $PROP$ and a set of actions $\mathcal{A}$. Moreover, assume a set of clocks C', with $C \cap C' = \emptyset$, a set of states S and a function $rel : S \to 2^{C'}$. Formulas of the real-time μ-calculus RTL_μ are defined as follows

$$A ::= \begin{array}{l} X \mid P \mid \neg P \mid A_1 \vee A_2 \mid A_1 \wedge A_2 \mid \omega \mid < \alpha > A \mid [\alpha]A \mid x.A \mid \\ \mu X\{U\}.A \mid \nu X\{U\}.A. \end{array}$$

where $\omega \in \Omega(C)$ is a clock constraint, $x \in C$ is a specification clock, $\alpha \in \mathcal{A}$ is an action, $P \in PROP$ is a propositional symbol, $X \in VAR$ is a logical variable and $U = \{(s,\omega) \mid s \in S, \omega \in \Omega(C \cup rel(s))\}$.

The modal operator $< \alpha >$ is the next time, which is equivalent in this framework to next state, while $[\alpha]$ is its dual. The quantifier $x.$ is called *reset* and is inspired by the freeze quantification (Alur and Henzinger, 1989). Formulas $\mu X\{\emptyset\}.A$ and $\nu X\{\emptyset\}.A$ stand respectively for the classical least and greatest fixpoint. In formulas $\mu X\{U\}.A$ and $\nu X\{U\}.A$, U is called the tag and represents a set of timed states over clocks $C' \cup C$. Note that the specification clocks of formulas and the clocks corresponding in timed states are required to be distinct for avoiding the reset of clock corresponding to locations.

Formulas are interpreted over a model $\mathcal{M} = ((TS, ts_o, \mathcal{A}, \mapsto), \gamma)$, where $\gamma : PROP \to 2^{TS}$ is the valuation function of propositional symbols and $(TS, ts_o, \mathcal{A}, \mapsto)$ is a labelled transition system, such that

- TS is a set of timed states (s, π), where $s \in S$ and $\pi : C \cup C' \to N \cup \{und\}$ is a partial assignment for the specification clocks $C' \cup C$, such that $\pi = \pi_1 \cup \pi_2$, with $\pi_1 : C \to N \cup \{und\}$, $\pi_2 : C' \to N$;
- $ts_0 = (s_0, \pi_0)$, where $\pi_0(x) = und$, for each $x \in C$;
- $\mapsto \subseteq TS \times \mathcal{A} \times TS$ is the next state relation such that, for each $(s_1, \pi_1) \overset{\alpha}{\mapsto} (s_2, \pi_2)$, $\pi_2(x) = \pi_1(x) + 1$, for $x \in C$ ($und + 1 \equiv und$).

The semantics of formulas with respect to a model $\mathcal{M} = ((TS, ts_o, \mathcal{A}, \mapsto), \gamma)$ and to a valuation $\rho : VAR \to 2^{TS}$ is defined as follows:

$$\begin{array}{l}
\| P \|_{\mathcal{M},\rho} = \gamma(P) \\
\| \neg P \|_{\mathcal{M},\rho} = TS \setminus \gamma(P) \\
\| \omega \|_{\mathcal{M},\rho} = \{(s,\pi) \in TS \mid \pi \models \omega\} \\
\| x.A \|_{\mathcal{M},\rho} = \{(s,\pi) \in TS \mid (s, \pi[x := 0]) \in \| A \|_{\mathcal{M},\rho}\} \\
\| < \alpha > A \|_{\mathcal{M},\rho} = \{(s,\pi) \in TS \mid (s,\pi) \overset{\alpha}{\mapsto} (s',\pi'), (s',\pi') \in \| A \|_{\mathcal{M},\rho}\} \\
\| [\alpha]A \|_{\mathcal{M},\rho} = \{(s,\pi) \in TS \mid \forall (s,\pi) \overset{\alpha}{\mapsto} (s',\pi'), (s',\pi') \in \| A \|_{\mathcal{M},\rho}\} \\
\| \mu X\{U\}.A \|_{\mathcal{M},\rho} = \mu V.(\| A \|_{\mathcal{M},\rho[V/X]} \setminus \| U \|_{\mathcal{M},\rho[V/X]}) \\
\| \nu X\{U\}.A \|_{\mathcal{M},\rho} = \nu V.(\| A \|_{\mathcal{M},\rho[V/X]} \cup \| U \|_{\mathcal{M},\rho[V/X]})
\end{array}$$

where $\| U \|_{\mathcal{M},\rho} = \{(s,\pi) \in TS \mid (s,\omega) \in U, \pi \models \omega\}$.

We use the notation $(s,\pi) \models_{\mathcal{M}} A$ for $(s,\pi) \in \| A \|_{\mathcal{M},\rho}$ for each valuation ρ. For a process p and an assignment π over $rel(p)$, we say that $(p,\pi) \models A$ iff $(p,\pi) \models_{(LTS(p),\gamma)} A$, where γ is the valuation $\gamma : Loc \rightarrow \mathcal{CTSP}$, such that $p \in \gamma(P)$ iff $P \in loc(p)$. Note that A must be a closed formula defined over propositional symbols Loc, actions $\mathcal{S}(Act)$, specification clocks C, with $C \cap Loc = \emptyset$, states $\mathcal{CTSP}$ and rel as defined in section 2. Moreover, for $\omega \in \Omega(C \cup rel(p))$, we say $p,\omega \models A$ iff, for each π with $\pi \models \omega$, $(p,\pi) \models A$.

It is important to stress that the logic RTL_μ subsumes the discrete real-time extension of CTL $RTCTL$ (Emerson et al., 1989). For instance the time-bounded property $E\Diamond_{\leq 3}A$ is expressed by the formula $x.\mu X.((A \wedge x \leq 3) \vee \langle - \rangle X)$, while $A\Box_{\leq 3}A$ is expressed by the formula $x.\nu X.(((A \wedge x \leq 3) \vee (x > 3)) \wedge [-]X)$.

4. The proof system

We introduce the proof system. We consider judgments such as $p,\omega \vdash A$, where $p \in \mathcal{CTSP}$, $\omega \in \Omega(rel(p)) \cup C)$ and A is a closed RTL_μ formula. We require in addition that for each $x + m\mathcal{R}y + m \in \omega$, $x,y \in C$. The obvious interpretation of $p,\omega \vdash A$ is that (p,π) satisfies A for each assignment $\pi \models \omega$. There are two classes of rules: logical rules that do not involve process operators and rules that work on the structure of the process for the next modalities. The proof system is developed in a goal directed fashion (top down).

We need some preliminary definitions. Assume $\omega \in \Omega(C)$. We denote by $\omega+\delta$ ($\omega-\delta$) for $\delta \in N$, the clock constraint $\omega[x+\delta/x]$ ($\omega[x-\delta/x]$), for each clock $x \in C$. For $C_1 \subseteq C$, we denote by $pro_{C_1}(\omega)$ the first order formula $\exists x_1,\ldots,x_n\ (\omega)$, where $C \setminus C_1 = \{x_1,\ldots x_n\}$. Let $\omega \in \Omega(C)$ and $C_1 \subseteq C$, $C_2 \cap C_1 = \emptyset$. We denote by $reset(\omega, C_1, C_2) \equiv pro_{C_1}(\omega) \wedge \bigwedge_{x \in C_2} x = 0$. It can be shown that $reset(\omega, C_1, C_2) \in \Omega(C_1 \cup C_2)$. Intuitively, $pro_{C_1}(\omega)$ and $reset(\omega, C_1, C_2)$ are the counterparts of the corresponding operations over clock assignments. For $\omega \in \Omega(C)$, we use the notation $\forall(\omega)$ for $\forall x_1,\ldots,\forall x_n(\omega)$, where $C = \{x_1,\ldots,x_n\}$.

4.1. Logical Rules

$$\frac{p,\omega \vdash P}{}\ prop_1 \quad P \in loc(p) \qquad\qquad \frac{p,\omega \vdash \neg P}{}\ prop_2 \quad P \notin loc(p)$$

$$\frac{p,\omega \vdash A_1 \wedge A_2}{p,\omega \vdash A_1 \quad p,\omega \vdash A_2}\ \wedge \qquad\qquad \frac{p,\omega \vdash A_1 \vee A_2}{p,\omega \vdash A_i}\ \vee$$

$$\frac{p,\omega_1 \vdash A}{p,\omega_2 \vdash A}\ thin \quad \models \forall(\omega_1 \supset \omega_2) \qquad\qquad \frac{p,\omega_1 \vee \omega_2 \vdash A}{p,\omega_1 \vdash A \quad p,\omega_1 \vdash A}\ disj$$

$$\frac{p,\omega_1 \vdash \omega_2}{}\ taut \quad \models \forall(\omega_1 \supset \omega_2)$$

$$\frac{p,\omega \vdash A}{}\ empty \quad \models \forall(\omega \wedge delay(p) \supset false)$$

$$\frac{p,\omega \vdash x.A}{p, reset(\omega, (rel(p) \cup C) \setminus \{x\}, x) \vdash A}\ reset$$

$$\frac{p,\omega \vdash \nu X\{U\}.A}{}\ \nu_1 \qquad \begin{array}{l}\{(p,\omega_i) \subseteq U\}_{i\in\{1,n\}} \\ \models \forall(\omega \supset \bigvee_{i\in\{1,n\}} \omega_i)\end{array}$$

$$\frac{p,\omega \vdash \nu X\{U\}.A}{p,\omega \vdash A[\nu X\{U \cup \{(p,\omega)\}\}.A/X]}\ \nu_2$$

$$\frac{p,\omega \vdash \mu X\{U\}.A}{p,\omega \vdash A[\mu X\{U \cup \{(p,\omega)\}\}.A/X]}\ \mu \qquad \begin{array}{l}\{(p,\omega_i) \subseteq U\}_{i\in\{1,n\}} \\ \forall i \in \{1,n\} \models \forall(\omega \wedge \omega_i \supset false)\end{array}$$

Rule *reset* is based on the semantics of quantifier $x.$ by exploiting the fact that $\pi \models \omega$ iff $\pi[x := 0] \models reset(\omega, (rel(p) \cup C) \setminus \{x\}, x)$. Rules ν_1, ν_2 and μ are obtained by generalizing to sets of timed states the rules for computing fixpoints locally (Winskel, 1991).

4.2. Rules for modal operators

We present the rules for establishing $< \alpha > A$ and $[\alpha]A$ in a compositional way. Let us explain the main ideas. Consider a process $p = \ell_1 :: \ell_2 :: [1,4]a.\ell_3 :: nil$ and a clock constraint $\omega \equiv \ell_2 = 3$. We have $p,\omega \models < a : \epsilon > \ell_3$, where $< a : \epsilon > \ell_3$ express the ability to perform the action $a : \epsilon$ and to reach a configuration containing location ℓ_3. Since p is not further decomposable we can immediately deduce from its syntax the transitions of its LTS. Therefore, we have the following proof.

$$\cfrac{\cfrac{\ell_1 :: \ell_3 :: nil, \ell_3 = 1 \vdash \ell_3}{\qquad} \; \ell_3 \in loc(\ell_1 :: \ell_3 :: nil)}{p, \omega \vdash < a : \epsilon > \ell_3} \models (\omega \supset 1 \leq \ell_2 \leq 4)$$

The side condition $\models (\omega \supset 1 \leq \ell_2 \leq 4)$ ensures that rule $non - idle$ is indeed applicable. Note that we consider the clock constraint $\ell_3 = 1$, since the clock assignment corresponding to $\ell_1 :: \ell_3 :: nil$ by rule $non - idle$ is $\{loc(\ell_3 :: nil) := 1\}$.

Consider a process $\ell :: t \bigtriangledown p$, where $t = \ell_1 :: [1,3]a.\ell_2 :: nil$ and $p = \ell_3 :: \ell_4 :: [0, \infty]b.\ell_5 :: nil$. We have $\ell :: t \bigtriangledown p, \omega \models < a : \epsilon > true$, for $\omega \equiv 1 \leq \ell_1 \leq 2$, since by rule $\bigtriangledown_1$ $(\ell :: t \bigtriangledown p, \pi) \overset{a:\epsilon}{\mapsto} (\ell :: \ell_2 :: nil, \ell_2 = 1)$, for each $\pi \models \omega$. Rule $\bigtriangledown_1$ suggests how to decompose the structure of the process. We obtain the following proof:

$$\cfrac{\cfrac{\ell :: t, \omega \vdash < a : \epsilon > true}{\cdots}}{\ell :: t \bigtriangledown p, \omega \vdash < a : \epsilon > true}\; type(a : \epsilon) = \epsilon$$

Note that the side condition $type(a : \epsilon) = \epsilon$ is necessary to obtain a sound rule, namely for ensuring that $\bigtriangledown_1$ is applicable.

We have $\ell :: t \bigtriangledown p, \omega \models < b : \epsilon > true$, since by rule $\bigtriangledown_2$ $(\ell :: t \bigtriangledown p, \pi) \overset{b:\epsilon}{\mapsto} (\ell :: t \bigtriangledown p', \ell_5 = 1 \wedge 2 \leq \ell_1 \leq 3)$ with $p' = \ell_4 :: \ell_5 :: nil$, for each $\pi \models \omega$. However, it would not be correct to deduce $\ell :: t \bigtriangledown p, \omega \vdash < b : \epsilon > true$ from $p, \omega \vdash < b : \epsilon > true$ as before.

In order to achieve *compositionality* we adapt a technique proposed by (Andersen et al., 1994), that permits to avoid the construction of the transition from $\ell :: t \bigtriangledown p$. The idea is that of using *extended assertions* such as $(A)_{(p,\omega)\bigtriangledown}$, with the semantics $\| (A)_{(p,\omega)\bigtriangledown} \| = \{(p', \pi') \mid (p \bigtriangledown p', \pi' \cup \pi \models A$, for each $\pi \models \omega\}$. Whit this understanding it can be obtained a proof as the following:

$$\cfrac{\cfrac{p, \omega_1 \vdash < b : \epsilon > (true)_{(\ell::t,\omega_2)\bigtriangledown}}{\cdots}}{\ell :: t \bigtriangledown p, \omega \vdash < b : \epsilon > true}\; \models \forall(\omega_2 \supset delay(\ell :: t))$$

where $\omega_1 \equiv true \equiv pro_{rel(p)}(\omega)$, $\omega_2 \equiv 2 \leq \ell_1 \leq 3 \equiv pro_{rel(\ell::t)}(\omega) - 1$. The constraint ω_2 is obtained by noting that $\pi \models pro_{rel(\ell::t)}(\omega)$ iff $\pi + 1 \models \omega_2$ as required in rule $\bigtriangledown_2$. The side condition $\models \forall(\omega_2 \supset delay(\ell :: t))$ guarantees that $\bigtriangledown_2$ is indeed applicable.

Now a proof of $p, \omega_1 \vdash < b : \epsilon > (true)_{(\ell::t,\omega_2)\nabla}$ can be obtained as follows.

$$\cfrac{\cfrac{p, \omega_1 \vdash < b : \epsilon > (true)_{(\ell::t,\omega_2)\nabla}}{p', \omega_1' \vdash (true)_{(\ell::t,\omega_2)\nabla}}\; shift}{\cfrac{\ell :: t \nabla p', \omega_2 \wedge \omega_1' \vdash true}{}}$$

where $\omega_1' \equiv \ell_5 = 1$.

A special rule called *shift rule* is applied for eliminating the operators from formulas. The resulting proof is compositional, namely, if we replace process t by t', this part of the proof is not affected except for the test concerning the delay. Similar techniques are applied for the other constructors.

We include in the syntax the following *extended assertions*:

$$(A)_{(\ell::t,\omega)\nabla} \mid (A)_{\ell\times(p,\omega)} \mid (A)_{\times+b} \mid (A)_{\times(\ell::t,\omega)\nabla} \mid (A)_{\times\nabla(p,\omega)} \mid (A)_{\times}$$

The semantics of these assertions is defined as follows:

- $\| (A)_{(\ell::t,\omega)\nabla} \| = \{(p, \pi) \mid (\ell :: t \nabla p, \pi \cup \pi') \models A, \forall \pi' \models \omega\}$;
- $\| (A)_{\ell\times(p,\omega)} \| = \{(p', \pi) \mid (\ell :: p' \times p, \pi \cup \pi') \models A, \forall \pi' \models \omega\}$;
- $\| (A)_{\times+b} \| = \{(\ell :: p_1 \times \ell_1 :: \ell_2 :: b', \pi) \mid (\ell :: p_1 \times \ell_1 :: \ell_2 :: b' + b, \pi) \models A\}$;
- $\| (A)_{\times(\ell::t,\omega)\nabla} \| = \{(\ell_1 :: p_1 \times p_2, \pi) \mid (\ell_1 :: p_1 \times \ell :: t \nabla p, \pi \cup \pi') \models A, \forall \pi' \models \omega\}$;
- $\| (A)_{\times\nabla(p,\omega)} \| = \{(\ell_1 :: p_1 \times \ell :: t, \pi) \mid (\ell_1 :: p_1 \times \ell :: t \nabla p, \pi \cup \pi') \models A, \forall \pi' \models \omega\}$;
- $\| (A)_{\times} \| = \{(\ell_1 :: \ell_2 :: (p_1 \times p_2) \times p_3, \pi) \mid (\ell_1 :: p_1 \times \ell_2 :: (p_2 \times p_3)) \models A\}$.

4.3. Shift Rules

$$\frac{p, \omega_1 \vdash (A)_{(\ell::t,\omega_2)\nabla}}{\ell :: t \nabla p, \omega_1 \wedge \omega_2 \vdash A} \qquad \frac{p_1, \omega_1 \vdash (A)_{\ell\times(p_2,\omega_2)}}{\ell :: p_1 \times p_2, \omega_1 \wedge \omega_2 \vdash A}$$

$$\frac{\ell :: p_1 \times \ell_1 :: \ell_2 :: b_1, \omega_1 \models (A)_{\times+b}}{\ell :: p_1 \times \ell_1 :: \ell_2 :: b_1 + b, \omega_1 \models A} \qquad \frac{\ell :: p_1 \times p_2, \omega_1 \vdash (A)_{\times(\ell_1::t,\omega_2)\nabla}}{\ell :: p_1 \times \ell_1 :: t \nabla p_2, \omega_1 \wedge \omega_2 \vdash A}$$

$$\frac{\ell :: p_2 \times \ell_1 :: t, \omega_1 \vdash (A)_{\times\nabla(p_1,\omega_2)}}{\ell :: p_2 \times \ell_1 :: t \nabla p_1, \omega_1 \wedge \omega_2 \vdash A} \qquad \frac{\ell :: \ell_1 :: (p_1 \times p_2) \times p_3, \omega_2 \vdash (A)_{\times}}{\ell :: p_1 \times \ell_1 :: (p_2 \times p_3), \omega_1 \wedge \omega_2 \vdash A}$$

These rules are trivially based on the previously defined semantics for extended assertions.

4.4. Rules for *nil*

$$\frac{\ell :: \ell_1 :: nil, \omega_1 \vdash < a : \bar{\epsilon} > A}{\ell :: \ell_1 :: nil, \omega_1 - 1 \vdash A} <> nil \quad a \in \#(*)$$

$$\frac{\ell :: \ell_1 :: nil, \omega_1 \vdash [a : \epsilon] A}{} []nil_1$$

$$\frac{\ell :: \ell_1 :: nil, \omega_1 \vdash [a : \bar{\epsilon}] A}{\ell :: \ell_1 :: nil, \omega_1 - 1 \vdash A} []nil_2 \quad a \in \#(*)$$

$$\frac{\ell :: \ell_1 :: nil, \omega_1 \vdash [a : \bar{\epsilon}] A}{} []nil_3 \quad a \notin \#(*)$$

4.5. Rules for timed prefixing

Let $\omega_2 \equiv reset(\omega_1, C, loc(t)) - 1$ and $\omega_3 \equiv reset(\omega_1 \wedge m \leq \ell_1 \leq M, C, loc(t)) - 1$.

$$\frac{\ell :: \ell_1 :: [m, M] a.t, \omega_1 \vdash < a : \epsilon > A}{\ell :: t, \omega_2 \vdash A} \tau <> \epsilon \qquad \models \forall(\omega_1 \supset m \leq \ell_1 \leq M)$$

$$\frac{\ell :: \ell_1 :: [m, M] a_1.t, \omega_1 \vdash < a_2 : \bar{\epsilon} > A}{\ell :: \ell_1 :: [m, M] a_1.t, \omega_1 - 1 \vdash A} \tau <> \bar{\epsilon} \qquad \begin{array}{l} (a_2 \in \#(*) \wedge \#(a_1) = \emptyset) \\ \vee (a_2 \in \#(a_1)) \\ \models \forall(\omega_1 - 1 \supset \ell_1 \leq M) \end{array}$$

$$\frac{\ell :: \ell_1 :: [m, M] a.t, \omega_1 \vdash [a : \epsilon] A}{\ell :: t, \omega_3 \vdash A} \tau[]\epsilon_1 \qquad \not\models \forall(\omega_1 \wedge m \leq \ell_1 \leq M \supset false)$$

$$\frac{\ell :: \ell_1 :: [m, M] a_1.t, \omega_1 \vdash [a_2 : \bar{\epsilon}] A}{\ell :: \ell_1 :: [m, M] a_1.t, \omega_1 - 1 \vdash A} \tau[]\bar{\epsilon}_1 \qquad \begin{array}{l} (a_2 \in \#(*) \wedge \#(a_1) = \emptyset) \vee \\ a_2 \in \#(a_1) \end{array}$$

$$\frac{\ell :: \ell_1 :: [m, M] a_1.t, \omega_1 \vdash [a_2 : \epsilon] A}{} \tau[]\epsilon_2 \quad a_1 \neq a_2$$

$$\frac{\ell :: \ell_1 :: [m, M] a_1.t, \omega_1 \vdash [a_2 : \bar{\epsilon}] A}{} \tau[]\bar{\epsilon}_2 \qquad \begin{array}{l} (a_2 \notin \#(*) \wedge \#(a_1) = \emptyset) \vee \\ ((a_2 \notin \#(a_1) \wedge \#(a_1) \neq \emptyset) \end{array}$$

$$\frac{\ell :: \ell_1 :: [m, M] a_1.t, \omega_1 \vdash [a_1 : \epsilon] A}{} \tau[]\epsilon_3 \quad \models \forall(\omega_1 \wedge m \leq \ell_1 \leq M \supset false)$$

These rules are based on rules $non - idle$ and $idle$ of the operational semantics by exploiting the fact that $\pi \models \omega_1$ iff $reset(\pi, C, rel(t)) + 1 \models \omega_2$ and analogously for ω_3.

4.6. Rules for choice

There are two sets of rules: rules applicable when the type of the action is non-idle and rules applicable when the type of the action is idle in order to distinguish whether either $choice_1$ or $choice_2$ is applicable.

1. Rules applicable iff $type(\alpha) = \epsilon$.

$$\frac{\ell :: \ell_1 :: b_1 + b_2, \omega_1 \vdash <\alpha> A}{\ell :: \ell_1 :: b_i, \omega_1 \vdash <\alpha> A} + <> \epsilon \qquad \frac{\ell :: \ell_1 :: b_1 + b_2, \omega_1 \vdash [\alpha]A}{\{\ell :: \ell_1 :: b_i, \omega_1 \vdash [\alpha]A\}_{i \in \{1,2\}}} + []\epsilon$$

2. Rules applicable iff $type(\alpha) = \bar{\epsilon}$.

$$\frac{\ell :: \ell_1 :: b_1 + b_2, \omega_1 \vdash <\alpha> A}{\{\ell :: \ell_1 :: b_i, \omega_1 \vdash <\alpha> true\}_{i \in \{1,2\}} \quad \ell :: \ell_1 :: b_1 + b_2, \omega_1 - 1 \vdash A} + <> \bar{\epsilon}$$

$$\frac{\ell :: \ell_1 :: b_1 + b_2, \omega_1 \vdash [\alpha]A}{\ell :: \ell_1 :: b_1 + b_2, \omega_1 - 1 \vdash A} + []\bar{\epsilon}_1 \qquad \frac{\ell :: \ell_1 :: b_1 + b_2, \omega_1 \vdash [\alpha]A}{\ell :: \ell_1 :: b_i, \omega_1 \vdash [\alpha]false} + []\bar{\epsilon}_2$$

4.7. Rules for refinement

Rules for refinement are similar to the ones for choice. However, in this case $rel(\ell :: t \bigtriangledown p) = rel(\ell :: t) \cup rel(p)$ and $rel(p) \cap rel(\ell :: t) = \emptyset$, by definition of well-formed process. Therefore, we consider the projections with respect to $rel(p)$ and $rel(\ell :: t)$. Let $\omega_2 \equiv pro_{rel(\ell::t)}(\omega_1) - 1$, $\omega_p \equiv pro_{C \cup rel(p)}(\omega_1)$ and $\omega_t \equiv pro_{C \cup rel(\ell::t)}(\omega_1)$.

1. Rules applicable iff $type(\alpha) = \epsilon$.

$$\frac{\ell :: t \bigtriangledown p, \omega_1 \vdash <\alpha> A}{\ell :: t, \omega_t \vdash <\alpha> A} \bigtriangledown <> \epsilon_1$$

$$\frac{\ell :: t \bigtriangledown p, \omega_1 \vdash [\alpha]A}{\ell :: t, \omega_t \vdash [\alpha]A \quad p, \omega_p \vdash [\alpha](A)_{(\ell::t,\omega_2)\bigtriangledown}} \bigtriangledown []\epsilon$$

$$\frac{\ell :: t \bigtriangledown p, \omega_1 \vdash <\alpha> A}{p, \omega_p \vdash <\alpha> (A)_{(\ell::t,\omega_2)\bigtriangledown}} \bigtriangledown <> \epsilon_2 \qquad \models \forall(\omega_2 \supset delay(\ell :: t))$$

2. Rules applicable iff $type(\alpha) = \bar{\epsilon}$.

$$\frac{\ell :: t \bigtriangledown p, \omega_1 \vdash < \alpha > A}{\ell :: t, \omega_t \vdash < \alpha > true \quad p, \omega_p \vdash < \alpha > true \quad \ell :: t \bigtriangledown p, \omega_1 - 1 \vdash A} \bigtriangledown <> \bar{\epsilon}$$

$$\frac{\ell :: t \bigtriangledown p, \omega_1 \vdash [\alpha]A}{\ell :: t \bigtriangledown p, \omega_1 - 1 \vdash A} \bigtriangledown []\bar{\epsilon}_1 \qquad \frac{\ell :: t \bigtriangledown p, \omega_1 \vdash [\alpha]A}{\ell :: t, \omega_t \vdash [\alpha]false} \bigtriangledown []\bar{\epsilon}_2$$

$$\frac{\ell :: t \bigtriangledown p, \omega_1 \vdash [\alpha]A}{p, \omega_p \vdash [\alpha]false} \bigtriangledown []\bar{\epsilon}_3$$

4.7.1. *Rules for recursion*

$$\frac{\ell :: rec\ x.t, \omega_1 \vdash \sigma A \quad \sigma \in \{< \alpha >, [\alpha]\}}{\ell :: t[rec\ x.t/x], \omega_1 \vdash \sigma A} rec$$

4.8. RULES FOR PARALLEL COMPOSITION

The case of parallel composition is the most complex. We would like to give a set of rules for proving $\ell :: p_1 \times p_2, \omega \vdash < \alpha > A$ from $\{\ell :: p_1 \times p_{2,j}, \omega_j \vdash < \alpha_j > A\}_{j \in \{1,n\}}$, where $p_{2,j}$ are components of p_2 and ω_j are obtained from ω by projection depending on the structure of p_2.

Suppose that $p_2 = \ell_1 :: \ell_2 :: [m, M]a.t$ and that $(\ell :: p_1 \times p_2, \pi_1 \cup \pi_2) \overset{\alpha}{\mapsto} (\ell :: p_1' \times p_2', \pi_1' \cup \pi_2')$, where $p_2' = \ell_1 :: t$ and $\pi_2' = \{loc(t) := 1\}$, for each $\pi_1 \cup \pi_2 \models \omega$. By definition of $\times$, there exists $(p_1, \pi_1) \overset{\alpha_1}{\mapsto} (p_1', \pi_1')$ with $\alpha \in \alpha_1 \times (a : \epsilon)$. Following this construction, we can derive $\ell :: p_1 \times p_2, \omega \vdash < \alpha > A$ from $p_1, \omega_1 \vdash < \alpha_1 > (A)_{\ell \times (p_2', \omega_2)}$, where $\omega_1 \equiv pro_{C \cup rel(p_1)}(\omega)$ and $\omega_2 \equiv loc(t) = 1$ under the condition that $\models \forall(\omega \supset m \leq \ell_1 \leq M)$. However, this method is not applicable in general. Suppose that $p_2 = \ell_1 :: \ell_2 :: b_1 + b_2$ and that one wants to prove $\ell :: p_1 \times p_2, \omega \vdash < \alpha > A$, where $type(\alpha) = \epsilon$. One might think of obtaining a proof from $\ell :: p_1 \times \ell_1 :: \ell_2 :: b_i, \omega \vdash < \alpha > A$ for some $i \in \{1, 2\}$. However, this kind of deduction is not correct in general. Actually, there exists $(\ell :: p_1 \times \ell_1 :: \ell_2 :: b_1 + b_2, \pi) \overset{\alpha}{\mapsto} (\ell :: p_1' \times p_2', \pi')$, iff by rule $\times$, $(p_1, \pi_1) \overset{\alpha_1}{\mapsto} (p_1', \pi_1')$ and $(\ell_1 :: \ell_2 :: b_1 + b_2, \pi_2) \overset{\alpha_2}{\mapsto} (p_2', \pi_2')$ such that $\alpha \in \alpha_1 \times \alpha_2$ and $\pi = \pi_1 \cup \pi_2$, $\pi' = \pi_1' \cup \pi_2'$. Now, if $type(\alpha_2) = \epsilon$, by rule $choice_1$, $(\ell_1 :: \ell_2 :: b_i, \pi_2) \overset{\alpha_2}{\mapsto} (p_2', \pi_2')$, for some $i \in \{1, 2\}$, and by rule $\times$, $(\ell :: p_1 \times \ell_1 :: \ell_2 :: b_i, \pi) \overset{\alpha}{\mapsto} (\ell :: p_1' \times p_2', \pi')$. In this case $\ell :: p_1 \times \ell_1 :: \ell_2 :: b_i, \omega \models < \alpha > A$ so that the suggested deduction is correct. On the other hand, if $type(\alpha_2) = \bar{\epsilon}$ then rule $choice_1$ is not applicable. In conclusion, it is necessary to know the type of α_2. Since $type(\alpha_2)$ cannot be deduced from $type(\alpha)$, we must adopt formulas with modal operators over *extended actions*.

DEFINITION 4. *Let Act be a set of basic actions. We define* $\mathcal{E}(Act)$ *as the set of* extended actions $a : \tau$, *where* $a \in Act$ *and* τ *is an* extended *type, such that:* $\epsilon, \bar{\epsilon}$ *are basic extended types and, if* τ_1 *is an extended type and* τ_2 *is a basic extended type, then* $\tau_1 \times \tau_2$ *is an extended type. For* $a : \tau \in \mathcal{E}(Act)$, *we define* $type^*(a : \tau) = \tau$, *if* τ *is basic, and* $type^*(a : \tau_1 \times \tau_2) = \epsilon$ *if* $type^*(a : \tau_i) = \epsilon$, *for some* $i \in \{1, 2\}$, *and* $type^*(a : \tau_1 \times \tau_2) = \bar{\epsilon}$, *otherwise.*

For $\alpha \in \mathcal{E}(Act)$, we say that $(p, \pi) \overset{\alpha'}{\mapsto} (p', \pi')$ is *compatible* with α iff

1. $\alpha = a : \tau$, $\alpha' = a : \tau'$ and $type(\alpha') = type^*(\alpha)$;

2. If $\alpha = a : \tau_1 \times \tau_2 \in \mathcal{E}(Act) \setminus \mathcal{S}(Act)$, then $(p, \pi) \overset{\alpha'}{\mapsto} (p', \pi')$ has been been obtained by rule $\times$ from $(p_i, \pi_i) \overset{\alpha'_i}{\mapsto} (p'_i, \pi'_i)$, such that $(p_i, \pi_i) \overset{\alpha'_i}{\mapsto} (p'_i, \pi'_i)$ is compatible with $\alpha_i = a_i : \tau_i$ and $\alpha'_i = a_i : \tau'_i$.

We can now define the semantics of formulas with extended actions.

- $\| < \alpha > A \| = \{(p, \pi) \mid \exists (p, \pi) \overset{\alpha'}{\mapsto} (p', \pi'), (p', \pi') \in \| A \|$, such that $(p, \pi) \overset{\alpha'}{\mapsto} (p', \pi')$ is compatible with α $\}$;

- $\| [\alpha]A \| = \{(p, \pi) \mid \forall (p, \pi) \overset{\alpha}{\mapsto} (p', \pi')$ that is compatible with α, $(p', \pi') \in \| A \| \}$.

Intuitively, $(p, \pi) \models < \alpha > A$ with $\alpha \in \mathcal{E}(Act) \setminus \mathcal{S}(Act)$ iff there exists $(p, \pi) \overset{\alpha'}{\mapsto} (p', \pi')$, where $(p', \pi') \models A$, such that $\alpha = a : \tau$, $\alpha' = a : \tau'$ and this transition is constructed consistently with τ. For instance, if $\tau = \epsilon \times \bar{\epsilon}$, then it must be obtained by rule $\times$ applied to premises $\{(p_i, \pi'_i) \overset{\alpha'_i}{\mapsto} (p'_i, \pi'_i)\}_{i \in \{1,2\}}$, where $type(\alpha'_1) = \epsilon$ and $type(\alpha'_2) = \bar{\epsilon}$.

We extend the operation $\times$ to extended actions. For $a_1 : \tau_1 \in \mathcal{E}(Act)$, $a_2 : \tau_2 \in \mathcal{S}(Act)$, $(a : \tau_1) \otimes (a : \tau_2) = \{a : \tau_1 \times \tau_2 \mid a \in a_1 \times a_2\} \subseteq \mathcal{E}(Act)$. For $\alpha = a : \tau_1 \times \tau_2 \in \mathcal{E}(Act)$, we define $proj_i(\alpha) = a : \tau_i$.

4.8.1. *Expansion rules*

Extended actions are introduced by the following rules.

$$\frac{\ell :: p_1 \times p_2, \omega_1 \vdash < \alpha > A}{\ell :: p_1 \times p_2, \omega_1 \vdash < \beta > A} \mathcal{E} <> \qquad \beta \in \mathcal{E}(\alpha)$$

$$\frac{\ell :: p_1 \times p_2, \omega_1 \vdash [\alpha]A}{\{\ell :: p_1 \times p_2, \omega_1 \vdash [\beta_i]A\}_{i \in \{1,n\}}} \mathcal{E}[] \qquad \{\beta_i\}_{i \in \{1,n\}} = \mathcal{E}(\alpha)$$

where for $a : \tau \in \mathcal{S}(Act)$, $\mathcal{E}(a : \tau) = \{a : \tau' \mid \tau \in \mathcal{E}(\tau)\}$, with $\mathcal{E}(\epsilon) = \{\epsilon \times \epsilon, \bar{\epsilon} \times \epsilon, \epsilon \times \bar{\epsilon}\}$ and $\mathcal{E}(\bar{\epsilon}) = \{\bar{\epsilon} \times \bar{\epsilon}\}$.

In other words, $\mathcal{E}(\tau)$ realizes the inverse of *type** and produces extended types consistent with the basic type τ.

4.8.2. *Rules for nil*
Let $p_2 = \ell_1 :: \ell_2 :: nil$ and $\omega_2 \equiv pro_{\{\ell_2\}}(\omega_1) - 1$, $\omega_{p_1} \equiv pro_{C \cup rel(p_1)}(\omega_1)$. Moreover, let $\{\beta_i\}_{i\in\{1,k\}} = \{\beta \mid \alpha \in \beta \otimes (b : \bar{\epsilon}), b \in \#(*)\}$.

$$\frac{\ell :: p_1 \times p_2, \omega_1 \vdash <\alpha> A}{p_1, \omega_{p_1} \vdash <\beta> (A)_{\ell\times(p_2,\omega_2)}} \times <> nil \qquad \begin{array}{l}\alpha \in \beta \otimes (b : \bar{\epsilon})\\ b \in \#(*)\end{array}$$

$$\frac{\ell :: p_1 \times p_2, \omega_1 \vdash [\alpha] A}{\{p_1, \omega_{p_1} \vdash [\beta_i](A)_{\ell\times(p_2,\omega_2)}\}_{i\in\{1,k\}}} \times [] nil$$

4.8.3. *Rules for prefixing*
Let $p_2 = \ell_1 :: \ell_2 :: [m, M]a.t$, $\omega_2 \equiv rel(t) = 1$, $\omega_3 \equiv pro_{rel(p_2)}(\omega_1) - 1$ and $\omega_{p_1} \equiv pro_{C \cup loc(p_1)}(\omega_1)$.

$$\frac{\ell :: p_1 \times p_2, \omega_1 \vdash <\alpha> A}{p_1, \omega_{p_1} \vdash <\beta> (A)_{\ell\times(\ell_1::t,\omega_2)}} \tau\times <>_1 \qquad \begin{array}{l}\alpha \in \beta \otimes (a : \epsilon)\\ \models \forall(\omega_1 \supset m \leq \ell_2 \leq M)\end{array}$$

$$\frac{\ell :: p_1 \times p_2, \omega_1 \vdash <\alpha> A}{p_1, \omega_{p_1} \vdash <\beta> (A)_{\ell\times(p_2,\omega_3)}} \tau\times <>_2 \qquad \begin{array}{l}\alpha \in \beta \times (b : \bar{\epsilon})\\ (b \in \#(*) \wedge \#(a) = \emptyset) \vee b \in \#(a)\\ \models \forall(\omega_3 \supset \ell_2 \leq M)\end{array}$$

Let $\{\alpha_i\}_{i\in\{1,n\}} = \{\alpha_i \mid \alpha \in \alpha_i \otimes (a : \epsilon)\}$ and $\{\beta_i\}_{i\in\{1,k\}} = \{\beta_i \mid \alpha \in \beta_i \otimes (b : \bar{\epsilon}), (b \in \#(*) \wedge \#(a) = \emptyset) \vee b \in \#(a)\}$. Moreover, let $\omega \equiv \omega_1 \wedge m \leq \ell \leq M$.

$$\frac{\ell :: p_1 \times p_2, \omega_1 \vdash [\alpha] A}{\begin{array}{c}\{p_1, \omega_{p_1} \vdash [\alpha_i](A)_{\ell\times(\ell_1::t,\omega_2)}\}_{i\in\{1,n\}}\\ \{p_1, \omega_{p_1} \vdash [\beta_i](A)_{\ell\times(p_2,\omega_3)}\}_{i\in\{1,k\}}\end{array}} \tau\times []_1 \qquad \not\models \forall(\omega \supset false)$$

$$\frac{\ell :: p_1 \times p_2, \omega_1 \vdash [\alpha] A}{\{p_1, \omega_{p_1} \vdash [\beta_i](A)_{\ell\times(p_2,\omega_3)}\}_{i\in\{1,k\}}} \tau\times []_2 \qquad \models \forall(\omega \supset false)$$

4.8.4. *Rules for choice*
1. Rules applicable iff $type(proj_2(\alpha)) = \epsilon$.

$$\frac{\ell :: p_1 \times \ell_1 :: \ell_2 :: b_1 + b_2, \omega_1 \vdash <\alpha> A}{\ell :: p_1 \times \ell_1 :: \ell_2 :: b_i, \omega_1 \vdash <\alpha> A} \times <> +\epsilon$$

$$\frac{\ell :: p_1 \times \ell_1 :: \ell_2 :: b_1 + b_2, \omega_1 \vdash [\alpha] A}{\{\ell :: p_1 \times \ell_1 :: \ell_2 :: b_i, \omega_1 \vdash [\alpha] A\}_{i\in\{1,2\}}} \times [] + \epsilon$$

2. Rules applicable iff $type(proj_2(\alpha)) = \bar{\epsilon}$.

$$\frac{\ell :: p_1 \times \ell_1 :: \ell_2 :: b_1 + b_2, \omega_1 \vdash <\alpha> A}{\{\ell :: p_1 \times \ell_1 :: \ell_2 :: b_i, \omega_1 \vdash <\alpha> (A)_{\times + b_j}\}_{i,j \in \{1,2\}, i \neq j}} \times <> +\bar{\epsilon}$$

$$\frac{\ell :: p_1 \times \ell_1 :: \ell_2 :: b_1 + b_2, \omega_1 \vdash [\alpha]A}{\{\ell :: p_1 \times \ell_1 :: \ell_2 :: b_i, \omega_1 \vdash [\alpha](A)_{\times + b_j}\}_{i,j \in \{1,2\}, i \neq j}} \times [] + \bar{\epsilon}_1$$

$$\frac{\ell :: p_1 \times \ell_1 :: \ell_2 :: b_1 + b_2, \omega_1 \vdash [\alpha]A}{\ell :: p_1 \times \ell_1 :: \ell_2 :: b_i, \omega_1 \vdash [\alpha]false} \times [] + \bar{\epsilon}_2$$

4.8.5. *Rules for refinement*

Let $\omega_p \equiv pro_{C \cup (rel(\ell :: p_1 \times p))}(\omega_1)$, $\omega_t \equiv pro_{C \cup rel(\ell :: p_1 \times \ell_1 :: t)}(\omega_1)$, $\omega_2 \equiv pro_{rel(\ell_1 :: t)}(\omega_1) - 1$ and $\omega_3 \equiv pro_{rel(p)}(\omega_1) - 1$.

1. Rules applicable iff $type(proj_2(\alpha)) = \epsilon$.

$$\frac{\ell :: p_1 \times \ell_1 :: t \bigtriangledown p, \omega_1 \vdash <\alpha> A}{\ell :: p_1 \times \ell_1 :: t, \omega_t \vdash <\alpha> A} \times <> \bigtriangledown \epsilon_1$$

$$\frac{\ell :: p_1 \times \ell_1 :: t \bigtriangledown p, \omega_1 \vdash <\alpha> A}{\ell :: p_1 \times p, \omega_p \vdash <\alpha> (A)_{\times(\ell_1 :: t, \omega_2)\bigtriangledown}} \times <> \bigtriangledown \epsilon_2 \qquad \models \forall(\omega_2 \supset delay(\ell_1 :: t))$$

$$\frac{\ell :: p_1 \times \ell_1 :: t \bigtriangledown p, \omega_1 \vdash [\alpha]A}{\ell :: p_1 \times p, \omega_p \vdash [\alpha](A)_{\times(\ell_1 :: t, \omega_2)\bigtriangledown} \quad \ell :: p_1 \times \ell_1 :: t, \omega_t \vdash [\alpha]A} \times [] \bigtriangledown \epsilon_2$$

2. Rules applicable iff $type(proj_2(\alpha)) = \bar{\epsilon}$.

$$\frac{\ell :: p_1 \times \ell_1 :: t \bigtriangledown p, \omega_1 \vdash <\alpha> A}{\begin{array}{l}\ell :: p_1 \times \ell_1 :: t, \omega_t \vdash <\alpha> (A)_{\times \bigtriangledown (p, \omega_3)} \\ \ell :: p_1 \times p, \omega_p \vdash <\alpha> (A)_{\times(\ell_1 :: t, \omega_2)\bigtriangledown}\end{array}} \times <> \bigtriangledown \bar{\epsilon}$$

$$\frac{\ell :: p_1 \times \ell_1 :: t \bigtriangledown p, \omega_1 \vdash [\alpha]A}{\begin{array}{l}\ell :: p_1 \times \ell_1 :: t, \omega_t \vdash [\alpha](A)_{\times \bigtriangledown (p, \omega_3)} \\ \ell :: p_1 \times p, \omega_p \vdash [\alpha](A)_{\times(\ell_1 :: t, \omega_2)\bigtriangledown}\end{array}} \times [] \bigtriangledown \bar{\epsilon}_1$$

$$\frac{\ell :: p_1 \times \ell_1 :: t \bigtriangledown p, \omega_1 \vdash [\alpha]A}{\ell :: p_1 \times \ell_1 :: t, \omega_t \vdash [\alpha]false} \times [] \bigtriangledown \bar{\epsilon}_2$$

$$\frac{\ell :: p_1 \times \ell_1 :: t \bigtriangledown p, \omega_1 \vdash [\alpha]A}{\ell :: p_1 \times p, \omega_p \vdash [\alpha]false} \times [] \bigtriangledown \bar{\epsilon}_3$$

4.8.6. *Rules for recursion*

$$\frac{\ell :: p_1 \times \ell_2 :: rec\, x.t, \omega_1 \vdash \sigma A \quad \sigma \in \{<\alpha>, [\alpha]\}}{\ell :: p_1 \times \ell_2 :: t[rec\, x.t/x], \omega_1 \vdash \sigma A} \times rec$$

4.8.7. *Rules for parallel composition*
Let $\{\beta_i\}_{i \in \{1,n\}} = left(\alpha)$.

$$\frac{\ell :: p_1 \times \ell_2 :: (p_3 \times p_4), \omega_1 \vdash <\alpha> A}{\ell :: \ell_2 :: (p_1 \times p_3) \times p_4, \omega_1 \vdash <\beta> (A)_\times} \times\times <> \quad \beta \in left(\alpha)$$

$$\frac{\ell :: p_1 \times \ell_2 :: (p_3 \times p_4), \omega_1 \vdash [\alpha] A}{\{\ell :: \ell_1 :: (p_1 \times p_3) \times p_4, \omega_1 \vdash [\beta_i](A)_\times\}_{i \in \{1,n\}}} \times \times []$$

where, for $a :: \tau_1 \times \tau_2 \in \mathcal{E}(Act)$, $left(a : \tau_1 \times \tau_2) = \{a : \tau\}$, such that $\tau = (\tau_1 \times \tau_{2,1}) \times \tau_{2,2}$, for $\tau_{2,1} \times \tau_{2,2} \in \mathcal{E}(\tau_2)$.

The idea is that it is sufficient to associate on the left by the associative property of parallel composition.

5. Soundness and Completeness

The proof system is sound in general, while it is complete for the class of *regular* processes (finite-state). A closed process p is *regular* iff it has guarded recursion and for each variable x, x does not occur free in a subprocess $\ell :: p_1 \times p_2$ and in process p_1, such that $\ell :: t \bigtriangledown p_1$ is a subprocess of p.

We exploit the following properties concerning the correspondence between assignments and constraints.

PROPOSITION 1. *Let C be a set of clocks and $C_1 \subseteq C$, $C_1 \cap C_2 = \emptyset$. For each assignment π over C and for each $\omega \in \Omega(C)$:*

1. *if $\pi \models \omega$ then $proc_{C_1}(\pi) \models proc_{C_1}(\omega)$. Moreover, if ω is satisfiable, then if $\pi' \models proc_{C_1}(\omega)$, there exists $\pi \models \omega$ and $\pi' = proc_{C_1}(\pi)$.*

2. *if $\pi \models \omega$, then $reset(\pi, C_1, C_2) \models reset(\omega, C_1 , C_2)$. Moreover, if ω is satisfiable and $\pi' \models reset(\omega, C_1, C_2)$, then there exists $\pi \models \omega$ and $reset(\pi, C_1, C_2,) = \pi'$.*

3. *$\pi \models \omega + \delta$ iff $\pi + \delta \models \omega$ and $\pi \models \omega - \delta$ iff $\pi - \delta \models \omega$.*

PROPOSITION 2. *Let $C_1 \cap C_2 = \emptyset$, $C_1 \subseteq C$ and $C_{i,1}, C_{i,2}$, such that $C_{j,i} \cup C_{k,i} = C_i$, for each $i, j, k \in \{1,2\}$ with $j \neq k$ and $C_{1,i} \cap C_{2,i} = \emptyset$. Let $\omega \in \Omega(C)$ such that, for each $x + n\mathcal{R}y + m \in \omega$, either $x, y \in C_{1,1}$ or $x, y \in C_{2,1}$. We have that $reset(\omega, C_1, C_2) \equiv reset(\omega, C_{1,1}, C_{1,2}) \wedge reset(\omega, C_{2,1}, C_{2,2})$.*

THEOREM 1. *Let p be a closed process, A be a closed RTL_μ formula and $\omega \in \Omega(C \cup rel(p))$. If $p, \omega \vdash A$, then $p, \omega \models A$.*

The proof of soundness is based on showing that each rule is backward sound. The soundness of fixpoints rules follows from the following reduction lemma that justifies the use of tagging fixpoints.

LEMMA 1. *Let ψ be a monotonic function on the lattice $\mathcal{P}(D)$. For $P \in \mathcal{P}(D)$,*

1. $P \subseteq \psi(\mu V.(\psi(V) \setminus P)) \Rightarrow P \subseteq \mu V.(\psi(V))$;
2. $P \subseteq \psi(\nu V.(\psi(V) \cup P)) \Leftrightarrow P \subseteq \nu V.(\psi(V))$.
3. *If $\mu V.\psi(V) = \psi^k(\emptyset)$ for some integer k, then there exists $\{P_i\}_{i \in \{1,n\}}$, such that $P = \bigcup_{i \in \{1,n\}} P_i$, and, for each $i \in \{1,n\}$, $P_i \subseteq \mu V.\psi(V) \Rightarrow P_i \subseteq \psi(\mu V.(\psi(V) \setminus P_i))$.*

THEOREM 2. *Let p be a regular process, A be a closed RTL_μ formula and $\omega \in \Omega(C \cup rel(p))$. If $p, \omega \models A$, then $p, \omega \vdash A$.*

A sketch of the proof of completeness can be found in the appendix A. The complete proofs can be found in (Levi, 1997).

6. Conclusions

In this paper we have considered a quite general discrete timed process language $\mathcal{TSP}$ which is parametric in the set of basic actions. An operator of process refinement and the use of locations permit to model in its framework a discrete version of timed statecharts with minimal and maximal delays associated to transitions. We have defined a compositional proof system for $\mathcal{TSP}$ based on a non-trivial adaptation of the techniques proposed in (Andersen et al., 1994) for a simple CCS-like language. The logic for expressing real-time properties is an extension of μ-calculus with freeze quantification and clock constraints over a discrete time domain.

We are currently interested in the realization of a tool for providing the user with mechanized assistance during the construction of the proofs. We are studying the embedding of our formal system in the proof assistant *Coq* (Cornes et al., 1995). This is very important since very simple proofs only can be done by hand.

One of the goals of a future research it would be to extend our approach to a dense time framework. The proposed model of timed statecharts is not adequate for modelling applications working whithin a dense environment, that may produce signals at arbitrary close instants of time. Dense timed versions of statecharts have been proposed in (Kesten and Pnueli, 1992; Peron and Maggiolo-Schettini, 1994). However, it is not clear if the synchrony hypothesis is still reasonable in this dense framework. For the verification we could for instance consider an extension of μ-calculus with dense time such as the one proposed by (Henzinger et al., 1994).

References

Alur, R.: 1991, 'Techniques for Automatic Verification of Real-time Systems'. Ph.D. thesis, University of Stanford.

Alur, R. and T. Henzinger: 1989, 'A Really Temporal Logic'. In: *Proceeding of the 30th Annual Symposium on Foundations of Computer Science.* pp. 164–169.

Alur, R. and T. Henzinger: 1990, 'Real-time logics: complexity and expressivness'. In: *Proc. Fifth IEEE Symp. on Logic In Computer Science.* pp. 390–401.

Alur, R. and T. Henzinger: 1992, 'Logics and models of real time: a survey'. In: *Proceedings of REX workshop 'Real-time Theory in Practice'*, Vol. 600 of *Lecture Notes in Computer Science.* pp. 74–106.

Andersen, H., C. Stirling, and G. Winskel: 1994, 'A Compositional Proof System for the Modal μ-Calculus'. In: *Proc. Ninth IEEE Symp. on Logic In Computer Science.* pp. 144–153.

Berry, G. and G. Gonthier: 1992, 'The ESTEREL Synchronous Programming Language: Design, Semantics, Implementation'. *Science of Computer Programming* **19**, 87–152.

Cornes, C., J. Courant, J. Fillatre, G. Huet, P. Manoury, C. Munoz, C. Murthy, C. Parent, C. Paulin-Mohring, A. Saibi, and B. Werner: 1995, 'The Coq Proof Assistant Reference Manual-Version 5.10'. Technical report, INRIA, Rocquenfort.

Emerson, E., A. Mok, A. Sistla, and J. Srinivasan: 1989, 'Quantitative temporal reasoning'. In: *First Annual Workshop on Computer-Aided Verification.*

Harel, D.: 1987, 'Statecharts: A Visual Formalism for Complex Systems'. *Science of Computer Programming* **8**, 231–274.

Harel, D., A. Pnueli, J. P. Schmidt, and R. Sherman: 1987, 'On the Formal Semantics of Statecharts'. In: *Proc. Second IEEE Symp. on Logic In Computer Science.* pp. 54–64.

Henzinger, T., X. Nicollin, J. Sifakis, and S. Yovine: 1994, 'Symbolic Model Checking for Real-Time Systems'. *Information and Computation* **111**, 193–244.

Kesten, Y. and A. Pnueli: 1992, 'Timed and Hybrid Statecharts and their Textual Representation'. In: *Formal Techniques in Real-time and Fault Tolerant Systems*, Vol. 571 of *Lecture Notes in Computer Science.* pp. 591–620.

Kozen, D.: 1983, 'Results on the Propositional mu-Calculus'. *Theoretical Computer Science* **27**, 333–354.
Levi, F.: 1996a, 'A Compositional μ-Calculus Proof System for Statecharts'. To appear on Theoretical Computer Science.
Levi, F.: 1996b, 'A Process Language for Statecharts'. In: *Analysis and Verification of Multiple-Agent Languages*, Vol. 1192 of *Lecture Notes in Computer Science*. pp. 388–403.
Levi, F.: 1997, 'Verification of Temporal and Real-Time Properties of Statecharts'. Ph.D. thesis, TD-6/97,Dipartimento di Informatica, Università di Pisa.
Maggiolo-Schettini, A., A. Peron, and S. Tini: 1996, 'Equivalences of Statecharts'. In: *Proc. of CONCUR 96*, Vol. 1119 of *Lecture Notes in Computer Science*. pp. 687–702.
Milner, R.: 1983, 'Calculi for Synchrony and Asynchrony'. *Theoretical Computer Science* **25**, 267–310.
Peron, A. and A. Maggiolo-Schettini: 1994, 'Transitions as Interrupts: A New Semantics for Timed Statecharts'. In: *Proceedings of International Symposium on Theoretical Aspects of Computer Software*. pp. 806–821.
Pnueli, A. and M. Shalev: 1991, 'What is in a Step'. In: *Theoretical Aspects of Computer Science*, Vol. 526 of *Lecture Notes in Computer Science*. pp. 244–464.
Prasad, K.: 1995, 'A calculus of broadcasting systems'. *Science of Computer Programming* **25**.
Pratt, V.: 1981, 'A decidable mu-calculus'. In: *Proceedings of 22nd. FOCS*. pp. 421–427.
Stirling, C. and D. Walker: 1991, 'Local Model Checking in the Modal mu-Calculus'. *Theoretical Computer Science* **89**, 161–177.
Winskel, G.: 1991, 'A Note on Model Checking the Modal Nu-calculus'. *Theoretical Computer Science* **83**, 157–187.

Appendix

A. Completeness of the proof system

In this appendix we show the main ideas underlying the proof of theorem 2. The proof is by induction on a relation $\prec$ over formulas that is well-founded for regular processes. For $p \in CTSP$, we denote by $\mathcal{R}_p$ the minimal subset of closed processes such that $p \in \mathcal{R}_p$, and if $p' \in \mathcal{R}_p$ and $(p', \pi') \overset{\alpha}{\mapsto} (p'', \pi'')$, then $p'' \in \mathcal{R}_p$. Let $\mathcal{S}_p$ be the minimal subset of closed processes such that, if p' is a subprocess of p, then $\mathcal{R}_{p'} \subseteq \mathcal{S}_p$. Regular processes are characterized by the fact that both $\mathcal{R}_p$ and $\mathcal{S}_p$ are finite. We show that there exists a finite quotient of $LTS(p)$ that is sufficient for establishing A by using a clock region technique (Henzinger et al., 1994; Alur, 1991).

DEFINITION 5. *For a closed process p and $\omega \in \Omega(rel(p) \cup C)$, we say that ω is a (p, A)-*constraint *iff the greatest constant of ω is k, where k is the greatest constant that occurs in p or in A. A set of timed states*

$\mathcal{R}$ *of* $LTS(p)$ *is a* (p, A)-region *iff,* $(p', \pi') \in \mathcal{R}$ *iff* $p' \in \mathcal{S}_p$ *and* $\pi' \models \omega$ *for some* (p, A)*-constraint* ω.

The basic observation is that, if p is regular, there are only finitely many (p, A)-regions. Moreover, the set of states satisfying a formula A is a (p, A)-region.

LEMMA 2. *For a closed process* p *and for a closed formula* A, $\| A \|$ *is a* (p, A)*-region.*

DEFINITION 6. *Let* p *be a closed process and let* A, A' *be closed formulas with tags* (p', ω), *where* ω *is a* (p, A)*-constraint. We define* $A' \prec A$ *iff either* A' *is a proper subformula of* A *or* $A = \sigma X\{U\}B$, *for* $\sigma \in \{\mu, \nu\}$, *and* $A' = B[\sigma X\{U \cup \{(p', \omega)\}\}B/X]$ *where* $\omega \in \Omega(rel(p') \cup C)$ *is a* (p, A)*-constraint, such that* $\| (p', \omega) \| \cap \| U \| = \emptyset$ *and* $p' \in \mathcal{S}_p$.

For a regular process p the set of (p, A)-regions is finite and by lemma 2 fixpoints can be computed whithin a finite number of steps so that $\prec$ is well-founded. Moreover, we exploit the following property.

LEMMA 3. *Let* p *be a closed process,* A *be a closed formula and* $\omega \in \Omega(rel(p) \cup C)$. *For* $\alpha \in \mathcal{E}(Act)$, *if* $p, \omega \models < \alpha > A$ *(resp.* $p, \omega \models [\alpha]A$

1. *there exist sets* $\{p_i \in \mathcal{S}_p\}_{i \in \{1,n\}}$ *and* $\{p'_j \in \mathcal{S}_p\}_{j \in \{1,k\}}$, *such that* $\{p_i, \omega_i \models A\}_{i \in \{1,n\}}$ *and* $\{p'_j, \omega'_j \models true\}_{j \in \{1,k\}}$;
2. $p, \omega \vdash < \alpha > A$ *(resp.* $p, \omega \vdash [\alpha]A$)) *can be derived from* $\{p_i, \omega_i \vdash A\}_{i \in \{1,n\}}$ *and* $\{p'_j, \omega'_j \vdash true\}_{j \in \{1,k\}}$.

Moreover, for each $i \in \{1, n\}$ *and* $\pi \models \omega$, *there exists* $(p, \pi) \overset{\alpha_i}{\mapsto} (p_i, \pi')$, *where* $\pi' \models \omega_i$, *such that it is consistent with* α.

Proof of theorem 2 is as follows.

Proof. f $p, \omega \models A$ and $\omega \wedge delay(p)$ is not satisfiable, by rule *empty*, $p, \omega \vdash A$. If $\omega \wedge delay(p)$ is satisfiable, it is sufficient to consider the case of $\models \forall(\omega \supset delay(p))$. The proof is by induction on $\prec$, that is well-founded for finite-state timed graphs.

1. The cases of $A \in \{P, \neg P, \omega, A_1 \vee A_2, A_1 \wedge A_2\}$ are trivial.
2. $A \equiv x.A_1$. By definition $p, \omega \models x.A_1$ iff, for each $\pi \models \omega$, $(s, \pi[x := 0]) \models A_1$. Since ω is satisfiable, by proposition 1 for each $\pi' \models reset(\omega, (C \cup rel(p)) \setminus \{x\}, x)$, there exists $\pi \models \omega$, such that $\pi' = pro_{(C \cup rel(p)) \setminus \{x\}}(\pi) \cup \{x := 0\}$. Since $x \in C$, then $\pi[x := 0] = \pi'$. Hence, $(s, \pi') \models A_1$, for each $\pi' \models reset(\omega, (C \cup rel(p)) \setminus \{x\}, x)$, i.e. $s, reset(\omega, (C \cup rel(p)) \setminus \{x\}, x) \models A_1$. Since $A_1 \prec A$, by induction hypothesis, $p, reset(\omega, (C \cup rel(p)) \setminus \{x\}, x) \vdash A_1$ and by rule *reset*, $p, \omega \vdash A$.

3. $A \equiv <\alpha> A_1$. By lemma 3 there exists $\{p_i, \omega_i \models A_1\}_{i\in\{1,n\}}$ and $\{p'_j, \omega'_j \models true\}_{j\in\{1,k\}}$, such that $p, \omega \vdash <\alpha> A_1$ is derivable from $\{p_i, \omega_i \vdash A_1\}_{i\in\{1,n\}}$ and $\{p'_j, \omega'_j \vdash true\}_{j\in\{1,k\}}$. By induction hypothesis proofs $\{p_i, \omega_i \vdash A_1\}_{i\in\{1,n\}}$ and $\{p'_j, \omega'_j \vdash true\}_{j\in\{1,k\}}$ indeed exist.

4. $A = [\alpha]A_1$. Similar to the previous case.

5. $A = \nu X\{U\}.B$. If $\| (p,\omega) \| \subseteq \| U \|$, then by ν_1, $p, \omega \vdash \nu X\{U\}.B$. Otherwise, by lemma 2, $\| \nu X\{U\}.B \|$ is a (p, A)-region, namely there exists a (p, A)-constraint $\omega' \in \Omega(rel(p) \cup C)$, such that $\models \forall(\omega \supset \omega')$ and $p, \omega' \models A$. We have $p, \omega \vdash A$ from $p, \omega' \vdash A$ by rule *thin*. By lemma 1, $(p, \omega') \in \| B[\nu X\{U \cup \{(p,\omega')\}.B/X] \cup U \|$. There are two cases:

   a) If $\| (p,\omega') \| \cap \| U \| = \emptyset$, then $p, \omega' \models B[\nu X\{U \cup \{(p,\omega')\}.B/X]$. Moreover, since $B[\nu X\{U \cup \{(p,\omega')\}.B/X] \prec A$, by induction hypothesis $p, \omega' \vdash B[\nu X\{U \cup \{(p,\omega')\}.B/X]$. We have $p, \omega' \vdash A$ by applying rule ν_2.

   b) If $\| (p,\omega') \| \cap \| U \| \neq \emptyset$, then there exists $\{(p, \omega_i) \in U\}_{i\in\{1,n\}}$, such that $\omega' \wedge \omega_i$ is satisfiable, for each $i \in \{1, n\}$. Consider $\omega'_1 \equiv \omega' \wedge (\bigvee_{i\in\{1,n\}} \omega_i)$ and $\omega'_2 \equiv \omega' \wedge \neg(\bigvee_{i\in\{1,n\}} \omega_i)$. It is obvious that $\omega' \equiv \omega'_1 \vee \omega'_2$. Hence, $p, \omega' \vdash A$ can be obtained by rule *disj* from $p, \omega'_1 \vdash A$ and $p, \omega'_2 \vdash A$. Moreover, $\models \forall(\omega'_1 \supset \bigvee_{i\in\{1,n\}} \omega_i)$. Hence, by rule ν_1, $p, \omega'_1 \vdash A$. On the other hand, $\| (p, \omega'_2) \| \cap \| U \| = \emptyset$. Therefore, $p, \omega'_2 \models B[\nu X\{U \cup \{(p,\omega'_2)\}.B/X]$ and, since $B[\nu X\{U \cup \{(p,\omega'_2)\}.B/X] \prec A$, by induction hypothesis $p, \omega'_2 \vdash B[\nu X\{U \cup \{(p,\omega'_2)\}.B/X]$. We have $p, \omega'_2 \vdash A$ by applying rule ν_2.

6. $A = \mu X\{U\}.B$. By lemma 1 and by lemma 2, there exist (p, A)-constraints $\omega_1, \ldots, \omega_k$, such that $\omega \equiv \bigvee_{i\in\{1,k\}} \omega_i$ and $p, \omega_i \models B[\mu X. \{U \cup \{(p,\omega_i)\}B/X]$. Moreover, by the semantics of tagged least fixpoints $\| (p,\omega_i) \| \cap \| U \| = \emptyset$. Therefore, $B[\mu X. \{U \cup \{(p,\omega_i)\}B/X] \prec A$ and by induction hypothesis, $p, \omega_i \vdash B[\mu X.\{U \cup \{(p,\omega_i)\}B/X]$, for each $i \in \{1, k\}$. By applying rules *disj* and rule μ, we have $p, \omega \vdash \mu X\{U\}.B$.

Temporal Logic for Stabilizing Systems

Yassine Lakhnech (yl@informatik.uni-kiel.d400.de)
Christian-Albrechts-Universität zu Kiel
Institut für Informatik und Praktische Mathematik
Preusserstrasse 1-9, 24105 Kiel, Germany

Michael Siegel (mis@wisdom.weizmann.ac.il)
Weizmann Institute of Science, Rehovot 76100, Israel
Dept. of Applied Mathematics and Computer Science

Abstract. This paper links two formerly disjoint research areas: temporal logic and stabilization. Temporal logic is a widely acknowledged language for the specification and verification of concurrent systems. Stabilization is a vitally emerging paradigm in fault tolerant distributed computing.

In this paper we give a brief introduction to stabilizing systems and present fair transition systems for their formal description. Then we give a formal definition of stabilization in linear temporal logic and provide a set of temporal proof rules specifically tailored towards the verification of stabilizing systems. By exploiting the semantical characteristics of stabilizing systems the presented proof rules are considerably simpler than the general temporal logic proof rules for program validity, yet we prove their completeness for the class of stabilizing systems.

These proof rules replace the hitherto informal reasoning in the field of stabilization and constitute the basis for machine-supported verification of an important class of distributed algorithms.

1. Introduction

This paper links two formerly disjoint research areas: temporal logic and stabilization. Temporal logic is a widely acknowledged language for the specification and verification of concurrent systems (Manna and Pnueli, 1991b; Manna and Pnueli, 1995). Stabilization is a concept used in various field of computer science s.a. databases and artificial intelligence (Gouda and Multari, 1991; Schneider, 1993) but has gained its importance as a vitally emerging paradigm in fault tolerant distributed computing (Schneider, 1993).

In this paper we give a brief introduction to stabilizing systems and present *fair transition systems* (Manna and Pnueli, 1991b) for their formal description. Then we give a formal definition of stabilization by means of *linear temporal logic* (LTL for short) and provide a set of temporal proof rules specifically tailored towards the verification of stabilizing systems. By exploiting the semantical characteristics of stabilizing systems the presented proof rules are considerably simpler than the general temporal logic proof rules for program validity, yet we prove

H. Barringer et al. (eds.), Advances in Temporal Logic, 71–90.

their completeness for the class of stabilizing systems. These proof rules replace the hitherto informal reasoning in the field of stabilization and constitute the basis for machine-supported verification of an important class of distributed algorithms.

The notion of stabilization has been introduced to the field of distributed computing by Edsger W. Dijkstra. In 1974, he presented several mutual exclusion algorithms for passing a token (Dijkstra, 1974) in a ring topology which all had an astonishing property: irrespectively of the initial number of token in the system, after a finite number of transitions the system converged to a global state where exactly one token was present in the ring and from then on passed between the nodes. Dijkstra called this behavior – convergence towards a set of predefined desired states – *stabilization*. Nine years later Leslie Lamport highlighted this so far little-known paper as a milestone in the work on fault-tolerant systems and as the most brilliant published work of Dijkstra (Lamport, 1984). Lamport observed that stabilizing systems show the remarkable property of being capable to automatically recover from transient errors, i.e. errors which do not continue to occur during the period of recovery (Schneider, 1993) and to remain correct thereafter. Concurrent systems with multiple tokens converge to a single token; graph markings become colorings; spanning forests become spanning trees; and inconsistent routing tables become consistent again. So the concept of stabilization allows to construct extremely robust fault-tolerant systems (Burns et al., 1993).

Unfortunately stabilization comes at a high price. Stabilizing algorithms are amongst the most complicated objects studied in the filed of distributed computing. Their intricacy stems from an extremely high degree of parallelism which is inherent to *all* stabilizing systems (see (Siegel, 1996) for a detailed discussion) combined with the standard problem in distributed computing that processes, cooperating in a network, have to perform actions taken on account of *local* information in order to accomplish a global objective. A lot of impossibility results, proving that a stabilizing solution for certain tasks and communication topologies cannot exist (Gouda et al., 1990; Lin and Simon, 1995) witness the complexity of stabilization as does the following citation, we quote Dijkstra (Dijkstra, 1974):

> " For more than a year—at least to my knowledge—it has been an open question whether nontrivial (self-)stabilizing systems could exist."

Most systems with distributed control are not stabilizing, i.e., once erroneously in an illegal state they generally never converge to a legal state again.

Since fault-tolerance is an increasingly important branch of distributed computing there is a well defined need for methods which support the design and verification of fault-tolerant systems based on stabilization, independent of technology, architecture, and application. The state-of-the-art so far are handwritten proofs on the basis of an intuitive understanding of the system –commonly described in pseudocode – and the desired properties. Undoubtly, handwritten proofs add confidence in the correctness of the algorithms. However, there is common consent that correctness proofs of complicated distributed systems are just as error-prone as the systems themselves (Owre et al., 1993). This insight resulted in an increasing interest in *automated* verification which comprise *algorithmic* techniques, known as model checking, e.g., (Alur et al., 1993; Burch et al., 1990), and *deductive* techniques, commonly referred to as theorem proving, e.g. (Boyer and Moore, 1986; Owre et al., 1992). The involvement of stabilizing systems makes tool support for their verification desirable if not indispensable. The prerequisite for tool-support is a formal basis comprising the following:

1. a formal model to describe stabilizing systems,
2. a formal specification language to describe properties of stabilizing systems, and
3. a formal framework to establish that a certain system obeys some specific properties.

This article provides such a formal basis. We use fair transition systems to describe stabilizing systems and advocate that LTL is an adequate specification language for this class of systems. The combination of temporal logic specifications and transition systems allows for algorithmic verification in the finite state case. To cover those cases where algorithmic verification is not applicable we investigate a temporal proof system for stabilizing systems. These proof rules are considerably simpler than the general proof rules for linear temporal logic (Manna and Pnueli, 1991a; Manna and Pnueli, 1991b; Manna and Pnueli, 1995). Nevertheless, we prove that the simple rules are just as complete as the general rules when considering stabilizing systems.

This article is organized as follows: In Section 2 we present fair transition systems and linear temporal logic. A brief introduction to stabilization as well as a formal definition by means of LTL is given in Section 3. The framework for formal reasoning about stabilization, consisting of a set of proof rules, is presented in Section 4. Some conclusions and prospects are given in Section 5.

Remark: Due to space limitations we only sketch some of the proofs. All proofs and technical details can be found in (Siegel, 1996).

2. Preliminaries

2.1. Fair Transition Systems

For the description of stabilizing systems, formalisms with all kinds of communication mechanisms (shared variables, synchronous and asynchronous communication) are used in the literature, and the execution models vary from pure interleaving to true parallelism. Therefore we use a generic abstract model, namely fair transition systems (Manna and Pnueli, 1991b), to formally capture stabilizing systems and their semantical characteristics. This choice guarantees broad applicability of our formalization as well as tool support since most existing analysis and verification tools are based on transition systems.

For the following presentation we assume a countable set *Var* of typed variables in which each variable is associated with a domain describing the possible values of that variable. A *state* s is a partial function $s : Var \longrightarrow Val$ assigning type-consistent values to a subset of variables in *Var*. By V_s we refer to the domain of s, i.e. the set of variables evaluated by s. The set of all states interpreting variables in $V \subseteq Var$ is denoted by Σ_V.

DEFINITION 1. *A* fair transition system $A = (V, \Theta, T, WF, SF)$ *consists of a finite set* $V \subseteq Var$ *of state variables, an assertion* Θ *characterizing* initial states, *a finite set* $T = \{t_1, \ldots, t_n\}$ *of* transitions *as well as* weak *and* strong fairness constraints *expressed by set* $WF \subseteq T$ *resp.* $SF \subseteq T$.

□

To refer to the components of a fair transition system (fts for short) A we use A as index; the state space of A is denoted by Σ_A and defined by $\Sigma_A \stackrel{\text{def}}{=} \Sigma_{V_A}$. The set of all fts's A with $V_A \subseteq Var$ is denoted by $\mathcal{S}$.

Each transition $t \in T$ is represented by a first order assertions $\rho_t(V, V')$, called a *transition predicate* (Manna and Pnueli, 1991b). A transition predicate relates the values of the variables V_s in a state s to their values in a successor state s' obtained by applying the transition to s. It does so by using two disjoint copies of set V_s. The occurrence of variable $v \in V_s$ refers to its value in state s while an occurrence of v' refers to the value of v in s'. Assuming that $\rho_t(V_s, V_s')$ is the transition predicate of transition t, state s' is called a *t-successor* of s iff $(s, s') \models \rho_t(V_s, V_s')$. A transition t is *enabled* in state s if there exists s' with $(s, s') \models \rho_t(V, V')$; otherwise it is *disabled.* Enabledness of a transition can be expressed by $en(t) \stackrel{\text{def}}{=} \exists V'.\rho_t(V, V')$. We require that T contains the *idling* transition t_{idle} whose transition relation is $\rho_{t_{idle}} : (V = V')$.

We use a standard linear semantics for fair transition systems (Manna and Pnueli, 1991a).

DEFINITION 2. *A* computation *of an fts* $A = (V, \Theta, T, WF, SF)$ *is an infinite sequence* $\sigma = \langle s_0, s_1, s_2, \ldots \rangle$ *of states* $s_i \in \Sigma_A$, *s.t.*

1. $s_0 \models \Theta$,
2. *for all* $i \in \mathbf{N}$ *state* s_{i+1} *is a t-successor of* s_i *for some transition* $t \in T$,
3. *if transition* $t \in WF$ *is continuously enabled from some point onwards in* σ, *there are infinitely many* $i \in \mathbf{N}$ *with* $(s_i, s_{i+1}) \models \rho_t(V, V')$,
4. *if transition* $t \in SF$ *is infinitely often enabled in* σ, *there are infinitely many* $i \in \mathbf{N}$ *with* $(s_i, s_{i+1}) \models \rho_t(V, V')$.

□

The set of computations generated by an fts A is denoted by $[\![A]\!]$. A finite sequence of states that satisfies the conditions 1. and 2. above is called *computation-prefix.*

For a given computation $\sigma = \langle s_0, s_1, s_2, \ldots \rangle$ and $i \in \mathbf{N}$ we define the prefix of σ up to index i by $\sigma_{\leq i} \stackrel{\text{def}}{=} \langle s_0, s_1, \ldots, s_i \rangle$. The suffix of σ starting at index i is the infinite sequence $\sigma_{\geq i} \stackrel{\text{def}}{=} \langle s_i, s_{i+1}, \ldots \rangle$. The set of computation *prefixes* of A is denoted by *Pref*(A) the set of computation *suffixes* by *Suff*(A). By σ_i we refer to the $(i+1)$-st state of computation σ.

We use $\{p\}\ t\ \{q\}$ as abbreviation of $(p \wedge \rho_t(V, V')) \rightarrow q'$, where q' results from q by replacing all variables of q by their primed version. For finite sets $T = \{t_1, \ldots, t_n\}$ of transitions we define $\{p\}\ T\ \{q\} \stackrel{\text{def}}{=} \bigwedge_{i=1}^{n} \{p\}\ t\ \{q\}$.

Finally, notice that each fts A is *machine closed* (Abadi and Lamport, 1991), that is, each computation-prefix of A is a prefix of a computation of A.

2.2. Linear Temporal Logic

As specification language we use a future fragment of Linear Temporal Logic (Manna and Pnueli, 1991b) without next-operator, referred to as LTL$^-$. Formulas are constructed from state formulas of some first order language $\mathcal{L}$, the temporal until operator U, and boolean operators $\wedge, \neg$ applied to formulas. Temporal formulas are denoted by $\varphi, \psi, \ldots$ and state formulas by $p, q, \ldots$. Temporal formulas are interpreted over infinite sequences of states (Manna and Pnueli, 1991b).

DEFINITION 3. *Given an infinite sequence* $\sigma = \langle s_0, s_1, s_2, \ldots \rangle$ *of states and* $\varphi \in LTL^-$. *We define that* σ satisfies φ, *denoted by* $\sigma \models \varphi$, *as:*

$$
\begin{array}{lll}
\sigma \models p & \textit{iff} & \sigma_0 \models p, \\
\sigma \models \varphi \wedge \psi & \textit{iff} & \sigma \models \varphi \ \ \textit{and} \ \ \sigma \models \psi, \\
\sigma \models \neg\varphi & \textit{iff} & \sigma \not\models \varphi, \\
\sigma \models \varphi U \psi & \textit{iff} & \exists i \geq 0.\ \sigma_{\geq i} \models \psi \ \ \textit{and} \ \ \forall j < i.\ \sigma_{\geq j} \models \varphi.
\end{array}
$$

□

We use the standard abbreviations $\Diamond\varphi \stackrel{\text{def}}{=} \mathit{true}U\varphi$ (eventually), $\Box\varphi \stackrel{\text{def}}{=} \neg\Diamond\neg\varphi$ (always) and $\varphi \Rightarrow \psi \stackrel{\text{def}}{=} \Box(\varphi \rightarrow \psi)$ (entails). An fts A satisfies a formula $\varphi \in \text{LTL}^-$, denoted by $A \models \varphi$, if all its computations satisfy φ. As notation for sequences we use $\circ$ to denote concatenation of sequences $last(seq)$ to return the last element of a (finite) sequence seq.

3. A Brief Introduction to Stabilization

Since stabilization has gained its importance as a paradigm for fault-tolerant computing, we address stabilization from this perspective.

3.1. Stabilization for Fault-Tolerance

Stabilization is besides *masking* and *detection/recovery* mechanisms the major paradigm in fault-tolerant computing (Arora, 1992). Stabilization is studied as an approach to cope with the *effect of arbitrary faults*, as long as some indispensable prerequisites for recovery, characterized by a so-called *fault-span*, are not violated. Such prerequisites may require that the actual code is not affected, or that the communication topology of the distributed system is not divided into unconnected parts. Fault-tolerance by stabilization ensures the continued availability of systems by correctly restoring the system state whenever the system exhibits incorrect behavior due to the occurrence of faults. Obviously, such a self-organizing behavior requires some assumptions about the occurrence of faults.

When talking about stabilization there is always an environment that acts as an adversary producing transient faults thus affecting the consistency of variables upon which the actual stabilization depends. In most publications on stabilization, see e.g. (Afek and Brown,

1993; Beauquier and Delaët, 1994; Dolev et al., 1993), the adversary is not explicitly modeled. The influence of faults is indirectly taken into account by considering *arbitrary initial states* in stabilizing systems. This models the situation that a fault-action just occurred and now time is given to the stabilizing system to reestablish a legal state.

From a more application oriented point of view stabilizing systems are constructed to be non-initializing because (Schneider, 1993):

- (re-)initialization of distributed systems is a very difficult task, cf. (Arora and Gouda, 1994), unless one constructs systems which can start in arbitrary states, and
- the main structuring principle to manage the complexity involved in the design and verification of these systems is to build complex stabilizing systems as a particular composition of simpler stabilizing systems, see e.g. (Arora and Gouda, 1994; Dolev et al., 1993; Gouda and Multari, 1991; Katz and Perry, 1993). In order for this composition to yield a stabilizing overall system, the simpler systems have to be non-initializing (see (Siegel, 1996) for details).

So, from now on we restrict our attention to the set $\mathcal{NI}$ of non-initializing systems, formally defined as $\mathcal{NI} \stackrel{\text{def}}{=} \{S \in \mathcal{S} | \models \Theta_S \leftrightarrow \mathit{true}\}$.

3.2. Formal Definition of Stabilization

Arora and Gouda advocate in (Arora, 1992; Arora and Gouda, 1993) a general and uniform definition of fault-tolerance based on the terms *convergence* and *closure*. These two properties characterize behavior of all stabilizing systems.

The observation which led to such a uniform definition is that there are two distinct kinds of behavior that a fault-tolerant system displays. As long as no fault occurs it performs the task that it is designed for, such as providing mutual exclusion, updating routing tables, or load balancing. As soon as a fault occurs a concerted effort of the processes causes the whole system to re-establish a so-called *legal* state (Dijkstra, 1974) from which point on it displays the original fault free behavior. So, in (Arora, 1992; Arora and Gouda, 1993) it is assumed that for a fault-tolerant system there exists a predicate *le* characterizing the set of legal system states which is invariant throughout fault-free system execution.

Furthermore Arora and Gouda state that faults can uniquely be represented as actions that upon execution perturb the system state (Cristian, 1985). Instead of anticipating various possible faults, it is only assumed that there exists a predicate *fs* weaker than *le*, called the *fault span* of the system (Arora, 1992), which defines the extent to which fault actions may perturb the legal states during system execution. Then they define that a system $S \in \mathcal{NI}$ is *stabilizing* w.r.t. *fs* and *le* iff it has the following two properties:

- *convergence*: Starting in an arbitrary state where *fs* holds, eventually a state is reached where *le* holds.
- *closure*: Predicates *fs* and *le* are invariant under execution of actions from S.

Based on this definition of stabilization Arora and Gouda define that a system $S \in \mathcal{NI}$ is *fault-tolerant* w.r.t. an adversary A and predicate *le* iff there exists a fault-span *fs* such that:

1. System S is stabilizing w.r.t. *fs* and *le*, and
2. predicate *fs* is invariant under execution of both actions from S and A.

Typical problems tackled in fault-tolerant computing by stabilization include communication protocols, clock synchronization, mutual exclusion, distributed reset, load balancing, spanning tree constructions and leader election (see (Siegel, 1996) for an extensive list of references).

The above informal description of stabilization can be formally captured by means of LTL$^-$ as follows:

DEFINITION 4. *Given fts* $S \in \mathcal{NI}$, *fault span fs and predicate le, with* $\models le \to fs$. *System* S *is* stabilizing *w.r.t. fs and le iff* S *satisfies the following properties:*

$$\begin{array}{ll} \textit{convergence:} & S \models \Box(fs \to \Diamond le) \\ \textit{closure:} & S \models \Box(fs \to \Box fs) \\ & S \models \Box(le \to \Box le) \end{array}$$

□

A formal definition of fault-tolerance w.r.t. an adversary and a set of legal states is straight-forward but requires a definition of parallel composition on fts's which we omitted due to space limitations (see (Siegel, 1996)). However it is worth noting that the forthcoming proof rules also apply in this case.

Properties of the form $\Box(p \to \Diamond q)$ are usually termed *response properties* in temporal logic (Manna and Pnueli, 1991b). We deviate from this convention and call them *convergence properties* since this is the term used in the stabilization community.

4. Proof Rules for Convergence and Closure

Despite the intricacy of stabilizing systems the mathematical foundations for their design and verification have hardly been investigated. In this section we present a collection of temporal proof rules which exploit the fact that stabilizing systems are non-initializing and that their fault-span is not left by system transitions, i.e. the fault-span is closed under system execution.

We give proof rules for program validity of formulas $\Box(p \rightarrow \Box p)$ and $\Box(p \rightarrow \Diamond q)$ for state predicates p, q. Repeated application of these rules gives a complete reduction of the program validity of these formulas into a set of assertional validities (Manna and Pnueli, 1991b). We prove relative completeness of our rules, i.e. we assume that there exists an oracle that provides proofs or otherwise verifies all generally valid assertions (Manna and Pnueli, 1991a); stated differently, we assume all valid assertions as axioms.

There exist complete proof rules for closure and convergence for arbitrary fts's (Manna and Pnueli, 1995). As explained, rather than simply recalling these very powerful rules, we adapt the general rules to deal with systems from set $\mathcal{NI}$.

In the following presentation we always state the general proof rule first, followed by the adapted rule for set $\mathcal{NI}$.

Note, that we do not have to deal with any past-formulas. There is nothing to prove about the past of a stabilizing systems since any of its states may have no past.

4.1. Proof Rules for Closure Properties

We start with a complete rule for proving $S \models \Box(p \rightarrow \Box p)$ for $S \in \mathcal{S}$ and some state predicate p. The rule for general invariance in (Manna and Pnueli, 1991a; Manna and Pnueli, 1995) refers to past modalities; in order to avoid the introduction of past-modalities we state a rule which only refers to future modalities.

$$\frac{\begin{array}{l} p \Rightarrow \varphi \\ \varphi \rightarrow p \\ \{\varphi\}\ T\ \{\varphi\} \end{array}}{\Box(p \rightarrow \Box p)} \qquad \text{GClos}$$

Concerning notation: In the following rules we assume that a fixed fts is given and refer to its components where necessary (e.g. to its set of transitions T in the rule above), cf. (Manna and Pnueli, 1991a). Recall, that operator $\Rightarrow$ denotes entailment $\varphi \Rightarrow \psi \stackrel{\text{def}}{=} \Box(\varphi \rightarrow \psi)$, whereas $\rightarrow$ denotes usual implication between assertions or formulas. Soundness and completeness of the above rule means that assertion p is closed in system S iff there exists a state formula φ such that

1. p implies φ in every reachable state of S,
2. φ implies p in every state, and
3. φ is invariant under transitions of S.

The first two premises imply that φ and p are equivalent on the set of reachable states of S. Typically, the additional state formula φ is used to characterize the set of p-states, i.e. states where predicate p holds, that are reachable in the system under investigation. Finding the appropriate assertion φ, i.e. coding enough reachability information into such an auxiliary predicate, is the intricate part in the application of rule GClos.

PROPOSITION 5. *Rule* GClos *is sound and complete for proving* $S \models \Box(p \rightarrow \Box p)$ *for* $S \in \mathcal{S}$.

$\Box$

Remark: Replacing in the second premise of rule GClos p by q actually yields a complete proof rule for the more general case $S \models \Box(p \rightarrow \Box q)$. Completeness of this rule is proven along the lines of the completeness proof for rule GClos which is sketched below.

Proof: As explained, we consider completeness of our proof rules relative to assertional validity. Hence it is sufficient to show that validity of the premises follows from validity of the conclusion. So, given a system S and predicate p such that $S \models \Box(p \rightarrow \Box p)$ we have to define predicate φ such that the premises of rule GClos becomes valid.

Assuming that the data domain of the assertion language is expressive enough to encode records of data and lists of records one can define a state assertion χ that holds in a state s iff s is reachable in S, i.e. appears in some computation of S (for technical details of this definition see (Manna and Pnueli, 1991a)).

Now, we consider as auxiliary predicate $\chi_p \stackrel{\text{def}}{=} \chi \wedge p$ and prove validity of the premises for predicate χ_p.

$S \models p \Rightarrow \chi_p$: Given a computation $\sigma \in [\![S]\!]$ and position i such that $\sigma_i \models p$. Obviously σ_i is reachable and satisfies p so $\sigma_i \models \chi_p$.

$\models \chi_p \rightarrow p$: By the definition of χ_p we have $\models \chi_p \rightarrow p$ because p is a conjunct of χ_p.

$\models \{\chi_p\}\ T\ \{\chi_p\}$: We prove that whenever state s satisfies predicate χ_p every t-successor s' of s, for $t \in T$, also satisfies χ_p.

$$\begin{array}{lll} & s \models \chi_p & \\ \text{iff} & s \models \chi \wedge p & \text{by definition of } \chi_p \\ \text{iff} & \exists \iota \in \mathit{Pref}(S).(\mathit{last}(\iota) = s \wedge s \models p) & \text{by definition of } \chi \end{array}$$

Let ι be a computation prefix leading to s, i.e. $\mathit{last}(\iota) = s$. We have that $\iota \circ \langle s' \rangle \in \mathit{Pref}(S)$ since s' is by assumption a t-successor of s for some $t \in T$ and since S is machine closed. Since $s \models p$ and $S \models \Box(p \rightarrow \Box p)$ we have $s' \models p$. We conclude $s' \models \chi \wedge p$, i.e. $s' \models \chi_p$.

□

The adapted complete rule for proving closure for systems $S \in \mathcal{NI}$ does not use any auxiliary predicate.

$\dfrac{\{p\}\ T\ \{p\}}{\Box(p \rightarrow \Box p)}$	Clos-$\mathcal{NI}$

The remaining premise $\{p\}\ T\ \{p\}$ is local in the sense that it does not require any more temporal reasoning; it is purely assertional. Note, that this rule is not complete if we consider arbitrary fts's.

Instead of proving directly soundness and completeness of rule Clos-$\mathcal{NI}$ for systems from $\mathcal{NI}$, we prove that $S \models \Box(p \rightarrow \Box p)$ (for $S \in \mathcal{NI}$) is derivable by rule GClos iff it is derivable by rule Clos-$\mathcal{NI}$. This reveals how we obtained the premise of rule Clos-$\mathcal{NI}$, namely by an explicit characterization of predicate φ in rule GClos and subsequent simplification of its premises.

PROPOSITION 6. *Given system $S \in \mathcal{NI}$. $S \models \Box(p \rightarrow \Box p)$ is derivable by rule* GClos *iff it is derivable by rule* Clos-$\mathcal{NI}$.

□

Proof: Given system $S \in \mathcal{NI}$ and predicate p. We just prove the more interesting direction: validity of the premises of rule GClos implies validity of the premise of rule Clos-$\mathcal{NI}$. So assume that there exists a predicate φ such that $S \models p \Rightarrow \varphi$, $\models \varphi \rightarrow p$ and $\models \{\varphi\}\ T\ \{\varphi\}$ holds. We have to prove that $\models \{p\}\ T\ \{p\}$ holds. We exploit the soundness and completeness result of rule GClos. From the assumptions we have by soundness of rule GClos that $S \models \Box(p \rightarrow \Box p)$. From the completeness proof for rule GClos we know that $\models \{\chi_p\}\ T\ \{\chi_p\}$ holds. Now it suffices to note that $\models \chi_p \leftrightarrow p$ for systems $S \in \mathcal{NI}$ since:

$$
\begin{array}{lll}
 & s \models \chi_p & \\
\text{iff} & s \models \chi \wedge p & \text{by definition of } \chi_p \\
\text{iff} & \exists \iota \in Pref(S).last(\iota) = s \wedge s \models p & \text{by definition of } \chi_p \\
\text{iff} & s \models p & \text{since } S \in \mathcal{NI}
\end{array}
$$

The last step is the reason why we can do without the auxiliary predicate φ in rule Clos-$\mathcal{NI}$: every p-state of S is reachable anyhow since $S \in \mathcal{NI}$ so predicate χ_p and p are equivalent.

From Proposition 5 and 6 we get:

COROLLARY 7. *Rule* Clos-$\mathcal{NI}$ *is sound and complete for proving* $S \models \Box(p \rightarrow \Box p)$ *for* $S \in \mathcal{NI}$.

□

Before we present proof rules for convergence we state a proposition which is important for the simplification of forthcoming proof rules.

PROPOSITION 8. *For all systems* $S \in \mathcal{NI}$ *and all temporal formulas* $\varphi \in LTL^-$ *the following holds:*

$$S \models \varphi \ \textit{iff} \ S \models \Box\varphi$$

□

The proof of this proposition is based on the observation that the set of computations of all systems $S \in \mathcal{NI}$ is suffix closed.

4.2. Proof Rules for Convergence

In the presentation of the proof rules for convergence we also consider subsets of $\mathcal{NI}$, consisting of all those systems which preserve a predefined state predicate p, i.e. sets of the form $\mathcal{NI}_p \stackrel{\text{def}}{=} \{S \in \mathcal{NI} \mid S \models \Box(p \rightarrow \Box p)\}$. The relevance of $\mathcal{NI}_p$ stems from the fact that we are interested in establishing the stabilization of a given system $S \in \mathcal{NI}$ w.r.t. some state predicates p, q. So we have to prove $S \models \Box(p \rightarrow \Diamond q) \wedge \Box(p \rightarrow \Box p) \wedge \Box(q \rightarrow \Box q)$. Proving $S \models \Box(p \rightarrow \Box p)$ first (by rule Clos-$\mathcal{NI}$), we have shown that S belongs to set $\mathcal{NI}_p$ and thus we can apply the rules for $\mathcal{NI}_p$ in order to establish the convergence of S.

The existing rules for proving convergence properties are usually partitioned into single-step convergence rules and extended convergence rules. We keep this pattern and mainly follow the presentation in (Manna and Pnueli, 1991a). Similar to the previous section we first state the general rule, followed by an adapted rule for set $\mathcal{NI}$ and finally the rule for $\mathcal{NI}_p$.

4.2.0.1. Single-step Convergence Rule under Weak Fairness

Single-step convergence rules are applicable in case that there exists at least one transition that accomplishes the desired convergence within one step. We obtain two slightly different rules, depending on whether this so-called helpful transition is executed weakly or strongly fair. The general proof rule as stated in (Manna and Pnueli, 1991a) is:

$$\frac{\begin{array}{l} p \Rightarrow (q \vee \varphi) \\ \{\varphi\}\ T\ \{q \vee \varphi\} \\ \{\varphi\}\ t\ \{q\} \\ \varphi \Rightarrow (q \vee en(t)) \end{array}}{\Box(p \rightarrow \Diamond q)} \qquad \textsf{WConv}$$

In this rule t identifies the helpful transition, contained in the set of weakly fair executed transitions, and $en(t)$ denotes the enabledness of t. The rule states that we have to find a state predicate φ such that p entails $q \vee \varphi$. Predicate φ has to be preserved by every transition in T unless q is established. Since the helpful transition t is enabled in φ-states if q is not yet established, we conclude that either q is established by a T-transition or, by weak fairness, finally t is executed which also establishes q.

When restricting our attention to systems in the set $\mathcal{NI}$ we get due to Proposition 8 the following simpler proof rule.

$$\frac{\begin{array}{l} p \rightarrow (q \vee \varphi) \\ \{\varphi\}\ T\ \{q \vee \varphi\} \\ \{\varphi\}\ t\ \{q\} \\ \varphi \rightarrow (q \vee en(t)) \end{array}}{\Box(p \rightarrow \Diamond q)} \qquad \textsf{WConv-}\mathcal{NI}$$

What looks like a minor change, replacing two times entailment by implication, constitutes in fact a considerable saving. Whereas the entailment properties of rule WConv require in general further invariants for their proof, in case of stabilizing systems we are done with proving ordinary implications!

PROPOSITION 9. *Given system $S \in \mathcal{NI}$. $S \models \Box(p \rightarrow \Diamond q)$ is derivable by rule* WConv *iff it is derivable by rule* WConv-$\mathcal{NI}$.

□

In case that the system under consideration belongs to set $\mathcal{NI}_p$ we get a further simplification.

$$\frac{\begin{array}{l}\{p \wedge \neg q\}\ t\ \{q\} \\ p \rightarrow (q \vee en(t))\end{array}}{\Box(p \rightarrow \Diamond q)} \qquad \text{WConv-}\mathcal{NI}_p$$

This simplification causes no loss of generality when considering systems in $\mathcal{NI}_p$.

PROPOSITION 10. *Given system* $S \in \mathcal{NI}_p$. $S \models \Box(p \rightarrow \Diamond q)$ *is derivable by rule* WConv *iff it is derivable by rule* WConv-$\mathcal{NI}_p$.

□

Proof: The proof of this proposition is based on the following two lemmata which are proved in (Siegel, 1996). These lemmata refer to a predicate χ_p^q which is a variant of predicate χ_p used in the completeness proof of Proposition 5. Predicate χ_p^q is defined such that $s \models \chi_p^q$ holds in a state $s \in \Sigma_S$ iff there exists a reachable p-state s' and a computation segment ι leading from s' to s, s.t. ι does not contain any q-states.

LEMMA 11. *Given system* $S \in \mathcal{S}$ *and predicates* p, q, φ. *If* $S \models p \Rightarrow (q \vee \varphi)$ *and* $\models \{\varphi\}\ T\ \{q \vee \varphi\}$ *holds then* $\models \chi_p^q \rightarrow \varphi$ *and* $\{\chi_p^q\}\ T\ \{q \vee \chi_p^q\}$.

□

The second lemma gives a characterization of predicate χ_p^q for systems $S \in \mathcal{NI}_p$.

LEMMA 12. *For systems* $S \in \mathcal{NI}_p$ *and predicates* p, q *we have* $\models \chi_p^q \leftrightarrow (p \wedge \neg q)$.

□

Using these lemmata it is not difficult to complete the proof for both directions of Proposition 10.

4.2.0.2. **Single-step Convergence Rule under Strong Fairness** The second single-step convergence rule relies on the existence of a helpful transition t in the set of strongly fair executed transitions. The general rule is:

$$\frac{\begin{array}{l}p \Rightarrow (q \vee \varphi) \\ \{\varphi\}\ T\ \{q \vee \varphi\} \\ \{\varphi\}\ t\ \{q\} \\ \varphi \Rightarrow \Diamond(q \vee en(t))\end{array}}{\Box(p \rightarrow \Diamond q)} \qquad \text{SConv}$$

As in rule WConv, t denotes the helpful transition, but now contained in the set of strongly fair executed transitions. Only the fourth premise is changed. We have to prove that φ entails that eventually $q \vee en(t)$ holds.

With the same justification as in the case of weak fairness we obtain the following two simplifications of this rule:

$$\frac{\begin{array}{l} p \rightarrow (q \vee \varphi) \\ \{\varphi\}\ T\ \{q \vee \varphi\} \\ \{\varphi\}\ t\ \{q\} \\ \varphi \rightarrow \Diamond(q \vee en(t)) \end{array}}{\Box(p \rightarrow \Diamond q)} \qquad \text{SConv-}\mathcal{NI}$$

We have the corresponding relative completeness result for rule SConv-$\mathcal{NI}$ w.r.t. rule SConv.

PROPOSITION 13. *Given system $S \in \mathcal{NI}$. $S \models \Box(p \rightarrow \Diamond q)$ is derivable by rule* SConv *iff it is derivable by rule* SConv-$\mathcal{NI}$.

□

In case that the system under consideration belongs to set $\mathcal{NI}_p$ we get:

$$\frac{\begin{array}{l} \{p \wedge \neg q\}\ t\ \{q\} \\ p \rightarrow \Diamond(q \vee en(t)) \end{array}}{\Box(p \rightarrow \Diamond q)} \qquad \text{SConv-}\mathcal{NI}_p$$

The expected result for $\mathcal{NI}_p$ in case of strong fairness is:

PROPOSITION 14. *Given system $S \in \mathcal{NI}_p$. $S \models \Box(p \rightarrow \Diamond q)$ is derivable by rule* SConv *iff it is derivable by rule* SConv-$\mathcal{NI}_p$.

□

As can be observed in most convergence proofs for stabilizing systems, *concerted effort* of several transitions is generally necessary to establish q after a p-state has been encountered. So more powerful rules for so called *extended* convergence are needed.

4.2.0.3. **Extended Convergence Rules** The general proof rule reduces extended convergence properties to a set of single-step convergence properties. These single-step convergence properties commonly serve to establish a well founded induction argument. We follow the presentation in (Manna and Pnueli, 1991a).

A binary relation $\preceq$ over set A is a pre-order if it is reflexive and transitive. If $a \preceq b$ but not $b \preceq a$ we say that a precedes b denoted by $a \prec b$. The irreflexive, asymmetric, and transitive ordering $(A, \prec)$ induced by $(A, \preceq)$ is *well-founded* if there does not exist an infinite sequence $\langle a_0, a_1, a_2, \ldots \rangle$ where $a_i \in A$ and $a_{i+1} \prec a_i$ for all $i \geq 0$. A pre-order $\preceq$ is called well-founded if its induced ordering $\prec$ is well-founded. Since we do not deal with past-operators we can use ranking functions which map states, rather than computation prefixes, to a well founded domain. So in the following rule from (Manna and Pnueli, 1991a) $\delta : \Sigma \mapsto A$ is a ranking function where Σ is the state space of the system under consideration, and A the domain of a well-founded pre-order $(A, \preceq)$. Then, the following rule can be used to prove convergence:

$$\frac{\begin{array}{l} p \Rightarrow (q \vee \varphi) \\ (\varphi \wedge \delta = a) \Rightarrow \Diamond(q \vee (\varphi \wedge \delta \prec a)) \end{array}}{\Box(p \rightarrow \Diamond q)} \qquad \textsf{EConv-1}$$

The first premise ensures that $q \vee \varphi$ holds at a position in a computation if p holds at that position. The second premise ensures that each reachable state where φ holds is eventually followed by a φ-state with a lower rank or a state where q holds.

PROPOSITION 15. (Manna & Pnueli). *Rule set* { EConv-1, WConv, SConv} *is sound and complete for proving* $S \models \Box(p \rightarrow \Diamond q)$ *for* $S \in \mathcal{S}$ *and predicates* p, q.

□

Simplifications of rule EConv-1 are obtained along the same lines as in the case of the single-step convergence rules.

$$\frac{\begin{array}{l} p \rightarrow (q \vee \varphi) \\ (\varphi \wedge \delta = a) \rightarrow \Diamond(q \vee (\varphi \wedge \delta \prec a) \end{array}}{\Box(p \rightarrow \Diamond q)} \qquad \textsf{EConv-1-}\mathcal{NI}$$

PROPOSITION 16. *Given system* $S \in \mathcal{NI}$. $S \models \Box(p \rightarrow \Diamond q)$ *is derivable by rule* EConv-1 *iff it is derivable by rule* EConv-1-$\mathcal{NI}$.

□

As corollary of Proposition 15 and 16 we get:

COROLLARY 17. *Rule set* { EConv-1-$\mathcal{NI}$, WConv-$\mathcal{NI}$, SConv-$\mathcal{NI}$} *is sound and complete for proving* $S \models \Box(p \rightarrow \Diamond q)$ *for* $S \in \mathcal{NI}$ *and predicates* p, q.

□

In case of set $\mathcal{NI}_p$ we get the following adapted rule:

$$\frac{(p \wedge \neg q \wedge \delta = a) \rightarrow \Diamond(q \vee \delta \prec a)}{\Box(p \rightarrow \Diamond q)} \qquad \textsf{EConv-1-}\mathcal{NI}_p$$

We have the following proposition.

PROPOSITION 18. *Given system* $S \in \mathcal{NI}_p$. $S \models \Box(p \rightarrow \Diamond q)$ *is derivable by rule* EConv-1 *iff it is derivable by rule* EConv-1-$\mathcal{NI}_p$.

□

From Proposition 15 and 18 we get the following soundness and completeness result for proving convergence properties of systems from $\mathcal{NI}_p$.

COROLLARY 19. *Rule set* { EConv-1-$\mathcal{NI}_p$, WConv-$\mathcal{NI}_p$, SConv-$\mathcal{NI}_p$} *is sound and complete for proving* $S \models \Box(p \rightarrow \Diamond q)$ *for* $S \in \mathcal{NI}_p$ *and predicates* p, q.

□

4.2.0.4. **Extended Convergence with Helpful Transitions** In (Manna and Pnueli, 1991a) a strategy is proposed to replace the premise concerning convergence in rule EConv-1 by applying a combination of single-step convergence rules. The resulting rule, stated below, uses the set $WF \cup SF$ containing all weakly or strongly fair executed transitions. Without loss of generality $WF \cap SF = \emptyset$ holds. Furthermore the rule assumes the existence of a set of state formulas $\{\varphi_1, \ldots, \varphi_n\}$, each φ_i corresponding to one $t_i \in WF \cup SF$, a well-founded pre-order $(A, \preceq)$, and a ranking function with range A. Let $\varphi \stackrel{\text{def}}{=} \bigvee_{i=1}^{n} \varphi_i$.

$$\frac{\begin{array}{l} p \Rightarrow (q \vee \varphi) \\ \{\varphi_i \wedge \delta = a\}\ T\ \{q \vee (\varphi \wedge \delta \prec a) \vee (\varphi_i \wedge \delta \preceq a)\} \\ \{\varphi_i \wedge \delta = a\}\ t_i\ \{q \vee (\varphi \wedge \delta \prec a)\} \\ \varphi_i \Rightarrow (q \vee en(t_i)) \quad \text{for } t_i \in WF \\ \varphi_i \Rightarrow \Diamond(q \vee en(t_i)) \quad \text{for } t_i \in SF \end{array}}{\Box(p \rightarrow \Diamond q)} \qquad \textsf{EConv-2}$$

In φ_i-states transition t_i is the helpful transition contained in $WF \cup SF$. The third premise guarantees that t_i, executed in a state where φ_i holds, either decreases the rank while preserving φ or establishes q. The other transitions in T either have to preserve these φ_i or have to lower the rank or establish q. The fourth and fifth premise guarantee that the t_i transitions are eventually executed when a φ_i-state has been encountered.

PROPOSITION 20. (Manna & Pnueli). *Rule* EConv-2 *is sound and complete for proving* $S \models \Box(p \rightarrow \Diamond q)$ *for* $S \in \mathcal{S}$ *and state predicates* p, q.

□

In case we restrict our attention to systems in set $\mathcal{NI}$ we obtain the following simplified rule:

$$\begin{array}{l} p \rightarrow (q \vee \varphi) \\ \{\varphi_i \wedge \delta = a\}\ T\ \{q \vee (\varphi \wedge \delta \prec a) \vee (\varphi_i \wedge \delta \preceq a)\} \\ \{\varphi_i \wedge \delta = a\}\ t_i\ \{q \vee (\varphi \wedge \delta \prec a)\} \\ \varphi_i \rightarrow (q \vee en(t_i)) \quad \text{for } t_i \in WF \\ \varphi_i \rightarrow \Diamond(q \vee en(t_i)) \quad \text{for } t_i \in SF \\ \hline \Box(p \rightarrow \Diamond q) \end{array} \qquad \text{EConv-2-}\mathcal{NI}$$

The corresponding proposition states:

PROPOSITION 21. *Given system* $S \in \mathcal{NI}$. $S \models \Box(p \rightarrow \Diamond q)$ *is derivable by rule* EConv-2 *iff it is derivable by rule* EConv-2-$\mathcal{NI}$.

□

From Proposition 20 and 21 we obtain:

COROLLARY 22. *Rule* EConv-2-$\mathcal{NI}$ *is sound and complete for proving* $S \models \Box(p \rightarrow \Diamond q)$ *for* $S \in \mathcal{NI}$ *and state predicates* p, q.

□

Finally, we get the following rule for extended convergence in case of $\mathcal{NI}_p$ where the φ_i are selected such that $\models (p \wedge \neg q) \rightarrow \bigvee_{i=1}^{n} \varphi_i$ holds.

$$\begin{array}{l} \{\varphi_i \wedge \delta = a\}\ T\ \{q \vee \delta \prec a \vee (\varphi_i \wedge \delta \preceq a)\} \\ \{\varphi_i \wedge \delta = a\}\ t_i\ \{q \vee \delta \prec a\} \\ \varphi_i \rightarrow (q \vee en(t_i)) \quad \text{for } t_i \in WF \\ \varphi_i \rightarrow \Diamond(q \vee en(t_i)) \quad \text{for } t_i \in SF \\ \hline \Box(p \rightarrow \Diamond q) \end{array} \qquad \text{EConv-2-}\mathcal{NI}_p$$

PROPOSITION 23. *Given system* $S \in \mathcal{NI}_p$. $S \models \Box(p \rightarrow \Diamond q)$ *is derivable by rule* EConv-2 *iff it is derivable by rule* EConv-2-$\mathcal{NI}_p$.

□

As in the previous case we get from Proposition 20 and 23:

COROLLARY 24. *Rule* EConv-2-$\mathcal{NI}_p$ *is sound and complete for proving* $S \models \Box(p \rightarrow \Diamond q)$ *for* $S \in \mathcal{NI}_p$ *and state predicates* p, q.

□

We have obtained rule EConv-2-$\mathcal{NI}_p$ by replacing φ in rule EConv-2 by $p \wedge \neg q$ and perform subsequent simplifications exploiting the closure of p. However, it is not possible to eliminate the auxiliary predicates φ_i in rule EConv-2-$\mathcal{NI}_p$; the construction of the φ_i's in the completeness proof of rule EConv-2 (Manna and Pnueli, 1991a) reveals that there does not exist a general characterization of the φ_i by means of the predicates p and q.

5. Conclusion

In this paper, we presented a formal framework for the analysis of stabilizing systems, which play an important role in the field of fault-tolerance. Temporal logic has been used to define the central notions underlying stabilization such as closure and convergence. Indeed, temporal logic turns out to be a natural formalism to define these notions. Then, we used the temporal approach to program verification to derive a set of proof rules which can be used to analyze stabilizing systems. Using the semantical features of these systems, such as the fact that they are non-initializing, we obtain a set of proof rules which are simpler than the general temporal proof rules. These proof rules provide a basis for computer aided verification of stabilizing systems.

In (Siegel, 1996) we have completed the list of proof rules for stabilizing systems by giving rules for *pseudo-stabilization* (Burns et al., 1993) and also some useful temporal tautologies which can be used in the verification of actual systems. Furthermore we present in (Siegel, 1996) a collection of design rules for the formal development of stabilizing systems based on the above presented verification rules.

Acknowledgments: We thank Amir Pnueli and the anonymous referees for valuable comments.

References

Abadi, M. and L. Lamport: 1991, 'The existence of refinement mappings'. *Theoretical Computer Science* **82**(2).

Afek, Y. and G. Brown: 1993, 'Self-stabilization over unreliable communication media'. *Distributed Computing* (7), 27–34.

Alur, R., T. Henzinger, and P. Ho: 1993, 'Automatic symbolic model checking of embedded systems'. In: *IEEE Real-Time Systems Symposium*.

Arora, A.: 1992, 'A Foundation of Fault Tolerant Computing'. Ph.D. thesis, The University of Texas at Austin.

Arora, A. and M. Gouda: 1993, 'Closure and convergence: a foundation of fault-tolerant computing'. *IEEE Transactions on Software Engineering* (19), 1015–1027.

Arora, A. and M. Gouda: 1994, 'Distributed reset'. *IEEE Transcations on Computers* (43), 1026–1038.

Beauquier, J. and S. Delaët: 1994, 'Probabilistic self-stabilizing mutual exclusion in uniform rings'. In: *PODC94 Proceedings of the Thirteenth Annual ACM Symposium on Principles of Distributed Computing.* p. 378.

Boyer, R. and J. Moore: 1986, 'Integrating decision procedures into heuristic theorem provers'. *Machine Intelligence* **11**.

Burch, J., E. Clarke, K. McMillan, D. Dill, and L. Hwang: 1990, 'Symbolic Model Checking: 10^{20} States and Beyond'. In: *Logic and Computer Science.*

Burns, J., M. Gouda, and R. Miller: 1993, 'Stabilization and pseudo-stabilization'. *Distributed Computing* **7**, 35–42.

Cristian, F.: 1985, 'A rigorous approach to fault-tolerant computing'. *IEEE Transactions on Software Engineering* **11**(1).

Dijkstra, E.: 1974, 'Self stabilizing systems in spite of distributed control'. *Communications of the ACM* **17**(11).

Dolev, S., A. Israeli, and S. Moran: 1993, 'Self-stabilization of dynamic systems assuming only read/write atomicity'. *Distributed Computing* **7**, 3–16.

Gouda, M., R. Howell, and L. Rosier: 1990, 'The instability of self-stabilization'. *Acta Informatica* **27**, 697–724.

Gouda, M. and N. Multari: 1991, 'Stabilizing communication protocols'. *IEEE Transactions on Computers* **40**, 448–458.

Katz, S. and K. Perry: 1993, 'Self-stabilizing extensions for message-passing systems'. *Distributed Computing* **7**, 17–26.

Lamport, L.: 1984, 'Solved problems, unsolved problems, and non-problems in concurrency'. In: *Proceedings of the 3rd Annual ACM Symposium on Principles of Distributed Computing.*

Lin, C. and J. Simon: 1995, 'Possibility and impossibility results for self-stabilizing phase clocks on synchronous rings'. In: *Proceedings of the Seconf Workshop on Self-Stabilizing Systems.* pp. 10.1–10.15.

Manna, Z. and A. Pnueli: 1991a, 'Completing the temporal picture'. *Theoretical Computer Science* **83**(1).

Manna, Z. and A. Pnueli: 1991b, *The Temporal Logic of Reactive and Concurrent Systems.* Springer Verlag.

Manna, Z. and A. Pnueli: 1995, *Temporal Verification of Reactive Systems.* Springer Verlag.

Owre, S., J. Rushby, and N. Shankar: 1992, 'PVS: a prototype verification system'. In: *11th Int Conf on Automated Deduction (CADE)*, Vol. 607 of *LNCS.* Springer Verlag.

Owre, S., J. Rushby, N. Shankar, and F. von Henke: 1993, 'Formal verification for fault-tolerant architectures: some lessons learned'. In: *FME 93: Industrial-strength Formal Methods*, Vol. 670 of *LNCS.* Springer Verlag.

Schneider, M.: 1993, 'Self-stabilization'. *ACM Computing Surveys* **25**, 45–67.

Siegel, M.: 1996, 'Phased Design and Verification of Stabilizing Systems'. Ph.D. thesis, University of Kiel.

Decidable Theories of ω-Layered Metric Temporal Structures

Angelo Montanari (montana@dimi.uniud.it),
Adriano Peron (peron@dimi.uniud.it) and
Alberto Policriti (policrit@dimi.uniud.it)
Dipartimento di Matematica e Informatica, Università di Udine
Via delle Scienze, 206 - 33100 Udine, Italy

Abstract. This paper focuses on decidability problems for metric and layered temporal logics. We prove the decidability of both the theory of metric temporal structures provided with an infinite number of arbitrarily coarse temporal layers and the theory of metric temporal structures provided with an infinite number of arbitrarily fine temporal layers. The proof for the first theory is obtained by reduction to the decidability problem of an extension of $S1S$ which is the logical counterpart of the class of ω-languages accepted by systolic tree automata. The proof for the second one is done through the reduction to the monadic second-order decidable theory of k successors SkS.

1. Introduction

This paper focuses on decidability problems for metric and layered temporal logics (Ciapessoni et al., 1993). The considered logics—suitable to model time granularity in various contexts—allow one to build granular temporal models by referring to the "natural scale" in any component of the model and properly constraining the interactions between differently grained components. As pointed out in (Benthem, 1995), the ability of providing and relating temporal representations at different 'grain levels' of the same reality is widely recognized as an important research theme for temporal logic and a major requirement for many applications (e.g., (Corsetti et al., 1991; Fiadeiro and Maibaum, 1992)).

In (Ciapessoni et al., 1993), Montanari et al. proposed a metric and layered temporal logic ($MLTL$ for short) for specifying granular real-time systems. Metric temporal logics, e.g., (Koymans, 1992; Montanari and de Rijke, 1995), extend propositional logic with a parameterized operator of relative temporal realization. $MLTL$ can be viewed as the combination of a number of differently-grained metric temporal logics. It replaces the flat temporal domain of metric temporal logics with a temporal universe consisting of a set of differently-grained temporal domains. A full account of syntax, semantics and axiomatization of $MLTL$ can be found in (Ciapessoni et al., 1993; Montanari and de Rijke, 1995); the basic expressive features of $MLTL$ are discussed

H. Barringer et al. (eds.), Advances in Temporal Logic, 91–108.

in Section 4, together with some examples of its use for the specification of relevant timing properties.

In order to guarantee the usefulness of metric and layered temporal logics as formal tools, it is necessary to show some basic decidability properties. The decidability problem for the pure metric (non-granular) fragment has been addressed by Alur and Henzinger in (Alur and Henzinger, 1993) which was, in fact, our starting point. They showed that, under suitable assumptions about the temporal domain and the associated operations, the validity and satisfiability problems for real-time logics extending propositional temporal logics with metric features are decidable. These problems can indeed be reduced to the decidability problem for a decidable theory: the well-known theory $S1S$. A first extension of their results, aiming at dealing with time granularity, has been presented in (Montanari and Policriti, 1996). Such an extension allows one to treat situations in which a finite number of coarsenings/refinements of the temporal domain is sufficient. The key idea to deal with the resulting finitely-layered metric temporal structures is to reformulate the decidability problem into an equivalent one relative to the finest metric component (layer). Hence, in both the original work by Alur and Henzinger and the above mentioned extension to finitely layered temporal structures, the basic tool for proving decidability properties is the theory $S1S$, and the basic engine is Büchi theorem on the decidability of regular ω-languages.

In the present work we deal with the more general case in which the underlying temporal structure consists of infinitely many temporal layers (ω-layered, k-refinable, metric temporal structures). We introduce the second-order language $\mathcal{L}^2_{\omega ML^k}$ for ω-layered (k-refinable) metric temporal structures, and show how to interpret it over different classes of structures. We first consider the case of temporal structures in which there is a finest temporal domain together with a infinite number of coarser and coarser domains (*upward unbounded layered structures*). To deal with such structures we use a more expressive theory, that we called $S1S^k$, which is a proper extension of $S1S$. The decidability of $S1S^k$ is shown by using a more powerful basic engine, namely the decidability of ω-languages recognized by k-ary Systolic Tree Automata. Such class of ω-languages has been recently proved to properly extend the class of regular ω-languages maintaining the same decidability properties (Monti and Peron, 1995a; Monti and Peron, 1995b). Since all the basic closure properties of regular ω-languages hold also for systolic tree ω-languages, a direct correspondence with the above mentioned second-order theory $S1S^k$ (properly extending $S1S$) can be established. From the one hand, upward unbounded layered structures provide an interesting example of application for the decidability of systolic tree ω-languages. On the other hand, we believe that the second-order theory $S1S$ is too weak to deal with infinitely coarsening domains, and that $S1S^k$ is a somehow "minimal" theory able to deal with such a case.

Successively, we deal with the problem of deciding infinitely refinable structures (*downward unbounded layered structures*). We prove that the decidability of the satisfiability problem for the theory of such structures can be reduced to the decidability of the satisfiability problem for SkS, the well-known monadic second-order decidable theory of k successors (Thomas, 1990). In this case the basic decidability engine is Rabin's Tree Theorem.

The paper is organized as follows. In Section 2, we define the theory $S1S^k$ and we recall the definition of the theory SkS. In Section 3, we formally define the theories of upward (resp. downward) unbounded layered structures, and prove that they are decidable. In Section 4, we show how the basic functionalities of $MLTL$ can be expressed in $\mathcal{L}^2_{\omega ML^k}$.

2. Preliminaries

2.1. THE THEORY $S1S^k$

In this section, we will define a (decidable) extension of the second-order theory of one successor $S1S$, called $S1S^k$, which is proved in (Montanari et al., 1996; Monti and Peron, 1995b) to be the logical counterpart of the operational definition of systolic ω-languages. The calculus $S1S^k$ extends the sequential calculus $S1S$ by adding a unary function symbol $\overset{k}{\leftarrow}$, called *power function*, and a unary predicate symbol L. For any natural number $x > 0$, the power function computes the natural number $x - x'$, where x' is the least power of k (with non-null coefficient) in the k-ary representation of x. The predicate L holds for a natural number x iff the least power of k (with non-null coefficient) in the k-ary representation of x has coefficient $k-1$.

DEFINITION 2.1. *The* power function $\overset{k}{\leftarrow}: \mathbb{N}^+ \rightarrow \mathbb{N}$ *is such that* $y = \overset{k}{\leftarrow}(x)$ *iff*

$$x = a_n k^n + a_{n-1}k^{n-1} + \ldots + a_m k^m,\ 0 \leq a_i \leq k-1, \text{ and } a_m \neq 0$$
$$y = a_n k^n + a_{n-1}k^{n-1} + \ldots + (a_m - 1)k^m.$$

The predicate L holds at x iff

$x = a_n k^n + a_{n-1}k^{n-1} + \ldots + a_m k^m$, $0 \leq a_i \leq k-1$, *and* $a_m = k - 1$.

The predicate "*is a power of k*" can be easily expressed as $0 = \overset{k}{\leftarrow}(x)$; hence, $S1S^k$ is at least as expressive as the well-known (decidable) extension of $S1S$ with the predicate "*is a power of k*".

As in the case of $S1S$, the model-theoretic structures for $S1S^k$ are ω-words over an alphabet Σ which, without loss of generality, we assume to be $\{0,1\}^n$. An ω-word α over Σ can be codified by a model

$$\underline{\alpha} = \langle \mathbb{N}, 0, succ, \leq, P_1, \ldots, P_n \rangle,$$

where $\mathbb{N}$ is the set of natural numbers, *succ* and $\leq$ are the successor function and the usual ordering of natural numbers, respectively, and $P_1, \ldots, P_n$ are subsets of $\mathbb{N}$ such that $j \in P_i$ iff the i-th element of $\alpha(j)$ is set to 1 (with $\alpha(j)$ denoting the j-th element of α).

DEFINITION 2.2. *The second-order language* $\mathcal{L}^2_{S1S^k}$ *is built up as follows:*

- terms *are freely constructed from individual variables by (zero or more) applications of the power function* $\overset{k}{\leftarrow}$;
- atomic formulae *are of the form* $L(t)$, $p(t)$, *and* $t \leq t'$, *where* t *and* t' *are terms and* p *is a first-order (or set) variable;*
- $S1S^k$-formulae *are freely constructed from atomic formulae by using the usual boolean connectives, and the quantifiers* $\exists$ *and* $\forall$ *acting on both individual and first-order variables.*

Note that the constant 0 and the successor function *succ* can be defined from $\leq$ and first-order quantification.

Let $\phi(p_1, \ldots, p_n)$ be a $S1S^k$-formula where at most the first-order variables $p_1, \ldots, p_n$ occur free, and $\underline{\alpha}$ be a ω-word model. We write $\underline{\alpha} \models \phi(p_1, \ldots, p_n)$ if ϕ is satisfied in $\underline{\alpha}$, with P_i as interpretation of p_i and where $\overset{k}{\leftarrow}$ and L are interpreted as in Definition 2.1.

The theory $S1S^k$ is a decidable extension of the theory $S1S$. As it is well known, the set of model-theoretic structures (i.e. ω-words) satisfying a formula ϕ of $S1S$ is a regular ω-language and the decidability of $S1S$ can be reduced to the (decidable) problem of emptiness for regular ω-languages (Büchi Theorem). The decidability of $S1S^k$ is proved in a similar way: in (Monti and Peron, 1995a) the class of ω-languages accepted by systolic tree automata is introduced. Such a class of languages properly extends the class of regular ω-languages, preserving all of its closure (w.r. to operations) and decidability properties. In (Montanari et al., 1996; Monti and Peron, 1995b), it has been proved that $S1S^k$ is the logical counterpart of systolic ω-languages, namely the set of ω-words satisfying a formula ϕ in $S1S^k$ is a systolic ω-language (an extension of Büchi Theorem). Therefore, the decidability of $S1S^k$ can be reduced to the (decidable) emptiness problem for systolic ω-languages. (The proof of the decidability of $S1S^2$ can be found in (Monti and Peron, 1995b). The (non-trivial) extension of the proof to any theory $S1S^k$, with $k \geq 2$, is given in the extended version of this paper (Montanari et al., 1996).)

2.2. The theory SkS

Leafless, perfectly balanced, k-ary labelled trees are the model-theoretic structures for interpreting formulae of SkS. Let Σ be the alphabet $\{0, \ldots, k-1\}^n$. An infinite, perfectly balanced k-ary tree t labelled over Σ, can be codified by a model

$$\underline{t} = \langle \{0, \ldots, k-1\}^*, \epsilon, succ_0, \ldots, succ_{k-1}, <_P, P_1, \ldots, P_n \rangle,$$

where $succ_0, \ldots, succ_{k-1}$ are the k successor functions over $\{0, \ldots, k-1\}^n$, with $succ_i(w) = w \cdot i$ (i.e. the concatenation of w and i) for $0 \leq i \leq k-1$, $<_P$ is the proper prefix relation over $\{0, \ldots, k-1\}^*$, and $P_1, \ldots, P_n$ are subsets of $\{0, \ldots, k-1\}^*$, where, for each $1 \leq i \leq n$, $w \in P_i$ iff the i-th component of the label of w equals 1.

DEFINITION 2.3. *The second-order language $\mathcal{L}^2_{SkS}$ is built up as follows:*

- terms *are freely constructed from individual variables and the constant* ϵ *by applying the successor functions* $succ_0, \ldots, succ_{k-1}$*;*

- atomic formulae *are of the forms* $t = t'$, $t <_P t'$ *and* $p(t)$, *where* t *and* t' *are terms and* p *is a first-order variable;*

- SkS-formulae *are freely constructed from atomic formulae by using the usual boolean connectives, and the quantifiers* $\exists$ *and* $\forall$ *acting on both individual and first-order variables.*

Let $\phi(p_1, \ldots p_n)$ be a SkS-formula, where at most the n first-order variables $p_1, \ldots, p_n$ occur free, and $\underline{t}$ be a tree model. We write $\underline{t} \models \phi(p_1, \ldots p_n)$ if ϕ is satisfied in $\underline{t}$, with P_i as interpretation of p_i (for $1 \leq i \leq n$). It is well known that the theory SkS is decidable (cf. (Thomas, 1990)).

3. Decidable theories of ω-layered temporal structures

In this section, we discuss the theory of ω-layered structures consisting of an infinite number of *arbitrarily coarse* infinite temporal domains (called upward unbounded layered structures), and the theory of ω-layered structures consisting of an infinite number of *arbitrarily fine* infinite temporal domains (called downward unbounded layered structures). Below, we first introduce the second-order language $\mathcal{L}^2_{\omega LM^k}$ for ω-layered, k-refinable, metric temporal structures; then we show how it can be interpreted over the class of upward unbounded layered structures as well as over the class of downward unbounded ones; finally, we prove the decidability of the two theories by reducing the former to the theory $S1S^k$ and the latter to the theory SkS.

Let $\mathcal{L}^2_{\omega LM^k}$ be the language including individual variables, the binary function symbol $\downarrow$, (uninterpreted) unary predicate symbols, the binary relational symbol $\leq$, and quantification of individual variables and (uninterpreted) unary predicate symbols. We restrict ourselves to formulae containing no free individual variables.

DEFINITION 3.1. (Basic language) *Let V and $\mathcal{V}$ be sets of individual and first-order variable symbols, respectively. Terms and formulae of $\mathcal{L}^2_{\omega LM^k}$ are built up as follows:*

- terms *are freely constructed from individual variables by applying the projection $\downarrow(\mathtt{j},\cdot)$, with $0 \leq j \leq k-1$;*
- atomic formulae *are of the form $\mathtt{p}(\mathtt{t})$ and $\mathtt{t} \leq \mathtt{t}'$, where $\mathtt{t}$ and $\mathtt{t}'$ are terms and $\mathtt{p} \in \mathcal{V}$;*
- formulae *are freely constructed from atomic formulae by using the usual boolean connectives, and the quantifiers $\exists$ and $\forall$ acting on both individual and first-order variables.*

3.1. Decidability of upward unbounded layered structures

We first define upward unbounded layered structures, and then show how to interpret $\mathcal{L}^2_{\omega LM^k}$ over them.

Let us denote by $\downarrow|_{T^i}$ the restriction of $\downarrow$ to T^i, and let $t < t'$ be a shorthand for $t \leq t'$ and $t' \not\leq t$.

DEFINITION 3.2. (Upward unbounded layered structure) *An upward unbounded ω-layered k-refinable metric temporal structure is a triplet $\langle \bigcup_{i\geq 0} T^i, \downarrow, \leq \rangle$, where*

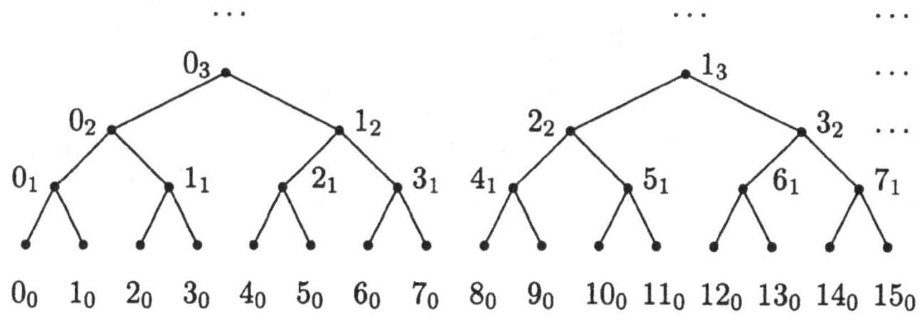

Figure 1. An upward unbounded (2-refinable) structure (i_j stands for $(+_j 1)^i(0_j)$).

- $\{T^i\}_{i\geq 0}$ *are pairwise disjoint enumerable sets;*
- $\downarrow$: $\{0,\dots,k-1\} \times \bigcup_{i\geq 1} T^i \rightarrow \bigcup_{i\geq 0} T^i$ *is a bijection such that for* $i > 0$, $\downarrow|_{T^i}$: $\{0,\dots,k-1\} \times T^i \rightarrow T^{i-1}$ *is a bijection;*
- $\leq$ *is the reflexive and transitive closure of the relation* $\tilde{\leq}$ *on* $\bigcup_{i\geq 0} T^i$ *such that*
  - $\downarrow(0,t)\tilde{\leq}t$ *and* $t\tilde{\leq}\downarrow(j,t)$, *for* $1 \leq j \leq k-1$;
  - $\downarrow(j,t)\tilde{\leq}\downarrow(j+1,t)$, *for* $1 \leq j \leq k-2$;
  - *if* $t\tilde{\leq}t'$ *and* $t' \notin\downarrow^n(t)$ *(for* $n \geq 0$*), then* $\downarrow(k-1,t)\tilde{\leq}t'$;
  - *if* $t\tilde{\leq}t'$ *and* $t \notin\downarrow^n(t')$ *(for* $n \geq 0$*), then* $t\tilde{\leq}\downarrow(0,t')$,
    where, for $t \in \bigcup_{i\geq 0} T^i$ *and* $n \in \mathbb{N}$,
    $\downarrow^n(t) \subseteq \bigcup_{i\geq 0} T^i$ *is the set such that* $\downarrow^0(t) = \{t\}$ *and*
    $\downarrow^n(t) = \{\downarrow(j,\bar{t}) : \bar{t} \in\downarrow^{n-1}(t), j \in \{0,\dots,k-1\}\}$.

Notice that, for each $t \in T^i$, with $i \geq 1$, and $0 \leq j \leq k-1$, $\downarrow(j,t)$ associates with t the $(j+1)$-th element of its k-decomposition with respect to T^{i-1}.

It can be shown that for each $i \geq 0$, T^i together with the restriction $\leq|_{T^i\times T^i}$ of $\leq$ to $T^i \times T^i$ is isomorphic to the natural numbers. For all $i \geq 0$ and $t \in T^i$, we denote the element $t' \in T^i$ such that $t < t'$ and $\neg\exists t''(t'' \in T^i \wedge t < t'' < t')$ by $+_i 1(t)$. An example of upward unbounded structure is given in Figure 1.

Let $\mathcal{I}$ be an interpretation of the language $\mathcal{L}^2_{\omega LM^k}$ over upward unbounded layered structures. As usual, let us denote by $\mathtt{c}^{\mathcal{I}}$ the element of the domain $\bigcup_{i\geq 0} T^i$ associated with the constant symbol $\mathtt{c}$ by $\mathcal{I}$. This notation is extended in a natural way to ground terms and atoms. We restrict ourselves to interpretations $\mathcal{I}$ that satisfy the following conditions:

$$(\downarrow(\mathtt{j},\mathtt{t}))^{\mathcal{I}} = \begin{cases} \downarrow(\mathtt{j},\mathtt{t}^{\mathcal{I}}) & \text{if } \mathtt{t}^{\mathcal{I}} \in \bigcup_{i>0} T^i \\ \bot & \text{otherwise} \end{cases} \qquad (\leq)^{\mathcal{I}} = \leq,$$

where $\perp$ stands for *undefined.*

DEFINITION 3.3. (Satisfiability relation) *Let $\mathcal{I}$ be an interpretation and $\mu : V \to \bigcup_{i\geq 0} T^i$ and $\nu : \mathcal{V} \to 2^{\bigcup_{i\geq 0} T^i}$ be the valuations of individual and first-order variables, respectively. We write $(\phi)^{\mathcal{I},\mu,\nu} = true$ to indicate that $\mathcal{I}$ satisfies ϕ under μ and ν. The satisfiability relation is defined as follows:*

- $(\mathtt{p}(\mathtt{t}))^{\mathcal{I},\mu,\nu} = true \Leftrightarrow \mathtt{t}^{\mathcal{I},\mu} \in \nu(\mathtt{p})$;
- $(\mathtt{t}_1 \leq \mathtt{t}_2)^{\mathcal{I},\mu,\nu} = true \Leftrightarrow (\mathtt{t}_1^{\mathcal{I},\mu}, \mathtt{t}_2^{\mathcal{I},\mu}) \in \leq^{\mathcal{I}}$;
- *boolean connectives and quantifiers are dealt with in the usual way.*

Notice that (atomic) formulae evaluate to false whenever at least one of their arguments evaluates to $\perp$. Given ϕ that contains no free individual variables, two interpretations may differ only in the values they assign to free first-order variables. Accordingly, an interpretation $\mathcal{I}$ for a formula with free predicate symbols $\mathtt{p}_1, \ldots, \mathtt{p}_m$ is given by m sets $\mathtt{p}_1^{\mathcal{I}}, \ldots, \mathtt{p}_m^{\mathcal{I}} \subseteq \bigcup_{i\geq 0} T^i$. We prove that the theory of upward unbounded layered structures is decidable, by defining a translation function τ that maps $\mathcal{L}^2_{\omega LM^k}$-formulae into equisatisfiable $S1S^k$-formulae. First of all, observe that any two upward unbounded layered structures are isomorphic. In fact, given $\mathcal{T}' = \langle \bigcup_{i\geq 0} T'^i, \downarrow', \leq' \rangle$ and $\mathcal{T}'' = \langle \bigcup_{i\geq 0} T''^i, \downarrow'', \leq'' \rangle$, the mapping $f : \bigcup_{i\geq 0} T'^i \to \bigcup_{i\geq 0} T''^i$ that associates the j-th element of T'^i with the j-th element of T''^i is a bijection which preserves projection and ordering. Therefore, it follows that a formula ϕ is satisfiable under an interpretation $\mathtt{p}_1^{\mathcal{I}}, \ldots \mathtt{p}_m^{\mathcal{I}} \subseteq \bigcup_{i\geq 0} \mathtt{T}'^i$ (where $\mathtt{p}_1, \ldots, \mathtt{p}_m$ are the free predicates symbols occurring in ϕ) if and only if ϕ is satisfiable under the interpretation $f(\mathtt{p}_1^{\mathcal{I}}), \ldots, f(\mathtt{p}_m^{\mathcal{I}}) \subseteq \bigcup_{i\geq 0} \mathtt{T}''^i$. This property allows us to replace the class of upward unbounded layered structures by a suitable single *concrete* structure which can be easily encoded into $S1S^k$. The *concrete* upward unbounded layered structure is $\mathcal{C} = \langle \bigcup_{i\geq 0} T^{i\mathcal{C}}, \downarrow^{\mathcal{C}}, \leq^{\mathcal{C}} \rangle$, where

- $T^{i\mathcal{C}} = \{k^i + nk^{i+1} : n \geq 0\}$;
- $\downarrow^{\mathcal{C}}$: $\{0, \ldots, k-1\} \times \bigcup_{i\geq 1} T^{i\mathcal{C}} \to \bigcup_{i\geq 0} T^{i\mathcal{C}}$ maps the pair $(j, k^i + nk^{i+1})$ into $k^{i-1} + (j + nk)k^i$, for $0 \leq j \leq k-1$ and $i \geq 0$;
- $\leq^{\mathcal{C}}$ is the restriction to $\bigcup_{i\geq 0} T^{i\mathcal{C}}$ of the usual ordering on natural numbers.

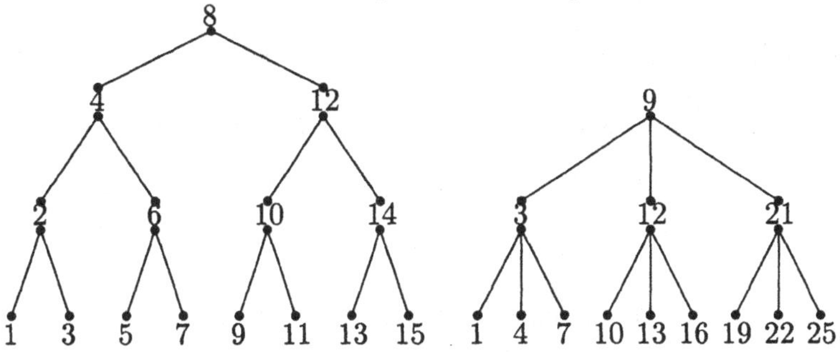

Figure 2. The trees associated with numbers 8 (for $k = 2$) and 9 (for $k = 3$).

Notice that the elements of the concrete structure are all (and only) the natural numbers having 1 as the least non-null coefficient in their k-ary representation. Notice also that, given a natural number

$$y = a_n k^n + a_{n-1}k^{n-1} + \ldots + a_m k^m, \text{ with } m > 0 \text{ and } a_m = 1,$$

the j-th child x of y (w.r. to the projection $\downarrow^{\mathcal{C}}$), with $0 \leq j \leq k-1$, is

$$x = a_n k^n + a_{n-1}k^{n-1} + \ldots + jk^m + 1k^{m-1}.$$

The trees associated with numbers 8 (for $k = 2$) and 9 (for $k = 3$) are shown in Figure 2. Now we show how the concrete projection function $\downarrow^{\mathcal{C}}$ can be expressed in terms of the power function $\overset{k}{\leftarrow}$ and the predicate L.

We first define $k-1$ auxiliary predicates L_j, with $1 \leq j \leq k-1$, such that $L_j(x)$ holds iff j is the least non-null coefficient in the k-ary representation of x (notice that $L_{k-1} = L$). For $j = 1, \ldots, k-1$, $\mathtt{L_j(x)}$ can be defined as follows:

$$\mathtt{L_j(x)} \text{ iff } \exists \mathtt{x_{k-1}, \ldots, x_j(L(x_{k-1}) \wedge x = x_j \wedge \bigwedge_{i=j+1}^{k-1} x_{i-1} = \overset{k}{\leftarrow} (x_i))}. \quad (1)$$

For any pair of natural numbers x and y such that both $L_1(x)$ and $L_1(y)$ hold, x is the 0-th child of y ($y \overset{0}{\rightarrow} x$ for short) iff $x = max\{w : w < y \wedge \overset{k}{\leftarrow}(w) = \overset{k}{\leftarrow}(y)\}$. Therefore, $y \overset{0}{\rightarrow} x$ can be defined as follows:

$$\begin{array}{c} \mathtt{L_1(y) \wedge L_1(x) \wedge x < y \wedge \overset{k}{\leftarrow} (y) = \overset{k}{\leftarrow} (x) \wedge} \\ \mathtt{\forall w((L_1(w) \wedge w < y \wedge \overset{k}{\leftarrow} (y) = \overset{k}{\leftarrow} (w)) \rightarrow w \leq x).} \end{array} \quad (2)$$

Analogously, for x and y such that both $L_1(x)$ and $L_1(y)$ hold, x is the j-th child of y ($y \overset{j}{\rightarrow} x$ for short), with $1 \leq j \leq k-1$, iff $L_j(\overset{k}{\leftarrow}(x)) \wedge$

$y = (\overset{k}{\leftarrow})^j(x)$. Therefore, $y \overset{j}{\to} x$, with $0 < j \leq k-1$) can be defined as follows

$$\begin{array}{l} \mathtt{L_1(y) \wedge L_1(x) \wedge (\exists x_j, \ldots, x_1(x_j =\overset{k}{\leftarrow} (x) \wedge} \\ \mathtt{L_j(x_j) \wedge x_1 = y \wedge \bigwedge_{i=2}^{j} x_{i-1} =\overset{k}{\leftarrow} (x_i))).} \end{array} \qquad (3)$$

The (inductively) defined map $\tau : \mathcal{L}^2_{\omega LM^k} \to S1S^k$ is as follows:

- if ϕ is an atomic formula devoid of any occurrence of terms of the form $\downarrow(\mathtt{j,t})$, then $\tau(\phi) = \phi$;
- if ϕ is an atomic formula and $\downarrow(\mathtt{j_1,x_1}), \ldots, \downarrow(\mathtt{j_n,x_n})$ are the n innermost occurrences of $\downarrow$ in ϕ, with $n \leq 2$, then
  $\tau(\phi) = \tau(\phi[\mathtt{z_1} \backslash \downarrow(\mathtt{j_1,x_1}), \ldots, \mathtt{z_n} \backslash \downarrow(\mathtt{j_n,x_n})]) \wedge$
  $\bigwedge_{1 \leq i \leq n} x_i \overset{j_i}{\to} z_i$,
  where $\mathtt{z_1}, \ldots, \mathtt{z_n}$ are fresh variables;
- if $\phi = \neg\psi$, then $\tau(\phi) = \neg\tau(\psi)$;
- if $\phi = \psi \wedge \theta$ (resp. $\phi = \psi \vee \theta$), then
  $\tau(\phi) = \tau(\psi) \wedge \tau(\theta)$ (resp. $\tau(\phi) = \tau(\psi) \vee \tau(\theta)$);
- if $\phi = \exists \mathtt{x}\psi$ (resp. $\forall \mathtt{x}\psi$), then
  $\tau(\phi) = \exists \mathtt{x}(\mathtt{L_1(x)} \wedge \tau(\psi))$ (resp. $\tau(\phi) = \forall \mathtt{x}(\mathtt{L_1(x)} \to \tau(\psi))$);
- if $\phi = \exists \mathtt{p}\psi$ (resp. $\forall \mathtt{p}\psi$), then
  $\tau(\phi) = \exists \mathtt{p}(\forall \mathtt{y}(\mathtt{p(y)} \to \mathtt{L_1(y)}) \wedge \tau(\psi))$
  (resp. $\tau(\phi) = \forall \mathtt{p}(\forall \mathtt{y}(\mathtt{p(y)} \to \mathtt{L_1(y)}) \to \tau(\psi))$).

LEMMA 3.4. *For any $\phi \in \mathcal{L}^2_{\omega LM^k}$, ϕ is satisfiable iff $\tau(\phi) \in \mathcal{L}^2_{S1S^k}$ is satisfiable.*

THEOREM 3.5. *The theory of upward unbounded layered structures is decidable.*

Proof.

It follows from Lemma 3.4 and the decidability of $S1S^k$ (cf. (Montanari et al., 1996)).

3.2. DECIDABILITY OF DOWNWARD UNBOUNDED LAYERED STRUCTURES

As for the upward case, we first formally define downward unbounded layered structures, and then show how to interpret $\mathcal{L}^2_{\omega LM^k}$ over them.

DEFINITION 3.6. (Downward unbounded layered structure) *A downward unbounded ω-layered k-refinable metric temporal structure is a triplet $\langle \bigcup_{i \geq 0} T^i, \downarrow, \leq \rangle$, where*

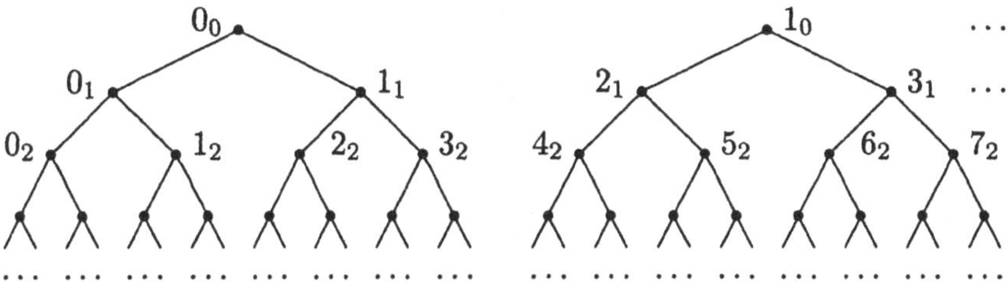

Figure 3. A downward unbounded (2-refinable) structure (i_j stands for $(+_j1)^i(0_j)$).

- $\{T^i\}_{i\geq 0}$ *are pairwise disjoint enumerable sets;*
- $\downarrow: \{0,\ldots,k-1\}\times\bigcup_{i\geq 0}T^i \rightarrow \bigcup_{i\geq 1}T^i$ *is a bijection such that for* $i\geq 0$, $\downarrow|_{T^i}: \{0,\ldots,k-1\}\times T^i \rightarrow T^{i+1}$ *is a bijection;*
- $\leq$ *is the reflexive and transitive closure of the relation* $\tilde{\leq}$ *on* $\bigcup_{i\geq 0}T^i$ *such that*
  - $\langle T^0, \tilde{\leq}|_{T^0\times T^0}\rangle$ *is isomorphic to* $\langle \mathbb{N},\leq\rangle$;
  - $t\tilde{\leq}\downarrow(j,t)$, *for* $0\leq j\leq k-1$;
  - $\downarrow(j,t)\tilde{\leq}\downarrow(j+1,t)$, *for* $0\leq j\leq k-2$;
  - *if* $t\tilde{\leq}t'$ *and* $t'\not\in\downarrow^n(t)$ *(for* $n\geq 0$*), then* $\downarrow(j,t)\tilde{\leq}t'$, *for* $0\leq j\leq k-1$.

As in the case of Definition 3.2, it can be shown that each layer is isomorphic to the natural numbers. An example of downward unbounded layered structure is shown in Figure 3.
Interpretations $\mathcal{I}$ of the language $\mathcal{L}^2_{\omega LM^k}$ over downward unbounded layered structures differ from interpretations over upward unbounded layered structures only in the definition of the semantic clause for $\downarrow$: $(\downarrow(\mathtt{j},\mathtt{t}))^{\mathcal{I}} = \downarrow(\mathtt{j},\mathtt{t}^{\mathcal{I}})$.
Unlike the case of upward unbounded layered structures, the interpretation of $\downarrow(\mathtt{j},\mathtt{t})$ over downward unbounded layered structures is indeed always defined. The satisfiability relation is defined exactly as in the case of upward unbounded layered structures.

We prove that the theory of downward unbounded layered structures is decidable, by showing that any formula $\phi\in\mathcal{L}^2_{\omega LM^k}$ can be transformed into an equisatisfiable formula of SkS. In particular, for each $0\leq j\leq k-1$, the function $\downarrow(j,.)$ acts as $succ_j$ (i.e. the j-th successor of SkS), and the ordering $\leq$ of $\mathcal{L}^2_{\omega LM^k}$ can be expressed by using the prefix order $\leq_P$ of SkS. In Figure 4, we show how a downward unbounded layered structure can be encoded into the domain of

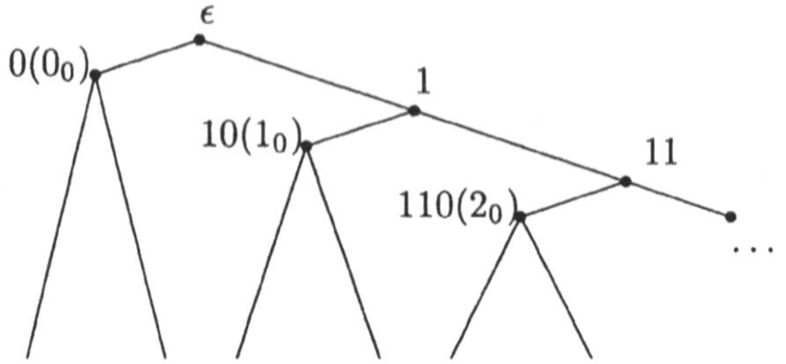

Figure 4. The encoding of a downward unbounded (2-refinable) structure into $\{0,1\}^*$.

interpretation of SkS (i.e., $\{0,\ldots,k-1\}^*$). The *concrete* downward unbounded layered structure is $\mathcal{R} = \langle \bigcup_{i\geq 0} T^{i\mathcal{R}}, \downarrow^{\mathcal{R}}, \leq^{\mathcal{R}}\rangle$, where

- $T^{i\mathcal{R}} = \{v \cdot 0 \cdot w : v \in \{k-1\}^*, w \in \{0,\ldots,k-1\}^*, |w| = i\}$;
- $\downarrow^{\mathcal{R}} (j, v) = v \cdot j (= succ_j(v))$, for all $v \in \bigcup_{i\geq 0} T^{i\mathcal{R}}$;
- $\leq^{\mathcal{R}}$ is the restriction to $\bigcup_{i\geq 0} T^{i\mathcal{R}}$ of the lexicographic ordering over $\{0,\ldots,k-1\}^*$.

PROPOSITION 3.7. *Let $\langle \bigcup_{i\geq 0} T^i, \downarrow, \leq\rangle$ be a downward unbounded layered structure. For each $i \geq 0$, $t \in T^i$, there exists $j \in \mathbb{N}$ such that $t = (+_i 1)^j(0_i)$. The map $cod : \bigcup_{i\geq 0} T^i \to \bigcup_{i\geq 0} T^{i\mathcal{R}}$ such that*

$$cod(t) = \begin{cases} (k-1)^j \cdot 0 & \text{if } i = 0; \\ (k-1)^q \cdot 0 \cdot r & \text{otherwise,} \end{cases}$$

where $j = k^i q + r$, with $0 \leq r < k$, is an isomorphism of downward unbounded layered structures.

In order to map a formula $\phi \in \mathcal{L}^2_{\omega LM^k}$ into an equisatisfiable formula of SkS, we define two auxiliary predicates $\mathtt{R}$ and $\mathtt{Dom}$ of SkS. These predicates are defined in such a way that $\mathtt{R}(\mathbf{x})$ holds iff $\mathbf{x}^{\mathcal{I}} \in \{k-1\}^*$, and $\mathtt{Dom}(\mathbf{x})$ holds iff $\mathbf{x}^{\mathcal{I}} \in \bigcup_{i\geq 0} T^{i\mathcal{R}}$ (which is a proper subset of $\{0,\ldots,k-1\}^* \setminus \{k-1\}^*$).
$\mathtt{R}(\mathbf{x})$ is a shorthand for the formula:

$$\exists p(p(\epsilon) \wedge p(x) \wedge \forall y,z(p(y) \to p(succ_{k-1}(y)) \wedge (p(y) \wedge \\ \bigvee_{i=0}^{k-1} succ_i(z) = y) \to p(z) \wedge (p(z) \wedge p(y)) \to \neg \bigvee_{i=0}^{k-2} succ_i(z) = y)),$$

while $\mathtt{Dom}(\mathbf{x})$ is a shorthand for the formula:

$$\exists p(p(x) \wedge \forall y,z(R(y) \to (\neg p(y) \wedge p(succ_0(y)) \wedge \\ \bigwedge_{j=1}^{k-1} \neg p(succ_j(y))) \wedge p(y) \to \bigwedge_{j=0}^{k-1} p(succ_j(y)) \wedge \\ (p(y) \wedge \neg R(z) \wedge \bigvee_{j=1}^{k-1} succ_j(z) = y) \to p(z))).$$

The reduction of a formula $\phi \in \mathcal{L}^2_{\omega LM^k}$ into a formula of SkS is as follows:

if $\phi = \mathtt{t} \leq \mathtt{t}'$, then
$\tau(\phi) = \mathtt{Dom}(\mathtt{t}_1) \wedge \mathtt{Dom}(\mathtt{t}_2) \wedge (\mathtt{t}_1 \leq_{\mathtt{P}} \mathtt{t}_2 \vee$
$\exists x (\bigvee_{l=0}^{k-1} (succ_l(x) \leq_P t_1 \wedge \bigvee_{s=l+1}^{k-1} succ_s(x) \leq_P t_2)))$,
where $\mathtt{t}_1$ and $\mathtt{t}_2$ are obtained from $\mathtt{t}$ and $\mathtt{t}'$, respectively, by replacing any occurrence of a term $\downarrow (j, t'')$ by $succ_j(t'')$;
if $\phi = \mathtt{p}(\mathtt{t})$, then $\tau(\phi) = \mathtt{Dom}(\mathtt{t}') \wedge \mathtt{p}(\mathtt{t}')$,
where $\mathtt{t}'$ is the term obtained from $\mathtt{t}$ by replacing any occurrence of a term $\downarrow (j, t'')$ by $succ_j(t'')$;
if $\phi = \neg\psi$, then $\tau(\phi) = \neg\tau(\psi)$;
if $\phi = \psi \wedge \theta$ (resp. $\phi = \psi \vee \theta$), then
$\tau(\phi) = \tau(\psi) \wedge \tau(\theta)$ (resp. $\tau(\phi) = \tau(\psi) \vee \tau(\theta)$);
if $\phi = \exists \mathtt{x}\psi$ (resp. $\forall \mathtt{x}\psi$), then
$\tau(\phi) = \exists \mathtt{x}(\mathtt{Dom}(\mathtt{x}) \wedge \tau(\psi))$ (resp. $\tau(\phi) = \forall \mathtt{x}(\mathtt{Dom}(\mathtt{x}) \rightarrow \tau(\psi))$;
if $\phi = \exists \mathtt{p}\psi$ (resp. $\forall \mathtt{p}\psi$), then
$\tau(\phi) = \exists \mathtt{p}(\forall \mathtt{y}(\mathtt{p}(\mathtt{y}) \rightarrow \mathtt{Dom}(\mathtt{y})) \wedge \tau(\psi))$
(resp. $\tau(\phi) = \forall \mathtt{p}(\forall \mathtt{y}(\mathtt{p}(\mathtt{y}) \rightarrow \mathtt{Dom}(\mathtt{y})) \rightarrow \tau(\psi)))$.

LEMMA 3.8. *For any $\phi \in \mathcal{L}^2_{\omega LM^k}$, ϕ is satisfiable iff $\tau(\phi) \in \mathcal{L}^2_{SkS}$ is satisfiable.*

THEOREM 3.9. *The theory of downward unbounded layered structures is decidable.*

Proof.
It follows from Lemma 3.8 and the decidability of SkS (cf. (Thomas, 1990)).

Remark. In both the downward and the upward case, the relation $\leq$ is defined in such a way that each pair $\langle T^i, \leq|_{T^i \times T^i}\rangle$, with $i \geq 0$, is isomorphic to the pair $\langle \mathbb{N}, \leq\rangle$. Moreover, in both cases the zero element for $\langle \bigcup_{i \geq 0} T^i, \leq\rangle$ coincides with the zero element for $\langle T^0, \leq|_{T^0 \times T^0}\rangle$. We cannot use the same definition of $\leq$ for both classes of structures: if we ordered downward unbounded structures in the same way of upward unbounded ones, we would loose the zero element, and vice versa. As a matter of fact, the topologies of upward and downward unbounded layered structures are intrinsically different. A number of natural orderings over the domain of both downward and upward unbounded layered structures can be easily defined on the basis of the given one. Significant examples are given in (Montanari et al., 1996).

4. High-level languages for ω-layered temporal structures

In this section, we demonstrate that the key features of $MLTL$ can actually be expressed in $\mathcal{L}^2_{\omega LM^k}$. To this end, we first define a high-level second-order language for the theory of upward (resp. downward) unbounded layered structures, including suitable primitives for modeling contextualization, projection, and displacement, and then show that the resulting language is in fact as expressive as the basic one ($\mathcal{L}^2_{\omega LM^k}$).

4.1. THE LANGUAGE FOR UPWARD UNBOUNDED LAYERED STRUCTURES

Let $\mathcal{UL}^2_{\omega LM^k}$ be the high-level second-order language for the theory of upward unbounded layered structures, including individual variables, the constant symbols $\mathtt{0_i}$ (local zero elements), the unary function symbols $\mathtt{+_i 1}$ (local successors), the binary function symbol $\downarrow$ (projection), (uninterpreted) unary predicate symbols, the unary predicate symbols $\mathtt{T^i}$ (vertical contextualizations or y-contextualizations) and $\mathtt{\Delta_i}$ (horizontal contextualizations or x-contextualizations), the binary relational symbols $\leq$ (ordering), $\preceq$ (approximate ordering) and $\equiv_{\mathtt{i,j}}$ (congruences), and quantification of individual variables and (uninterpreted) unary predicate symbols (with $i \geq 0$). We restrict ourselves to formulae that contain no free individual variables.

An interpretation $\mathcal{I}$ for the language $\mathcal{L}^2_{\omega LM^k}$ over upward unbounded layered temporal structures is also an interpretation for the high-level language $\mathcal{UL}^2_{\omega LM^k}$ if and only if it satisfies the following further conditions:

$\mathtt{0}_\mathtt{i}^{\mathcal{I}} = t$, where t is the (unique) element of T^i such that $\forall t'(t' \in T^i \rightarrow t \leq t')$;

$$(\mathtt{+_i 1(t)})^{\mathcal{I}} = \begin{cases} +_i 1(t^{\mathcal{I}}) & \text{if } t^{\mathcal{I}} \in T^i \\ \perp & \text{otherwise,} \end{cases}$$

$(\mathtt{T^i})^{\mathcal{I}} = T^i$; $(\mathtt{\Delta_i})^{\mathcal{I}} = \{(+_j 1)^i(0_j^{\mathcal{I}}) : j \geq 0\}$;

$(\preceq)^{\mathcal{I}} = \{(t, t') : ancestor(t, t') \vee ancestor(t', t) \vee t \leq t'\}$, where, for all $t, t' \in \bigcup_{i \geq 0} T^i$, $ancestor(t, t')$ if and only if $\exists j(t =\downarrow (j, t') \vee \exists j(t'' =\downarrow (j, t') \wedge ancestor(t, t'')))$;

$(\equiv_{\mathtt{i,j}})^{\mathcal{I}} = \{(t, t') : t \in T^i \wedge t' \in T^i \wedge t = (+_i 1)^m(0_i^{\mathcal{I}}) \wedge t' = (+_i 1)^n(0_i^{\mathcal{I}}) \wedge m \equiv_j n\}$.

For each $i \geq 0$, 0_i denotes the origin of the domain T^i and $+_i 1$ is interpreted as the successor function over T^i. y-contextualizations $T^i(t)$ and x-contextualizations $\Delta_i(t)$ respectively restrict the range of possible values of the term t, constraining it to be interpreted over the domain T^i and over the set of elements at distance i from the origin of the domain they belong to. For each $i \geq 0$, the set of elements satisfying Δ_i, together with (the obvious restriction of) $\leq$, is isomorphic to $\langle \mathbb{N}, \leq \rangle$, so x-contextualization and y-contextualization act in a perfectly symmetric way. For each $t, t' \in \bigcup_{i \geq 0} T^i$, $\preceq$ relates t to t' whenever $t \leq t'$ holds or either t' belongs to the projection of t over the domain t' belongs to or t belongs to the projection of t' over the domain t belongs to. Finally, for all $i \geq 0$ and $j \geq 2$, $\equiv_{i,j}$ denotes the congruence modulo-j over T^i.

EXAMPLE 4.1. We describe a property whose specification exploits the expressive power of the operators of x-contextualization and projection of upward unbounded layered structures, and cannot be expressed in the theory of finitely-layered ones. For the sake of simplicity, we take $k = 2$. Consider a process of decay, where the elapsing time between two successive occurrences of a given phenomenon exponentially increases. It can be modeled in $\mathcal{UL}^2_{\omega LM^k}$ as follows. Let p and q be two predicates that respectively detect occurrences and not occurrences of the considered phenomenon. With respect to each temporal domain, we assume that the phenomenon occurs at the first two time instants (that is, with delay 0); the phenomenon is not detected at the next time instant; it occurs again at the fourth instant (that is, with delay 2^0); then it is not detected for 2 time instants, but it surely happens before the end of an interval of length 2 (that is, with delay greater than or equal to 2^1 and less than 2^2); then it is not detected for 4 time instants, but it surely happens before the end of an interval of length 4 (that is, with delay greater than or equal to 2^2 and less than 2^3); and so on. If an occurrence of the phenomenon is detected at a time instant $i \in T^j$, then it occurs at least in its rightmost child, while if it is not detected at $i \in T^j$, then it is not detected at each of its children.

Formally, the process is specified by the following $\mathcal{UL}^2_{\omega LM^k}$-formula:

$$\forall x((\Delta_0(x) \vee \Delta_1(x)) \rightarrow p(x) \wedge \Delta_2(x) \rightarrow q(x) \wedge p(x) \leftrightarrow \neg q(x) \wedge \\ \wedge q(x) \rightarrow (q(\downarrow(0, x)) \wedge q(\downarrow(1, x))) \wedge p(x) \rightarrow p(\downarrow(1, x))).$$

It is easy to verify that each model of the above formula associates with each layer T^j an ω-sequence of the form $s_0 \cdot s_1 \cdot \ldots \cdot s_i \cdot \ldots$, where $s_0 = \{p\}$ and

$$s_i = \{p\} \cdot \{q\}^{2^{i-1}} \cdot w_i, \text{ with } w_i \in \{p, q\}^*, |w_i| = 2^{i-1} - 1.$$

It is possible to prove that the set of ω-sequences fulfilling the above condition is a non-regular ω-language over the alphabet $\Sigma = \{p, q\}$.

The notion of satisfaction of a formula $\phi \in \mathcal{UL}^2_{\omega LM^k}$ by an interpretation $\mathcal{I}$ is a straightforward generalization of the corresponding notion for $\mathcal{UL}^2_{\omega LM^k}$.

In (Montanari et al., 1996) we prove that $\mathcal{UL}^2_{\omega LM^k}$ is as expressive as $\mathcal{L}^2_{\omega LM^k}$. More precisely, it is possible to show that , for each $i \geq 0$, the constant $\mathtt{0_i}$, the successor function $+_{\mathtt{i}}1$, the vertical and horizontal contextualizations $\mathtt{T^i}$ and $\mathtt{\Delta_i}$, the congruences $\equiv_{\mathtt{i,j}}$, with $j \geq 2$, and the approximate ordering $\preceq$ can be expressed in terms of $\downarrow$ and $\leq$.

4.2. THE LANGUAGE FOR DOWNWARD UNBOUNDED LAYERED STRUCTURES

The formulae of the high-level second-order language $\mathcal{DL}^2_{\omega LM^k}$ for the theory of downward unbounded layered structures differ from the ones of $\mathcal{UL}^2_{\omega LM^k}$ only in the fact that atomic formulae having the form $\mathtt{\Delta_i(t)}$ are replaced by atomic formulae having the form $\mathtt{\Delta_{i,j}(t)}$, for $i, j \geq 0$. Interpretations $\mathcal{I}$ of $\mathcal{DL}^2_{\omega LM^k}$ differ from $\mathcal{UL}^2_{\omega LM^k}$-interpretations only in the definition of the semantic clause for $\downarrow$, and in the replacement of the semantic clause for Δ_i by the following one:

$$(\mathtt{\Delta_{i,j}})^{\mathcal{I}} = \{(+_k 1)^i((\downarrow)^k(0, (+_0 1)^j(0_0^{\mathcal{I}}))) : k \geq 0\}.$$

This semantic clause states that for any $k \geq 0$ and any $t \in T^k$, $\Delta_{i,j}$ holds at t iff t is at distance i from the (unique) element $t' \in T^k$ belonging to the leftmost branch of the tree rooted at the (unique) element $t'' \in T^0$ which is at distance j from 0_0. It is easy to see that the predicate Δ_i of $\mathcal{UL}^2_{\omega LM^k}$ is equivalent to $\Delta_{i,0}$. The satisfiability relation is defined exactly as in the case of $\mathcal{UL}^2_{\omega LM^k}$, except for the obvious replacement of Δ_i by $\Delta_{i,j}$.

EXAMPLE 4.2. Let $\mathtt{p}$ and $\mathtt{q}$ be two predicates (denoting a pair of events or of states). We show how the properties of downward unbounded layered structures can be exploited to constrain $\mathtt{p}$ and $\mathtt{q}$ to be locally indistinguishable (resp. distinguishable). We say that two predicates $\mathtt{p}$ and $\mathtt{q}$ are *locally indistinguishable* w.r. to a time instant t if both $\mathtt{p}$ and $\mathtt{q}$ hold at t, and there exists a child t' of t such that $\mathtt{p}$ and $\mathtt{q}$ are locally indistinguishable w.r. to t'. Two predicates are *locally distinguishable* w.r. to a time instant t if they are not locally indistinguishable w.r. to it. The condition that $\mathtt{p}$ and $\mathtt{q}$ are locally indistinguishable w.r. to (a time instant t denoted by) x can be expressed as follows:

$$\exists \mathtt{r}(\mathtt{path}(\mathtt{x}, \mathtt{r}) \wedge \forall \mathtt{y}(\mathtt{r}(\mathtt{y}) \rightarrow \mathtt{p}(\mathtt{y}) \wedge \mathtt{q}(\mathtt{y}))),$$

where $\mathtt{path(x,r)}$ stands for:

$$\begin{gathered}\mathtt{r(x)} \wedge \forall \mathtt{y}((\mathtt{r(y)} \rightarrow \bigvee_{\mathtt{i=0}}^{\mathtt{k-1}} \mathtt{r}(\downarrow(\mathtt{i},\mathtt{y})))) \wedge \forall \mathtt{i},\mathtt{j}(\mathtt{r}(\downarrow(\mathtt{i},\mathtt{y})) \wedge \\ \mathtt{r}(\downarrow(\mathtt{j},\mathtt{y})) \rightarrow \mathtt{i}=\mathtt{j}) \wedge \forall \mathtt{z}(\mathtt{y} \neq \mathtt{x} \wedge \mathtt{r(y)} \wedge \bigvee_{\mathtt{i=0}}^{\mathtt{k-1}} \downarrow(\mathtt{i},\mathtt{z}) = \mathtt{y} \rightarrow \mathtt{r(z)}) \wedge \\ (\bigvee_{\mathtt{i=0}}^{\mathtt{k-1}} \downarrow(\mathtt{i},\mathtt{y}) = \mathtt{x} \rightarrow \neg \mathtt{r(y)}) \wedge \\ (\neg \exists \mathtt{z} \bigvee_{\mathtt{i=0}}^{\mathtt{k-1}} \downarrow(\mathtt{i},\mathtt{z}) = \mathtt{y} \wedge \mathtt{y} \neq \mathtt{x} \rightarrow \neg \mathtt{r(y)})),\end{gathered}$$

($\forall \mathtt{i},\mathtt{j}(\mathtt{r}(\downarrow(\mathtt{i},\mathtt{y})) \wedge \mathtt{r}(\downarrow(\mathtt{j},\mathtt{y})) \rightarrow \mathtt{i}=\mathtt{j})$ is the obvious shorthand). The condition that $\mathtt{p}$ and $\mathtt{q}$ are locally distinguishable w.r. to (a time instant t denoted by) x can be expressed as follows:

$$\forall \mathtt{r}(\mathtt{path(x,r)} \rightarrow \exists \mathtt{y}(\mathtt{r(y)} \wedge ((\mathtt{p(y)} \wedge \neg \mathtt{q(y)})$$

$$\vee(\neg \mathtt{p(y)} \wedge \mathtt{q(y)}) \vee (\neg \mathtt{p(y)} \wedge \neg \mathtt{q(y)})))).$$

As in the upward unbounded case, it is possible to show that $\mathcal{DL}^2_{\omega LM^k}$ is as expressive as $\mathcal{L}^2_{\omega LM^k}$.

Remark. In this section we have shown that the basic functionalities of $MLTL$ can be expressed in $\mathcal{L}^2_{\omega LM^k}$. We want to point out that, however, there exist significant properties of ω-layered structures that cannot be expressed in $\mathcal{L}^2_{\omega LM^k}$. As an example, it is not possible to define a binary predicate *same_layer* such that, for all $t, t' \in \bigcup_{i \geq 0} T^i$, *same_layer*$(t,t')$ iff $\exists i(t \in T^i \wedge t' \in T^i)$. As a matter of fact, extending $\mathcal{L}^2_{\omega LM^k}$ with the predicate *same_layer* would make the theory of downward unbounded layered structures undecidable, a result which follows from the undecidability of the extension of SkS with the *equal_level* predicate E given by $E(u,v)$ iff $|u| = |v|$, with $u, v \in \{0,1\}^*$ (cf. (Läuchli and Savoiz, 1987)). It is possible to show that a similar undecidability result holds for the theory of upward unbounded ones.

Conclusions and further developments

In this paper we proved the decidability of theories of metric and layered temporal structures provided with an infinite number of either arbitrarily coarse or arbitrarily fine layers. We are currently exploring the natural generalization to layered structures which are both upward and downward unbounded. The questions whether we can decide upward unbounded layered structures using the theory $S1S$ (instead of $S1S^k$) and whether we can decide downward unbounded layered structures using $S1S^k$ (instead of SkS) are still open. Our conjecture is that the proposed reductions are actually the minimal ones. Finally, notice that the above results directly hold for metric and layered temporal logics non-axiomatically defined. Since they are decidable, one can list all their theorems and hence axiomatic completeness trivially follows, even though the axioms are not produced explicitly. An explicit axiomatic counterpart could be obtained extending a simplified variant of $TPTL$ (real-time propositional temporal logic), where state variables are replaced by time variables and $\bigcirc$ is interpreted as the successor over time, with contextual and projection operators of $MLTL$.

References

Alur, R. and T. Henzinger: 1993, 'Real-time logics: complexity and expressiveness'. *Information and Computation* **104**, 35–77.

Benthem, J. v.: 1995, 'Temporal Logic'. In: D. Gabbay, C. Hogger, and J. Robinson (eds.): *Handbook of Logic in Artificial Intelligence and Logic Programming, Vol. 4*. Oxford University Press, pp. 241–350.

Ciapessoni, E., E. Corsetti, A. Montanari, and P. San Pietro: 1993, 'Embedding Time Granularity in a Logical Specification Language for Synchronous Real-Time Systems'. *Science of Computer Programming* **20**, 141–171.

Corsetti, E., A. Montanari, and E. Ratto: 1991, 'Dealing with Different Time Granularities in Formal Specifications of Real-Time Systems'. *The Journal of Real-Time Systems* **3**, 191–215.

Fiadeiro, J. and T. Maibaum: 1992, 'Sometimes "Tomorrow" is "Sometimes" - Action Refinement in a Temporal Logic of Objects'. In: *Proc. ICTL'94, LNAI 827.* pp. 48–66.

Koymans, R.: 1992, 'Specifying Message Passing and Time-Critical Systems with Temporal Logic'. In: *LNCS 651.*

Läuchli, H. and C. Savoiz: 1987, 'Monadic Second-Order Definable Relations on the Binary Tree'. *Journal of Symbolic Logic* **52**, 219–226.

Montanari, A. and M. de Rijke: 1995, 'Two-Sorted Metric Temporal Logic'. Technical Report CS-R9577, CWI, University of Amsterdam. To appear in Theoretical Computer Science.

Montanari, A., A. Peron, and A. Policriti: 1996, 'Decidable theories of ω-layered metric temporal structures'. Technical Report ML-96-07, Institute for Logic Language and Information, University of Amsterdam.

Montanari, A. and A. Policriti: 1996, 'Decidability Results for Metric and Layered Temporal Logics'. *Notre Dame Journal of Formal Logic* **37**, 260–282.

Monti, A. and A. Peron: 1995a, 'Systolic Tree ω-Languages'. In: *Proceedings of STACS-95, LNCS 900.* pp. 131–142.

Monti, A. and A. Peron: 1995b, 'Systolic Tree ω-Languages: The Operational and the Logical View'. Technical Report SI-95/11, Dipartimento di Scienze dell'Informazione, Università di Roma "La Sapienza".

Thomas, W.: 1990, 'Automata on Infinite Objects'. In: J. van Leeuwen (ed.): *Handbook of Theoretical Computer Science, Vol. B.* Elsevier Science Publishers, pp. 133–191.

Synthesis with Incomplete Informatio
*

Orna Kupferman[†]
EECS Department, UC Berkeley, Berkeley CA 94720-1770, U.S.A.
orna@eecs.berkeley.edu, http://www-cad.eecs.berkeley.edu/~orna

Moshe Y. Vardi[‡]
Rice University, Department of Computer Science, Houston, TX 77251-1892, U.S.A.
vardi@cs.rice.edu, http://www.cs.rice.edu/~vardi

Abstract. In program synthesis, we transform a specification into a system that is guaranteed to satisfy the specification. When the system is open, then at each moment it reads input signals and writes output signals, which depend on the input signals and the history of the computation so far. The specification considers all possible input sequences. Thus, if the specification is linear, it should hold in every computation generated by the interaction, and if the specification is branching, it should hold in the tree that embodies all possible input sequences.

Often, the system cannot read all the input signals generated by its environment. For example, in a distributed setting, it might be that each process can read input signals of only part of the underlying processes. Then, we should transform a specification into a system whose output depends only on the readable parts of the input signals and the history of the computation. This is called *synthesis with incomplete information*. In this work we solve the problem of synthesis with incomplete information in its full generality. We consider linear and branching settings with complete and incomplete information. We claim that *alternation* is a suitable and helpful mechanism for coping with incomplete information. Using *alternating tree automata*, we show that incomplete information does not make the synthesis problem more complex, in both the linear and the branching paradigm. In particular, we prove that independently of the presence of incomplete information, the synthesis problems for CTL and CTL* are complete for EXPTIME and 2EXPTIME, respectively.

1. Introduction

In *program synthesis*, we transform a specification into a program that is guaranteed to satisfy the specification. Earlier works on synthesis consider *closed systems*. There, a program that meets the specification can be extracted from a constructive proof that the formula is satisfiable (Manna and Waldinger, 1980; Emerson and Clarke, 1982). As argued in (Dill, 1989; Pnueli and Rosner, 1989; Abadi et al., 1989), such synthesis paradigms are not of

* Part of this work was done in Bell Laboratories during the DIMACS Special Year on Logic and Algorithms.

† Supported in part by the ONR YIP award N00014-95-1-0520, by the NSF CAREER award CCR-9501708, by the NSF grant CCR-9504469, by the AFOSR contract F49620-93-1-0056, by the ARO MURI grant DAAH-04-96-1-0341, by the ARPA grant NAG2-892, and by the SRC contract 95-DC-324.036.

‡ Supported in part by the National Science Foundation grants CCR-9628400 and CCR-9700061, and by a grant from the Intel Corporation.

H. Barringer et al. (eds.), Advances in Temporal Logic, 109–127.

much interest when applied to *open systems*, which interact with an environment. Consider for example a scheduler for a printer that serves two users. The scheduler is an open system. Each time unit it reads the input signals $J1$ and $J2$ (a job sent from the first or the second user, respectively), and it writes the output signals $P1$ and $P2$ (print a job of the first or the second user, respectively). The scheduler should be designed so that jobs of the two users are not printed simultaneously, and whenever a user sends a job, the job is printed eventually. Of course, this should hold no matter how the users send jobs.

We can specify the requirement for the scheduler in terms of a *linear temporal logic* (LTL) formula ψ (Pnueli, 1981). Satisfiability of ψ does not imply that a required scheduler exists. To see this, observe that ψ is satisfied in every structure in which the four signals never hold. In addition, an evidence to ψ's satisfiability is not of much help in extracting a correct scheduler. Indeed, while such an evidence only suggests a scheduler that is guaranteed to satisfy ψ for *some* input sequence, we want a scheduler that satisfies ψ for *all* possible scripts of jobs sent to the printer. We now make this intuition more formal. Given sets I and O of input and output signals, respectively, we can view a program as a *strategy* $P : (2^I)^* \rightarrow 2^O$ that maps a finite sequence of sets of input signals into a set of output signals. When P interacts with an environment that generates infinite input sequences, it associates with each input sequence an infinite computation over $2^{I \cup O}$. Given an LTL formula ψ over $I \cup O$, *realizability* of ψ is the problem of determining whether there exists a program P all of whose computations satisfy ψ. Correct synthesis of ψ then amounts to constructing such P (Pnueli and Rosner, 1989).

The linear paradigm for realizability and synthesis is closely related to *Church's solvability problem* (Church, 1963). There, we are given a regular relation $R \subseteq (2^I)^\omega \times (2^O)^\omega$ and we seek a function $f : (2^I)^\omega \rightarrow (2^O)^\omega$, generated by a strategy, such that for all $x \in (2^I)^\omega$, we have $R(x, f(x))$. We can view the relation R as a linear specification for the program: it defines all the permitted pairs of input and output sequences. A function f as above then maps every possible input sequence into a permitted output sequence, and can be therefore viewed as a correct program. The solutions to Church's problem and the LTL synthesis problem are similar (Rabin, 1970; Pnueli and Rosner, 1989), and consist of a reduction to the nonemptiness problem of *tree automata* (an earlier and more complicated solution can be found in (Büchi and Landweber, 1969)).

Though the program P is deterministic, it induces a computation tree. The branches of the tree correspond to external nondeterminism, caused by different possible inputs. Thus, the tree has a fixed branching degree $|2^I|$, and it embodies all the possible inputs (and hence also computations) of P. When we synthesize P from an LTL specification ψ, we require ψ to hold in all the paths of P's computation tree. Consequently, we cannot impose possibility requirements on P (Lamport, 1980; Emerson and Halpern, 1986).

In the scheduler example, while we can require, for instance, that for every infinite sequence of inputs a job of the first user is eventually printed, we cannot require that every finite sequence of inputs *can be* extended so that a job of the first user is eventually printed. In order to express possibility properties, we should specify P using *branching temporal logics*, which enable both universal and existential path quantification (Emerson, 1990). Given a branching specification ψ over $I \cup O$ (we consider here specifications given in terms of CTL or CTL* formulas), realizability of ψ is the problem of determining whether there exists a program P whose computation tree satisfies ψ. Correct synthesis of ψ then amounts to constructing such P. We note that this problem is different from the supervisor-synthesis problem considered in (Antoniotti, 1995). There, a given structure needs to be restricted (by disabling some of its transitions) in order to satisfy a given branching specification.

So far, we considered the case where the specifications (either linear or branching) refer solely to signals in I and O, both are known to P. This is called synthesis with *complete information.* Often, the program does not have complete information about its environment. For example, in a distributed setting, it might be that each process can read input signals of only part of the underlying processes. Let E be the set of input signals that the program can not read. In the scheduler example, we take $E = \{B1, B2\}$, where $B1$ holds when the job sent to the printer by the first user is a paper containing a bug, and similarly for $B2$. Unfortunately, while the scheduler can see whether the users send jobs, it cannot trace bugs in their papers.

Since P cannot read the signals in E, its activity is independent of them. Hence, it can still be viewed as a strategy $P : (2^I)^* \rightarrow 2^O$. Nevertheless, the computations of P are now infinite words over $2^{I \cup E \cup O}$. Similarly, embodying all the possible inputs to P, the computation tree induced by P now has a fixed branching degree $|2^{I \cup E}|$ and it is labeled by letters in $2^{I \cup E \cup O}$. Note that different nodes in this tree may have, according P's incomplete information, the same "history of inputs". In the scheduler example, P cannot distinguish between two nodes that the input sequences leading to them differ only in the values of $B1$ and $B2$ in some points along them.

Often, programs need to satisfy specifications that refer to signals they cannot read. For example, following several events, it was decided to change the specification for the printer scheduler so that if the first user sends to the printer a buggy paper, then the paper is never printed. The new scheduler needs to satisfy the specification even though it cannot trace bugs. This problem, of synthesis with *incomplete information*, is the subject of this work. Formally, given a specification ψ over the sets I, E, and O of readable input, unreadable input, and output signals, respectively, synthesis with incomplete information amounts to constructing a program $P : (2^I)^* \rightarrow 2^O$, which is independent of E, and which realizes ψ (that is, if ψ is linear then all the computations of P satisfy ψ, and if ψ is branching then the computation tree of P satisfies ψ).

It is known how to cope with incomplete information in the linear paradigm. In particular, the approach used in (Pnueli and Rosner, 1989) can be extended to handle LTL synthesis with incomplete information. Essentially, nondeterminism of the automata can be used to guess the missing information, making sure that no guess violates the specification (Vardi, 1995). Similarly, methods for control and synthesis in other linear paradigms (e.g., when specifying terminating programs by regular languages) have been extended to handle incomplete information (Kumar and Garg, 1995; Kumar and Shayman, 1995).

Coping with incomplete information is more difficult in the branching paradigm. The methods used in the linear paradigm are not applicable here. To see why, let us consider first realizability with complete information. There, recall, we are given a branching specification ψ over $2^{I \cup O}$, and we check whether there exists a program P whose computation tree satisfies ψ. In other words, we check whether we can take the *I-exhaustive* tree (i.e., the 2^I-labeled tree that embodies all input sequences) and annotate it with outputs so that the resulted $2^{I \cup O}$-labeled tree satisfies ψ. This problem can be easily solved using tree automata. Essentially, we first construct a tree automaton $\mathcal{A}_\psi$ that accepts exactly all $2^{I \cup O}$-labeled trees that satisfy ψ (Emerson and Sistla, 1984; Vardi and Wolper, 1986). Then, we construct a tree automaton $\mathcal{A}_P$ that accepts all the "potential" computation trees (i.e., all $2^{I \cup O}$-labeled trees obtained by annotating the I-exhaustive tree with outputs). Finally, we check that the intersection of the two automata is nonempty. Can we follow the same lines in the presence of incomplete information? Now, that ψ is defined over $I \cup E \cup O$, the automaton $\mathcal{A}_\psi$ should accept all the $2^{I \cup E \cup O}$-labeled trees that satisfy ψ. This causes no difficulty. In addition, since the environment produces signals in both I and E, the automaton $\mathcal{A}_P$ should consider annotations of the $(I \cup E)$-exhaustive tree. Unlike, however, in the case of complete information, here not all annotations induce potential computation trees. Since P is independent of signals in E, an annotation induces a potential computation tree only if every two nodes that have the same history (according to P's incomplete information) are annotated with the same output! This consistency condition is non-regular and cannot be checked by an automaton. It is this need, to restrict the set of candidate computation trees to trees that meet some non-regular condition, that makes incomplete information in the branching paradigm so challenging.

In this paper we solve the problem of synthesis with incomplete information for the branching paradigm, which we show to be a proper extension of the linear paradigm and the complete-information paradigm. We claim that *alternation* is a suitable and helpful mechanism for coping with incomplete information. Using *alternating tree automata*, we show that incomplete information does not make the synthesis problem more complex, in both the linear and the branching paradigm. In fact, as alternating tree automata shift all the combinatorial difficulties of the synthesis problem to the nonemptiness test, the automata-based algorithms that we describe are as simple as the known automata-based algorithms for the satisfiability problem. In particular, we prove that independently of the presence of incomplete information, the synthesis problems for CTL and CTL* are complete for EXPTIME and 2EXPTIME, respectively. These results join the 2EXPTIME-complete bound for LTL synthesis in both settings (Pnueli and Rosner, 1989; Rosner, 1992; Vardi, 1995). Keeping in mind that the satisfiability problems for LTL, CTL, and CTL* are complete for PSPACE, EXPTIME, and 2EXPTIME (Emerson, 1990), it follows that while the transition from closed to open systems dramatically increases the complexity of synthesis in the linear paradigm, it does not influence the complexity in the branching paradigm.

2. Preliminaries

Given a finite set Υ, an Υ*-tree* is a set $T \subseteq \Upsilon^*$ such that if $x \cdot \upsilon \in T$, where $x \in \Upsilon^*$ and $\upsilon \in \Upsilon$, then also $x \in T$. When Υ is not important or clear from the context, we call T a tree. The elements of T are called *nodes*, and the empty word ϵ is the *root* of T. For every $x \in T$, the nodes $x \cdot \upsilon \in T$ where $\upsilon \in \Upsilon$ are the *children* of x. Each node x of T has a *direction* in Υ. The direction of the root is υ^0, for some designated $\upsilon^0 \in \Upsilon$, called the *root direction.* The direction of a node $x \cdot \upsilon$ is υ. We denote by $dir(x)$ the direction of node x. An Υ-tree T is a *full infinite tree* if $T = \Upsilon^*$. Unless otherwise mentioned, we consider here full infinite trees. A *path* π of a tree T is a set $\pi \subseteq T$ such that $\epsilon \in \pi$ and for every $x \in \pi$ there exists a unique $\upsilon \in \Upsilon$ such that $x \cdot \upsilon \in \pi$. For a path π and $j \geq 0$, let π_j denote the node of length j in π. Each path $\pi \subseteq T$ corresponds to a word $T(\pi) = dir(\pi_0) \cdot dir(\pi_1) \cdots$ in Υ^ω.

Given two finite sets Υ and Σ, a Σ*-labeled* Υ*-tree* is a pair $\langle T, V\rangle$ where T is an Υ-tree and $V : T \rightarrow \Sigma$ maps each node of T to a letter in Σ. When Υ and Σ are not important or clear from the context, we call $\langle T, V\rangle$ a labeled tree. Each path π in $\langle T, V\rangle$ corresponds to a word $V(\pi) = V(\pi_0) \cdot V(\pi_1) \cdots$ in Σ^ω. For a Σ-labeled Υ-tree $\langle T, V\rangle$, we define the *x-ray* of $\langle T, V\rangle$, denoted $xray(\langle T, V\rangle)$, as the $(\Upsilon \times \Sigma)$-labeled Υ-tree $\langle T, V'\rangle$ in which each node is labeled by both its direction and its labeling in $\langle T, V\rangle$. Thus, for every $x \in T$, we have $V'(x) = \langle dir(x), V(x)\rangle$. Essentially, the labels in $xray(\langle T, V\rangle)$ contain information not only about the surface of $\langle T, V\rangle$ (its labels) but also about its skeleton (its nodes).

Let $\Upsilon_2 = \{0, 1\}$ and $\Upsilon_4 = \{00, 01, 10, 11\}$. When $T = \Upsilon_2^*$, we say that T is a full infinite binary tree (*binary tree*, for short). We refer to the child $x \cdot 0$ of x as the left child and refer to the child $x \cdot 1$ as the right child. When $T = \Upsilon_4^*$, we say that T is a full infinite 4-ary tree (*4-ary tree*, for short). We refer to $x \cdot 00$ as the leftmost child, $x \cdot 01$ is its brother to the right, then comes $x \cdot 10$, and $x \cdot 11$ is the rightmost child (see Figure below).

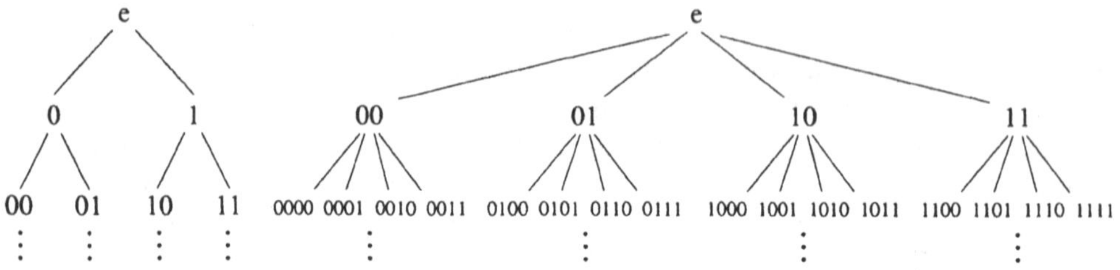

For a node $x \in \Upsilon_4^*$, let $hide(x)$ be the node in Υ_2^* obtained from x by replacing each letter c_1c_2 by the letter c_1. For example, the node 0010 of the the 4-ary tree in the figure corresponds, by *hide*, to the node 01 of the binary tree. Note that the nodes 0011, 0110, and 0111 of the 4-ary tree also correspond, by *hide*, to the node 01 of the binary tree. We extend the hiding operator to paths $\pi \subset \Upsilon_4^*$ in the straightforward way. That is, the path $hide(\pi) \subset \Upsilon_2^\omega$ is obtained from π by replacing each node $x \in \pi$ by the node $hide(x)$.

Let Z be a finite set. For a Z-labeled Υ_2-tree $\langle \Upsilon_2^*, V\rangle$, we define the *widening* of $\langle \Upsilon_2^*, V\rangle$, denoted $wide(\langle T, V\rangle)$, as the Z-labeled Υ_4-tree $\langle \Upsilon_4^*, V'\rangle$ where for every $x \in \Upsilon_4^*$, we have $V'(x) = V(hide(x))$. Note that for every node $x \in \Upsilon_4^*$, the children $x \cdot 00$ and $x \cdot 01$ of x agree on their label in $\langle \Upsilon_4^*, V'\rangle$. Indeed, they are both labeled with $V(hide(x) \cdot 0)$. Similarly, the children $x \cdot 10$ and $x \cdot 11$ are both labeled with $V(hide(x) \cdot 1)$. The essence of widening is that for every path π in $\langle \Upsilon_2^*, V\rangle$ and for every path $\rho \in hide^{-1}(\pi)$, the path ρ exists in $\langle \Upsilon_4^*, V'\rangle$ and $V(\pi) = V'(\rho)$. In other words, for every two paths ρ_1 and ρ_2 in $\langle \Upsilon_4^*, V'\rangle$ such that $hide(\rho_1) = hide(\rho_2) = \pi$, we have $V'(\rho_1) = V'(\rho_2) = V(\pi)$.

Given finite sets X and Y, we generalize the operators *hide* and *wide* to Z-labeled $(X \times Y)$-trees by parameterizing them with the set Y as follows (the operators defined above correspond to the special case where $X = Y = \Upsilon_2$). The operator $hide_Y : (X \times Y)^* \to X^*$ replaces each letter $x \cdot y$ by the letter x. Accordingly, the operator $wide_Y$ maps Z-labeled X-trees to Z-labeled $(X \times Y)$-trees. Formally, $wide_Y(\langle X^*, V\rangle) = \langle (X \times Y)^*, V'\rangle$, where for every node $w \in (X \times Y)^*$, we have $V'(w) = V(hide_Y(w))$.

3. The Problem

Consider a program P that interacts with its environment. Let I and E be the sets of input signals, readable and unreadable by P, respectively, and let O be the set of P's output signals. We can view P as a *strategy* $P : (2^I)^* \to 2^O$. Indeed, P maps each finite sequence of sets of readable input signals into a set of output signals. Note that the information of P on its input is incomplete and it does not depend on the unreadable signals in E. We assume the following interaction between P and its environment.

The interaction starts by P outputting $P(\epsilon)$. The environment replies with some $\langle i_1, e_1\rangle \in 2^I \times 2^E$, to which P replies with $P(i_1)$. Interaction then continues step by step, with an output $P(i_1 \cdot i_2 \cdots i_j)$ corresponding to a sequence $\langle i_1, e_1\rangle \cdot \langle i_2, e_2\rangle \cdots \langle i_j, e_j\rangle$ of inputs. Thus, P associates with each infinite input sequence $\langle i_1, e_1\rangle \cdot \langle i_2, e_2\rangle \cdots$, an infinite computation $[P(\epsilon)] \cdot [i_1 \cup e_1 \cup P(i_1)] \cdot [i_2 \cup e_2 \cup P(i_1 \cdot i_2)] \cdots$ over $2^{I \cup E \cup O}$. We note that our choice of P starting the interaction is for technical convenience only.

Though the program P is deterministic, it induces a computation tree. The branches of the tree correspond to external nondeterminism, caused by different possible inputs. Thus, P is a $2^{I \cup E \cup O}$-labeled $2^{I \cup E}$-tree, where each node with direction $i \cup e \in I \cup E$ is labeled by $i \cup e \cup o$, where o is the set of output signals that P assigns to the sequence of readable inputs leading to the node. Formally, we obtain the computation tree of P by two transformations on the 2^O-labeled tree $\langle (2^I)^*, P\rangle$, which represents P. First, while $\langle (2^I)^*, P\rangle$ ignores the signals in E and the extra external nondeterminism induced by them, the computation tree of P, which embodies all possible computations, takes them into account. For that, we define the 2^O-labeled tree $\langle (2^{I \cup E})^*, P'\rangle = wide_{(2^E)}(\langle (2^I)^*, P\rangle)$. By the definition of the *wide* operator, each two nodes in $(2^{I \cup E})^*$ that correspond, according to P's incomplete information, to the same input sequence are labeled by P' with the same output. Now, as the signals in I and E are represented in $\langle (2^{I \cup E})^*, P'\rangle$ only in its nodes and not in its labels, we define the computation tree of P as $\langle (2^{I \cup E})^*, P''\rangle = xray(\langle (2^{I \cup E})^*, P'\rangle)$. Note that, as I, E, and O are disjoint, we refer to $wide_{(2^E)}(\langle (2^I)^*, P\rangle)$ as a $2^{I \cup E}$-tree, rather than a $(2^I \times 2^E)$-tree. Similarly, $xray(\langle (2^{I \cup E})^*, P'\rangle)$ is a $2^{I \cup E \cup O}$-labeled tree, rather than a $(2^{I \cup E} \times 2^O)$-labeled tree.

Given a CTL* formula ψ over the sets $I \cup E \cup O$ of atomic propositions, the problem of branching realizability with incomplete information is to determine whether there is a program P whose computation tree satisfies ψ. The synthesis problem requires the construction of such P.

It is easy to see that problem of branching synthesis with incomplete information is a proper extension of the problem of branching synthesis with complete information. We claim that it is also a proper extension of the linear synthesis problem with incomplete (and hence also complete) information (Pnueli and Rosner, 1989; Pnueli and Rosner, 1990; Vardi, 1995). For that, we show that synthesis for an LTL formula φ can be reduced to synthesis for the CTL* formula $A\varphi$. Moreover, φ and $A\varphi$ are realized by the same programs. Consider a program P. Assume first that P realizes $A\varphi$. Then, clearly, P associates with every input sequence a computation that satisfies φ, and thus it realizes φ as well. The second direction seems less immediate, but it follows easily from the fact that P is a (deterministic) strategy. Indeed, if P does not realize $A\varphi$, then there must be an input sequence with which P does no associate a computation satisfying φ.

In the next section we solve the general synthesis problem and we show that generalization comes at no cost. In more detail, we show that CTL* synthesis with incomplete information can be done in 2EXPTIME, matching the 2EXPTIME lower bound for LTL synthesis with complete information.

4. The Solution

Church's solvability problem and the LTL synthesis problem are solved, in (Rabin, 1970; Pnueli and Rosner, 1989), using nondeterministic tree automata. Both solutions follow the same lines: we can translate the linear specification for the program into a word automaton, we can determinize the automaton an extend it to a tree automaton, and we can check the nonemptiness of this tree automaton when restricted to trees that embodies all possible inputs. Such method cannot be easily extended to handle synthesis with incomplete information in the branching paradigm. Here, we need to check the nonemptiness of a tree automaton when restricted not only to trees that embody all possible inputs, but also to trees in which the labels of the output signals are consistent. This condition is non-regular and cannot be checked by an automaton. In order to solve this difficulty, we need the framework of alternating tree automata.

4.1. Alternating Tree Automata

Alternating tree automata run on labeled trees. They generalize nondeterministic tree automata and were first introduced in (Muller and Schupp, 1987). For simplicity, we refer first to automata over infinite binary trees. Consider a nondeterministic tree automaton $\mathcal{A}$ with a set Q of states and transition function δ. The transition function δ maps an automaton state $q \in Q$ and an input letter $\sigma \in \Sigma$ to a set of pairs of states. Each such pair suggests a nondeterministic choice for the automaton's next configuration. When the automaton is in a state q and is reading a node x labeled by a letter σ, it proceeds by first choosing a pair $\langle q_1, q_2\rangle \in \delta(q,\sigma)$ and then splitting into two copies. One copy enters the state q_1 and proceeds to the node $x \cdot 0$ (the left child of x), and the other copy enters the state q_2 and proceeds to the node $x \cdot 1$ (the right child of x).

Let $\mathcal{B}^+(\Upsilon_2 \times Q)$ be the set of positive Boolean formulas over $\Upsilon_2 \times Q$; i.e., Boolean formulas built from elements in $\Upsilon_2 \times Q$ using $\wedge$ and $\vee$, where we also allow the formulas **true** and **false** and, as usual, $\wedge$ has precedence over $\vee$. For a set $S \subseteq \Upsilon_2 \times Q$ and a formula $\theta \in \mathcal{B}^+(\Upsilon_2 \times Q)$, we say that S *satisfies* θ iff assigning **true** to elements in S and assigning **false** to elements in $(\Upsilon_2 \times Q) \setminus S$ makes θ true. We can represent δ using $\mathcal{B}^+(\Upsilon_2 \times Q)$. For example, $\delta(q,\sigma) = \{\langle q_1, q_2\rangle, \langle q_3, q_1\rangle\}$ can be written as $\delta(q,\sigma) = (0,q_1) \wedge (1,q_2) \vee (0,q_3) \wedge (1,q_1)$.

In nondeterministic tree automata, each conjunction in δ has exactly one element associated with each direction. In alternating automata on binary trees, $\delta(q,\sigma)$ can be an arbitrary formula from $\mathcal{B}^+(\Upsilon_2 \times Q)$. We can have, for instance, a transition

$$\delta(q,\sigma) = (0,q_1) \wedge (0,q_2) \vee (0,q_2) \wedge (1,q_2) \wedge (1,q_3).$$

The above transition illustrates that several copies may go to the same direction and that the automaton is not required to send copies to all the directions. Formally, a finite alternating automaton on infinite binary trees is a tuple $\mathcal{A} = \langle \Sigma, Q, \delta, q_0, \alpha \rangle$ where Σ is the input alphabet, Q is a finite set of states, $\delta : Q \times \Sigma \to \mathcal{B}^+(\Upsilon_2 \times Q)$ is a transition function, $q_0 \in Q$ is an initial state, and α specifies the acceptance condition (a condition that defines a subset of Q^ω).

A *run of an alternating automaton* $\mathcal{A}$ on an input binary labeled tree $\langle T, V \rangle$ is a tree $\langle T_r, r \rangle$ in which the root is labeled by q_0 and every other node is labeled by an element of $\Upsilon_2^* \times Q$. Unlike T, in which each node has exactly two children, the tree T_r may have nodes with many children and may also have *leaves* (nodes with no children). Thus, $T_r \subset \mathbb{N}^*$ and a path in T_r may be either finite, in which case it contains a leaf, or infinite. Each node of T_r corresponds to a node of T. A node in T_r, labeled by (x, q), describes a copy of the automaton that reads the node x of T and visits the state q. Note that many nodes of T_r can correspond to the same node of T; in contrast, in a run of a nondeterministic automaton on $\langle T, V \rangle$ there is a one-to-one correspondence between the nodes of the run and the nodes of the tree. The labels of a node and its children have to satisfy the transition function. Formally, $\langle T_r, r \rangle$ is a Σ_r-labeled tree where $\Sigma_r = \Upsilon_2^* \times Q$ and $\langle T_r, r \rangle$ satisfies the following:

1. $\epsilon \in T_r$ and $r(\epsilon) = (\epsilon, q_0)$.

2. Let $y \in T_r$ with $r(y) = (x, q)$ and $\delta(q, V(x)) = \theta$. Then there is a (possibly empty) set $S = \{(c_0, q_0), (c_1, q_1), \ldots, (c_{n-1}, q_{n-1})\} \subseteq \Upsilon_2 \times Q$, such that the following hold:

   – S satisfies θ, and

   – for all $0 \leq i < n$, we have $y \cdot i \in T_r$ and $r(y \cdot i) = (x \cdot c_i, q_i)$.

For example, if $\langle T, V \rangle$ is a binary tree with $V(\epsilon) = a$ and $\delta(q_0, a) = ((0, q_1) \vee (0, q_2)) \wedge ((0, q_3) \vee (1, q_2))$, then the nodes of $\langle T_r, r \rangle$ at level 1 include the label $(0, q_1)$ or $(0, q_2)$, and include the label $(0, q_3)$ or $(1, q_2)$. Note that if $\theta =$ **true**, then y need not have children. This is the reason why T_r may have leaves. Also, since there exists no set S as required for $\theta =$ **false**, we cannot have a run that takes a transition with $\theta =$ **false**. Each infinite path ρ in $\langle T_r, r \rangle$ is labeled by a word in Q^ω. Let $inf(\rho)$ denote the set of states in Q that appear in $r(\rho)$ infinitely often. A run $\langle T_r, r \rangle$ is accepting iff all its infinite paths satisfy the acceptance condition. In alternating *Rabin* tree automata, $\alpha \subseteq 2^Q \times 2^Q$, and an infinite path ρ satisfies an acceptance condition $\alpha = \{\langle G_1, B_1\rangle, \ldots, \langle G_m, B_m\rangle\}$ iff there exists $1 \leq i \leq m$ for which $inf(\rho) \cap G_i \neq \emptyset$ and $inf(\rho) \cap B_i = \emptyset$. In alternating *Büchi* tree automata, $\alpha \subseteq Q$, and an infinite path ρ satisfies α iff $inf(\rho) \cap \alpha = \emptyset$.

For an acceptance condition α over the state space Q and a given set S, the *adjustment* of α to the state space $Q \times S$, denoted $\alpha \times S$, is obtained from α by replacing each set F participating in α by the set $F \times S$.

As with nondeterministic automata, an automaton accepts a tree iff there exists an accepting run on it. We denote by $\mathcal{L}(\mathcal{A})$ the language of the automaton $\mathcal{A}$; i.e., the set of all labeled trees that $\mathcal{A}$ accepts. We say that $\mathcal{A}$ is *nonempty* iff $\mathcal{L}(\mathcal{A}) \neq \emptyset$. We say that $\mathcal{A}$ is a *nondeterministic* tree automaton iff all the transitions of $\mathcal{A}$ have only disjunctively related atoms sent to the same direction; i.e., if the transitions are written in DNF, then every disjunct contains at most one atom of the form $(0, q)$ and one atom of the form $(1, q)$.

Alternating tree automata over Υ-trees, for an arbitrary finite set Υ, are defined similarly, with a transition function $\delta : Q \times \Sigma \to \mathcal{B}^+(\Upsilon \times Q)$. Accordingly, a run of an alternating automaton over an Υ-tree is a Σ_r-labeled tree with $\Sigma_r = \Upsilon^* \times Q$. In particular, alternating word automata can be viewed as special case of alternating tree automata, running over Υ-trees with a singleton Υ. The transition function of an alternating word automaton is $\delta : Q \times \Sigma \to \mathcal{B}^+(Q)$. That is, we omit the unique direction from the atoms in δ.

We define the *size* $|\mathcal{A}|$ of an alternating automaton $\mathcal{A} = \langle \Sigma, Q, \delta, q_0, \alpha\rangle$ as $|Q| + |\alpha| + |\delta|$, where $|Q|$ and $|\alpha|$ are the respective cardinalities of the sets Q and α, and where $|\delta|$ is the sum of the lengths of the satisfiable (i.e., not **false**) formulas that appear as $\delta(q, \sigma)$ for some $q \in Q$ and $\sigma \in \Sigma$ (the restriction to satisfiable formulas is to avoid an unnecessary $|Q| \cdot |\Sigma|$ minimal size for δ). Note that $\mathcal{A}$ can be stored in space $O(|\mathcal{A}|)$.

4.2. Automata-Theoretic Constructions

The first step in our solution is to translate the input CTL* formula into an alternating tree automaton. Here, alternation enables us to define automata that are exponentially smaller than automata with no alternations.

THEOREM 4.1. (Bernholtz et al., 1994) *Given a CTL* formula ψ over a set AP of atomic propositions and a set Υ of directions, there exists an alternating Rabin tree automaton $\mathcal{A}_{\Upsilon,\psi}$ over 2^{AP}-labeled Υ-trees, with $2^{O(|\psi|)}$ states and two pairs, such that $\mathcal{L}(\mathcal{A}_{\Upsilon,\psi})$ is exactly the set of trees satisfying ψ.*

The computation tree of P is a $2^{I \cup E \cup O}$-labeled $2^{I \cup E}$-tree. Given an automaton $\mathcal{A}$, which accepts a set of such computation trees, we construct, in Theorem 4.2 below, an automaton $\mathcal{A}'$ that accepts the projection of the trees in $\mathcal{L}(\mathcal{A})$ on the 2^O element of their labels.

THEOREM 4.2. *Given an alternating tree automaton $\mathcal{A}$ over $(\Upsilon \times \Sigma)$-labeled Υ-trees, we can construct an alternating tree automaton $\mathcal{A}'$ over Σ-labeled Υ-trees such that*

(1) *$\mathcal{A}'$ accepts a labeled tree $\langle \Upsilon^*, V\rangle$ iff $\mathcal{A}$ accepts $xray(\langle \Upsilon^*, V\rangle)$.*

(2) *$\mathcal{A}'$ and $\mathcal{A}$ have the same acceptance condition.*

(3) $|\mathcal{A}'| = O(|\mathcal{A}|)$.

Proof: Let $\mathcal{A} = \langle \Upsilon \times \Sigma, Q, \delta, q_0, \alpha\rangle$. Then, $\mathcal{A}' = \langle \Sigma, Q \times \Upsilon, \delta', \langle q_0, \upsilon^0\rangle, \alpha \times \Upsilon\rangle$, where for every $q \in Q$, $\upsilon \in \Upsilon$, and $\sigma \in \Sigma$, the transition $\delta'(\langle q, \upsilon\rangle, \sigma)$ is obtained from $\delta(q, \langle \upsilon, \sigma\rangle)$ by replacing each atom (υ', q') by the atom $(\upsilon', \langle q', \upsilon'\rangle)$.

That is, a state $\langle q, \upsilon\rangle$ in $\mathcal{A}'$ corresponds to a state q in $\mathcal{A}$ that reads only nodes with direction υ. Thus, the Υ-element in the states of $\mathcal{A}'$ provides the "missing" Υ-element of the alphabet of $\mathcal{A}$. It is easy to see that $\mathcal{A}'$ indeed accepts exactly all the trees whose x-rays are accepted by $\mathcal{A}$. □

Consider a $2^{I \cup E \cup O}$-labeled $2^{I \cup E}$-tree. Recall that we cannot define an automaton that accepts the language of $\mathcal{A}$ when restricted to consistent trees. Instead, we define, in Theorem 4.3 below, an automaton $\mathcal{A}'$ that accepts a $2^{I \cup E \cup O}$-labeled 2^O-tree iff its 2^E-widening is accepted by $\mathcal{A}$. Each such 2^E-widening is guaranteed to be consistent! Thus, $\mathcal{A}'$ is nonempty iff $\mathcal{A}$ when restricted to consistent trees is nonempty.

THEOREM 4.3. *Let X, Y, and Z be finite sets. Given an alternating tree automaton $\mathcal{A}$ over Z-labeled $(X \times Y)$-trees, we can construct an alternating tree automaton $\mathcal{A}'$ over Z-labeled X-trees such that*

(1) *$\mathcal{A}'$ accepts a labeled tree $\langle X^*, V\rangle$ iff $\mathcal{A}$ accepts $wide_Y(\langle X^*, V\rangle)$.*

(2) *$\mathcal{A}'$ and $\mathcal{A}$ have the same acceptance condition.*

(3) $|\mathcal{A}'| = O(|\mathcal{A}|)$.

Proof: Let $\mathcal{A} = \langle Z, Q, \delta, q_0, \alpha\rangle$. Then, $\mathcal{A}' = \langle Z, Q, \delta', q_0, \alpha\rangle$, where for every $q \in Q$ and $z \in Z$, the transition $\delta'(q, z)$ is obtained from $\delta(q, z)$ by replacing each atom $(\langle x, y\rangle, q')$ by the atom (x, q'). We prove the correctness of our construction. Thus, we prove that for every Z-labeled X-tree $\langle X^*, V\rangle$, we have

$$\langle X^*, V\rangle \in \mathcal{L}(\mathcal{A}') \text{ iff } wide_Y(\langle X^*, V\rangle) \in \mathcal{L}(\mathcal{A}).$$

Assume first that $wide_Y(\langle X^*, V\rangle) \in \mathcal{L}(\mathcal{A})$. We prove that $\langle X^*, V\rangle \in \mathcal{L}(\mathcal{A}')$. Let $\langle T_r, r\rangle$ be an accepting run of $\mathcal{A}$ on $wide_Y(\langle X^*, V\rangle)$. By the definition of a run, we have that $r : T_r \to ((X \times Y)^* \times Q)$ maps a node of the run-tree to a pair consisting of a node in $(X \times Y)^*$ and a state of $\mathcal{A}$ that reads this node.

Consider the labeled tree $\langle T_r, r'\rangle$ where for every $w \in T_r$ with $r(w) = \langle u, q\rangle$ we have $r'(w) = \langle hide(u), q\rangle$. It is easy to see that $\langle T_r, r'\rangle$ is an accepting run of $\mathcal{A}'$ on $\langle X^*, V\rangle$.

Assume now that $\langle X^*, V\rangle \in \mathcal{L}(\mathcal{A}')$. We prove that $wide_Y(\langle X^*, V\rangle) \in \mathcal{L}(\mathcal{A})$. Since $hide^{-1}(u)$, for $u \in X^*$, is not a singleton, this direction is harder. We first define an alternating tree automaton $\mathcal{A}''$. The automaton $\mathcal{A}''$ is identical to the automaton $\mathcal{A}'$, only that it annotates the states with a direction from Y that maintains the information that gets lost when we obtain δ' from δ. Formally, $\mathcal{A}'' = \langle Z, Q \times Y, \delta'', \langle q_0, y^0\rangle, \alpha \times Y\rangle$, where y^0 is the root direction of Y, and for every $q \in Q$, $y \in Y$, and $z \in Z$, the transition $\delta''(\langle q, y\rangle, z)$ is obtained from $\delta(q, z)$ by replacing each atom $(\langle x, y'\rangle, q')$ by the atom $(x, \langle q', y'\rangle)$.

It is easy to see that $\mathcal{L}(\mathcal{A}'') = \mathcal{L}(\mathcal{A}')$. We now prove that if $\langle X^*, V\rangle \in \mathcal{L}(\mathcal{A}'')$, then $wide_Y(\langle X^*, V\rangle) \in \mathcal{L}(\mathcal{A})$. Let $\langle T_r, r''\rangle$ be an accepting run of $\mathcal{A}''$ on $\langle X^*, V\rangle$. By the definition of a run, we have that $r'' : T_r \to (X^* \times (Q \times Y))$ maps a node of the run to a pair consisting of a node in X^* and a state of $\mathcal{A}''$ that reads this node. Consider the labeled tree $\langle T_r, r\rangle$ with $r : T_r \to ((X \times Y)^* \times Q)$, where

- $r(\epsilon) = r''(\epsilon)$, and
- for every $wc \in T_r$ with $r(w) = \langle u, q'\rangle$ and $r''(wc) = \langle ux, \langle q, y\rangle\rangle$, we have $r(wc) = \langle u\langle x, y\rangle, q\rangle$.

It is easy to see that $\langle T_r, r\rangle$ is an accepting run of $\mathcal{A}$ on $wide_Y(\langle X^*, V\rangle)$. □

Given an alternating tree automaton $\mathcal{A}$, let $cover(\mathcal{A})$ and $narrow_Y(\mathcal{A})$ denote the corresponding automata $\mathcal{A}'$ constructed in Theorems 4.2 and 4.3 (for a set Y of directions), respectively.

4.3. Automata-Theoretic Solution

We can now use Theorems 4.1, 4.2, and 4.3, to solve the general synthesis problem.

THEOREM 4.4. *For every CTL* formula ψ over the sets I, E, and O of readable input, unreadable input, and output signals, ψ is realizable iff $narrow_{(2^E)}(cover(\mathcal{A}_{(2^{I\cup E}),\psi}))$ is nonempty.*

Proof: Let $\mathcal{A} = \mathcal{A}_{(2^{I\cup E}),\psi}$. Recall that $\mathcal{A}$ is an alternating tree automaton over $2^{I\cup E\cup O}$-labeled $2^{I\cup E}$-trees. Hence, the automaton $cover(\mathcal{A})$ runs over 2^O-labeled $2^{I\cup E}$-trees and the automaton $narrow_{(2^E)}(cover(\mathcal{A}))$ runs over 2^O-labeled 2^I-trees.

Assume first that ψ is realizable. Then, by Theorem 4.1, there exists a program $P : (2^I)^* \to 2^O$ such that $xray(wide_{(2^E)}(\langle (2^I)^*, P\rangle))$ is accepted by $\mathcal{A}$.

Hence, by Theorem 4.2, $wide_{(2^E)}(\langle(2^I)^*, P\rangle)$ is accepted by $cover(\mathcal{A})$. Hence, by Theorem 4.3, $\langle(2^I)^*, P\rangle$ is accepted by $narrow_{(2^E)}(cover(\mathcal{A}))$, which is therefore nonempty.

Assume now that $narrow_{(2^E)}(cover(\mathcal{A}))$ is nonempty. Then, there exists a tree $\langle(2^I)^*, P\rangle$ accepted by $narrow_{(2^E)}(cover(\mathcal{A}))$. Then, by Theorems 4.2 and 4.3, $xray(wide_{(2^E)}(\langle(2^I)^*, P\rangle))$ is accepted by $\mathcal{A}$. Hence, by Theorem 4.1, P realizes ψ.

□

We say that P is a *finite-state program* iff there exists a deterministic finite-state word automaton $\mathcal{U}_P$ over the alphabet 2^I, such that the set of states of $\mathcal{U}_P$ is $Q \times 2^O$ for some finite set Q, and for every word $x \in (2^I)^*$, the run of $\mathcal{U}_P$ on x terminates in a state in $Q \times \{P(x)\}$. Thus, finite-state programs are generated by finite-state automata. We sometimes refer also to finite state strategies, for arbitrary Σ-labeled Υ-trees. It is shown in (Pnueli and Rosner, 1989) that for every realizable LTL specification we can synthesize a finite-state program. In the theorem below, we show that this holds also for the general synthesis problem.

THEOREM 4.5. *Given an alternating tree automaton $\mathcal{A}$ over Σ-labeled Υ-trees, the following are equivalent:*

1. *$\mathcal{A}$ is nonempty.*
2. *There exists a finite-state strategy $f : \Upsilon^* \to \Sigma$ such that $\langle\Upsilon^*, f\rangle \in \mathcal{L}(\mathcal{A})$.*

Furthermore, the nonemptiness algorithm for $\mathcal{A}$ can be extended. within the same complexity bounds to produce a finite-state strategy.

Proof: Rabin proved the theorem for nondeterministic Rabin tree automata (Rabin, 1970). Muller and Schupp translated alternating tree automata to nondeterministic tree automata of the same acceptance condition (Muller and Schupp, 1995). □

THEOREM 4.6. *The synthesis problem for LTL and CTL***, with either complete or incomplete information, is 2EXPTIME-complete.*

Proof: We reduced the synthesis problem for a CTL* formula ψ to the nonemptiness problem of an alternating Rabin tree automaton with exponentially many states and linearly many pairs. The upper bounds then follow from the known translation of alternating Rabin tree automata to nondeterministic Rabin tree automata (Muller and Schupp, 1995) and the complexity of the nonemptiness problem for the latters (Emerson and Jutla, 1988; Pnueli and Rosner, 1989). The lower bounds follow from the known lower bounds to the realizability problem for LTL and the satisfiability problems for CTL* (Vardi and Stockmeyer, 1985; Pnueli and Rosner, 1989; Rosner, 1992). □

4.4. Handling CTL Specifications More Efficiently

As CTL is a subset of CTL*, the algorithm described in Theorem 4.4 suggests a solution for the synthesis problem of CTL. Nevertheless, though CTL is simpler than CTL* (in particular, when ψ is a CTL formula, the automaton $\mathcal{A}_{\Upsilon,\psi}$ described in Theorem 4.1 is an alternating Büchi tree automaton with $O(|\psi|)$ states), the time required for executing our algorithm for a CTL formula is double-exponential in the length of the formula, just as the one for CTL*. To see this, note that the number of states in $cover(\mathcal{A}_{\Upsilon,\psi})$ depends not only on the number of states in $\mathcal{A}_{\Upsilon,\psi}$, but also on the size of Υ. Since $cover(\mathcal{A}_{\Upsilon,\psi})$ is an alternating tree automaton, checking it (or its narrowing) for nonemptiness is exponential in the number of its states, and therefore also in Υ. Hence, as the size of Υ is exponential in the number of input signals, independent of the type of ψ, we end-up with a double-exponential-time algorithm. In this section we describe an exponential-time algorithm for solving the synthesis problem for CTL. The idea is to minimize the use of alternation: employ it when it is crucial (checking that candidate trees are consistent), and give it up when it is a luxury (checking that candidate trees are I-exhaustive). For that, we need to change the order in which things are checked. That is, we first check consistency, and then check I-exhaustiveness.

We first need some definitions and notations. We assume that the reader is familiar with the logic CTL. We consider here CTL formulas in a positive normal form. Thus, a CTL formula is one of the following:

- **true**, **false**, p, or $\neg p$, where $p \in AP$.
- $\varphi_1 \vee \varphi_2$, $\varphi_1 \wedge \varphi_2$, $EX\varphi_2$, $AX\varphi_2$, $E\varphi_1 U\varphi_2$, $A\varphi_1 U\varphi_2$, $EG\varphi_2$, or $AG\varphi_2$, where φ_1 and φ_2 are CTL formulas.

As we have already mentioned informally, we say that a $2^{I\cup O}$-labeled 2^I-tree $\langle(2^I)^*, V\rangle$ is *I-exhaustive* iff for every node $w \in (2^I)^*$ we have $dir(w) \cap I = V(w) \cap I$. Recall that the operator $wide_{(2^E)}$ maps a $2^{I\cup O}$-labeled 2^I-tree $\langle(2^I)^*, V\rangle$ to a $2^{I\cup O}$-labeled $2^{I\cup E}$-tree $\langle(2^{I\cup E})^*, V'\rangle$ such that for every node $w \in (2^{I\cup E})^*$, we have $V'(w) = V(hide_{(2^E)}(w))$. We define a variant of the operator $wide_{(2^E)}$, called $fat_{(2^E)}$. Given a $2^{I\cup O}$-labeled 2^I-tree $\langle(2^I)^*, V\rangle$, the operator $fat_{(2^E)}$ maps $\langle(2^I)^*, V\rangle$ into a set of $2^{I\cup E\cup O}$-labeled $2^{I\cup E}$-trees such that $\langle(2^{I\cup E})^*, V'\rangle \in fat_{(2^E)}(\langle(2^I)^*, V\rangle)$ iff the following hold.

1. $V'(\varepsilon) \cap 2^{I\cup O} = V(\varepsilon)$, and
2. for every $w \in (2^{I\cup E})^+$, we have $V'(w) = V(hide_{(2^E)}(w)) \cup (dir(w) \cap E)$.

That is, $fat_{(2^E)}(\langle(2^I)^*, V\rangle)$ contains $2^{|E|}$ trees, which differ only on the label of their roots. The trees in $fat_{(2^E)}(\langle(2^I)^*, V\rangle)$ are very similar to the tree $wide_{(2^E)}(\langle(2^I)^*, V\rangle)$, with each node labeled, in addition to its label in

$wide_{(2^E)}(\langle(2^I)^*, V\rangle)$, also with its direction, projected on E. An exception is the root, which is labeled, in addition to its label in $wide_{(2^E)}(\langle(2^I)^*, V\rangle)$, also with some subset of E. Among all the trees in $fat_{(2^E)}(\langle(2^I)^*, V\rangle)$, of special interest to us is the tree with root labeled with $V(\varepsilon)$; that is, the tree in which the E element of the label of the root is the empty set (the root direction of 2^E). We call this tree $wide'_{(2^E)}(\langle(2^I)^*, V\rangle)$. Note that when $\langle(2^I)^*, V\rangle$ is I-exhaustive, then $wide'_{(2^E)}(\langle(2^I)^*, V\rangle)$ is equal to $xray(wide_{(2^E)}(\langle(2^I)^*, V\rangle))$. Indeed, both are $2^{I\cup E\cup O}$-labeled $2^{I\cup E}$-trees with each node $w \in 2^{I\cup E}$ labeled by $V(hide_{(2^E)}(w)) \cup dir(w)$.

We can now turn to the constructions analogous to *cover* and *narrow*. We start with an alternating tree automaton that checks candidate trees for satisfaction of ψ, making sure that attention is restricted to consistent trees.

THEOREM 4.7. *Given a CTL formula ψ over the sets I, E, and O of readable input, unreadable input, and output signals, there exists an alternating Büchi tree automaton $\mathcal{A}_\psi$ over $2^{I\cup O}$-labeled 2^I-trees, with $O(|\psi|)$ states, such that $\mathcal{L}(\mathcal{A}_\psi)$ is exactly the set of trees $\langle T, V\rangle$ for which $wide'_{(2^E)}(\langle T, V\rangle)$ satisfies ψ.*

Proof: We define $\mathcal{A}_\psi = \langle 2^{I\cup O}, Q, \delta, q_0, \alpha\rangle$, where

- $Q = \{q_0\} \cup (cl(\psi) \times \{\exists, \forall\})$. Typically, when the automaton is in state $\langle\varphi, \exists\rangle$, it accepts all trees $\langle T, V\rangle$ for which there exists a tree in $fat_{(2^E)}(\langle T, V\rangle)$ that satisfies φ. When the automaton is in state $\langle\varphi, \forall\rangle$, it accepts all trees $\langle T, V\rangle$ for which all the trees in $fat_{(2^E)}(\langle T, V\rangle)$ satisfy φ. We call $\exists$ and $\forall$ the *mode* of the state.

- The transition function $\delta : Q \times 2^{I\cup O} \to \mathcal{B}^+(2^I \times Q)$ is defined by means of a function $\delta' : cl(\psi) \times 2^{I\cup E\cup O} \to \mathcal{B}^+(2^I \times Q)$ explained and defined below.

  The function δ' is essentially the same function used in the original alternating tree automata framework for CTL model checking in (Bernholtz et al., 1994). The only change is that δ' attributes the states with modes. For all $\sigma \in 2^{I\cup E\cup O}$, we define δ' as follows.

  - For $p \in I \cup E \cup O$, we have
    - $\delta'(p, \sigma) = \textbf{true}$ if $p \in \sigma$.
    - $\delta'(p, \sigma) = \textbf{false}$ if $p \notin \sigma$.
    - $\delta'(\neg p, \sigma) = \textbf{true}$ if $p \notin \sigma$.
    - $\delta'(\neg p, \sigma) = \textbf{false}$ if $p \in \sigma$.
  - $\delta'(\varphi_1 \vee \varphi_2, \sigma) = \delta'(\varphi_1, \sigma) \vee \delta'(\varphi_2, \sigma)$.
  - $\delta'(\varphi_1 \wedge \varphi_2, \sigma) = \delta'(\varphi_1, \sigma) \wedge \delta'(\varphi_2, \sigma)$.
  - $\delta'(EX\varphi_2, \sigma) = \bigvee_{v\in 2^I}(v, \langle\varphi_2, \exists\rangle)$.
  - $\delta'(AX\varphi_2, \sigma) = \bigwedge_{v\in 2^I}(v, \langle\varphi_2, \forall\rangle)$.

- $\delta'(E\varphi_1 U\varphi_2, \sigma) = \delta'(\varphi_2, \sigma) \vee [\delta'(\varphi_1, \sigma) \wedge \bigvee_{\upsilon \in 2^I} (\upsilon, \langle E\varphi_1 U\varphi_2, \exists\rangle)]$.
- $\delta'(A\varphi_1 U\varphi_2, \sigma) = \delta'(\varphi_2, \sigma) \vee [\delta'(\varphi_1, \sigma) \wedge \bigwedge_{\upsilon \in 2^I} (\upsilon, \langle E\varphi_1 U\varphi_2, \forall\rangle)]$.
- $\delta'(EG\varphi_2, \sigma) = \delta'(\varphi_2, \sigma) \wedge \bigvee_{\upsilon \in 2^I} (\upsilon, \langle EG\varphi_2, \exists\rangle)$.
- $\delta'(AG\varphi_2, \sigma) = \delta'(\varphi_2, \sigma) \wedge \bigwedge_{\upsilon \in 2^I} (\upsilon, \langle EG\varphi_2, \forall\rangle)$.

Now, for all $\varphi \in cl(\psi)$ and $\upsilon \in 2^{I \cup O}$, we define δ as follows.

- $\delta(\langle\varphi, \exists\rangle, \upsilon) = \bigvee_{\tau \in 2^E} \delta'(\varphi, \upsilon \cup \tau)$.
- $\delta(\langle\varphi, \forall\rangle, \upsilon) = \bigwedge_{\tau \in 2^E} \delta'(\varphi, \upsilon \cup \tau)$.

In addition, as we fixed the root direction of 2^E to $\emptyset$, we define $\delta(q_0, \upsilon) = \delta'(\psi, \upsilon)$.

Note that for some forms for φ, that either do not depend in the satisfaction of signals in E in the present, or depend in nothing but their satisfaction in the present, we can simplify δ as follows (we use $*$ to denote either $\exists$ or $\forall$).

- For $p \in I \cup O$, we have
  - $\delta(\langle p,^* \rangle, \upsilon) =$ **true** if $p \in \upsilon$.
  - $\delta(\langle p,^* \rangle, \upsilon) =$ **false** if $p \notin \upsilon$.
  - $\delta(\langle \neg p,^* \rangle, \upsilon) =$ **true** if $p \notin \upsilon$.
  - $\delta(\langle \neg p,^* \rangle, \upsilon) =$ **false** if $p \in \upsilon$.
- For $p \in E$, we have
  - $\delta(\langle p, \exists\rangle, \upsilon) = \delta(\langle \neg p, \exists\rangle, \upsilon) =$ **true**.
  - $\delta(\langle p, \forall\rangle, \upsilon) = \delta(\langle \neg p, \forall\rangle, \upsilon) =$ **false**.
- $\delta(\langle EX\varphi_2,^* \rangle, \upsilon) = \delta'(EX\varphi_2, \upsilon)$.
- $\delta(\langle AX\varphi_2,^* \rangle, \upsilon) = \delta'(AX\varphi_2, \upsilon)$.

– The set α of accepting states consists of all the G-formulas in $cl(\psi)$ with either modes; that is, states of the form $\langle EG\varphi_2, \exists\rangle$, $\langle EG\varphi_2, \forall\rangle$, $\langle AG\varphi_2, \exists\rangle$, or $\langle AG\varphi_2, \forall\rangle$.

□

Not all the trees $\langle T, V\rangle$ accepted by $\mathcal{A}_\psi$ correspond to strategies that realize ψ. In order to be such a strategy, $\langle T, V\rangle$ should be I-exhaustive. This, however, can be checked independently, by a nondeterministic tree automaton.

THEOREM 4.8. *Given finite sets I and O of readable input and output signals, there exists a nondeterministic Büchi tree automaton $\mathcal{A}_{exh}$ over $2^{I \cup O}$-labeled 2^I-trees, with $2^{|I|}$ states, such that $\mathcal{L}(\mathcal{A}_{exh})$ is exactly the set of I-exhaustive trees.*

Proof: We define $\mathcal{A}_{exh} = \langle 2^{I\cup O}, 2^I, \delta, \emptyset, 2^I \rangle$, where for every $\tau \in 2^I$ and $\sigma \in 2^{I\cup O}$, we have

$$\delta(\tau, \sigma) = \left[\begin{array}{ll} \bigwedge_{\upsilon \in 2^I}(\upsilon, \upsilon) & \text{if } \tau = \sigma \cap I, \\ \textbf{false} & \text{otherwise.} \end{array} \right.$$

□

THEOREM 4.9. *For every CTL formula ψ over the sets I, E, and O of readable input, unreadable input, and output signals, ψ is realizable iff $\mathcal{L}(\mathcal{A}_\psi) \cap \mathcal{L}(\mathcal{A}_{exh})$ is nonempty.*

Proof: Assume first that ψ is realizable. Then, there exists a strategy $P : (2^I)^* \to 2^O$ that realizes ψ. We claim that the $2^{I\cup O}$-labeled 2^I-tree $\langle (2^I)^*, P' \rangle = xray(\langle (2^I)^*, P \rangle)$ is in $\mathcal{L}(\mathcal{A}_\psi) \cap \mathcal{L}(\mathcal{A}_{exh})$. First, as its 2^I labels are obtained by the operator x-ray, the tree $\langle (2^I)^*, P' \rangle$ is I-exhaustive, and is therefore in $\mathcal{L}(\mathcal{A}_{exh})$. Since P realizes ψ, we know, by the definition of realizability, that $xray(wide_{(2^E)}(\langle (2^I)^*, P \rangle))$ satisfies ψ. In addition, as $\langle (2^I)^*, P' \rangle$ is I-exhaustive, we have that $wide'_{(2^E)}(\langle (2^I)^*, P' \rangle)$ is equal to $xray(wide_{(2^E)}(\langle (2^I)^*, P' \rangle))$, which is, by the definition of P' and the operator x-ray, equal to $xray(wide_{(2^E)}(\langle (2^I)^*, P \rangle))$. Hence, $wide'_{(2^E)}(\langle (2^I)^*, P' \rangle)$ satisfies ψ, and belongs, by Theorem 4.7, to $\mathcal{L}(\mathcal{A}_\psi)$.

Assume now that $\mathcal{L}(\mathcal{A}_\psi) \cap \mathcal{L}(\mathcal{A}_{exh})$ is nonempty. Then, there exists a $2^{I\cup O}$-labeled 2^I-tree $\langle (2^I)^*, P \rangle \in \mathcal{L}(\mathcal{A}_\psi) \cap \mathcal{L}(\mathcal{A}_{exh})$. By Theorem 4.7, the labeled tree $wide'_{(2^E)}(\langle (2^I)^*, P \rangle)$ satisfies ψ. In addition, by Theorem 4.8, the labeled tree $\langle (2^I)^*, P \rangle$ is I-exhaustive and hence, $wide'_{(2^E)}(\langle (2^I)^*, P \rangle)$ is equal to $xray(wide_{(2^E)}(\langle (2^I)^*, P \rangle))$. Finally, the tree $xray(wide_{(2^E)}(\langle (2^I)^*, P' \rangle))$, where P' is the strategy obtained from P by projecting its labels on 2^O is equivalent to $xray(wide_{(2^E)}(\langle (2^I)^*, P \rangle))$. Hence, by the definition of realizability, P' realizes ψ. □

THEOREM 4.10. *The synthesis problem for CTL, with either complete or incomplete information, is EXPTIME-complete.*

Proof: We reduced the synthesis problem for a CTL formula ψ to the nonemptiness problem of an intersection of two automata. The first automaton, $\mathcal{A}_\psi$, is an alternating Büchi tree automaton with $O(|\psi|)$ states. The second, $\mathcal{A}_{exh}$, is a nondeterministic Büchi automaton with $2^{O(|\psi|)}$ states. By (Muller and Schupp, 1995), we can translate $\mathcal{A}_\psi$ to an alternating Büchi tree automaton with $2^{O(|\psi|)}$ states. By (Vardi and Wolper, 1986), we can therefore check the nonemptiness of the product of $\mathcal{A}_\psi$ and $\mathcal{A}_{exh}$ in time exponential in $|\psi|$. The lower bound follows from the known lower bound to the satisfiability problem for CTL (Fischer and Ladner, 1979). □

5. Discussion

We suggest a framework for the synthesis problem in its full generality. The difficulties caused by incomplete information require the use of alternating tree automata. While the synthesis problem for open systems is different and more general than the satisfiability problem, alternating tree automata suggest very similar solutions to these problems. Both solutions are based on the translation of a temporal logic formula ψ into an alternating tree automaton $\mathcal{A}_\psi$, and on the nonemptiness test for $\mathcal{A}_\psi$. For the synthesis problem, the automaton $\mathcal{A}_\psi$ needs to pass a simple transformation ("*cover*") before the nonemptiness test. In the presence of incomplete information, it needs to pass an additional simple transformation ("*narrow*"). Our framework handles, as a special case, the classical LTL synthesis problem and suggest a simpler solution for this case. In particular, it avoids the determinization procedure required in (Pnueli and Rosner, 1989). Typically, alternating tree automata shift all the combinatorial difficulties of the synthesis problem, in both the linear and the branching framework, to the nonemptiness test.

References

Abadi, M., L. Lamport, and P. Wolper: 1989, 'Realizable and Unrealizable Concurrent Program Specifications'. In: *Proc. 16th Int. Colloquium on Automata, Languages and Programming*, Vol. 372. pp. 1–17.

Antoniotti, M.: 1995, 'Synthesis and verification of discrete controllers for robotics and manufacturing devices with temporal logic and the Control-D system'. Ph.D. thesis, New York University, New York.

Bernholtz, O., M. Vardi, and P. Wolper: 1994, 'An Automata-Theoretic Approach to Branching-Time Model Checking'. In: D. L. Dill (ed.): *Computer Aided Verification, Proc. 6th Int. Conference*, Vol. 818 of *Lecture Notes in Computer Science*. Stanford, pp. 142–155.

Büchi, J. and L. Landweber: 1969, 'Solving sequential conditions by finite-state strategies'. *Trans. AMS* **138**, 295–311.

Church, A.: 1963, 'Logic, arithmetics, and automata'. In: *Proc. International Congress of Mathematicians, 1962.* pp. 23–35.

Dill, D.: 1989, *Trace theory for automatic hierarchical verification of speed independent circuits.* MIT Press.

Emerson, A. and A. Sistla: 1984, 'Deciding full Branching Time Logics'. *Information and Control* **61**(3), 175–201.

Emerson, E.: 1990, 'Temporal and modal logic'. *Handbook of theoretical computer science* pp. 997–1072.

Emerson, E. and E. Clarke: 1982, 'Using Branching Time Logic to Synthesize Synchronization Skeletons'. *Science of Computer Programming* **2**, 241–266.

Emerson, E. and J. Halpern: 1986, 'Sometimes and Not Never Revisited: On Branching Versus Linear Time'. *Journal of the ACM* **33**(1), 151–178.

Emerson, E. and C. Jutla: 1988, 'The Complexity of Tree Automata and Logics of Programs'. In: *Proceedings of the 29th IEEE Symposium on Foundations of Computer Science.* White Plains, pp. 368–377.

Fischer, M. and R. Ladner: 1979, 'Propositional dynamic logic of regular programs'. *J. of Computer and Systems Sciences* **18**, 194–211.

Kumar, R. and V. Garg: 1995, *Modeling and control of logical discrete event systems*. Kluwer Academic Publishers.

Kumar, R. and M. Shayman: 1995, 'Supervisory control of nondeterministic systems under partial observation and decentralization'. *SIAM Journal of Control and Optimization.*

Lamport, L.: 1980, 'Sometimes is sometimes "Not never" - on the temporal logic of programs'. In: *Proceedings of the 7th ACM Symposium on Principles of Programming Languages.* pp. 174–185.

Manna, Z. and R. Waldinger: 1980, 'A deductive approach to program synthesis'. *ACM Transactions on Programming Languages and Systems* **2**(1), 90–121.

Muller, D. and P. Schupp: 1987, 'Alternating automata on infinite trees'. *Theoretical Computer Science* **54**,, 267–276.

Muller, D. and P. Schupp: 1995, 'Simulating Aternating tree automata by nondeterministic automata: New results and new proofs of theorems of Rabin, McNaughton and Safra'. *Theoretical Computer Science* **141**, 69–107.

Pnueli, A.: 1981, 'The Temporal Semantics of Concurrent Programs'. *Theoretical Computer Science* **13**, 45–60.

Pnueli, A. and R. Rosner: 1989, 'On the Synthesis of a Reactive Module'. In: *Proceedings of the Sixteenth ACM Symposium on Principles of Programming Languages.* Austin.

Pnueli, A. and R. Rosner: 1990, 'Distributed Reactive Systems are hard to Synthesize'. In: *Proc. 31st IEEE Symposium on Foundation of Computer Science.* pp. 746–757.

Rabin, M.: 1970, 'Weakly definable relations and special automata'. In: *Proc. Symp. Math. Logic and Foundations of Set Theory.* pp. 1–23.

Rosner, R.: 1992, 'Modular Synthesis of Reactive Systems'. Ph.D. thesis, Weizmann Institute of Science, Rehovot, Israel.

Vardi, M.: 1995, 'An automata-theoretic approach to fair realizability and synthesis'. In: P. Wolper (ed.): *Computer Aided Verification, Proc. 7th Int'l Conf.*, Vol. 939 of *Lecture Notes in Computer Science.* pp. 267–292.

Vardi, M. and L. Stockmeyer: 1985, 'Improved upper and lower bounds for modal logics of programs'. In: *Proc 17th ACM Symp. on Theory of Computing.* pp. 240–251.

Vardi, M. and P. Wolper: 1986, 'Automata-Theoretic Techniques for Modal Logics of Programs'. *Journal of Computer and System Science* **32**(2), 182–221.

Deductive Verification of Parameterized Fault-Tolerant Systems: A Case Study
*

Nikolaj S. Bjørner, Uri Lerner and Zohar Manna
(nikolaj|uri|manna@cs.stanford.edu)
Computer Science Department, Stanford University

Abstract. We present a methodology and a formal toolset for verifying fault-tolerant systems, based upon the temporal verification system STeP. Our test case is the modeling and verification of a parameterized fault-tolerant leader-election algorithm recently proposed in (Garavel and Mounier, 1996).

Our methods settle the general N-process correctness for the algorithm, which had been previously verified only for $N = 3$. We formulate the notion of *Uniform Compassion* to model progress in faulty systems more faithfully, and combine it with the more standard notions of fairness. We also show how the correctness proofs generalize to different channel models by a reduction to a simple channel model.

1. Introduction

The analysis of distributed algorithms can be rather complex when a large and intertwined set of interactions is considered. If the algorithms have only a finite set of states this analysis can be carried out automatically using *model checking* tools (McMillan, 1993). This yields a practical way to debug parameterized systems by checking different instantiations of the parameters. However, model checking runs quickly out of steam as the instantiated parameters take larger values.

Recently, a new *leader-election* algorithm was proposed as a challenging verification problem in (Garavel and Mounier, 1996). Leader-election algorithms are used in a distributed environment to select one station that would serve as a coordinator in performing tasks needed by other stations in the system. In its general form the algorithm involves N independent processes and N communication links, each capable of holding k messages, yielding $O((N + kN)^N)$ states. The state explosion problem limited the model checking verification in (Garavel and Mounier, 1996) to $N = 3$ stations with only $k = 1$ messages in a link.

Contrary to model checking methods, deductive methods can handle infinite-state and parameterized systems. The Stanford Temporal

* This research was supported in part by the National Science Foundation under grant CCR-95-27927, the Defense Advanced Research Projects Agency under NASA grant NAG2-892, ARO under grant DAAH04-95-1-0317, ARO under MURI grant DAAH04-96-1-0341, and by Army contract DABT63-96-C-0096 (DARPA).

H. Barringer et al. (eds.), Advances in Temporal Logic, 129–148.

Prover, STeP (Bjørner et al., 1996; Bjørner et al., 1995), provides a mechanized deductive framework for verifying linear-time temporal properties of reactive and concurrent systems. We argue that the use of transition systems for system modeling, linear-time temporal logic for requirements specification and STeP for verification is general and effective. We claim that the metric for the usefulness of formal validation should be the sum of the effort invested in these three areas: modeling, specification and verification. The proof of the correctness of the algorithm was presented to us as an open problem. We found and verified the proof within one week using STeP.

Using our parameterized approach we generalize the algorithm, which was originally presented only for links with capacity 1. We investigate the general case of links holding up to k messages, and show how this case can be reduced to links with no buffering capacity. We even investigate links that can store an unbounded number of messages, and show how our proof generalizes to this case as well.

To model progress more faithfully we introduce the new notion of *Uniform Compassion*, which is a stronger assumption than the standard compassion. We claim that Uniform Compassion captures the behavior of a faulty system correctly, and simplifies the formalization and proofs.

Although the analysis of parameterized systems is in general undecidable (Apt and Kozen, 1986), a growing number of publications report sound, and in some cases complete, analysis methods for some restricted classes of parameterized systems (Emerson and Namjoshi, 1996; Ip and Dill, 1996; Clarke et al., 1995; Lesens et al., 1997). The algorithm we analyze in this paper does not belong to the categories handled by these methods.

In Section 2 we review the computational model of transition systems and temporal specifications. We also introduce Uniform Compassion and define a new proof rule to account for it. In Section 3 we model the leader-election algorithm as a transition system. In Section 4 we show that our modeling and analysis can be reused for different versions of the algorithm where network stations are linked by channels of various capacities. The verification of the safety and liveness requirements is finally given in Sections 5 and 6.

2. Computational Model

2.1. Fair Transition systems

In the style of (Manna and Pnueli, 1995), a fair transition system $\mathcal{S}$ is of the form $\langle \mathcal{V}, \Theta, \mathcal{T}, \mathcal{J}, \mathcal{C} \rangle$, where $\mathcal{V}$ is a finite set of system variables,

Θ is the initial condition, and $\mathcal{T}$ is a finite set of transitions. The vocabulary $\mathcal{V}$ contains data variables, control variables and auxiliary variables. The set of *states* (interpretations) over $\mathcal{V}$ is denoted by Σ. A *φ-state* is a state s where assertion (first-order formula) φ holds. The initial condition Θ is an assertion over $\mathcal{V}$. A transition τ maps each state $s \in \Sigma$ into a (possibly empty) set of τ-successor states, $\tau(s) \subseteq \Sigma$. The mapping associated with τ is defined by an assertion $\rho_\tau(\overline{x}, \overline{x}')$, called the *transition relation*, which relates the values $\overline{x}$ of the variables in state s and the values $\overline{x}'$ in a successor state $s' \in \tau(s)$. A transition τ is *enabled* at state s if $\tau(s) \neq \emptyset$. The assertion $En(\tau)$ characterizes the states where τ is enabled. Transition τ is *taken* from s to s' if $s' \in \tau(s)$. The Hoare-triple $\{\varphi\}\tau\{\psi\}$ denotes the verification condition $\varphi(\overline{x}) \land \rho_\tau(\overline{x}, \overline{x}') \rightarrow \psi(\overline{x}')$. With $\{\varphi\}\mathcal{T}\{\psi\}$ we associate the conjunction $\bigwedge_{\tau \in \mathcal{T}} \{\varphi\}\tau\{\psi\}$. An infinite sequence of states $\sigma = s_0, s_1, \ldots$ is a *run* over system $\mathcal{S}$ if

Initiality The initial state s_0 satisfies Θ.

Consecution For each state s_i either $s_{i+1} = s_i$ (stuttering) or there is a transition $\tau \in \mathcal{T}$ such that $s_{i+1} \in \tau(s_i)$.

A run σ is a *computation* if the following fairness constraints are satisfied:

Justice For any just transition $\tau \in \mathcal{J}$: Transition τ cannot be continuously enabled without being taken.

Compassion For any compassionate transition $\tau \in \mathcal{C}$: If τ is enabled infinitely often it is taken infinitely often.

We use standard linear-time temporal logic (Manna and Pnueli, 1991) to specify requirements of reactive systems. For example the formula $\Box\varphi$ is true if φ holds on all states, and $\Diamond\varphi$ is true if φ holds some time in the future. We use $\varphi \Rightarrow \psi$ as shorthand for $\Box(\varphi \rightarrow \psi)$. A temporal formula that is satisfied by all computations of $\mathcal{S}$ is *$\mathcal{S}$-valid*.

Deductive verification of $\mathcal{S}$-validity relies on a library of verification rules that reduce the verification of temporal formulas into first-order premises. For example, the *basic invariance* verification rule `B-INV` reduces the proof of $\Box p$ to the verification of the first-order premises $\Theta \rightarrow p$ and $\{p\}\mathcal{T}\{p\}$. A complete set of inference rules for standard fair transition systems is presented in (Manna and Pnueli, 1989).

2.2. Uniform Compassion

To model liveness in fault-tolerant systems we introduce the notion of *Uniform Compassion*. Uniform Compassion captures the assumption that random faults cannot happen selectively only in certain states. For example, we want to rule out the case where there is an infinite number of messages from stations i and j, messages from station i are always lost and messages from station j are not always lost.

When making a transition uniformly compassionate we assume that it is taken uniformly on all the states where it is enabled without special preference to some states. Next to the justice and compassion sets $\mathcal{J}$ and $\mathcal{C}$ in fair transition systems we add the set of uniformly compassionate transitions $\mathcal{UC}$ and restrict computations to satisfy the extra condition:

Uniform compassion For any uniformly compassionate transition $\tau \in \mathcal{UC}$ and assertion φ: If τ is enabled infinitely often at φ-states, then it is taken infinitely often at a φ-state.

The notion of Uniform Compassion is closely related to the *extreme-fairness* found in (Pnueli, 1983), which is used in a slightly different model. Under this model a transition may have a few branches. When the transition is taken, a branch is selected nondeterministically. Under extreme fairness either a transition is taken only finitely often from a φ-state or every branch of the transition is taken infinitely often from a φ-state.

Note that for finite domains Uniform Compassion can be simulated by splitting uniformly compassionate transitions into a set of transitions, each corresponding to one possible state of the entire state space. However, even in this case Uniform Compassion offers an exponentially more succinct (and a more natural) way to represent systems.

We will use the verification rule below for transitions in $\mathcal{UC}$.

RESP–UC (Response under Uniform Compassion)	
For $\tau_h \in \mathcal{UC}$, assertions p, q, ψ and φ	
$U1$	$p \rightarrow \varphi \vee \psi \vee q$
$U2$	$\{\varphi\}\mathcal{T}\{q \vee \varphi \vee \psi\}$
$U3$	$\psi \Rightarrow \diamondsuit(\varphi \vee q)$
$U4$	$\{\varphi\}\tau_h\{q\}$
$U5$	$\varphi \rightarrow En(\tau_h)$
	$p \Rightarrow \diamondsuit q$

THEOREM 2.1. *Rule RESP-UC is sound.*

Proof: From $U1 - U3$ we get that $p \Rightarrow (\Box \diamondsuit \varphi \vee \diamondsuit q)$. If $p \Rightarrow \diamondsuit q$ we are done. Assume $p \Rightarrow \Box \diamondsuit \varphi$. From U5 and using $\tau_h \in \mathcal{U}_C$ we get that

τ_h will be taken infinitely often from a φ-state. From $U4$ this implies $\Box\Diamond q$.

2.3. STeP

The STeP system was used to formally check the proofs of the safety and liveness requirements of the leader-election algorithm. This helped reveal several misguided proof attempts and finally led to a formally verified proof.

As input STeP accepts a state-based encoding of a reactive system as a fair transition system and a temporal requirements specification. Fair transition systems are encoded using a UNITY-style (Chandy and Misra, 1988) programming language. Each transition τ is labeled by its name, auxiliary parameters $[i]$, and a fairness constraint. The parameters describe a set of transitions, one for each different instance of the parameters. The body of a transition is a guarded command consisting of an enabling condition *guard* and a (simultaneous) state assignment $\overline{x} := e$. Variables not mentioned explicitly in the state assignment are unchanged. The transition relation associated with transition $\tau[i]$ is $guard \wedge \overline{x}' = e \wedge \overline{y}' = \overline{y}$, where $\overline{y}$ are the state variables not included in $\overline{x}$. An example of such an encoding is presented in Figure 2.

Verification rules, verification diagrams (Manna and Pnueli, 1994), and decision procedures for propositional temporal logic are used to reduce the verification of temporal formulas to first-order verification conditions. STeP then provides an interactive Gentzen-style environment for establishing first-order formulas. A tightly integrated suite of decision procedures combined with first-order reasoning (Bjørner et al., 1997) facilitates this part of the verification considerably. We will give system and specification descriptions in STeP syntax to show how the formalization is presented to our tool.

3. The fault-tolerant leader-election algorithm

3.1. The leader-election algorithm

We are given a ring-formed network of stations indexed $1..N$ that share a common resource R. The stations can send messages to each other via channels in a clockwise fashion to determine which station can access R. Figure 1 shows the network configuration for three stations with links L_1, L_2 and L_3 and the shared resource in the middle. Simple algorithms for this distributed leader-election problem have been proposed in (Chang and Roberts, 1979), whereas (Garavel and Mounier, 1996) gives an algorithm that addresses the issues of faulty channels and dead stations.

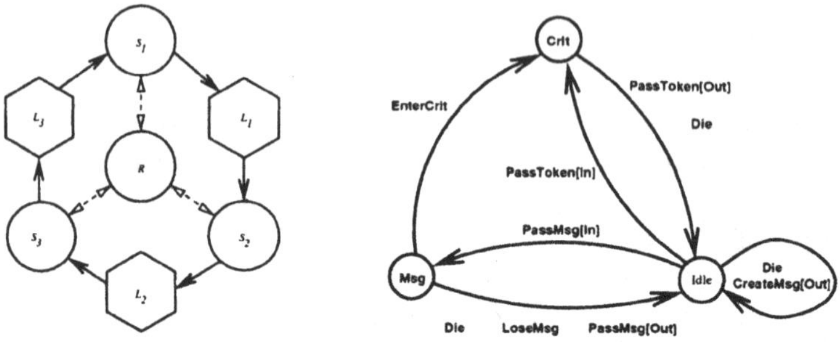

Figure 1. Ring network and leader-election algorithm transitions

In the fault-tolerant leader-election algorithm each station can reside at three control locations: an idling location `Idle`, a message-passing location `Msg`, and a critical location `Crit`, where the station has exclusive access to the resource. We further assume that a station may die.

A station enters its critical section when it receives the token message from its predecessor. When it exits the critical section it sends the token message to its successor. We assume that the token message can get lost when it is sent (unreliable links). To overcome this we introduce a leader-election protocol. Every living station may send a claim message, meaning it assumes the token is lost and wants to create a new one. The claim message includes the station index and an extra bit, whose role will be discussed later.

When station i gets a claim message, it passes the message iff i is bigger than the index of station that created the claim. A dead station passes the messages it receives unless it created the message. This behaviour guarantees that only the lowest living station will get its own message back, so it cannot happen that two different stations will become the leader. Claim messages can get lost, just like token messages.

When a station gets its own message back it still cannot necessarily enter the critical section, because there may be a token in the network. As we prove in Section 5, in this case the token would have been received by the station exactly once between the time a message was sent and the time a message was received. Therefore it is enough to keep one extra bit which changes whenever a station gets the token. When a station creates a claim message it adds the bit to the message. When a station gets its own message back it creates a new token iff the bit in the message is equal to the local bit in the station. Dead stations always delete their own messages.

The somewhat peculiar requirement that dead stations should be able to detect their own messages seems somewhat ad-hoc, but is necessary to guarantee liveness. Suppose that dead stations did not delete their messages, then we could have the following scenario: Station 1 sends $N-1$ messages and then dies. This implies that there is only one station in `Idle` mode throughout the network. Since no station can delete any one of the $N-1$ messages no two consecutive stations will ever be together in `Idle` mode, so a new claim message will never be created and liveness will be violated.

It is possible to change the algorithm by adding an extra field to the messages. The extra field is altered by live stations and is used to make sure that the same claim cannot go through the same live station twice. [1]

3.2. The formal model

In the original presentation of the algorithm, each control location has an analogous location in the case the station had died. We model these states by adding an extra bit `alive` to each station, indicating whether it is alive or dead.

In addition to `alive` each station saves another three variables. The integer `J` and the flag `b0` are the station index and extra bit fields in the message the station has received. The flag `b` is the local bit of the station. We represent these four local variables as four arrays of size N, where station i may access only entry i in the arrays.

The transition diagram on the right of Figure 1 summarizes the transitions that a station can take between its locations. The algorithm is encoded as a uniformly fair transition system in Figure 2. The encoding assumes that the links have zero capacity. The communication between the stations therefore proceeds by a synchronous handshake. Section 4 demonstrates that this choice is not a limitation.

The `PassToken` and `PassMsg` transitions send a token (claim message) to the next station that receives it (that is, the synchronous version combines `PassToken[Out]` and `PassToken[In]` into a single transition). `LoseToken` and `LoseMsg` are similar, but the token (claim message) gets lost due to the unreliable channels. `LoseMsg` is also used when a station deletes its own message (because it has died or the bit

[1] Ironically, liveness can be guaranteed without a dead station deleting its own message if we let the `LoseMsg` transition be compassionate. This represents a network that is less reliable than our model, since it guarantees that if we pass an infinite number of messages an infinite number of messages will be lost.

is in the wrong state). `CreateMsg` is used to create a claim message. `EnterCrit` is used to create a new token when a station gets its own message back. Finally, `Die` simulates the death of a station by moving to `Idle` and setting `alive` to *false*.

The key requirements for the algorithm are:

Safety (Mutual Exclusion): At most one station will be at the critical section at any given time:

$$\mathtt{Crit}(i) \land \mathtt{Crit}(j) \Rightarrow i = j$$

Liveness: Every living station gets access to its critical section infinitely often:

$$\square \diamondsuit (\neg \mathtt{alive}[i] \lor \mathtt{Crit}(i))$$

4. Modeling channels

The algorithm in Figure 2 is formulated for networks where message-passing proceeds synchronously. The links between two consecutive stations have capacity 0. We refer to the N-station instance of this algorithm as $\mathcal{S}[0, N]$. In general, let $\mathcal{S}[k, N]$ and $\mathcal{S}[\infty, N]$ be the versions where the links have buffering capacities k and ∞ respectively. The version model checked in (Garavel and Mounier, 1996) is $\mathcal{S}[1, 3]$.

In this section we demonstrate that our analysis of $\mathcal{S}[0, N]$ can be reused to verify correctness for the other versions of the algorithm, thereby producing correctness proofs "for free".

4.1. Simulating $\mathcal{S}[k, N]$ using $\mathcal{S}[0, N(k+1)]$

We now establish that we can simulate $\mathcal{S}[k, N]$ using $\mathcal{S}[0, N(k+1)]$ preserving certain properties. The idea is to simulate an asynchronous link between two stations by adding k dead stations between them. Dead stations behave as links, by passing on messages from other stations.

Not every execution sequence of $\mathcal{S}[0, N(k+1)]$ corresponds to an execution of $\mathcal{S}[k, N]$ with the dead-station-as-link interpretation. For example, consider the case where four dead stations simulate a four-message capacity link, and there are two messages in the link. When a capacity four buffer holds two messages, the actions of placing a new message on the buffer and taking a message from the buffer should both be enabled. However, in the configuration presented in Figure 3 a new message cannot be sent or received.

```
type loc = { lCrit, lIdle, lMsg }
in    N : int where N > 1
local l : array [1..N] of loc where Forall i : [1..N] . l[i] = lIdle
local J : array [1..N] of [1..N]
local b,b0 : array [1..N] of bool where Forall i : [1..N] . b[i]
local alive : array [1..N] of bool where Forall i : [1..N] . alive[i]

macro Crit(i:[1..N]) = l[i] = lCrit
macro Idle(i:[1..N]) = l[i] = lIdle
macro Msg(i:[1..N]) = l[i] = lMsg
macro next(i:[1..N]) = if i = N then 1 else i + 1 (* i mod N + 1 *)

Transition LoseToken [i : [1..N]] :
  enable Crit(i) /\ Idle(next(i))
  assign (l[i],b[i]) := (lIdle,!b[i])

Transition PassToken [i : [1..N]] Compassionate :
  enable Crit(i) /\ Idle(next(i))
  assign (l[i],b[i],l[next(i)]) := (lIdle,!b[i],lCrit)

Transition CreateMsg [i : [1..N]] Compassionate :
  enable Idle(i) /\ Idle(next(i)) /\ alive[i]
  assign (l[next(i)],b0[next(i)],J[next(i)]) := (lMsg,b[i],i)

Transition EnterCrit [i : [1..N]] Just :
  enable Msg(i) /\ b0[i] = b[i] /\ J[i] = i /\ alive[i]
  assign l[i] := lCrit

Transition LoseMsg [i : [1..N]] Just :
  enable Msg(i) /\ !(b0[i] = b[i] /\ J[i] = i /\ alive[i]) /\
        (alive[i] /\ i < J[i] \/ i = J[i] \/ Idle(next(i)))
  assign l[i] := lIdle

Transition PassMsg [i : [1..N]] Uniformly Compassionate :
  enable Msg(i) /\ (alive[i] --> i > J[i]) /\ Idle(next(i))
        /\ i != J[i]
  assign (l[i],l[next(i)],b0[next(i)],J[next(i)])
        := (lIdle,lMsg,b0[i],J[i])

Transition Die [i : [1..N]] :
  enable alive[i]
  assign (alive[i],l[i]) := (false,lIdle)
```

Figure 2. Synchronous leader-election algorithm as a STeP transition system

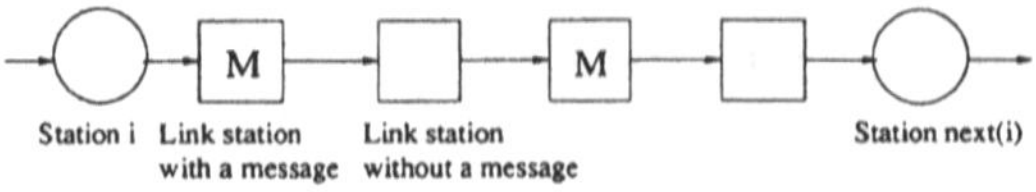

Figure 3. Illegal link behavior

Fortunately, all that is needed for our purposes is a mapping from any $\mathcal{S}[k,N]$-computation σ to some $\mathcal{S}[0,N(k+1)]$-computation σ' that preserves the temporal properties of interest. The way a single state in $\mathcal{S}[k,N]$ is associated with a state in $\mathcal{S}[0,N(k+1)]$ is illustrated in Figure 4.

The mapping makes sure that for every state s in the run of $\mathcal{S}[k,N]$ there is a matching state $m(s)$ in the run in $\mathcal{S}[0,N(k+1)]$. We call the $m(s)$ states *genuine states.* The rest of the states in $\mathcal{S}[0,N(k+1)]$ are the *artificial states.*

In a genuine state, station i in the asynchronous version corresponds closely to station $f(i)$ in the synchronous version. The two stations are in the same state, their local variables have the same values, and a transition is enabled in one iff it is enabled in the other.

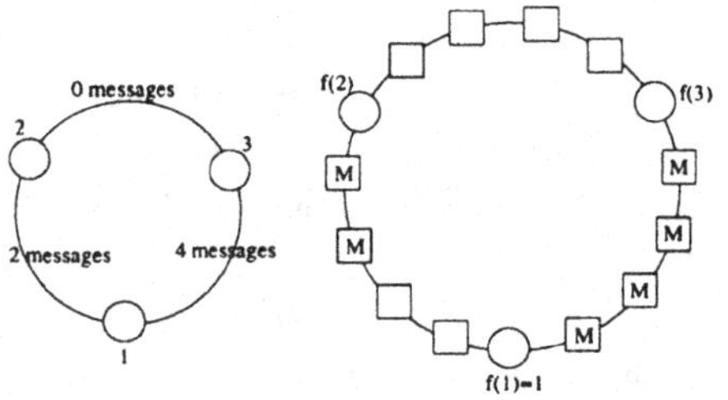

Figure 4. Matching states in $\mathcal{S}[4,3]$ and $\mathcal{S}[0,15]$

A transition in the asynchronous system is simulated by a sequence of transitions in the synchronous system to maintain the state correspondence. The sequence of steps needed for this correspondence is illustrated in Figure 5. Initially, the synchronous version makes sure to kill the stations designated as links.

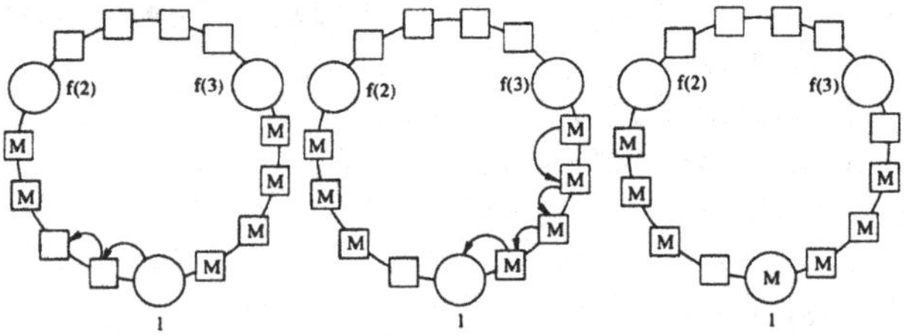

Figure 5. Sending and receiving a message

4.2. PROVING MUTUAL EXCLUSION FOR $\mathcal{S}[k, N]$

THEOREM 4.1. *If Mutual Exclusion holds in $\mathcal{S}[0, N]$ for every N then it must hold in $\mathcal{S}[k, M]$ for every k, M.*

Proof: Assume Mutual Exclusion does not hold for $\mathcal{S}[k, M]$ for certain k, M. This implies the existence of a computation that leads to a state s where two different stations have a token. We simulate this computation with a $\mathcal{S}[0, N(k+1)]$ system. In state $m(s)$ in $\mathcal{S}[0, N]$ two different stations will have a token, therefore Mutual Exclusion is violated for the synchronous case. ∎

It should be obvious that we can modify this theorem to many other safety properties, assuming we can formalize them both in $\mathcal{S}[k, M]$ and in $\mathcal{S}[0, N]$.

4.3. LIVENESS

Establishing that the simulation using $\mathcal{S}[0, N(k+1)]$ preserves the Liveness property of $\mathcal{S}[k, N]$ is more complicated than Mutual Exclusion, since we have to take the fairness constraints of transitions into account. Uniform Compassion is especially troublesome as the set of assertions that hold for the different systems are incompatible. For example, we can compose a formula φ stating : "There is a link in illegal mode and a `CreateMsg` transition is enabled for station k". This formula may hold infinitely often, but the transition would not be taken according to the simulation definition, violating Uniform Compassion for the $\mathcal{S}[0, N(k+1)]$ system.

A key observation in overcoming this difficulty is that the deductive verification that will be presented in Section 6 only appeals to a restricted set $\mathcal{U} = \{\varphi_1, \varphi_2, \ldots, \varphi_n\}$ of assertions when using the verification rule for Uniform Compassion. Liveness not only holds for all computations, but also for all runs that satisfy Justice, Compassion and Uniform Compassion with respect to $\mathcal{U}$. Therefore, it is enough to show that if the Liveness requirements fail for $\mathcal{S}[k, N]$ then they fail for a run that satisfies the Uniform Compassion requirements with respect to $\mathcal{U}$.

THEOREM 4.2. *If Liveness holds in $\mathcal{S}[0, N]$ for every N then it must hold in $\mathcal{S}[k, M]$ for every k, M.*

Proof: Define $\mathcal{U}$ to be the set of state formulas used in the Liveness proof. As we shall see in Section 6 we have:

$\mathcal{U} = \{\texttt{SmallestAlive}(k) \wedge \texttt{HasMsg}(k, j) \wedge \texttt{Idle}(\texttt{next}(j)) \mid 1 \le k, j \le N\}$

where $\texttt{SmallestAlive}(k)$ is *true* iff station k is alive and no lower ranking station is alive.

Assume there is a computation of $A = \mathcal{S}[k, M]$ where Liveness does not hold. We simulate this computation by a run on $B = \mathcal{S}[0, M(k+1)]$ (where Liveness is violated). It is enough to show that this run satisfies Justice, Compassion and Uniform Compassion with respect to $\mathcal{U}$ (it does not necessarily satisfy Uniform Compassion with respect to every state formula). To establish this we consider the fair transitions in our system.

- Just and Compassionate Transitions

  We distinguish between links and original stations. It is fairly easy to show that in our simulation fairness is not violated by link stations. We concentrate on transitions for original stations. If such a transition is enabled continuously (infinitely often) in B it is also enabled continuously (infinitely often) in the genuine states of B. Therefore, if Justice (Compassion) is not violated in A it is not violated in B.

- Uniformly Compassionate Transitions

  The only relevant transition is **PassMsg**. This transition is always taken for the dead station representing a link, so again we restrict ourselves to original stations.

  If $\varphi \in \mathcal{U}$ holds infinitely often, then it holds infinitely often in the genuine states, so we can restrict ourselves to these states (note that this is not true in general for every assertion φ). If $\varphi \in \mathcal{U}$ holds in $m(s) \in B$ then a similar formula must hold in $s \in A$ where we replace the predicate $\texttt{Idle}(\texttt{next}(k))$ by a predicate $\neg\texttt{FullLink}(k)$. Therefore if Uniform Compassion is not violated in A it is not violated in B with respect to $\mathcal{U}$. ∎

4.4. Unbounded links

Our reduction considered only bounded asynchronous links, but we can also use it for unbounded links.

COROLLARY 4.3. *If Mutual Exclusion holds for $\mathcal{S}[0, N]$ for every N, then it must hold for $\mathcal{S}[\infty, M]$ for every M.*

Proof: Assume Mutual Exclusion does not hold in $\mathcal{S}[\infty, M]$. Then there must exist a computation σ with a finite prefix σ' in which mutual exclusion is violated. Let k be the (finite) number of messages sent in

σ'. Then σ' must also be a legal prefix of a computation in $\mathcal{S}[k, M]$. From Theorem 4.1 we get that Mutual Exclusion must also be violated for some N in $\mathcal{S}[0, N]$. ∎

We cannot use a similar argument for the Liveness property, since the computation segment in which the property may be violated is not finite. Indeed, we can have an infinite sequence of states in $\mathcal{S}[\infty, N]$ in which Liveness is violated. The only transitions taken are sending claim messages from stations on the network. We cannot give a similar sequence of states in $\mathcal{S}[0, N]$.

It is possible, however, to reuse the Liveness proof for the unbounded case with some local modifications. We actually can drop Part-1 of the proof (lemmass **liveness-A**, **liveness-B** and **liveness-C**). Lemmas **liveness-D** and **liveness-F** will be modified by the use of ranking functions that represent the number of messages waiting in the link. See Section 6 for more details.

5. Safety proof

The verification of the safety requirement proceeds in two steps, by establishing two invariants. The invariants are stated in Figure 6.

First we establish the auxiliary invariant `safety-A`: if station m has a message from station k, where $k \geq m$, then all stations with rank lower than m must have died before. This property holds because if station t with a rank lower than m is alive, it will not pass a message from station k, therefore m cannot receive the message.

The main specification is stated as the assertion `Mux`, appearing as the first conjunct of invariant `safety-B`. `Mux` states that at no time there is more than one token in the system. However, this conjunct by itself is not inductive since it does not prevent arbitrary stations to create tokens. We use the second conjunct of `safety-B` to obtain an inductive assertion.

We focus on stations m and k, where m has a message from k, and on the location of the token (if one exists). We examine two basic cases. The first case is where a token resides between k and m (including m, not including k). Note that if the the message can get from m to k a new token may be created and mutual exclusion will be violated. The reason this cannot happen is that in this case we would have $b[k] \neq b0[m]$. Therefore, we state that if there is a token between k and m then $b[k] \neq b0[m]$.

The second case is where there is a token between m and k (including k, not including m). We also include in this case the possibility of a token being created between m and k (this is possible if a living station gets its own message back and creates a new token). In this case we must have $b[k] = b0[m]$, so when the token passes through k we go to the previous case. We can state this as follows: if we are not in the first case, and still $b[k] \neq b0[m]$, then there is no token between m and k and none can be created.

Due to space constraints we do not show the actual proof of the invariance in this paper.

```
macro Mux = Forall x,y : [1..N] . Crit(x) /\ Crit(y) --> x=y
macro NoToken = Forall z : [1..N] . !Crit(z)
macro Between(x:[1..N],i:[1..N],j:[1..N]) =
    if i < j then (i<x /\ x<=j) else (i<x \/ x<=j)
macro TokenBetween(k:[1..N],m:[1..N]) =
    Exists j : [1..N] . Between(j,k,m) /\ Crit(j)
macro HasMsg(k:[1..N],m:[1..N]) = Msg(m) /\ J[m]=k
macro NoPreviousToken(k:[1..N],m:[1..N]) =
  Forall r : [1..k], n : [1..N] .
    HasMsg(r,n) /\ alive[r] /\ b[r]=b0[n] --> Between(n,r,m)

PROPERTY safety-A :
  [] Forall m,k,t : [1..N] .
     HasMsg(k,m) /\ t < m /\ m <= k --> !alive[t]

PROPERTY safety-B :
  [] (Mux /\
    Forall k, m : [1..N] . alive[k] /\ HasMsg(k,m) -->
     if TokenBetween(k,m) then
       b[k]!=b0[m]
     else
       (b[k]!=b0[m] --> NoPreviousToken(k,m) /\ NoToken))
```

Figure 6. Safety specification in STeP

6. Liveness proof

In this Section we present the proof of the liveness requirement. We present the proof in an hierarchical form going from general high-level lemmas to the concrete formalization. Due to space constraints we only present the detailed proof of selected lemmas of expository interest.

6.1. Proof overview

We now give an informal overview of the proof. We decompose the proof into smaller parts and give an intuitive explanation why each part is correct.

In the proof we concentrate on the lowest living station. If there is no such station, then all the stations are dead and no progress will be made. In general the proof has three major parts:

Part–1 Show that the lowest living station will be able to send a claim message. A key lemma in this part is that messages from a dead station will eventually disappear from the network (established using **liveness-A,B,C** which are presented below).

Part–2 Show that if there is no token in the network then the lowest living station will create one. We do this by showing that the lowest living station will get its own message back (**liveness-D,E**).

Part–3 Show that every station will eventually get the token, assuming that there is at least one living station (**liveness-F,G**).

6.2. The Main Lemmas

The lemmas used in this proof are as follows:

liveness-A Every station gets to `Idle`: $\square \diamondsuit \texttt{Idle}(k)$.

To prove this property we first show that at least one station is in `Idle`. We then claim that if station i is in `Idle` then eventually station $\texttt{pred}(i)$ will get to `Idle`. By induction we get that every station will eventually get to `Idle`.

liveness-B Messages of dead stations disappear:
$\neg\texttt{alive}[k] \Rightarrow \diamondsuit \neg\texttt{HasMsg}(j, k)$.

Assume station k is dead and cannot create new claim messages. We want to show that eventually there will not be any messages from k in the network. This is true since every message from k will be lost along the network or will eventually get to k and then be deleted by the dead station.

liveness-C Smallest alive can send a claim or gets the token:
`SmallestAlive`$(k) \Rightarrow \Diamond(\neg$`alive`$[k] \vee$ `Crit`$(k) \vee En($`CreateMsg`$[k]))$.

This property is implied by the last two properties. Let k be the lowest living station. By **liveness-B** eventually all the messages from lower ranking stations will disappear. After that, by **liveness-A** its successor will eventually get to `Idle`. If k is in `Msg` then it is either a message from itself or a message from a higher ranking station. In any case k will either enter `Crit` or delete the message and get to `Idle`. Since `next`(k) must stay in `Idle` we will have infinitely often that both k and `next`(k) will be in `Idle`.

liveness-D Smallest alive gets its claim:
`SmallestAlive`$(k) \Rightarrow \Diamond(\neg$`alive`$[k] \vee$ `Crit`$(k) \vee En($`EnterCrit`$[k]))$.

We show by induction that every station will eventually get the claim from the smallest alive. The base case was established in **liveness-C**. For the induction step we use Uniform Compassion, with the assertion φ stating that a station has a message from the lowest alive station. Since this happens infinitely often, the message will be passed to the next station.

liveness-E Smallest alive creates a token:
`SmallestAlive`$(k) \Rightarrow \Diamond(\neg$`alive`$[k] \vee$ `Crit`$(k))$.

This is implied directly by **liveness-D**, assuming there is no token in the network that can change the bit in the station.

liveness-F Smallest alive sends the token to other stations:
`SmallestAlive`$(k) \Rightarrow \Diamond(\neg$`alive`$[k] \vee$ `Crit`$(j))$.

This lemma is established by induction much like **liveness-D** (although here we use only standard Compassion and not Uniform Compassion). The base case is established by **liveness-E**, and the induction step is a straightforward application of the compassion rule.

liveness-G Every alive station gets the token: $\Box \Diamond(\neg$`alive`$[k] \vee$ `Crit`$(k))$.

This is implied directly by **liveness-F**. If there is a living station then there is a smallest alive station, so either all the living stations will die or they will get the token.

In the next Sections we illustrate how the formally checked proof of lemmas **liveness-A** and **liveness-D** is carried out with STeP. The other cases are verified in a similar way. We have formally checked all lemmas. In this process we found bugs in preliminary informal proofs, and used the formal tool to improve the proof-structure.

6.3. THE STEP PROOF OF **liveness-A**

To establish that every station reaches its `Idle` state infinitely often we use the auxiliary lemmas in Figure 7. Property `A.1`, proved using rule B-INV, establishes that at any time some station resides at `Idle`.

```
macro dist(k1,k2) = (k1-k2) mod N
macro pred(i) = if i = 1 then N else i - 1

PROPERTY A.1 : [] Exists k : [1..N] . Idle(k)
PROPERTY A.2 : Idle(k) ==> <> Idle(pred(k))
PROPERTY A.3 : Forall m : [0..N-1], k1, k2 : [1..N] .
    m=dist(k1,k2) --> (Idle(k1) ==> <>Idle(k2))
PROPERTY A : []<> Idle(k)
```

Figure 7. **liveness-A**

The proof of `A.2` is done using a verification diagram, presented in Figure 6.3. The diagram contains the different cases for process A and process $B = \texttt{pred}(A)$. Nodes labeled by φ_3 and φ_2 treat the cases where B is in the `Msg` state. It can then either go directly to the `Idle` state or enter the critical section in node φ_1. From φ_1 the transition `PassToken` is enabled since A is in the `Idle` state. Thus, taking the transition will lead to the node labeled by φ_0. STeP checks this argument formally by assigning the corresponding first-order verification conditions to each node and departing transition. Verification diagrams are described in more detail in (Manna and Pnueli, 1994).

To prove `A.3` we apply the mathematical induction rule, where `A.2` is used for the induction step.

Finally, `A` follows from `A.1` and `A.3`. We use STeP's interactive Gentzen-style prover to split temporal operators, skolemize and instantiate `A.3` so the goal sequents can be decided using decision procedures. STeP combines propositional temporal decision procedure with ground decision procedures for linear arithmetic and other ground-decidable theories. This is helpful since instantiations of quantifiers give subgoals that are not purely propositionally valid. For instance $\Box\Diamond(k < 0 \wedge a \leq k) \rightarrow \Box\Diamond(a < 0)$ can be proved automatically, since all counter-models must contain the unsatisfiable state assignment $\langle k < 0 : \mathit{true},\ a \leq k : \mathit{true},\ a < 0 : \mathit{false}\rangle$.

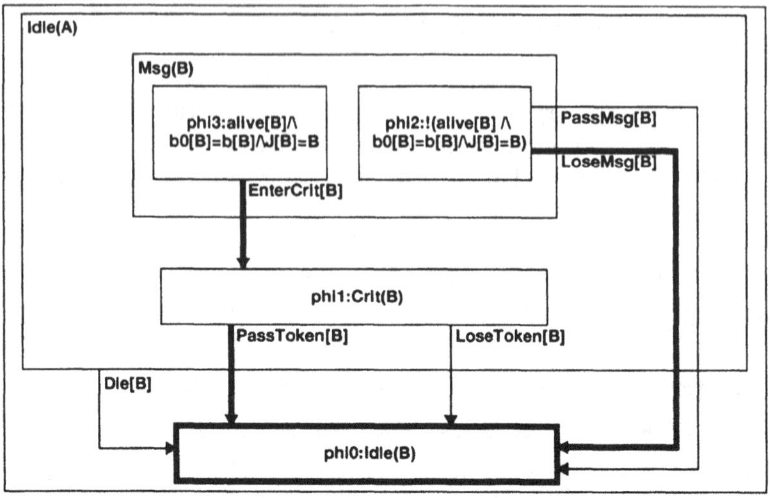

Figure 8. Verification diagram for `A.2`

6.4. The STeP proof of **liveness-D**

```
macro HasBMsg(k,j) = HasMsg(k,j) /\ b0[j] = b[k]

PROPERTY D.1 :
  SmallestAlive(k) ==>
      <>(!alive[k] \/ Crit(k) \/ SmallestAlive(k)/\
           HasBMsg(k,next(k)))
PROPERTY D.2 :
  SmallestAlive(k) ==>
      <>(!alive[k] \/ Crit(k) \/ SmallestAlive(k)/\
           HasBMsg(k,j))
PROPERTY D :
  SmallestAlive(k) ==> <>(!alive[k] \/ Crit(k) \/
           Enabled(EnterCrit[k]))
```

Figure 9. Proof of property `Liveness-D`

The first step in establishing that a message eventually circles the entire network, is the verification of property `D.1`. It states that the neighboring station to k, where `SmallestAlive(`k`)`, must obtain a claim from k. Property D.1 follows as **liveness-C** ensures that the compassionate transition $\texttt{CreateMsg}[k]$ is enabled infinitely often if $\texttt{SmallestAlive}(k) \wedge \neg\ \texttt{Crit}(k)$ holds continuously.

Property D.2 is proved by induction on $\texttt{dist}(k, j)$. It establishes that j must obtain a claim from k infinitely often. The base of the induction is D.1. For the induction step, we assume

$$\texttt{SmallestAlive}(k) \Rightarrow$$
$$\Diamond(\neg\texttt{alive}[k] \vee \texttt{Crit}(k) \vee \texttt{SmallestAlive}(k) \wedge \texttt{HasBMsg}(k, j))$$

and show

$$\texttt{SmallestAlive}(k) \Rightarrow$$
$$\Diamond(\neg\texttt{alive}[k] \vee \texttt{Crit}(k) \vee \texttt{SmallestAlive}(k) \wedge \texttt{HasBMsg}(k, \texttt{next}(j)))$$

using rule RESP-UC with intermediary assertions ψ : $\texttt{SmallestAlive}(k)$, φ : $\texttt{SmallestAlive}(k) \wedge \texttt{HasBMsg}(k, j) \wedge \texttt{Idle}(\texttt{next}(j))$, and the helpful transition τ_h : $\texttt{PassMsg}[k]$.

The first-order verification conditions generated by rule RESP-UC are verified using STeP's Gentzen prover using standard first-order reasoning aided by STeP's decision procedures. The verification rule also produces the temporal premise $U3$, which follows from the induction hypothesis, together with **liveness-A** and some simple safety properties that constrain j's movements from the Msg location.

7. Conclusion

We presented a formal proof of a generalized version of a fault-tolerant leader-election algorithm recently proposed in (Garavel and Mounier, 1996). Safety was verified using standard verification rules for invariants. To analyze liveness under fault-tolerance, we developed a notion of Uniform Compassion to get a realistic model. Our approach allowed us to use existing tools already present in STeP.

This paper did not investigate fairness under fault-tolerancy from a model checking and proof-theoretical perspective in full depth. We feel there is still room to explore these issues.

ACKNOWLEDGEMENTS

We thank Xavier Boyen, Bernd Finkbeiner, Henny B. Sipma and Tomás E. Uribe for their most valuable comments and for insightful discussions.

References

Apt, K. and D. Kozen: 1986, 'Limits for automatic verification of finite-state concurrent systems'. *Information Processing Letters* **22**(6).

Bjørner, N., A. Browne, E. Chang, M. Colón, A. Kapur, Z. Manna, S. Sipma, and T. Uribe: 1995, 'STeP: The Stanford Temporal Prover, User's Manual'. Technical Report STAN-CS-TR-95-1562, Computer Science Department, Stanford University.

Bjørner, N., A. Browne, E. Chang, M. Colón, A. Kapur, Z. Manna, S. Sipma, and T. Uribe: 1996, 'STeP: Deductive-algorithmic verification of reactive and real-time systems'. In: *Proc. 8th Intl. Conf. on Computer Aided Verification*, Vol. 1102 of *LNCS.*

Bjørner, N., M. Stickel, and T. Uribe: 1997, 'A practical combination of first-order reasoning and decision procedures'. In: *14th Intl. Conf. on Automated Deduction.*

Chandy, K. M. and J. Misra: 1988, *Parallel Program Design.* Addison-Wesley.

Chang, E. and R. Roberts: 1979, 'An improved algorithm for decentralized extrema-finding in circular configurations of processes'. *Communications of the ACM* **5**(22), 281–283.

Clarke, E., O. Grumberg, and S. Jha: 1995, 'Verifying parameterized networks using abstraction and regular languages'. In: I. Lee and S. Smolka (eds.): *CONCUR '95: Concurrency Theory, 6th Intl. Conf.*

Emerson, E. and K. Namjoshi: 1996, 'Automatic verification of parameterized synchronous systems'. In: R. Alur and T. Henzinger (eds.): *Proc.* 8th *Intl. Conference on Computer Aided Verification.* pp. 87–98.

Garavel, H. and L. Mounier: 1996, 'Specification and verification of various distributed leader election algorithms for unidirectional ring networks'. Technical Report Rapport de recherche 2986, INRIA, Rhone-Alpes, France.

Ip, C. and D. Dill: 1996, 'Verifying systems with replicated components in Murφ'. In: R. Alur and T. Henzinger (eds.): *Proc.* 8th *Intl. Conference on Computer Aided Verification.* pp. 147–158.

Lesens, D., N. Halbwachs, and P. Raymond: 1997, 'Automatic verification of parameterized linear networks of processes'. In: 24th *ACM Symp. Princ. of Prog. Lang.* pp. 346–357.

Manna, Z. and A. Pnueli: 1989, 'Completing the temporal picture'. In: G. Ausiello, M. Dezani-Ciancaglini, and S. R. D. Rocca (eds.): *Proc. 16th Int. Colloq. Aut. Lang. Prog.* pp. 534–558. Also in Theoretical Computer Science.

Manna, Z. and A. Pnueli: 1991, *The Temporal Logic of Reactive and Concurrent Systems: Specification.* Springer-Verlag.

Manna, Z. and A. Pnueli: 1994, 'Temporal verification diagrams'. In: *Proc. Int. Symp. on Theoretical Aspects of Computer Software*, Vol. 789 of *LNCS.* pp. 726–765.

Manna, Z. and A. Pnueli: 1995, *Temporal Verification of Reactive Systems: Safety.* Springer-Verlag.

McMillan, K.: 1993, *Symbolic Model Checking.* Kluwer Academic Pub.

Pnueli, A.: 1983, 'On the extremely fair treatment of probabilistic algorithms'. In: *Proc. 15th ACM Symp. Theory of Comp.* pp. 278–290.

Using Otter for Temporal Resolution

Clare Dixon (C.Dixon@doc.mmu.ac.uk) *
Department of Computing, Manchester Metropolitan University, Manchester M1 5GD, United Kingdom

Abstract. This paper describes how to use OTTER, a resolution based theorem prover for classical propositional and first-order logics to implement Fisher's temporal resolution method. Fisher's clausal temporal resolution method involves both classical-style 'step' resolution within states and 'temporal' resolution over states. As the translation to a normal form removes most of the temporal operators, and the resolution within states is very similar to resolution in classical logics, most of the inferences required for the step resolution phase can be performed by OTTER after a simple translation into classical propositional logic. However, the application of the temporal resolution rule is more complex, requiring a search for sets of formulae that together imply an invariant property, for resolution with an eventuality formula. We suggest how OTTER can be used to fulfil the obligations of a particular algorithm that performs this search. The correctness of the method is discussed and we conclude by outlining how it has helped develop a prototype theorem prover based on Fisher's method.

Keywords: temporal logics, theorem-proving, resolution

1. Introduction

Temporal logics have been useful for the specification and verification of reactive systems (Barringer, 1987; Hailpern, 1982; Manna and Pnueli, 1992; Owicki and Lamport, 1982). Verification that a given temporal property, for example liveness or mutual exclusion, holds for a program or specification written in temporal logic involves formal proof. However few automated theorem provers exist for such logics. With the availability of efficient theorem provers for classical logics, for example OTTER (McCune, 1990), it is tempting to simply translate temporal logic formulae into first-order classical logic, add the necessary information to capture the temporal structures we are dealing with, and carry out the proof within this framework. However using this approach we may move from a propositional temporal logic that is decidable into first-order classical logic which is not. Further, this approach will not work for the more powerful temporal logics that allow the next operator, for example PTL, the logic used in this paper, which incorporates a variety of induction. Finally, by translating to classical logic we lose

* This work was supported partially by an EPSRC PhD Studentship and partially by EPSRC Research Grant GR/K57282

H. Barringer et al. (eds.), Advances in Temporal Logic, 149–166.

any structure or clarity that may be gained from working within the temporal logic itself.

In this paper we suggest a mechanism for using OTTER (McCune, 1990) a resolution theorem prover for propositional and first-order classical logic, to implement a resolution based theorem prover for linear time propositional temporal logic suggested by Fisher (Fisher, 1991). The method involves the translation to a normal form, classical resolution within states (known as *step resolution*) and *temporal resolution* over states between $\Diamond\neg p$ (*sometime* $\neg p$) and formulae that together imply $\Box p$ (*always* p). While different algorithms for implementing the temporal resolution rule have been identified (Dixon, 1996), we here concentrate on one of these algorithms that is the most suitable for implementation using OTTER. The translation to a normal form removes most of the temporal operators except '$\bullet$' (*in the previous moment*) and '$\Diamond$' (*sometime in the future*). To perform step resolution all we must do is choose a suitable encoding of our temporal formulae so that resolution is carried out between the correct formulae, i.e. those referring to the same moment in time. To implement temporal resolution we repeatedly use OTTER to fulfil the obligations of a search algorithm that detects sets of formulae whose combination implies a $\Box$-formula.

The paper is organised as follows. The logic we use, PTL, a description of Fisher's temporal resolution method and an algorithm for implementing the temporal resolution rule are given in §2. In §3 we describe how OTTER may be used to implement each of the parts of the temporal resolution method. Correctness issues are discussed in §4. Results, related work and conclusions are examined in §5, §6 and §7 respectively.

2. The Temporal Resolution Method

Here we summarise the syntax and semantics of the logic used and then describe Fisher's temporal resolution method. Finally, an algorithm for applying the temporal resolution rule, known as Breadth-First Search, is given.

2.1. Syntax and Semantics

The logic used in this paper is Propositional Temporal Logic (PTL), in which we use a linear, discrete model of time with finite past and infinite future. PTL may be viewed as a classical propositional logic augmented with both future-time and past-time temporal operators. Future-time temporal operators include '$\Diamond$' (*sometime in the future*), '$\Box$' (*always in the future*), '$\bigcirc$' (*in the next moment in time*), '$\mathcal{U}$' (*until*), '$\mathcal{W}$' (*unless* or *weak until*), each with a corresponding past-time operator. Since our temporal models assume a finite past, for convenience, we also use an operator **start** which only holds at the beginning of time.

Models for PTL consist of a sequence of *states*, representing moments in time, i.e.,

$$\sigma = s_0, s_1, s_2, s_3, \ldots$$

Here, each state, s_i, contains those propositions satisfied in the ith moment in time. As formulae in PTL are interpreted at a particular moment, the satisfaction of a formula f is denoted by

$$(\sigma, i) \models f$$

where σ is the model and i is the state index at which the temporal statement is to be interpreted. For any well-formed formula f, model σ and state index i, then either $(\sigma, i) \models f$ or $(\sigma, i) \not\models f$. For example, a proposition symbol, 'p', is satisfied in model σ and at state index i if, and only if, p is one of the propositions in state s_i, i.e.,

$$(\sigma, i) \models p \quad \text{iff} \quad p \in s_i.$$

The semantics of the temporal connectives used in the normal form or the resolution rule are defined as follows

$$\begin{array}{l}
(\sigma, i) \models \textbf{start} \text{ iff } i = 0; \\
(\sigma, i) \models \bullet A \text{ iff } i > 0 \text{ and } (\sigma, i-1) \models A; \\
(\sigma, i) \models \Diamond A \text{ iff there exists a } j \geqslant i \text{ s.t. } (\sigma, j) \models A; \\
(\sigma, i) \models \Box A \text{ iff for all } j \geqslant i \text{ then } (\sigma, j) \models A; \\
(\sigma, i) \models A \mathcal{U} B \text{ iff there exists a } k \geqslant i \text{ s.t. } (\sigma, k) \models B \\
\qquad\qquad \text{and for all } i \leqslant j < k \text{ then } (\sigma, j) \models A; \\
(\sigma, i) \models A \mathcal{W} B \text{ iff } (\sigma, i) \models A \mathcal{U} B \text{ or } (\sigma, i) \models \Box A.
\end{array}$$

The full syntax and semantics of PTL will not be presented here, but can be found in (Fisher, 1991).

2.2. A Normal Form for PTL

Formulae in PTL can be transformed to a normal form, Separated Normal Form (SNF) (Fisher, 1991; Fisher, 1992), which is the basis of the resolution method used in this paper. We note that the transformation into SNF preserves satisfiability so any contradiction generated from the formula in SNF implies a contradiction in the original formula. Formulae in SNF are of the general form

$$\Box \bigwedge_i R_i$$

where each R_i is known as a *rule* and must be one of the following forms.

$$\begin{array}{lcll} \textbf{start} & \Rightarrow & \bigvee_{b=1}^{r} l_b & \text{(an } \textit{initial} \ \square\text{-rule)} \\ \bigcirc \bigwedge_{a=1}^{g} k_a & \Rightarrow & \bigvee_{b=1}^{r} l_b & \text{(a } \textit{global} \ \square\text{-rule)} \\ \textbf{start} & \Rightarrow & \Diamond l & \text{(an } \textit{initial} \ \Diamond\text{-rule)} \\ \bigcirc \bigwedge_{a=1}^{g} k_a & \Rightarrow & \Diamond l & \text{(a } \textit{global} \ \Diamond\text{-rule)} \end{array}$$

Here k_a, l_b, and l are literals. The outer '$\square$' operator, that surrounds the conjunction of rules is usually omitted. Similarly, for convenience the conjunction is dropped and we consider just the set of rules R_i.

SNF rules may be combined by conjoining the left hand sides of rules and conjoining the right hand sides and then rewriting into DNF. Such rules are known as merged-SNF (SNF_m) (Fisher, 1991) and are required to apply the temporal resolution rule (see §2.4).

2.3. Step Resolution

'Step' resolution consists of the application of standard classical resolution rule to formulae representing constraints at a particular moment in time, together with simplification rules for transferring contradictions within states to constraints on previous states. Simplification and subsumption rules are also applied.

Pairs of initial $\square$-rules, or global $\square$-rules, may be resolved using the following (step resolution) rule where $\mathcal{L}_1$ and $\mathcal{L}_2$ are both last-time formulae or both **start**.

$$\begin{array}{rcl} \mathcal{L}_1 & \Rightarrow & A \vee r \\ \mathcal{L}_2 & \Rightarrow & B \vee \neg r \\ \hline (\mathcal{L}_1 \wedge \mathcal{L}_2) & \Rightarrow & A \vee B \end{array}$$

Once a contradiction within a state is found using step resolution, the following rule can be used to generate extra global constraints.

$$\begin{array}{rcl} \bigcirc P & \Rightarrow & \textbf{false} \\ \hline \textbf{start} & \Rightarrow & \neg P \\ \bigcirc \textbf{true} & \Rightarrow & \neg P \end{array}$$

This rule states that if, by satisfying P in the last moment in time a contradiction is produced, then P must never be satisfied in *any* moment in time. The new constraints therefore represent $\square \neg P$.

Subsumption and simplification also form part of the step resolution process. For example the SNF rules $\bigcirc \mathbf{false} \Rightarrow A$ and $P \Rightarrow \mathbf{true}$ can be removed during simplification as they represent valid subformulae and therefore cannot contribute to the generation of a contradiction. The first rule is valid as $\bigcirc \mathbf{false}$ can never be satisfied, and the second is valid as **true** is always satisfied.

The step resolution process terminates when either no new resolvents are derived, or **false** is derived in the form of one of the following rules.

$$\begin{array}{rcl} \mathbf{start} & \Rightarrow & \mathbf{false} \\ \bigcirc \mathbf{true} & \Rightarrow & \mathbf{false} \end{array}$$

2.4. Temporal Resolution

During temporal resolution the aim is to resolve a $\Diamond$–rule, $\mathcal{L} \Rightarrow \Diamond l$, where $\mathcal{L}$ may be either a last-time formula or **start**, with a set of rules that together imply $\Box \neg l$, for example a set of rules that together have the effect of $\bigcirc A \Rightarrow \Box \neg l$. However the interaction between the '$\bigcirc$' and '$\Box$' operators in PTL makes the definition of such a rule non-trivial and further the translation from PTL to SNF will have removed all but the outer level of $\Box$–operators. So, resolution will be between a $\Diamond$–rule and a *set* of rules that together imply a $\Box$–formula which will contradict the $\Diamond$–rule. Thus, given a set of rules in SNF, then for every rule of the form $\mathcal{L} \Rightarrow \Diamond l$ temporal resolution may be applied between this $\Diamond$–rule and a set of global $\Box$–rules, which taken together force $\neg l$ always to be satisfied.

The temporal resolution rule[1] is given by the following

$$\begin{array}{rcl} \bigcirc A_0 & \Rightarrow & B_0 \\ \ldots & & \ldots \\ \bigcirc A_n & \Rightarrow & B_n \\ \mathcal{L} & \Rightarrow & \Diamond l \\ \hline \mathcal{L} & \Rightarrow & (\bigwedge_{i=0}^{n} \neg A_i)\, \mathcal{W}\, l \end{array}$$

with side conditions

$$\left\{ \begin{array}{rcl} \text{for all } 0 \le i \le n & \vdash & B_i \Rightarrow \neg l \\ \text{and} & \vdash & B_i \Rightarrow \bigvee_{j=0}^{n} A_j \end{array} \right\}$$

[1] In previous presentations of this rule two resolvents are given. However only the resolvent given here is necessary for completeness, so for simplicity we omit the other resolvent.

where the side conditions ensure that the set of SNF_m rules $\bigcirc A_i \Rightarrow B_i$ together imply $\square\neg l$. So if any of the A_i are satisfied then $\neg l$ will be *always* be satisfied. Such a set of rules are known as a *loop* in $\neg l$.

2.5. The Temporal Resolution Method

After the translation to the normal form, step and temporal resolution are carried out, any new resolvents being translated into SNF. This process continues until either a contradiction has been detected or no new resolvent can be generated.

2.6. Breadth-First Search

Several different algorithms have been developed to implement the temporal resolution rule (Dixon, 1996) as it is significantly different from classical resolution. Here we concentrate on the one that lends itself to implementation using OTTER, known as Breadth-First Search.

2.6.1. *Breadth-First Search Algorithm*

For each rule of the form $\mathcal{L} \Rightarrow \Diamond l$ carry out the following.

1. Search for all the rules of the form $\bigcirc X_k \Rightarrow \neg l$, for $k = 0$ to b (called *start rules*), disjoin the left hand sides and make the *top node* H_0 equivalent to this, i.e.

$$H_0 \Leftrightarrow \bigvee_{k=0}^{b} X_k.$$

   Simplify H_0. If $\vdash H_0 \Leftrightarrow \mathbf{true}$ we terminate having found a loop.

2. Given node H_i, build node H_{i+1} for $i = 0, 1, \ldots$ by looking for rules or combinations of rules of the form $\bigcirc A_j \Rightarrow B_j$, for $j = 0$ to m where
   $\vdash B_j \Rightarrow H_i$ and $\vdash A_j \Rightarrow H_0$. Disjoin the left hand sides so that

$$H_{i+1} \Leftrightarrow \bigvee_{j=0}^{m} A_j$$

   and simplify as previously.

3. Repeat (2) until

   a) $\vdash H_i \Leftrightarrow \mathbf{true}$. We terminate having found a Breadth-First loop and return **true**.

b) $\vdash H_i \Leftrightarrow H_{i+1}$. We terminate having found a Breadth-First loop and return the DNF formula H_i.

c) The new node is empty. We terminate without having found a loop.

If the formula H_i is returned then this represents $\bigcirc H_i \Rightarrow \Box\neg l$ which can be used in the application of the temporal resolution rule. The correctness of the Breadth-First Search Algorithm has been shown in (Dixon, 1995).

3. Using Otter to Implement Temporal Resolution

Here we describe how OTTER may be used to implement both step and temporal resolution. To apply the temporal resolution rule we use OTTER to fulfil the proof obligations of the Breadth-First Search algorithm.

OTTER (Organized Techniques for Theorem-proving and Effective Research) is a resolution theorem prover, for classical propositional and first-order logics, allowing the use of resolution inference rules and strategies. Implemented in C, input and output to OTTER is via files that set and clear flags relating to these inference rules and strategies, enable the input of the list of clauses (or formulae) and identify a Set of Support (SoS). In OTTER a clause $a \vee \neg b$ is written as $a \mid -b$. and the set of clauses that are not in the SoS are known as *usable*.

We use OTTER rather than any other theorem prover for classical logics as it is efficient, well known, widely used and resolution based. Using other theorem provers for classical logics may also be possible, however a resolution based approach is preferable as it is then easier to see the relationship between step resolution inferences in the temporal system and resolution inferences in the translated system.

3.1. Step Resolution

As step resolution is similar to classical resolution a careful encoding of the SNF rules allows most of the inferences to be carried out. Step resolution is carried out between pairs of initial or global $\Box$-rules so the translation into classical propositional logic we require is limited to these $\Box$-rules. The translation must preserve the left hand side of rules while allowing no resolution between them, but must rewrite the right hand side so that resolution may be performed only between pairs of initial $\Box$-rules or pairs of global $\Box$-rules. The translation we propose prefixes the propositions on the left hand side with *slast_neg* or *slast* dependent on whether they are preceded by a negation or not. For example, if a global $\Box$-rule had x on its left hand side this would be translated as *slast_x* where as $\neg y$ on the left hand side of such a rule would be translated as *slast_neg_y*. The right hand side of initial $\Box$-rules are translated simply as a disjunction of literals whereas the propositions on the right hand side of global $\Box$-rules are prefixed by -*s* or *s* if they were preceded by a negation or not. The translation is given in more detail below.

$$\begin{array}{rclcl}
\mathbf{start} & \Rightarrow & \bigvee_{j=1}^{n} l_j & \longrightarrow & R_1(l_1) \mid \ldots \mid R_1(l_n). \\
\bigcirc\mathbf{true} & \Rightarrow & \bigvee_{j=1}^{n} l_j & \longrightarrow & R_2(l_1) \mid \ldots \mid R_2(l_n). \\
\bigcirc \bigwedge_{i=1}^{m} k_i & \Rightarrow & \bigvee_{j=1}^{n} l_j & \longrightarrow & L(k_1) \mid \ldots \mid L(k_m) \mid R_2(l_1) \mid \\
 & & & & \ldots \mid R_2(l_n).
\end{array}$$

Where the literals on the left hand side of rules[2] are rewritten as propositions as follows

$$\begin{array}{rcl}
L(l) & \longrightarrow & slast_l \\
L(\neg l) & \longrightarrow & slast_neg_l
\end{array}$$

and literals on the right hand side of initial and global $\Box$-rules are rewritten respectively as follows.

$$\begin{array}{rclcrcl}
R_1(l) & \longrightarrow & l & \qquad & R_2(l) & \longrightarrow & s_l \\
R_1(\neg l) & \longrightarrow & -l & \qquad & R_2(\neg l) & \longrightarrow & -s_l
\end{array}$$

Resolution can now take place between pairs of initial $\Box$-rules, for example

$$\begin{array}{rcl}
\mathbf{start} & \Rightarrow & a \vee \neg b \\
\mathbf{start} & \Rightarrow & c \vee b \\
\hline
\mathbf{start} & \Rightarrow & a \vee c
\end{array}
\qquad \text{becomes} \qquad
\begin{array}{c|c}
a & -b. \\
c & b. \\
\hline
a & c.
\end{array}$$

or pairs of global rules, for example

$$\begin{array}{rcl}
\bigcirc x & \Rightarrow & a \vee \neg b \\
\bigcirc \neg y & \Rightarrow & c \vee b \\
\hline
\bigcirc (x \wedge \neg y) & \Rightarrow & a \vee c
\end{array}$$

[2] Initially we used $-L(k_i)$ here to mimic the translation of implication into disjunction but as all literals on the left hand sides of rules have the same sign in OTTER syntax, so they don't get resolved together, the "−" sign was removed.

becomes

$$\begin{array}{r|c|l} slast\_x & s\_a & -s\_b. \\ slast\_neg\_y & s\_c & s\_b. \\ \hline slast\_x \mid slast\_neg\_y & s\_a & s\_c. \end{array}$$

Subsumption should take place as required. For example we might have the rule $\bigcirc(a \wedge b) \Rightarrow p \vee q$ which is subsumed by $\bigcirc a \Rightarrow p$. In OTTER format the clause will be $slast\_a$ $|slast\_b \mid p \mid q$ which is subsumed by $slast\_a \mid p$. Some simplification takes place, for example rules of the form $slast\_x \mid s\_a \mid -s\_a$. representing $\bigcirc x \Rightarrow a \vee \neg a$ (which is simplified to $\bigcirc x \Rightarrow \mathbf{true}$), will be removed. However OTTER will not be able to detect and remove rules of the form $slast\_x \mid slast\_neg\_x \mid s\_a$. i.e. $\bigcirc(x \wedge \neg x) \Rightarrow a$ (which is simplified to $\bigcirc \mathbf{false} \Rightarrow a$) so this must be carried out separately. Termination can be detected in OTTER. Pairs of resolvents e.g. **start** $\Rightarrow a$ and **start** $\Rightarrow \neg a$ which may be resolved to give **start** $\Rightarrow$ **false**, will now be the clauses a and $-a$ in OTTER which when resolved give us the empty clause. Similarly $\bigcirc \mathbf{true} \Rightarrow a$ and $\bigcirc \mathbf{true} \Rightarrow \neg a$ will now be clauses s_a and $-s\_a$ in OTTER again producing the empty clause.

The rule that translates rules of the form $\bigcirc P \Rightarrow \mathbf{false}$ into extra global constraints cannot be replicated using OTTER. Instead when no further resolution step can be carried out we search for rules of this form and translate them into the required format. Then these new rules are input as the set of support.

The use of OTTER for step resolution can be incorporated into the main processing algorithm as follows.

3.1.1. *The Full Algorithm using Otter for Step Resolution (OTRES)*
Set the lists *usable* and *sos* to empty.

1. Input formula, translate it to SNF and store the resulting initial and global $\square$-rules in *sos*.

2. Output lists *usable* and *sos* in OTTER syntax.

3. Generate input for and run OTTER with the two lists of rules *usable* and *sos*.

4. Parse OTTER output searching for

   a) a proof by detecting the text `---->` `UNIT CONFLICT` ; the procedure terminates having detected a contradiction; or

   b) the new list of usable rules.

5. Set *usable* to be the rules output from OTTER, remove any rules that can be simplified to $\bigcirc$**false** $\Rightarrow X$ and set *sos* to be empty.

6. a) If *usable* contains rules of the form $\bigcirc P \Rightarrow$ **false** delete from *usable*, add the rules **start** $\Rightarrow \neg P$ and $\bigcirc$**true** $\Rightarrow \neg P$ to *sos* and continue processing at part 2;

   b) otherwise perform temporal resolution on *usable*.

7. a) If no new resolvents have been returned from temporal resolution then we terminate having found the formula satisfiable;

   b) otherwise translate the new resolvents into SNF, store in *sos* and continue processing at part 2.

3.2. Using Otter for Breadth-First Search

Here we use OTTER as a subroutine that performs a proof for each node constructed in the Breadth-First Search Algorithm. Given that we have constructed the node N_i the main part of the Breadth-First Search Algorithm requires us to find a set of SNF_m rules $\bigcirc A_j \Rightarrow B_j$ such that $\vdash B_j \Rightarrow N_i$ and $\vdash A_j \Rightarrow N_0$. To achieve the former we put the set of global $\Box$-rules as the set of *usable* rules in OTTER and resolve with the rule $\bigcirc$**true** $\Rightarrow \neg N_i$. We must then search for rules of the form $\bigcirc C_j \Rightarrow$ **false**. To achieve the latter we can simply combine each C_j with each disjunct from the top node (N_0). The next node, N_{i+1} is the disjunction of these formulae.

3.2.1. *The Breadth-First Search Algorithm using Otter.*

Assume we wish to resolve with $\Diamond\neg l$, i.e. we want to find a loop in $\neg l$.

1. Given a set of global $\Box$-rules R add the rule $\bigcirc$**true** $\Rightarrow l$ to the set of support and resolve. If we obtain a contradiction, (as R contains the rule $\bigcirc$**true** $\Rightarrow \neg l$) then return the loop **true**. Otherwise look for rules of the form $\bigcirc X_k \Rightarrow$ **false** for $k = 0$ to b and let

$$H_0 \Leftrightarrow \bigvee_{k=0}^{b} X_k$$

   If no such rules exist then no loop can be found.

2. Given H_i to build H_{i+1} add the rule $\bigcirc$**true** $\Rightarrow \neg H_i$ to the set of support[3] and resolve with R. If we obtain a contradiction go to step

[3] Actually as $\neg H_i$ is in CNF there will be one rule for each conjunct of $\neg H_i$.

3(a). Otherwise search for rules of the form $\bigcirc C_j \Rightarrow$ **false** and let

$$H_{i+1} = \bigvee_{j=0}^{m} \left(\bigvee_{k=0}^{b} (C_j \wedge X_k) \right).$$

3. Termination occurs when either

   a) a contradiction has been detected, i.e. return the loop **true**;

   b) $H_i \Leftrightarrow H_{i+1}$ return the DNF formula H_i;

   c) there are no rules of the form $\bigcirc C_j \Rightarrow$ **false** in the set after performing resolution i.e. no loops have been found.

Note that in step 2 H_{i+1} is not in its simplest form but when we negate and add $\bigcirc$**true** $\Rightarrow \neg H_{i+1}$ to the set of support for resolution with OTTER will remove some of the formulae that could be simplified at this stage by subsumption and simplification.

4. Correctness

To establish soundness, completeness and termination of the temporal resolution method using OTTER for both step and temporal resolution, (OTRES), we relate each inference made in OTRES to one in Fisher's algorithm. We assume the correctness of OTTER itself.

THEOREM 1. *A PTL formula φ is unsatisfiable if and only if it has a refutation by OTRES.*

Proof. (Outline) Fisher's method has been shown correct in (Peim, 1994) and proofs are carried out in relation to this. We must show that OTTER performs every inference from Fisher's method (and no more). It can be shown, using the translation suggested, every inference from Fisher's method is carried out by OTTER and that every inference carried out by OTTER is required by Fisher's method. The correctness of the OTTER version of Breadth-First Search is established from the correctness results for the Breadth-First Search algorithm (Dixon, 1995) and that OTTER implements this algorithm directly. Finally, we must show that the use of the set of support in the main algorithm does not restrict the system and cause incompleteness. Initially, all the rules are placed in the set of support as we try to derive all possible resolvents (but allowing subsumption, simplification etc). Subsequently only new rules, either the resolvents from temporal resolution or the translation of rules of the form $\bigcirc P \Rightarrow$ **false**, are placed into the set of support.

This will not restrict resolution as all the rules in the set of usable rules have had as much step resolution as possible performed on them. As the newly derived rules are part of the set of support resolution can take place between one of them and any other rule so all resolution inferences can be made.

THEOREM 2. *For any PTL formula φ the OTRES algorithm performed on φ terminates.*

Proof. (Outline) Termination of Fisher's algorithm is shown in (Peim, 1994) so similarly the top level algorithm will terminate. We must ensure that each of the temporal and step resolution components still terminates. For temporal resolution the Breadth-First Search algorithm has been shown to terminate in (Dixon, 1995) so as OTTER is used to implement each step of this algorithm this part of the method still terminates. We must check that the cycle performing step resolution terminates. During a cycle of step resolution (steps 2–6 of the algorithm in §3.1) either a contradiction is detected (step 4(a)) and termination is ensured or we keep looking for rules of the form $\bigcirc P \Rightarrow \mathbf{false}$ and running OTTER again. If no rules of the form $\bigcirc P \Rightarrow \mathbf{false}$ are found again the step resolution cycle terminates. We must ensure that we cannot keep generating the *same* rule $\bigcirc P \Rightarrow \mathbf{false}$, deleting it and adding $\bigcirc\mathbf{true} \Rightarrow \neg P$ and $\mathbf{start} \Rightarrow \neg P$ to the set of support. In this situation if $\bigcirc P \Rightarrow \mathbf{false}$ is *again* detected, by deleting it and *again* adding $\bigcirc\mathbf{true} \Rightarrow \neg P$ and $\mathbf{start} \Rightarrow \neg P$ to the set of support we know that either these rules or rules that subsume them will be in the set of usable rules. As we set the flag that processes clauses on input, `process_input`, these clauses will be deleted and the cycle will terminate.

5. A Prototype OTRES System

Here we give a summary of the results obtained from testing the OTRES system on examples and some comments on its operation.

5.1. Summary of Results

A prototype implementation of Fisher's temporal resolution theorem prover was constructed using SICStus Prolog (Carlsson and Widen, 1991). This proved all but 6 of the 49 valid temporal formulae given in (Manna and Pnueli, 1981) on a SPARCstation 1. The system ran out of memory for the 6 unproved examples. The OTRES system not only proved the 6 remaining examples but also proved much larger problems for example, the mutual exclusion property for Peterson's Algorithm (Peterson, 1981; Pnueli, 1984) which was not previously possible.

We note however that for some of the smaller examples in (Manna and Pnueli, 1981) the timings were greater using OTRES that for the Prolog version. This is due to the overhead required for the interface between the controlling system and OTTER. For this reason we feel that timings are not a good comparison of the systems. Before such results can be usefully compared an improved interface between the controlling system and OTTER is required. However, in Appendix A we give a breakdown of the number of inputs to OTTER required and the number of rules generated for the 6 examples mentioned above.

5.2. Comments on Operation

First, the termination of the step resolution cycle mentioned in §4 is illustrated with an example. Then we consider how the repetition of some inferences may be made using OTTER to implement Breadth-First Search.

Example: Given the rule-set

$$\begin{array}{llll} 1 & \bigcirc a \Rightarrow \neg z & 3 & \bigcirc b \Rightarrow a \\ 2 & \bigcirc b \Rightarrow z & 4 & \bigcirc a \Rightarrow b \end{array}$$

this will be used as the SoS for input to OTTER. Output from OTTER (translated into the usual syntax and annotated with numbers) will be rules 1–4 plus the following new rule.

$$5 \quad \bigcirc a \wedge b \Rightarrow \mathbf{false} \quad [1,2]$$

Rules 1–4 will be input as the usable rules. Rule 5 will be deleted and rewritten as the following two rules which will be re-input to OTTER as the SoS.

$$\begin{array}{llll} 6 & \bigcirc\mathbf{true} \Rightarrow \neg a \vee \neg b & [5] \\ 7 & \mathbf{start} \Rightarrow \neg a \vee \neg b & [5] \end{array}$$

After the second input to OTTER we obtain rules 1–4, 6, 7 as output plus

$$\begin{array}{lll} 8 & \bigcirc a \Rightarrow \neg a & [6,4] \\ 9 & \bigcirc b \Rightarrow \neg b & [6,3] \\ 10 & \bigcirc a \wedge b \Rightarrow \mathbf{false} & [8,3] \end{array}$$

and have derived $\bigcirc a \wedge b \Rightarrow \mathbf{false}$ for the second time. However, on the next input to OTTER rules 1–4 and 6–9 will be the usable rules and the SoS will have rules that are rewrites of rule 10 (and are identical to rules 6 and 7).

$$
\begin{array}{llcll}
11 & \bigcirc\,\mathbf{true} & \Rightarrow & \neg a \vee \neg b & [10] \\
12 & \mathbf{start} & \Rightarrow & \neg a \vee \neg b & [10]
\end{array}
$$

When the input to OTTER is processed rules 11 and 12 are subsumed by rules 6 and 7 making the SoS empty, no resolution can take place and the rules output are 1–4, 6–9. No further inputs to OTTER are attempted as the rules output do not contain any rules of the form $\bigcirc P \Rightarrow \mathbf{false}$.

5.2.1. *Repetition of work for loop detection.*
Using OTTER for loop detection we may deduce information that satisfies the conditions for the Breadth-First Search algorithm but even if it forms part of a loop resolvents from it will not provide any new information.

6. Related Work

Resolution systems based on translations from modal logics into first-order classical logics are given in (Auffray and Enjalbert, 1989; Chan, 1987; Nonnengart, 1996; Ohlbach, 1988). Auffray and Enjalbert (Auffray and Enjalbert, 1989) translate modal formulae into a first-order classical logic known as *Path Logic* and different properties of the accessibility relation are captured by sets of equations. The system described by Ohlbach (Ohlbach, 1988) translates modal formulae into a first-order classical logic known as *P-logic.* Here different properties of the modal accessibility relation are captured by special unification algorithms. The system described in (Chan, 1987) is similar but deals with only the propositional S4 system. Nonnengart (Nonnengart, 1996) uses a semi-functional translation into first-order logic where modal formulae are translated into first-order formulae and a background theory is added dependent on which logic is being considered. Although both modal and temporal logics are considered, the approach cannot deal with temporal logics using the next operator.

A related approach to the above is given in (Frisch and Scherl, 1990). Here modal formulae are translated into a constraint logic. Deduction is again resolution based but also involves special processing mechanisms to deal with the constraints that represent relations between worlds.

All of the above are *global* approaches, i.e. involve the translation into a first-order logic and then deduction within this framework. We merely use OTTER as an engine to perform step resolution and then to apply the Breadth-First Search algorithm. It can be thought of as a procedure call from the main program.

7. Conclusions

We have suggested how to implement Fisher's temporal resolution method using OTTER to perform both step and temporal resolution. The implementation allows us to complete larger examples than previously. As step resolution is almost the same as classical resolution, the use of OTTER on the translated rules performs almost all of the required inferences. The implementation of the temporal resolution step however is different. We use OTTER to perform the Breadth-First Search *algorithm* rather than trying to get OTTER to generate a proof directly.

We avoid problems of undecidability that occur in translating temporal logics into into first-order classical logics by translating from PTL into classical propositional logic. We can do this as the translation to the normal form has removed most of the temporal operators and during step resolution we only apply inference rules to the initial and global $\square$-rules so we only require the translation of these rules. Further, during temporal resolution the application of Breadth-First Search algorithm is only to the global $\square$-rules. Finally, although we have developed a method to remove rules prior to the search for loops (Dixon, 1997) little work has been done to guide the search for a temporal proof. Using OTTER may help us experiment with classical strategies in this temporal setting.

We implement the Breadth-First Search algorithm rather than the others described in (Dixon, 1996) as the structure we build is simply a series of formulae in DNF that satisfy certain properties. The other two methods suggested construct much more complex structures so it is not so obvious that OTTER could be used to implement them at all.

Although the system has allowed us to attempt more challenging examples we note the following limitations. Firstly the overhead of input and output to OTTER does increase the overall timings on smaller examples. Secondly, parts of the step resolution method cannot be carried out using OTTER and it is rather clumsy having to move in and out of OTTER repeatedly. Further, as illustrated above, some sets of rules may require an extra input to OTTER before we can detect the end of the step resolution part of the algorithm. Ideally we would prefer a translation (and the clever use of OTTER) that would perform at least *all* of the possible step resolution inferences in one attempt. In particular that could detect and rewrite rules of the form $\bigcirc P \Rightarrow$ **false** as $\bigcirc$**true** $\Rightarrow \neg P$ and **start** $\Rightarrow \neg P$. Finally during temporal resolution there may be some repetition of the inferences carried out during step resolution producing formulae that satisfy the criteria for Breadth-First Search but whose resolvents provide no new information.

In conclusion this approach does provide a way of utilising OTTER a well developed, efficient theorem prover for classical logics to prove temporal theorems.

Acknowledgements

Thank to Martin Peim for the interface to OTTER and suggesting that OTTER could be used for loop search. Thanks also to Howard Barringer and Graham Gough and to Michael Fisher for his comments on an earlier draft of this paper.

References

Y. Auffray and P. Enjalbert. Modal Theorem Proving: An Equational Viewpoint. In *Proceedings of the International Joint Conference on Artificial Intelligence (IJCAI)*, pages 441–445, Detroit, USA, 1989. Morgan Kaufmann.

H. Barringer. Using Temporal Logic in the Compositional Specification of Concurrent Systems. In A. P. Galton, editor, *Temporal Logics and their Applications*, chapter 2, pages 53–90. Academic Press Inc. Limited, London, December 1987.

M.-C. Chan. The Recursive Resolution Method for Modal Logic. *New Generation Computing*, 5:155–183, 1987.

M. Carlsson and J. Widen. *SICStus Prolog User's Manual.* Swedish Institute of Computer Science, Kista, Sweden, September 1991.

C. Dixon. *Strategies for Temporal Resolution.* PhD thesis, Department of Computer Science, University of Manchester, 1995.

C. Dixon. Search Strategies for Resolution in Temporal Logics. In M. A. McRobbie and J. K. Slaney, editors, *Proceedings of the Thirteenth International Conference on Automated Deduction (CADE)*, volume 1104 of *Lecture Notes in Artificial Intelligence*, pages 672–687, New Brunswick, New Jersey, July/August 1996. Springer-Verlag.

C. Dixon. Temporal Resolution: Removing Irrelevant Information. In *Proceedings of TIME-97 the Fourth International Workshop on Temporal Representation and Reasoning*, Daytona Beach, Florida, May 1997.

M. Fisher. A Resolution Method for Temporal Logic. In *Proceedings of the Twelfth International Joint Conference on Artificial Intelligence (IJCAI)*, Sydney, Australia, August 1991. Morgan Kaufman.

M. Fisher. A Normal Form for First-Order Temporal Formulae. In *Proceedings of Eleventh International Conference on Automated Deduction (CADE)*, volume 607 of *Lecture Notes in Computer Science*, Saratoga Springs, New York, June 1992. Springer-Verlag.

A. Frisch and R. Scherl. A constraint logic approach to modal deduction. *Lecture Notes in Computer Science*, 478:234–250, 1990.

B. T. Hailpern. *Verifying Concurrent Processes Using Temporal Logic*, volume 129 of *Lecture Notes in Computer Science*. Springer-Verlag, 1982.

W. W. McCune. *OTTER 2.0 Users Guide*. Argonne National Laboratory, 9700 South Cass Avenue, Argonne, Illinois 60439-4801, March 1990. ANL-90/9.

Z. Manna and A. Pnueli. Verification of Concurrent Programs: The Temporal Framework. In Robert S. Boyer and J. Strother Moore, editors, *The Correctness Problem in Computer Science*, pages 215–273. Academic Press, London, 1981.

Z. Manna and A. Pnueli. *The Temporal Logic of Reactive and Concurrent Systems: Specification*. Springer-Verlag, New York, 1992.

A. Nonnengart. Resolution-Based Calculi for Modal and Temporal Logics. In M. A. McRobbie and J. K. Slaney, editors, *Proceedings of the Thirteenth International Conference on Automated Deduction (CADE)*, volume 1104 of *Lecture Notes in Artificial Intelligence*, pages 598–612, New Brunswick, New Jersey, July/August 1996. Springer-Verlag.

H.-J. Ohlbach. A Resolution Calculus for Modal Logics. *Lecture Notes in Computer Science*, 310:500–516, May 1988.

S. Owicki and L. Lamport. Proving Liveness Properties of Concurrent Programs. *ACM Transactions on Programming Languages and Systems*, 4(3):455–495, July 1982.

M. Peim. Propositional Temporal Resolution Over Labelled Transition Systems. Unpublished Technical Note, Department of Computer Science, University of Manchester, 1994.

G. L. Peterson. Myths about the Mutual Exclusion Problem. *Information Processing Letters*, 12(3):115–116, 1981.

A. Pnueli. In Transition From Global to Modular Temporal Reasoning about Programs. In Krysztof Apt, editor, *Logics and Models of Concurrent Systems*, pages 123–144, La Colle-sur-Loup, France, October 1984. NATO, Springer-Verlag.

Appendix

A. Analysis of Some Examples

The examples (to be shown unsatisfiable) we consider are as follows.

1 $((\bigcirc w_1 \mathcal{U} (\bigcirc w_2)) \wedge \neg \bigcirc (w_1 \mathcal{U} w_2)) \vee$
$(\bigcirc (w_1 \mathcal{U} w_2) \wedge \neg(\bigcirc w_1 \mathcal{U} \bigcirc w_2))$

2 $(\bigcirc((\neg w_1 \vee w_2) \wedge (w_1 \vee \neg w_2)) \wedge \neg((\neg \bigcirc w_1 \vee \bigcirc w_2)$
$\wedge(\neg \bigcirc w_2 \vee \bigcirc w_1)))\vee$
$(((\neg \bigcirc w_1 \vee \bigcirc w_2) \wedge (\neg \bigcirc w_2 \vee \bigcirc w_1)) \wedge \neg \bigcirc((\neg w_1 \vee w_2)$
$\wedge(w_1 \vee \neg w_2)))$

3 $(((w_1 \wedge w_2) \mathcal{U} w_3) \wedge \neg((w_1 \mathcal{U} w_3) \wedge (w_2 \mathcal{U} w_3)))\vee$
$(((w_1 \mathcal{U} w_3) \wedge (w_2 \mathcal{U} w_3)) \wedge \neg((w_1 \wedge w_2) \mathcal{U} w_3))$

4 $((w_1 \mathcal{U} (w_2 \vee w_3)) \wedge \neg((w_1 \mathcal{U} w_2) \vee (w_1 \mathcal{U} w_3)))\vee$
$((w_1 \mathcal{U} w_2) \vee (w_1 \mathcal{U} w_3)) \wedge \neg(w_1 \mathcal{U} (w_2 \vee w_3))$

5 $(\Diamond w_1 \wedge \Diamond w_2) \wedge \neg(\Diamond(w_1 \wedge \Diamond w_2) \vee \Diamond(w_2 \wedge \Diamond w_1))$

6 $((w_1 \mathcal{U} w_2) \wedge \neg(w_2 \vee (w_1 \wedge \bigcirc(w_1 \mathcal{U} w_2))))\vee$
$((w_2 \vee (w_1 \wedge \bigcirc(w_1 \mathcal{U} w_2))) \wedge \neg(w_1 \mathcal{U} w_2))$

Table I shows the number of inputs to OTTER required for each round of step and temporal resolution and the amount of rules generated after each cycle of step resolution for the above examples. For example problem 6 required one input to OTTER during step resolution (generating 51 rules) and three inputs to OTTER for temporal resolution. After a further two inputs to OTTER (generating 324 rules) a contradiction was detected. The figures in bold represent how far in the proof the non-OTTER system reached for example, problem 6 failed during the second round of step resolution.

Table I. Use of OTTER in Some Examples

	Number of inputs to OTTER						Number of rules generated					
	1	2	3	4	5	6	1	2	3	4	5	6
Step 1	**2**	**2**	1	1	1	1	192	93	60	75	22	51
Temp. 1	3	-	3	3	2	3						
Step 2	3	-	**1**	**3**	3	**2**	489	-	138	314	41	324
Temp. 2	4	-	-	3	4	-						
Step 3	2	-	-	3	**1**	-	834	-	-	482	55	-
Temp. 3	-	-	-	3	-	-						
Step 4	-	-	-	1	-	-	-	-	-	508	-	-

Guiding Clausal Temporal Resolution

Michael Fisher (M.Fisher@doc.mmu.ac.uk)
Department of Computing, Manchester Metropolitan University, Manchester M1 5GD, United Kingdom

Clare Dixon (C.Dixon@doc.mmu.ac.uk)
Department of Computing, Manchester Metropolitan University, Manchester M1 5GD, United Kingdom

Abstract. The effective mechanisation of temporal logic is vital to the application of temporal reasoning in many fields, for example the verification of reactive systems, the implementation of temporal query languages, and temporal logic programming. Consequently, a variety of proof methods have been developed, implemented and applied. While *clausal* temporal resolution has been successfully employed for a range of problems, a number of improvements are still required. One particular drawback for certain applications is the restriction that temporal resolution operations must occur only after *all* relevant non-temporal resolution steps have been carried out. It is this restriction that we consider in this paper, where we introduce, justify and apply a new temporal resolution rule. This rule, which may be seen as a generalisation of the existing temporal resolution rule, can be applied at *any* time during the refutation, thus providing the possibility for much greater interleaving between temporal and non-temporal operations. In addition, the use of this temporal resolution rule can provide information that is useful in guiding any subsequent search.

Keywords: temporal logic, theorem-proving, clausal resolution, linear-time

1. Introduction

The effective mechanisation of temporal logic is vital to the application of temporal reasoning in many fields, for example the verification of reactive systems (Manna and Pnueli, 1992), the implementation of temporal query languages (Chomicki and Niwinski, 1995), and temporal logic programming (Abadi and Manna, 1989). Consequently, a range of proof methods have been developed, implemented and applied. In addition to well-known tableau (Wolper, 1985) and automata-theoretic (Sistla et al., 1987) methods, there has been a resurgence in interest in resolution-based methods (Abadi and Manna, 1990; Cavali and Fariñas del Cerro, 1984; Venkatesh, 1986; Fisher, 1991).

Although *clausal* temporal resolution (Fisher, 1991) has been developed and implemented (Dixon, 1995; Dixon, 1996) and has been used successfully in a variety of applications, experiments have revealed a practical problem with the resolution rule originally formulated in (Fisher, 1991). This is related to the restriction that temporal resolution operations generally occur only after *all* relevant non-temporal

H. Barringer et al. (eds.), Advances in Temporal Logic, 167–184.

resolution steps have been carried out. In effect, the formulae must be maximally combined before any search for candidates for temporal resolution can be made. While this ensures that the phases of the method remain separate and simplifies the temporal resolution rule, it also means that examples where a large amount of non-temporal resolution can *potentially* occur may be very expensive.

As the non-temporal resolution we use (termed *step resolution*) is effectively a form of classical resolution, we would like to utilise standard strategies, such as Set of Support (Wos et al., 1965), during this phase. In particular, we would like to use information from the temporal resolution step to derive the set of support for step resolution. However, the problem is that the current temporal resolution rule requires that all the relevant step resolution rules have been applied so that the formulae are in a suitable form. Thus, the current implementation applies step resolution between *all* clauses (unless an immediate contradiction is generated), before temporal resolution can be applied. Thus, in applications where a large amount of step resolution *can* potentially occur, yet a contradiction may only be derived after temporal resolution has been carried out, this rigidly separated approach is expensive. Not only is this slow, but the majority of the resolvents derived are likely to be irrelevant to further (temporal) resolution. This can be seen particularly in examples (such as representations of protocols or mutual exclusion algorithms) where large finite-state machines are effectively encoded within the temporal formulae (e.g. Peterson's mutual exclusion algorithm (Peterson, 1981) encoded in PTL in (Pnueli, 1984)).

Since the introduction of this resolution method, research into its implementation has centered around the development of algorithms for generating candidates for temporal resolution operations. Indeed, the search for these candidates is the most expensive part of the whole process. However, as such a search is certain to be required for any refutation that uses temporal resolution, our concern here is to avoid *unnecessary* step resolution operations wherever possible. Thus, in this paper, we consider a revised temporal resolution rule which enables the temporal and step resolution phases to be interleaved to a greater extent. This not only allows us to generate temporal resolvents earlier (leaving full temporal resolution checks until later), but also allows the resolvents derived in the temporal resolution operation to be used to guide further step resolution (for example, by using the Set of Support strategy). Thus, while the new rule represents a variation on the original version, its use also has implications concerning the order in which temporal and step resolution operations occur.

The structure of the paper is as follows. In §2 we define the form of temporal logic considered, namely Propositional Temporal Logic

(PTL) (Gabbay et al., 1980). For simplicity, we consider this future-time temporal logic rather than that defined in (Fisher, 1991), which incorporated past-time operators. However, the refined resolution rule can be transferred to the past-time framework if necessary. In §3, we review the original temporal resolution method and, in particular, the temporal resolution rule, while in §4, we introduce, justify and apply the new temporal resolution rule. The implications of this refined rule for the method in general, and the loop search component in particular, are considered in §5 . Finally, in §6, we provide conclusions and discuss future work in this area.

2. Temporal Logic

In this section, we present the syntax and semantics of the temporal logic we consider, namely PTL (Gabbay et al., 1980). This generalises classical propositional logic, and thus it contains the standard propositional connectives $\neg$ (not) and $\vee$ (or); the remaining connectives are assumed to be introduced as abbreviations in the usual way. We use temporal connectives that can refer to the *future*, namely $\bigcirc$ (for 'next') and $\mathcal{U}$ (for 'until'). We explain these connectives in detail below. The temporal connectives are interpreted over a *flow of time* that is linear, discrete, bounded in the past, and infinite in the future. An obvious choice for such a flow of time is $(\mathbb{N}, <)$, i.e., the natural numbers ordered by the usual 'less than' relation.

2.1. Syntax

We now formally present the syntax of PTL.

THEOREM 1. *The language of PTL contains the following symbols:*

1. *A set* $\Phi = \{p, q, r, \ldots\}$ *of* primitive propositions*;*
2. **true** *and* **false***;*
3. *The binary propositional connective* $\vee$ *(or), and unary propositional connective* $\neg$ *(not);*
4. *The nullary temporal connective* **start***, the unary temporal connective* $\bigcirc$ *(next) and binary temporal connective* $\mathcal{U}$ *(until).*

THEOREM 2. *The set* WFF *of well-formed formulae of PTL is defined by the following rules:*

$\langle M,u\rangle \models \mathbf{true}$		
$\langle M,u\rangle \not\models \mathbf{false}$		
$\langle M,u\rangle \models \mathbf{start}$	iff	$u=0$
$\langle M,u\rangle \models q$	iff	$\pi_p(u,q)=T$ (where $q\in\Phi$)
$\langle M,u\rangle \models \neg\varphi$	iff	$\langle M,u\rangle \not\models \varphi$
$\langle M,u\rangle \models \varphi\vee\psi$	iff	$\langle M,u\rangle \models \varphi$ or $\langle M,u\rangle \models \psi$
$\langle M,u\rangle \models \bigcirc\varphi$	iff	$\langle M,u+1\rangle \models \varphi$
$\langle M,u\rangle \models \varphi\,\mathcal{U}\,\psi$	iff	$\exists v\in\mathbb{N}$ such that $(u\leq v)$ and $\langle M,v\rangle\models\psi$, and $\forall w\in\mathbb{N}$, if $(u\leq w<v)$ then $\langle M,w\rangle\models\varphi$

Figure 1. Semantics of PTL

1. *(Primitive propositions are formulae): if $p\in\Phi$ then $p\in$* WFF;
2. *(Nullary connectives):* **true, false, start** $\in$ WFF;
3. *(Unary connectives): if $\varphi\in$ WFF then $\neg\varphi\in$ WFF, $\bigcirc\varphi\in$ WFF, and $(\varphi)\in$ WFF;*
4. *(Binary connectives): if $\varphi,\psi\in$ WFF, then $\varphi\vee\psi\in$ WFF, and $\varphi\,\mathcal{U}\,\psi\in$ WFF.*

2.2. SEMANTICS

We define a model, M, for PTL as a structure $\langle\mathcal{D},\pi_p\rangle$ where

- $\mathcal{D}$ is the temporal domain, i.e. the Natural Numbers ($\mathbb{N}$), and
- $\pi_p:\mathcal{D}\times\Phi\rightarrow\{T,F\}$ is a function assigning T or F to each atomic proposition at each moment in time.

As usual, we define the semantics of the language via the satisfaction relation '$\models$'. For PTL, this relation holds between pairs of the form $\langle M,u\rangle$ (where M is a model and $u\in\mathbb{N}$), and well-formed PTL formulae. The rules defining the satisfaction relation are given in Figure 1. Satisfiability and validity in PTL are defined in the usual way.

Other standard temporal connectives are introduced as abbreviations, in terms of $\mathcal{U}$, e.g.,

$$\begin{aligned} \Diamond\varphi &\stackrel{\text{def}}{=} \mathbf{true}\,\mathcal{U}\,\varphi \\ \Box\varphi &\stackrel{\text{def}}{=} \neg\Diamond\neg\varphi \\ \varphi\mathcal{W}\psi &\stackrel{\text{def}}{=} \varphi\mathcal{U}\psi \vee \Box\varphi \end{aligned}$$

We now informally consider the meaning of the temporal connectives. First, consider the two basic connectives: $\bigcirc$ and $\mathcal{U}$. The $\bigcirc$ connective means 'at the next time'. Thus $\bigcirc\varphi$ will be satisfied at some time if φ is satisfied at the *next* moment in time. The $\mathcal{U}$ connective means 'until'. Thus $\varphi\mathcal{U}\psi$ will be satisfied at some time if ψ is satisfied at either the present time or some time in the future, and φ is satisfied at all times until the time that ψ is satisfied. Of the derived connectives, $\Diamond\varphi$ will be satisfied at some time if φ is satisfied either at the present moment or at some future time. The formula $\Box\varphi$ will be satisfied at some time if φ is satisfied at the present and all future times. The binary $\mathcal{W}$ connective means 'unless'. Thus $\varphi\mathcal{W}\psi$ will be satisfied at some time if either φ is satisfied until such time as ψ is satisfied, or else φ is always satisfied. Note that $\mathcal{W}$ is similar to, but weaker than, the $\mathcal{U}$ connective; for this reason it is sometimes called 'weak until'. Finally, a temporal operator that takes no arguments is defined which is true only at the first moment in time: this operator is '**start**'.

3. A Resolution-Based Proof Method for PTL

Before describing the resolution method in detail, we outline the motivation for the approach adopted. First, we recap the problems associated with clausal resolution in non-classical logics. The main problem with extending resolution to temporal logics, such as PTL, is that literals cannot generally be moved across temporal contexts. In particular, if T is a temporal operator, p and $\mathsf{T}\neg p$ cannot generally be resolved. Thus, the only inferences that can be made occur in particular temporal contexts. For example, both p and $\neg p$ can be resolved, as, for certain types of temporal operator, can $\mathsf{T}p$ and $\mathsf{T}\neg p$.

The clausal resolution method introduced in (Fisher, 1991) addresses this problem by utilising a normal form, called Separated Normal Form (SNF), which separates out complex formulae from their contexts through the use of *renaming* (Plaisted and Greenbaum, 1986), and a new temporal resolution rule introduced specifically for formulae in the normal form.

The resolution method consists of the following cycle of steps. To determine whether a formula, $\varphi \in$ WFF, is unsatisfiable

1. Rewrite φ into SNF

2. Repeat

    a) apply *step* resolution until either a contradiction is generated, or no further step resolution can be carried out

    b) rewrite any new resolvents into SNF

    c) apply simplification and subsumption rules

    d) apply the temporal resolution rule

    e) rewrite new resolvents into SNF

    until either **false** is derived or no more rules can be applied.

We briefly review each of these items, starting with the normal form itself.

3.1. SEPARATED NORMAL FORM

This resolution method (Fisher, 1991), depends on formulae being rewritten into a normal form, called Separated Normal Form (SNF) (Fisher, 1997). In this section, we review SNF but do not consider the transformation procedure that takes an arbitrary formula of PTL and rewrites it into SNF (for further details, see (Fisher, 1997)).

A formula in SNF is of the form:

$$\Box \bigwedge_{i=1}^{n} (\varphi_i \Rightarrow \psi_i)$$

where each of the '$\varphi_i \Rightarrow \psi_i$' (called *rules*) is one of the following.

$$\begin{array}{lcll}
\textbf{start} & \Rightarrow & \bigvee_{k=1}^{r} l_k & \text{(an \textit{initial} rule)} \\
\bigwedge_{j=1}^{q} m_j & \Rightarrow & \bigcirc \bigvee_{k=1}^{r} l_k & \text{(an \textit{always} rule)} \\
\bigwedge_{j=1}^{q} m_j & \Rightarrow & \Diamond l & \text{(a \textit{sometime} rule)}
\end{array}$$

where each m_j, l_k or l is a literal.

3.2. Step Resolution

The step resolution rules are simply versions of the classical resolution rule rewritten in two ways. First, the *initial* step resolution rule:

$$\frac{\begin{array}{rcl}\textbf{start} & \Rightarrow & \psi_1 \vee l \\ \textbf{start} & \Rightarrow & \psi_2 \vee \neg l\end{array}}{\begin{array}{rcl}\textbf{start} & \Rightarrow & \psi_1 \vee \psi_2\end{array}}$$

Then the *global* step resolution rule:

$$\frac{\begin{array}{rcl}\varphi_1 & \Rightarrow & \bigcirc(\psi_1 \vee l) \\ \varphi_2 & \Rightarrow & \bigcirc(\psi_2 \vee \neg l)\end{array}}{\begin{array}{rcl}(\varphi_1 \wedge \varphi_2) & \Rightarrow & \bigcirc(\psi_1 \vee \psi_2)\end{array}}$$

3.3. Simplification

The simplification rules used are similar to the classical case, consisting of both simplification and subsumption rules, and so will not be duplicated here. An additional rule is required when a contradiction in a state is produced, i.e.,

$$\frac{\varphi \Rightarrow \bigcirc\textbf{false}}{\begin{array}{rcl}\textbf{start} & \Rightarrow & \neg\varphi \\ \textbf{true} & \Rightarrow & \bigcirc\neg\varphi\end{array}}$$

This shows that, if a particular formula leads to a contradiction, then that formula should not be satisfied either in the initial state or in any subsequent state.

3.4. Temporal Resolution

Rather than describe the temporal resolution rule in detail, we refer the interested reader to (Fisher, 1991). The basic idea is to resolve one sometime rule with a *set* of always rules as follows.

$$\left.\frac{\begin{array}{rcl}\varphi_1 & \Rightarrow & \bigcirc\psi_1 \\ \varphi_2 & \Rightarrow & \bigcirc\psi_2 \\ \vdots & \vdots & \vdots \\ \varphi_n & \Rightarrow & \bigcirc\psi_n \\ \chi & \Rightarrow & \Diamond\neg l\end{array}}{\begin{array}{rcl}\chi & \Rightarrow & (\neg\bigvee_{i=1}^{n}\varphi_i)\,\mathcal{W}\,\neg l\end{array}}\right\} \text{ where } \bigwedge_{i=1}^{n}(\varphi_i \Rightarrow \bigcirc(l \wedge \bigvee_{j=1}^{n}\varphi_j))$$

The side condition ensures that the set of $\varphi_i \Rightarrow \bigcirc\psi_i$ rules together imply

$$\bigvee_{i=1}^{n} \varphi_i \Rightarrow \bigcirc \Box l .$$

Such a set of rules is known as a *loop* in l (see §4.1 for details). The resolvent states that, once χ has occurred, none of the φ_i must occur while the eventuality (i.e. $\Diamond\neg l$) is outstanding. This resolvent must again be translated into SNF.

3.5. Termination

Finally, if either **start** $\Rightarrow$ **false** or **true** $\Rightarrow$ $\bigcirc$**false** are produced, the original formula is unsatisfiable and the resolution process terminates.

3.6. Example

Consider the following example where the original formula to be checked for satisfiability has been transformed into the SNF rules numbered 1 to 8 below.

1.	a	$\Rightarrow$	$\bigcirc l$
2.	b	$\Rightarrow$	$\bigcirc l$
3.	a	$\Rightarrow$	$\bigcirc b$
4.	b	$\Rightarrow$	$\bigcirc a$
5.	x	$\Rightarrow$	$\Diamond\neg l$
6.	**start**	$\Rightarrow$	a
7.	**start**	$\Rightarrow$	x
8.	**start**	$\Rightarrow$	l

Now, no step resolution is possible here but we can apply temporal resolution to 5 together with rules 1 to 4. These first four rules together characterise a loop in l, namely

$$(a \vee b) \Rightarrow \bigcirc \Box l$$

and the temporal resolution operation produces the following resolvent.

$$x \Rightarrow (\neg a \wedge \neg b)\, \mathcal{W} \neg l$$

One of the SNF rules produced by rewriting this resolvent is

9. **start** $\Rightarrow$ $\neg x \vee \neg l \vee \neg a$

from which a contradiction can be generated by initial step resolution with rules 6, 7 and 8.

3.7. Correctness

The soundness and (refutation) completeness of the original temporal resolution method have been established in (Fisher, 1991; Peim, 1994; Dixon, 1995).

4. Refining the Resolution Method

4.1. Motivation

As described earlier, the problem with the resolution method as it was defined in (Fisher, 1991) is that step resolution (essentially classical resolution) must be applied until either a contradiction is generated or no further step resolution operations can take place. In certain examples, this means that the method is particularly slow as it has to carry out excessive step resolution before even considering temporal resolution. For example:

$$\begin{array}{lrcl}
1. & a & \Rightarrow & \bigcirc l \\
2. & a & \Rightarrow & \bigcirc(a \vee \neg c) \\
3. & a & \Rightarrow & \bigcirc c \\
4. & d & \Rightarrow & \bigcirc(\neg c \vee \neg p \vee \neg q) \\
5. & e & \Rightarrow & \bigcirc(\neg c \vee p \vee \neg q) \\
6. & (f \wedge p) & \Rightarrow & \bigcirc(\neg c \vee \neg p \vee q) \\
7. & (g \wedge q) & \Rightarrow & \bigcirc(\neg c \vee p \vee q) \\
8. & \mathbf{start} & \Rightarrow & d \\
9. & \mathbf{start} & \Rightarrow & a \\
10. & d & \Rightarrow & \Diamond \neg l
\end{array}$$

Here, if step resolution is carried out fully, then rules 3 to 7 can potentially generate a large number of new rules. To derive a contradiction, all we need do is resolve 2 and 3 and then apply temporal resolution to the resolvent together with rules 1 and 10. This will quickly lead to a contradiction.

Thus, the main property we desire of a refined resolution rule is to be able to apply it at an *earlier* stage in the refutation. In particular, we need to be able to carry out temporal resolution between a sometime rule and a set of always rules which *partially* characterise a $\square$ formula (also known as a *loop* (Dixon, 1995)).

4.2. An Improved Temporal Resolution Rule

The essential idea behind the new temporal resolution rule is that, rather than insisting that we already have a set of rules that exactly characterise a loop (i.e the $\varphi_i \Rightarrow \bigcirc\psi_i$ in §3.4), we derive a more complex resolvent that allows for the possibility that such a loop does not exist. If such a loop does exist, then the new resolvent turns out to be equivalent to the original form. More importantly, if such a loop is not immediately obvious, the resolvents produced (once transformed into SNF) can be used to guide further step resolution in such a way that a loop will be detected if possible.

This new temporal resolution rule is

$$\begin{array}{ccc}
\varphi_1 & \Rightarrow & \bigcirc\psi_1 \\
\varphi_2 & \Rightarrow & \bigcirc\psi_2 \\
\vdots & \vdots & \vdots \\
\varphi_n & \Rightarrow & \bigcirc\psi_n \\
\chi & \Rightarrow & \Diamond\neg l \\
\hline
\end{array}$$

$$\chi \Rightarrow \left[(\Diamond\neg \bigwedge_{i=1}^{n}(\varphi_i \Rightarrow \bigcirc(l \wedge \bigvee_{j=1}^{n}\varphi_j))) \vee (\neg \bigvee_{i=1}^{n}\varphi_i)\mathcal{W}\neg l \right]$$

Note that there is now no side-condition; this has been incorporated into the more complex resolvent. In fact the $\Diamond\neg\ldots$ subformula is derived directly from the side condition, as discussed in §4.4. Thus, rather than being *sure* that

$$(\bigvee_{i=1}^{n}\varphi_i) \Rightarrow \bigcirc\Box l$$

and deriving resolvents that stop any φ_i occurring while $\Diamond\neg l$ is outstanding, the new resolvent states that *either* we must stop any φ_i from occurring while the eventuality has yet to be satisfied (as in the original resolvent), or it must be the case that, at some point in the future

$$\neg \bigwedge_{i=1}^{n}(\varphi_i \Rightarrow \bigcirc(l \wedge \bigvee_{j=1}^{n}\varphi_j))$$

and so $\bigcirc\Box l$ will not necessarily follow from satisfying one of the φ_i. The intuition behind this resolvent is that it disjoins the resolvent from the original rule together with a formula that will be unsatisfiable if an appropriate loop exists. In this sense, the $\Diamond$ subformula represents a check that the derivation of the $\mathcal{W}$ subformula was a sound thing to do. Importantly, not only can the verification of this check be delayed until later in the refutation, but, if it is true, the verification process will be trivial.

Before considering the correctness and implementation of the new rule, we will examine some simple examples of its use.

4.3. Examples

4.3.1. *Simple Loop Example*

We first consider the simple example given in §3.6. Recall that the initial set of rules is as follows.

1.	a	$\Rightarrow$	$\bigcirc l$
2.	b	$\Rightarrow$	$\bigcirc l$
3.	a	$\Rightarrow$	$\bigcirc b$
4.	b	$\Rightarrow$	$\bigcirc a$
5.	x	$\Rightarrow$	$\Diamond \neg l$
6.	**start**	$\Rightarrow$	a
7.	**start**	$\Rightarrow$	x
8.	**start**	$\Rightarrow$	l

First, let us imagine that we chose the correct combination of literals that characterise the loop in l, i.e. $a \lor b$. In this case, the resolvent we generate is

$$x \Rightarrow \Diamond\neg((a \lor b) \Rightarrow \bigcirc(l \land (a \lor b))) \lor (\neg a \land \neg b)\,\mathcal{W}\,\neg l \qquad (\dagger)$$

When this is rewritten into SNF, some of the rules produced are as follows, where t and s are new proposition symbols.

9.	**start**	$\Rightarrow$	$\neg x \lor t \lor \neg l \lor \neg a$
10.	t	$\Rightarrow$	$\Diamond s$
11.	**start**	$\Rightarrow$	$\neg s \lor a \lor b$
12.	**true**	$\Rightarrow$	$\bigcirc(\neg s \lor a \lor b)$
13.	s	$\Rightarrow$	$\bigcirc(\neg l \lor \neg a)$
14.	s	$\Rightarrow$	$\bigcirc(\neg l \lor \neg b)$

N.B., all of the above are derived from the renaming and expansion of ($\dagger$). Now refutation proceeds as follows where the *temporal check* in step 22 is discussed below.

15.	$(s \land b)$	$\Rightarrow$	$\bigcirc$**false**	[Global SRES: 13, 2, 4]
16.	$(s \land a)$	$\Rightarrow$	$\bigcirc$**false**	[Global SRES: 14, 1, 3]
17.	**start**	$\Rightarrow$	$\neg s \lor \neg b$	[Simplification: 15]
18.	**start**	$\Rightarrow$	$\neg s \lor \neg a$	[Simplification: 16]
19.	**true**	$\Rightarrow$	$\bigcirc(\neg s \lor \neg b)$	[Simplification: 15]
20.	**true**	$\Rightarrow$	$\bigcirc(\neg s \lor \neg a)$	[Simplification: 16]
21.	**true**	$\Rightarrow$	$\bigcirc \neg s$	[Glob. SRES: 12, 19, 20]
22.	**start**	$\Rightarrow$	$\neg t \lor s$	[Temp. Check: 21, 10]
23.	**start**	$\Rightarrow$	**false**	[Initial SRES: 6, 7, 8, 9, 18, 22]

Note that the *temporal check* used in deriving rule 30 is actually a form of temporal resolution, but one where only a trivial search is required. In fact, in this, we only consider matching rules of the form $\textbf{true} \Rightarrow \bigcirc\neg\varphi$ with $\chi \Rightarrow \Diamond\varphi$. Although the new resolution rule introduces a new eventuality to be resolved, this only represents a check to establish that the looping condition is satisfied. If it is, then the check will succeed; if an appropriate loop is not present then $\textbf{true} \Rightarrow \bigcirc\neg\varphi$ will not be generated and the check will fail.

Regarding the above example, we also note that

1. if we replace $a \Rightarrow \bigcirc b$ (i.e. rule 3) by the pair of rules $a \Rightarrow \bigcirc(b \vee \neg c)$ and $a \Rightarrow \bigcirc c$ the refutation using the new resolution operation still succeeds, and,
2. if we completely remove $a \Rightarrow \bigcirc b$, then no refutation can be found.

4.3.2. *Avoiding Excess Step Resolution*
Consider the example given in §4.1:

1.	a	$\Rightarrow$	$\bigcirc l$
2.	a	$\Rightarrow$	$\bigcirc(a \vee \neg c)$
3.	a	$\Rightarrow$	$\bigcirc c$
4.	d	$\Rightarrow$	$\bigcirc(\neg c \vee \neg p \vee \neg q)$
5.	e	$\Rightarrow$	$\bigcirc(\neg c \vee p \vee \neg q)$
6.	$(f \wedge p)$	$\Rightarrow$	$\bigcirc(\neg c \vee \neg p \vee q)$
7.	$(g \wedge q)$	$\Rightarrow$	$\bigcirc(\neg c \vee p \vee q)$
8.	**start**	$\Rightarrow$	d
9.	**start**	$\Rightarrow$	a
10.	d	$\Rightarrow$	$\Diamond\neg l$
11.	**start**	$\Rightarrow$	l

If we apply the temporal resolution rule to 10, taking $\{\varphi_i\}$ simply as $\{a\}$, we derive the resolvent $d \Rightarrow \Diamond\neg(a \Rightarrow \bigcirc(l \wedge a)) \vee (\neg a)\mathcal{W}(\neg l)$.

As in the previous example, we rename and expand this into appropriate SNF rules as follows, where w, x and y are new proposition symbols.

12.	**start**	$\Rightarrow$	$\neg d \vee x \vee \neg l \vee \neg a$
13.	**start**	$\Rightarrow$	$\neg d \vee x \vee \neg l \vee w$
14.	**true**	$\Rightarrow$	$\bigcirc(\neg d \vee x \vee \neg l \vee \neg a)$
15.	**true**	$\Rightarrow$	$\bigcirc(\neg d \vee x \vee \neg l \vee w)$
16.	w	$\Rightarrow$	$\bigcirc(\neg l \vee \neg a)$
17.	w	$\Rightarrow$	$\bigcirc(\neg l \vee w)$
18.	x	$\Rightarrow$	$\Diamond y$

$$
\begin{array}{lrcl}
19. & \textbf{start} & \Rightarrow & \neg y \lor a \\
20. & \textbf{true} & \Rightarrow & \bigcirc(\neg y \lor a) \\
21. & y & \Rightarrow & \bigcirc(\neg l \lor \neg a)
\end{array}
$$

The refutation proceeds as follows

$$
\begin{array}{lrcll}
22. & (a \land y) & \Rightarrow & \bigcirc\textbf{false} & \text{[Glob. SRES: } 1,2,3,21] \\
23. & \textbf{start} & \Rightarrow & \neg a \lor \neg y & \text{[Simplification: } 22] \\
24. & \textbf{true} & \Rightarrow & \bigcirc(\neg a \lor \neg y) & \text{[Simplification: } 22] \\
25. & \textbf{true} & \Rightarrow & \bigcirc(\neg y) & \text{[Global SRES: } 20,24] \\
26. & \textbf{start} & \Rightarrow & \neg x \lor y & \text{[Temp. Check: } 18,25] \\
27. & \textbf{start} & \Rightarrow & \textbf{false} & \text{[Initial SRES: } 8,9,11, \\
 & & & & \qquad 12,23,26]
\end{array}
$$

4.4. CORRECTNESS

The refined resolution rule can be derived directly from the original temporal resolution rule as follows. The original rule is

$$
\left.\begin{array}{c}
\begin{array}{rcl}
\varphi_1 & \Rightarrow & \bigcirc\psi_1 \\
\varphi_2 & \Rightarrow & \bigcirc\psi_2 \\
\vdots & \vdots & \vdots \\
\varphi_n & \Rightarrow & \bigcirc\psi_n \\
\chi & \Rightarrow & \Diamond\neg l
\end{array} \\
\hline
\chi \Rightarrow (\neg \bigvee_{i=1}^{n} \varphi_i)\,\mathcal{W}\,\neg l
\end{array}\right\} \text{ if the side-condition } \textbf{looping} \text{ holds}
$$

where **looping** is

$$
\bigwedge_{i=1}^{n}(\varphi_i \Rightarrow \bigcirc(l \land \bigvee_{j=1}^{n} \varphi_j))
$$

Moving the side condition (**looping**) to the conclusion of the rule, we get

$$
\begin{array}{c}
\begin{array}{rcl}
\varphi_1 & \Rightarrow & \bigcirc\psi_1 \\
\varphi_2 & \Rightarrow & \bigcirc\psi_2 \\
\vdots & \vdots & \vdots \\
\varphi_n & \Rightarrow & \bigcirc\psi_n \\
\chi & \Rightarrow & \Diamond\neg l
\end{array} \\
\hline
\Box\textbf{looping} \Rightarrow (\chi \Rightarrow (\neg \bigvee_{i=1}^{n} \varphi_i)\,\mathcal{W}\,\neg l)
\end{array}
$$

Note that, as the side condition must be valid for the original rule to be used (alternatively, recall that SNF incorporates an '$\Box$' operator outside each rule), it is $\square$**looping** rather than just **looping** that is moved to the conclusion of the rule.

With a little classical reasoning, we can manipulate this to

$$\frac{\begin{array}{ccc} \varphi_1 & \Rightarrow & \bigcirc\psi_1 \\ \varphi_2 & \Rightarrow & \bigcirc\psi_2 \\ \vdots & \vdots & \vdots \\ \varphi_n & \Rightarrow & \bigcirc\psi_n \\ \chi & \Rightarrow & \Diamond\neg l \end{array}}{\chi \Rightarrow (\square\textbf{looping} \Rightarrow (\neg\bigvee_{i=1}^{n}\varphi_i)\,\mathcal{W}\,\neg l)}$$

which gives us the refined rule. The resolvent can be further transformed to $\chi \Rightarrow \Diamond\neg$**looping** $\vee \ldots$

THEOREM 1. *The new resolution rule is sound.*

Proof. (Outline) This follows directly from the above derivation and the soundness of the original temporal resolution rule.

THEOREM 2. *As long as only a finite number of potential loops are chosen, the application of the resolution rule terminates.*

Proof. (Outline) Each potential loop generates a new eventuality when temporal resolution is applied. However, since only trivial verification (not full loop checking) is required for these eventualities, they do not generate any further resolvents. Thus, if only a finite number of loops are examined, then the process terminates.

THEOREM 3. *The new resolution rule is complete.*

Proof. (Outline) This follows from the fact that if the side condition is satisfied in the original rule (which is complete) then we know that $\vdash$ **looping** and hence $\vdash$ $\square$**looping**. Combining this (using only classical reasoning) with the resolvent from the refined rule gives us the resolvent from the original rule. Together with the termination of the process, this provides completeness.

5. Implementation and Usage

Given this new resolution rule, several questions are raised regarding its usage and practical implementation occur. We consider each of these below.

1. *As the choice of φ_i seems to be crucial, how is this choice made? What happens if the wrong φ_i's are chosen?*

For the initial choice of φ_i's, we either use heuristics to examine the original formulae and thus suggest potential loops, or apply breadth-first loop search (Dixon, 1996). If the full breadth-first search fails to find a loop, we use each of the conjunctions in the last search node constructed as a guess for the loop (on the basis that the search might have been heading in the right direction, but just needed a little more step resolution).

If we choose the wrong φ_i's, then either some of the φ_i's are correct or none of them are. In the latter case, then because of the correctness of the resolution method, the search eventually fails. If we have chosen a *partially correct* set of φ_i's, then the process of verifying the temporal check will lead us towards the correct loop. For example, consider

$$\begin{array}{lrcl}
1. & a & \Rightarrow & \bigcirc(l \vee x) \\
2. & b & \Rightarrow & \bigcirc\neg x \\
3. & a & \Rightarrow & \bigcirc b \\
4. & b & \Rightarrow & \bigcirc a \\
5. & \mathbf{start} & \Rightarrow & d \\
6. & \mathbf{start} & \Rightarrow & a \\
7. & \mathbf{start} & \Rightarrow & b \\
8. & d & \Rightarrow & \Diamond\neg l \\
9. & \mathbf{start} & \Rightarrow & l
\end{array}$$

Here the actual loop that will derive a contradiction is $(a \wedge b) \Rightarrow \bigcirc \Box l$. Now imagine that we choose a as the φ_i, guessing that the loop is $a \Rightarrow \bigcirc \Box l$. We generate a resolvent, including a temporal check, as before and continue step resolution. When we come to verify the check, rather than finding

$$\mathbf{true} \Rightarrow \bigcirc\mathbf{looping}$$

we actually generate

$$(a \wedge b) \Rightarrow \bigcirc\mathbf{looping}$$

Thus, in the case where the check fails, a new choice for a loop will be generated. In the above case, we select $a \wedge b$ as the φ_i and continue.

2. *How can we be sure that if there is an appropriate loop, the* temporal check *will only involve searching for* **true** $\Rightarrow \ldots$*? Also, how much step resolution is involved in this check?*

Recall that the temporal check is of the form $\Diamond\neg$**looping** where **looping** is the side condition for the original resolution rule. If we have chosen the correct loop, then **looping** is valid as, by generalisation is $\square$**looping**. Once we rename $\Diamond\neg$**looping** to $\Diamond x$ and $x \Rightarrow \neg$**looping**, then step resolution should give us $x \Rightarrow \bigcirc$**false** and thus **true** $\Rightarrow \neg x$.

The amount of step resolution required in this last phase is constrained in that we only resolve $x \Rightarrow \neg$**looping** together with rules that actually construct the loop in order to generate $x \Rightarrow \bigcirc$**false** and then only step resolution between rules of the form **true** $\Rightarrow \ldots$ is required in order to generate **true** $\Rightarrow \neg x$.

3. *Given that the temporal resolution rule can now be applied at any time in the refutation, what is the general strategy for its use?*

The basic cycle we propose is

a) Choose a potential loop for an eventuality, e.g. via breadth-first search.

b) If breadth-first search succeeds, use the original temporal resolution rule and take resolvent generated as part of the set of support.

   If breadth-first search fails, use the new resolution rule, with the choice of loop being based upon the progress the breadth-first search process made, and use the resolvent generated as part of the set of support.

c) Apply step resolution, utilising the set of support identified above.

d) If a contradiction is generated, or if no new resolvents are produced, terminate.

   Otherwise, go to (a).

6. Conclusions and Future Work

In this paper, we have introduced, justified and applied a refined temporal resolution rule. This not only allows greater flexibility in the search for a refutation in clausal temporal resolution, but provides guidance for further search once the resolvents produced are used as the set of support.

Although space precludes the inclusion of larger examples and timings, initial results are encouraging. This resolution rule is now being incorporated into the development of CLATTER, a temporal theorem-prover based on the clausal resolution method (Fisher, 1991), its basic implementation (Dixon, 1996) and further refinement (Dixon, 1997).

Further work is required on both the heuristics that are required if breadth-first search fails completely, and the refinement of the set of support derived from the resolvents produced.

Acknowledgements

This work was partially supported by EPSRC under Research Grant GR/K57282.

References

M. Abadi and Z. Manna. Temporal Logic Programming. *Journal of Symbolic Computation*, 8: 277–295, 1989.

M. Abadi and Z. Manna. Nonclausal Deduction in First-Order Temporal Logic. *ACM Journal*, 37(2):279–317, April 1990.

A. Cavali and L. Fariñas del Cerro. A Decision Method for Linear Temporal Logic. In R. E. Shostak, editor, *Proceedings of the 7th International Conference on Automated Deduction*, pages 113–127. LNCS 170, 1984.

J. Chomicki and D. Niwinski. On the Feasibility of Checking Temporal Integrity Constraints. *Journal of Computer and System Sciences*, 51(3):523–535, December 1995.

C. Dixon. *Strategies for Temporal Resolution.* PhD thesis, Department of Computer Science, University of Manchester, Manchester M13 9PL, U.K., December 1995.

C. Dixon. Search Strategies for Resolution in Temporal Logics. In *Proceedings of the Thirteenth International Conference on Automated Deduction (CADE)*, Lecture Notes in Computer Science. Springer-Verlag, August 1996.

C. Dixon. Temporal Resolution: Removing Irrelevant Information. In *Proceedings of International Workshop on Temporal Reasoning (TIME)*, Daytona Beach, Florida, May 1997.

M. Fisher. A Resolution Method for Temporal Logic. In *Proceedings of the Twelfth International Joint Conference on Artificial Intelligence (IJCAI)*, Sydney, Australia, August 1991. Morgan Kaufman.

M. Fisher. A Normal Form for Temporal Logic and its Application in Theorem-Proving and Execution. *Journal of Logic and Computation*, 7(4), August 1997.

D. Gabbay, A. Pnueli, S. Shelah, and J. Stavi. The Temporal Analysis of Fairness. In *Proceedings of the Seventh ACM Symposium on the Principles of Programming Languages*, pages 163–173, Las Vegas, Nevada, January 1980.

Z. Manna and A. Pnueli. *The Temporal Logic of Reactive and Concurrent Systems: Specification.* Springer-Verlag, New York, 1992.

M. Peim. Propositional Temporal Resolution Over Labelled Transition Systems. Unpublished Technical Note, Department of Computer Science, University of Manchester, 1994.

G. L. Peterson. Myths about the Mutual Exclusion Problem. *Information Processing Letters*, 12(3):115–116, 1981.

D. A. Plaisted and S. A. Greenbaum. A Structure-Preserving Clause Form Translation. *Journal of Symbolic Computation*, 2(3):293–304, September 1986.

A. Pnueli. In Transition from Global to Modular Temporal Reasoning about Programs. In *Advanced Institute on Logics and Models for Verification and Specification of Concurrent Systems*, La Colle Sur Loupe, October 1984. Springer-Verlag.

A. P. Sistla, M. Vardi, and P. Wolper. The Complementation Problem for Büchi Automata with Applications to Temporal Logic. *Theoretical Computer Science*, 49:217–237, 1987.

G. Venkatesh. A Decision Method for Temporal Logic based on Resolution. *Lecture Notes in Computer Science*, 206:272–289, 1986.

P. Wolper. The Tableau Method for Temporal Logic: An overview. *Logique et Analyse*, 110–111:119–136, June-Sept 1985.

L. Wos, G. Robinson, and D. Carson. Efficiency and Completeness of the Set of Support Strategy in Theorem Proving. *ACM Journal*, 12:536–541, October 1965.

Determinism and the Origins of Temporal Logic

Torben Braüner
Centre for Philosophy and Science-Theory
Aalborg University
Langagervej 6
9220 Aalborg East, Denmark
E-mail: tor@brics.dk

Per Hasle
Information and Media Science
Aarhus University
Niels Juelsgade 84
8200 Aarhus N, Denmark
E-mail: phasle@imv.aau.dk

Peter Øhstrøm
Department of Communication
Aalborg University
Langagervej 8
9220 Aalborg East, Denmark
E-mail: poe@hum.auc.dk

Abstract.
The founder of symbolic temporal logic, A. N. Prior was to a great extent motivated by philosophical concerns. The philosophical problem with which he was most concerned was determinism versus free will. The aim of this paper is to point out some crucial interrelations between this philosophical problem and temporal logic. First, we sketch how Prior's personal reasons for studying the problems related to determinism were philosophical - initially, indeed theological. Second, we discuss his reconstruction of the classical Master Argument, which has since Antiquity been considered a strong argument for determinism. Furthermore, the treatment of determinism in two of Prior's proposed temporal systems, namely the Ockhamistic and the Peircean systems, is investigated. Third, we illustrate the fundamental role of the very same issue in more recent discussions of some tempo-modal systems: The 'Leibniz-system' based on ideas of Nishimura (1979) as well as Belnap and Green's argument (! 1994), to which we add some necessary revisions.

1. Introduction

Temporal logic is now being studied within (at least) three different academic communities:

1. The study of expressions of time within natural language understanding,

2. Various applications and other studies of temporal logic within computer science,

H. Barringer et al. (eds.), Advances in Temporal Logic, 185–206.

3. The study of time and modality within philosophical logic.

It is clear that while 1. and 2. have not been established as proper research fields before our century, 3. can be traced back to ancient philosophy. For centuries philosophers have been doing important research within what is now called temporal logic. It is also clear that in our century, at least some of those who have established 1. and 2. were aware of important contributions of the philosophers who have studied the logic of time and modality. The philosophical study of temporal logic has the oldest tradition compared to the other main communities working with temporal logic. It is also evident that at least some important innovations within symbolic temporal logic stem directly from philosophical ideas.

A.N. Prior (1914-1969) is the philosopher who has done most to collect material from old studies of temporal logic and to transform the material into a modern form. Prior also established a number of fundamental results within temporal logic. In fact, he was the first modern logician to make systematic studies of the logic of time and modality. His main contribution was the formal logic of tenses. For this reason it is fair to call Prior the founding father of modern temporal logic. From his work it seems clear that the critical study of arguments for determinism and fatalism was a major motivation for the formulation of temporal logic. Most of the systems proposed by him were transformations of philosophical ideas into the modern language of formal logic.

Prior's personal reasons for studying the problems related to determinism were philosophical - initially, indeed theological. In the first section of this paper this kind of intellectual reasons are presented. We also intend to present an outline of some of Prior's critical studies of the arguments for determinism in order to demonstrate how such studies have given rise to a number of interesting systems of temporal logic. This outline falls in two parts which constitute the second and third section of the paper. In the second section, we discuss his reconstruction of the classical Master Argument, which was considered a convincing argument for determinism in Antiquity. Furthermore, we investigate the treatment of determinism in two of Prior's proposed temporal systems, namely the Ockhamistic and the Peircean systems, respectively (some of the earliest branching-time logics). Herein we see a clear example of how the issue of determinism was related to tempo-modal logic from th! e outset. In the third section, we investigate the fundamental role of the very same issue in more recent discussions of some tempo-modal systems: The 'Leibniz-system' based on ideas of Nishimura (1979) as well as Belnap and Green's argument (1994), to which we add some necessary revisions.

2. Prior on Determinism

In his memorial paper on Prior, (Kenny, 1970), A. J. P. Kenny summed up his life and work with these words:

> Prior's greatest scholarly achievement was undoubtedly the creation and development of tense-logic. But his research and reflection on this topic led him to elaborate, piece by piece, a whole metaphysical system of an individual and characteristic stamp. He had many different interests at different periods of his life, but from different angles he constantly returned to the same central and unchanging themes. Throughout his life, for instance, he worked away at the knot of problems surrounding determinism: first as a predestinarian theologian, then as a moral philosopher, finally as a metaphysician and logician. ((Kenny, 1970), p. 348)

By now, it is recognized that with the construction of temporal logic Prior made a highly original and lasting contribution to philosophy and logic. Considerable changes in approach notwithstanding, one can indeed trace motivations and considerations of a theological nature underlying later formal and philosophical achievements. So far, however, little has been done in order to investigate the relation between his theological work and his later work on temporal logic. One reason for this is the simple fact that much of the early work is all but inaccessible, a significant part of it indeed unpublished. Apart from sheer historical interest it seems to us that the logic community will be well served by spelling out this relation in somewhat greater detail. The aim of this section is to point out a few major aspects of this relation.

PREDESTINATION

Prior was brought up as a Methodist, but at the age of 18 he became a Presbyterian. The Presbyterian denomination is Calvinist. Now, the central insight of the Reformation was that man could not save himself through deeds, but rather salvation was pure grace, a gift from God, demanding only faith. However, this immediately raises the question whether faith, then, is something man is free to accept or reject, or whether some are 'elected' to be believers - receiving passively the gift of faith - while others are not given this gift. The reformers differed on this point, but Calvin, at any rate, took a firm and consequent stand: indeed there is no such thing as free choice with respect to faith; every person is predestined either to belief or disbelief, and hence, to salvation or damnation. The most marked feature of Presbyterianism, then, is its teaching concerning predestination.

One remarkable defence of predestination and determinism is given in Prior's paper 'Determinism in philosophy and theology', (Prior, 1944). Here, the doctrine of predestination is thematically compared with philosophical determinism, respectively, indeterminism. The paper is difficult to date, but it was probably written in the mid-fourties, see (Hasle, 1997). In what follows, we shall have a look at a few of the main themes.

The paper opens with the observation that in "modern discussions", determinism is often associated with a "scientific creed" as opposed to the idea of free will, which is considered to be religious. But this perception is immediately countered:

> It is exceedingly rare for philosophers to pay any great attention to the fact that a whole line of Christian thinkers, running from Augustine (to trace it back no further) through Luther and Calvin and Pascal to Barth and Brunner in our own day, have attacked free will in the name of religion. ((Prior, 1944), p. 1)

Furthermore, Prior argued that the ordinary ideas of free will, when understood as moral accountability and general indeterminism, are at least as absurd as the idea of predestination:

> We are guilty of that which we are totally helpless to alter; and to God alone belongs the glory of what we do when we are truly free. - Absurd as these doctrines appear, *they are in the end no more so than the ordinary non-Augustinian concept of "moral accountability"*... (our italics) ((Prior, 1944), p. 2)

Finally, Prior goes to describe how certain human experiences are actually compatible with the notion of predestination, observing that

> Even those of us who accept a straightforward determinism have to give some account of men's *feeling* of freedom, and their *feeling* of guilt; and it is at least conceiveable that the "absurdities" of Augustinianism contain a more accurate psychological description of the state of mind concerned, than does the "absurdity" of the ordinary non-Augustinian concept of "moral accountability". ((Prior, 1944), p. 3)

Various strikingly and elegant arguments are offered to underpin the plausibility of Augustinianism in the face of human experience.

FROM PREDESTINATION TO INDETERMINISM - AND THE DISCOVERY OF TEMPORAL LOGIC

Prior's stance on determinism was to change from the early fifties and onwards, a development closely related to his discovery of temporal logic. Throughout the fourties, he was interested in logic - mainly traditional non-symbolic logic - but apparently even more interested in philosophical and historical issues within theology. This began to change about 1949, and his first interest in modal logic was aroused around 1951. At this time he also developed into an adherent of indeterminism, and indeed, of free will. In (Copeland, 1996) Jack Copeland describes how

> ...Aristotle speaks of some propositions about the future - namely, those about such events as are not already predetermined - as being neither true nor false when they are uttered... This appealed to Prior, once a Barthian Calvinist but now [ca. 1950/51] on the side of indeterminism and free will. There can be no doubt that Prior's interest in tense logic was bound up with his belief in the existence of real freedom. ((Copeland, 1996), p. 16)

Clearly, he must have been revising his former attempts to defend (and understand) the doctrine of predestination. In 1951, Prior read the paper (Findlay, 1941) by J. N. Findlay, where, in a footnote, the possibility of developing a formal calculus of tenses is suggested. In 'The Craft of Formal Logic', (Prior, 1951), Prior wrote as follows:

> In Peter of Spain [Medieval logician, d. 1277], the word 'mode' is used in a broad sense for any sort of qualification of anything, adjectival or adverbial; but more strictly for adverbial qualifications, and most strictly of all for qualifications of the manner of connection between the subject and the predicate, which is what he takes the ordinary modes of necessity, etc., to be. Other adverbial modes express time... (The grammatical word 'mood' comes from the Latin 'modus' as well as the logical word 'mode'. That there should be a modal logic of time-distinctions has been suggested in our own day by Professor Findlay [In Findlay's 'Time: A Treatment of Some Puzzles'] ...). ((Prior, 1951), p. 750)

He decided to take up Findlay's challenge, see (Øhrstrøm and Hasle, 1993). Major sources for him were also Łukasiewicz' discussion of future contingents in Aristotle's *De Interpretatione*, (Łukasiewicz, 1970), and the Diodorean "Master Argument", which he came to study via a paper by Benson Mates on 'Diodorean Modalities', (Mates, 1949). In both of these problem sets - future contingents and the Master Argument - the logic of time is strongly interwoven with the discussion of determinism versus indeterminism. Thus from the very outset of Prior's development of temporal logic, the problem of determinism was dealt with in parallel with the logic of time. Moreover, it is clear that the determinism-issue has roots in the problem of predestination, and that Prior's dealing with it was a natural continuation of his earlier preoccupation with such matters. Of course, at the same time there is also a breach in the very approach to these problems. The emphasis on time and change is it! self a marked departure from the peculiarly atemporal spirit of the Calvinist teaching on predestination.

In the second half of the fifties, Prior increasingly took up the notion of (Divine) Foreknowledge, which is obviously related to the determinism and predestination issues. There is a straightforward logical relation between the notions of predestination (of an event) and Divine Foreknowledge (of the event): The former implies the latter. His studies led him to consider the classical Christian belief in Divine Foreknowledge as unacceptable (except in a very restricted form, see (Prior, 1962a)). In 'Some Free Thinking About Time', (Prior, 1996), he stated his belief in indeterminism as well as the resulting limitations on Divine Foreknowledge very clearly:

> I believe that what we see as a progress of events *is* a progress of events, a *coming to pass* of one thing after another, and not just a timeless tapestry with everything stuck there for good and all... ((Prior, 1996), p. 47)

> This belief of mine... is bound up with a belief in real freedom. One of the big differences between the past and the future is that once something has become past, it is, as it were, out of our reach - once a thing has happened, nothing we can do can make it not to have happened. But the future is to some extent, even though it is only to a very small extent, something we can make for ourselves.... if something is the work of a free agent, then it wasn't going to be the case until that agent decided that it was. ((Prior, 1996), p. 48)

> I would go further than Duns Scotus and say that there are things about the future that God doesn't yet know because they're not yet to be known, and to talk about knowing them is like saying that we can know falsehoods. ((Prior, 1996), p. 48)

So, in parallel with his discovery of temporal logic he became a firm believer in indeterminism and free will. Furthermore, his logical studies led him to consider such a belief to be inconsistent with what he regarded as indispensable parts of the Christian faith, namely, the doctrines of Predestination and Divine Foreknowledge. In fact, he had become an agnostic, although not an atheist.

3. From Determinism to Temporal Logic

The Master Argument of Diodorus Cronus

The idea that there could be a logic of time distrinctions was, as mentioned above, first brought to Prior's attention in 1951 when he read a paper by John Findlay, cf. (Copeland, 1996), p. 15 ff. However, his first serious work in the field was in 1953/1954, when he studied a short article by Benson Mates, (Mates, 1949). The paper was concerned with Diodorean logic. Diodorus Cronus (ca. 340-280 B.C.) was a philosopher of the Megarian school, who in Antiquity achieved wide fame as a logician and a formulator of philosophical paradoxes. The most well-known of these paradoxes is the so-called Master Argument which in Antiquity was understood as an argument designed to prove the truth of fatalism. Prior realised that it might be possible to relate Diodorus' ideas to contemporary works on modality by developing a calculus which included temporal operators analogous to the operators of modal logic. Therefore he attempted a reconstruction of this logic. This work became a cornerst! one in his development of a modern temporal logic.

The Master Argument is a trilemma. According to Stoic philosopher Epictetus (ca. 55-135), Diodorus argued that the following three propositions cannot all be true, (Mates, 1953), p. 38:

(D1) Every proposition true about the past is necessary.

(D2) An impossible proposition cannot follow from (or after) a possible one.

(D3) There is a proposition which is possible, but which neither is nor will be true.

Having established this incompatibility, and taking (D1) and (D2) to be immediately plausible, Diodorus could justify that (D3) is false. Assuming (D1) and (D2), he went on to explain (or define) possibility and necessity as follows:

(D$\Diamond$) The possible is that which either is or will be true.

(D$\Box$) The necessary is that which, being true, will not be false.

Unfortunately, only the premises and the conclusion of the argument are known. During the last few decades various philosophers and logicians have tried to reconstruct the argument as it might have been. The reconstruction of the Master Argument certainly constitutes a genuine problem within the history of logic. It should, however, be noted that the argument is worth studying for reasons other than historical. First of all, the Master Argument has been read as an argument for determinism. Secondly, the Master Argument can be regarded as an attempt to clarify the conceptual relations between time and modality. When seen in this perspective any attempted reconstruction of the argument is important also from a systematic point of view, and this is obviously true for any version of the argument, even if it is historically incorrect.

In the following we shall discuss Prior's reconstruction of the Master Argument. In 1954 he wrote the paper 'Diodoran Modalities', (Prior, 1955), wherein the essence of his reconstruction is already present. However, in this reconstruction he brought in Łukasiewicz' three-valued logic in order to make certain points about indeterminateness. Later, he could make his points without having to refer to three-valued logic. His reconstruction was thus improved in (Prior, 1967), and it is to this reconstruction we shall refer in the following.

It is assumed that tensed propositions are propositional functions, with times as arguments, (Prior, 1967), p. 32 ff. Prior uses tense- and modal operators in his reconstruction, and interprets the logical consequence involved in (D2) as what is in modal logic usually called 'strict implication'. The strict implication $p \rightarrow q$ is defined as $\Box(p \Rightarrow q)$. In what follows, the tense operators P and F are to be thought of as meaning respectively 'it has been the case that ...' and 'it will be the case that ...'. We define Hp and Gp as respectively $\neg P \neg p$ and $\neg F \neg p$.

On these assumptions it is possible to symbolise (D1), (D2) and (D3) in the following way:

(Dl) $Pq \Rightarrow \Box Pq$

(D2) $((p \rightarrow q) \wedge \Diamond p) \Rightarrow \Diamond q$

(D3) $\Diamond r \wedge \neg r \wedge \neg Fr$, for some proposition r

However, these three premises are not sufficient for reconstructing the argument (when thus formalised). In addition Prior needs two extra premises, namely the following:

(D4) $(q \wedge Gq) \Rightarrow PGq$

(D5) $q \rightarrow HFq$

We are now ready to state Prior's reconstruction. It is proved by reductio ad absurdum that the three Diodorean premises (D1), (D2), and (D3) are inconsistent given (D4) and (D5). The proof goes as follows:

1. $\Diamond r \wedge \neg r \wedge \neg Fr$ (from D3)

2. $\Diamond r$ (from 1)

3. $r \rightarrow HFr$ (from D5)

4. $\Diamond HFr$ (from D2, 2 and 3)

5. $\neg r \wedge G\neg r$ (from 1)

6. $PG\neg r$ (from 5 and D4)

7. $\Box PG\neg r$ (from 6 and D1)

8. $\neg\Diamond HFr$ (from 7; contradicts 4)

From a historical point of view, Prior's addition of (D4) and (D5) is problematic (even though the argument thus reconstructed is interesting in its own right). In fact, the statement (D4) is rather complicated and not as innocuous as it may seem at first glance. It may be seen as related to the question regarding denseness of time. It is not very likely that Diodorus would involve such an assumption without making it an explicit premise in the Master Argument. As regards (D5), we know that Diodorus used the Master Argument as a case for the definitions (D$\Diamond$) and (D$\Box$). That is, in the argument itself $\Box$ (and $\Diamond$ as well) should be regarded as primitive. It is hard to believe that Diodorus would involve a premise about $\Box$ without stating it explicitly.

The Master Argument as an Argument for Determinism

It is very likely that the Master Argument was originally designed to prove fatalism or determinism. Because of the plausibility of (D1) and (D2), the argument was understood as a rather strong case against (D3). The denial of (D3) is equivalent to the view that if a proposition is possible, then either it is true now or it will be true at some future time. So in a nutshell the conclusion of the argument is that an event which will never happen and is not happening now cannot be possible. It should be clear that the argument is interesting not only for historical reasons. Its systematical content is entirely relevant for a modern discussion of determinism, too.

The present-day philosopher wanting to argue against fatalism and determinism should relate to all known versions of the Master Argument, directly or indirectly. From Prior's work it turns out that in a modern context there is no need to interpret the conclusion as a statement of determinism or fatalism. This point can be made clear if we consider the structure of time as branching rather than linear. In this kind of structure, any given 'now' gives rise to a number of possible and different futures - sometimes called the 'forking paths into the future'. An adequate conception of the notion of 'possibility' can then be captured by the formula

$$\Diamond p \Leftrightarrow (p \vee Fp)$$

This amounts to the definitions (D$\Diamond$) and (D$\Box$) being adopted. In fact, the very use of the idea of 'possible futures' can be understood as an acceptance of the conclusion of the Master Argument, that is, the denial of (D3), which means that p is true only if p is true now of in the future ($\Diamond p \Rightarrow (p \vee Fp)$). Thus it is evident that if time is branching then any possibility must belong to some possible future or the 'now'. So when we investigate the Master Argument from the perspective of the historical development of the logical study of time, the argument turns out to be a demonstration of a fundamental relationship between time and modality rather than a case for fatalism or determinism. In 'Tense-Logic and the Continuity of Time', (Prior, 1962b), Prior himself summed up his investigation into Diodorus as follows:

> The aim of the Master Argument, as I conceive it, was to refute the Aristotelian view that while it is now beyond the power of men or gods to affect the past, there are alternative futures between which choice is possible. Against this, Diodorus held that the possible is simply what either is or will be true. ((Prior, 1962b), p. 138)

In order to formalise the notion of branching time, we need a set $TIME$ equipped with a relation $<$ together with a function T which assigns a truth value $T(t,p)$ to each pair consisting of an element t of $TIME$ and a propositional letter p. The elements of $TIME$ are to be considered as instants and $<$ as the before-relation. To account for the branchingness of time, it is assumed that $<$ is transitive and furthermore satisfies the condition

$$(t < t' \wedge t'' < t') \Rightarrow (t < t'' \vee t'' < t \vee t = t'')$$

which is called backwards linearity. By induction, we extend the range of the valuation operator T from propositional letters to arbitrary formulae as follows:

$$\begin{array}{rll} T(t, p \wedge q) & \text{iff} & T(t,p) \wedge T(t,q) \\ T(t, \neg p) & \text{iff} & \neg T(t,p) \\ T(t, Pp) & \text{iff} & \exists t' < t.\, T(t',p) \\ T(t, Fp) & \text{iff} & \exists t' > t.\, T(t',p) \end{array}$$

A formula p is said to be valid if and only if p is true in any structure $(TIME, <, T)$, that is, we have $T(t,p)$ for any instant t.

The conceptual price for involving branching time as described above is that the notion of the future is conflated with the notion of possibility. This means we have to say that any contingent future event is in fact going to be (and in fact also not going to be). Prior was not ready to pay such a price. He clearly felt that there is more to the notion of truth about the future than the mere idea of future possibilities. So although he accepted the idea of branching time he would not like to do it in the manner expressed in the semantics described above, realising that if he wanted to avoid the conclusion of the Master Argument, at least one of the premises has to be denied. According to Prior, the obvious candidates for such a denial are (D1) and (D5).

One could deny (D1), that is, the statement $Pq \Rightarrow \Box Pq$, since it is not reasonable to view a true proposition about the future as necessary, just because it is formulated as a prophecy stated in the past. Such a proposition is about the past only in a spurious sense. The statement

It has been that Dion never will be here

which can be symbolised as $P\neg Fr$, should not be counted as necessary even if it is true. So even if we accept $\neg r$, $\neg Fr$, and $P\neg Fr$, there is no a priori reason for excluding the conceptual possibility of Dion's being here at some future time, or his 'always been going to be here', that is, $\Diamond Fr$ and $\Diamond PGr$. Therefore, it would be reasonable to question (D1). William of Ockham (ca. 1285 - 1349) did indeed take a similar position; he maintained that we must distinguish between sentences which are properly about the past, and those which are only spuriously so, that is, a distinction between (proper) past and pseudo-past, see (William of Ockham, 1983). In that case (D1) should be restricted to propositions where F does not occur in q; this in turn destroys the line of thought in the Diodorean (and similar) arguments when seen as arguments for determinism. Prior developed this solution further in what he called the Ockhamistic system after this medieval! logican.

Nevertheless, Prior himself accepted (D1) without Ockham's restriction, but questioned the validity of (D5), that is, the statement $p \rightarrow HFp$. If we understand 'will be' as 'determinately will be', then (D5) can certainly be denied. He showed how the denial of this premise could give rise to a temporal logical system, which he called the Peircean system after the logician Charles Sanders Peirce (1839 - 1914). This was the system which Prior himself preferred.

In view of Prior's indeterminism, it may seem surprising that he accepted (D1). However, in his first work on Diodorus (1954/55) he followed Łukasiewicz' ideas on three-valued logic and thus considered the truth-value of statements about the contingent future to be indeterminate (that is, 1/2, rather than 0 or 1); already on these grounds, determinism would not follow from the acceptance of (D1). In 1967, he had himself adopted the Peirce-solution to be presented below. Again, (D1) does not lead to determinism on such a supposition.

In the following we shall discuss Prior's Ockhamistic and Peircean systems. In addition we shall argue that there are at least two more possibilities which Prior did not take into consideration.

THE OCKHAMISTIC SYSTEM

In order to describe the semantics for the new tempo-modal system which is called the Ockhamistic system, we need again a set $TIME$ equipped with a transitive and backwards linear relation $<$ together with a function T which assigns a truth value to each pair consisting of an instant and a propositional letter. The novel feature of the semantics is a notion of temporal 'routes' or 'temporal branches', cf. (Prior, 1967), p. 126 ff. Branches are maximal linear subsets in $(TIME, <)$. The set of branches induced by $(TIME, <)$ will be denoted C. Note that the conditions of transitivity and backwards linearity makes each branch c backwards-closed, that is, if $t \in c$ and $t' < t$ then $t' \in c$.

We shall now give a valuation operator where truth is relative to an instant as well as to a branch to which the instant belongs. So $Ock(t, c, p)$ is to be thought of as p being true at the instant t in the branch c. We define the valuation operator Ock as follows:

$$\begin{array}{rll} Ock(t,c,p) & \text{iff} & T(t,p), \text{ where } p \text{ is a propositional letter} \\ Ock(t,c,p \wedge q) & \text{iff} & Ock(t,c,p) \wedge Ock(t,c,q) \\ Ock(t,c,\neg p) & \text{iff} & \neg Ock(t,c,p) \\ Ock(t,c,Pp) & \text{iff} & \exists t' < t.\ Ock(t',c,p) \\ Ock(t,c,Fp) & \text{iff} & \exists t' > t.\ t' \in c \wedge Ock(t',c,p) \\ Ock(t,c,\Box p) & \text{iff} & \forall c' \in C.\ t \in c' \Rightarrow Ock(t,c',p) \end{array}$$

A formula p is said to be Ockham-valid if and only if p is Ockham-true in any structure $(TIME, <, T)$, that is, we have $Ock(t, c, p)$ for any instant t and branch c such that $t \in c$.

It may be doubted whether Prior's Ockhamistic system is in fact an accurate representation of the temporal logical ideas propagated by William of Ockham. According to Ockham, God knows the contingent future, so it seems that he would accept an idea of absolute truth, also when regarding a statement Fq about the contingent future - and not only what Prior has called "prima-facie assignments" like $Ock(t, c, Fq)$, cf. (Prior, 1967), p. 126. That is, such a proposition can be made true 'by fiat' simply by constructing a concrete structure which satisfies it. But Ockham would say that Fq can be true at t without being relativised to any branch. In the next part of this paper we shall show that it is possible to establish a system which seems a bit closer to Ockham's original ideas (namely the system which we call 'Leibnizian').

It should be noted that this difference with respect to the notion of truth is mainly philosophical. Prior's Ockhamistic system appears to comprehend at least all the theorems which should be accepted according to Ockham's original ideas. Let us, for instance, consider the formula $q \Rightarrow HFq$. From the above definitions it is obvious that $Ock(t, c, q \Rightarrow HFq)$ for any t and any c with $t \in c$. Therefore $q \Rightarrow HFq$ is valid in this system. Likewise, the formula $q \Rightarrow \Box q$, where the tense operator F does not occur in q, is valid. This formula is in good accordance with the medieval dictum "unumquodque, quando est, oportet esse" ("Anything, when it is, is necessary"), see (Rescher and Urquhart, 1971), p. 191.

The Peircean System

Now, let us turn to another tempo-modal system which Prior studied carefully, the so-called Peircean system. In this system four different operators can be considered, F, G, f, and g, all of which are related to the future. These operators can be translated into an Ockhamistic formulation in the following way:

$$\begin{array}{ccc} F & \mapsto & \Box F \\ f & \mapsto & \Diamond F \\ G & \mapsto & \Box G \\ g & \mapsto & \Diamond G \end{array}$$

Operators regarding the past are left unaltered. It should be noted that there is no Peircean expression which translates into the Ockhamistic tense operator F. Using the translation, validity within the Peirce system can be defined in terms of validity in Ockhamistic structures: We simply define a formula to be Peirce-valid if and only if its translation is Ockham-valid.

The advantage of the Ockhamistic system could be said to be its property of making a genuine distinction between the following three types of statement:

(i) Necessarily, Mr. Smith will commit suicide.

(ii) Possibly, Mr. Smith will commit suicide.

(iii) Mr. Smith will commit suicide.

However, in the Peirce-system the type of future statement seen in (iii) will have to be interpreted as meaning either (i) or (ii). There is no 'plain future' in this system. Of course, that is a deliberate and philosophically motivated choice cf. (Øhrstrøm and Hasle, 1995), chapter 2.2. Charles S. Peirce himself wrote:

> A certain event either will happen or will not. There is nothing now in existence to constitute the truth of its being about to happen, or of its being not about to happen, unless it be certain circumstances to which only a law or uniformity can lend efficacy. But that law or uniformity, the nominalists say, has no real being, it is only a mental representation. If so, neither the being about to happen nor the being about not to happen has any reality at present ... If, however, we admit that the law has a real being, and of the mode of being an individual, but even more real, then the future necessary consequent of a present state of things is as real and true as the present state of things itself. ((Peirce, 1 58), 6.368)

Peirce saw himself as a realist. Truth and reality were for him objective, albeit in a sense differing from classical 'naive realism'. As argued by for instance Harry R. Klocker, (Klocker, 1968), p. 80 ff, according to Peirce, truth is that viewpoint upon which everybody examining the state of things eventually agree, and reality is the object represented by this viewpoint.

Given this analysis one must admit that the Ockhamistic system cannot a priori be preferred on philosophical grounds; but on linguistic grounds at least, it seems clear that (iii) should be distinguished from (i) and (ii). It is hard to see how one can do without the notion of plain future.

Let us consider the formula $q \Rightarrow HFq$. When this formula is translated into the Ockhamistic language, we get the following formula:

$$q \Rightarrow H\Box Fq$$

Clearly, an Ockham-structure can be found where $Ock(t, c, q \Rightarrow H\Box Fq)$ is false for some t and some c with $t \in c$. Therefore, the formula $q \Rightarrow HFq$, which is a consequence of (D5), is not valid in Prior's Peircean system.

Now, the truth-operator in the Peircean system does not have to be defined in terms of the Ockhamistic operator. It would have been possible to present it independently. But since we want to compare the two systems, the above definitions are appropriate. We can immediately verify the most interesting feature of Prior's definition of Peircean truth:

$$Pei(t, c, Fp) \quad \text{iff} \quad \forall c' \in C.\ t \in c' \Rightarrow \exists t' > t.\ t' \in c' \wedge Pei(t', c', p)$$

This appears to be in good accordance with the ideas of C. S. Peirce, since he rejected the very idea that statements regarding the contingent future could be true.

4. The Development after Prior

Leibnizian Tense Logic

In the article (Nishimura, 1979a) Hirokazu Nishimura formulated a new temporal model which is slightly different from the Ockham-model which Prior had considered. Nishimura's model involves not only instants, but also 'histories' defined as maximal linear subsets of the set of instants. In fact, Nishimura's model of time can be viewed as a union of disjoint histories. According to the model, the tenses (past, present, future) are always relative to a history. Clearly, Nishimura's model has to involve some relation of "identity" of histories before certain events. For instance, relative to one possible history H1 there is going to be a sea-fight tomorrow, and relative to another history H2 there is not going to be a sea-fight tomorrow. Let us for the sake of argument suppose that in H2 there is never going to be a sea-fight. Then H1 and H2 may be identical in all respects up till tomorrow, except that "there will be a sea-fight" is true today (and earlier) with respect to H! 1, but at any time false with respect to H2. The situation can be represented by the following structure:

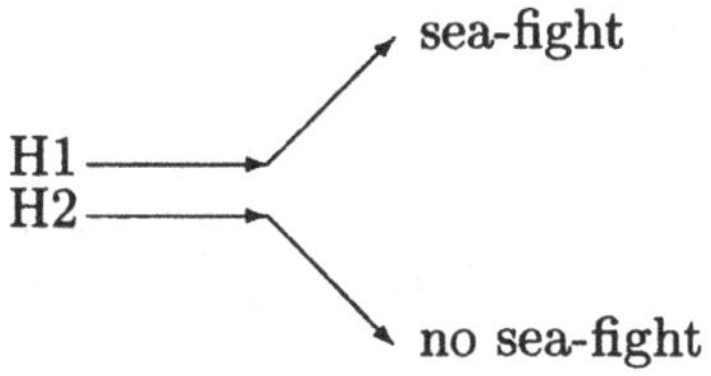

Therefore, in order to establish such an identity relation, future statements must be disregarded. (The identity relation of course only applies to the 'sections' of H1 and H2 up till tomorrow.) In dealing with the model, it is natural to consider the full set of histories as pre-defined. This view is similar to the concept of creation of a temporal world, which G. W. Leibniz (1646-1716) defended in his philosophy.

In the remarks above, we have disregarded the fact that Nishimura's model is metric. With the aim of comparing it to Prior's Ockham-model, we shall recast it in a non-metric fashion, that is, we shall give a formal non-metric branching time model incorporating Nishimura's ideas. The resulting model will be called the Leibniz-model. It is in most respects very close to the Ockham-model. In fact, any Leibniz-valid formula is Ockham-valid, but not vice versa.

In order to give a formal account of the Leibnizian model, we need a set $TIME$ equipped with two relations $<$ and $\approx$ together with a function T which assigns a truth value to each pair consisting of an instant and a propositional letter. It is assumed that $<$ is transitive as well as backwards and forwards linear. Also, it is assumed that $\approx$ is an equivalence relation. Furthermore, the relations $<$ and $\approx$ have to interact as stipulated by the condition

$$(t \approx t' \wedge t'' < t) \Rightarrow \exists t''' < t'.\, t'' \approx t'''$$

It is assumed that the valuation operator T respects the $\approx$ relation in the sense that

$$t \approx t' \Rightarrow T(t,p) = T(t',p)$$

for every propositional letter p. Note that, by induction, the range of this statement can be extended from propositional letters to arbitrary formulae not involving the tense operator F.

We extend the range of the valuation operator T to arbitrary formulae as follows:

$$\begin{array}{rcl} T(t, p \wedge q) & \text{iff} & T(t,p) \wedge T(t,q) \\ T(t, \neg p) & \text{iff} & \neg T(t,p) \\ T(t, Pp) & \text{iff} & \exists t' < t.\, T(t',p) \\ T(t, Fp) & \text{iff} & \exists t' > t.\, T(t',p) \\ T(t, \Box p) & \text{iff} & \forall t' \approx t.\, T(t',p) \end{array}$$

A formula p is said to be Leibniz-valid if and only if p is true in any Leibniz-structure $(TIME, <, \approx, T)$, that is, we have $T(t,p)$ for any instant t.

It should be noted that an equivalence relation on $TIME$ can be obtained by defining a pair of instants t and t' to be related if and only if

$$t < t' \vee t' < t \vee t = t'$$

The equivalence classes induced by this equivalence relation constitute the histories of the structure. Also, note that a history is backwards closed as well as forwards closed.

In what follows, we shall prove that a Leibniz-valid formula is Ockham-valid by giving a construction which to a structure $(TIME, <, T)$ assigns a Leibniz-structure $(TIME', <', \approx', T')$ such that any formula is Ockham-true in $(TIME, <, T)$ if and only if it is true in the structure $(TIME', <', \approx', T')$. The set $TIME'$ is defined as follows:

$$TIME' = \{(t,c) \in TIME \times C \mid t \in c\}$$

Furthermore, the relations $<'$ and $\approx'$ as well as the function T' are defined as follows:

$$\begin{array}{rcl} (t,c) <' (t',c') & \text{iff} & t < t' \wedge c = c' \\ (t,c) \approx' (t',c') & \text{iff} & t = t' \\ T'((t,c),p) & \text{iff} & T(t,p), \text{ where } p \text{ is a propositional letter} \end{array}$$

Intuitively, what is going on is that the histories of $(TIME', <')$ are obtained by splitting up the branches of $(TIME, <)$. The relation $\approx'$ and the valuation operator T' are then defined as appropriate. Using the definitions, it is straightforward to check that $(TIME', <', \approx', T')$ is indeed a Leibniz-structure. By induction, it is now straightforward to check that whenever $t \in c$ we have $Ock(t, c, p)$ if and only if $T'((t, c), p)$ for any formula p. Hence, a formula is Ockham-true in the structure $(TIME, <, T)$ if and only if it is true in the Leibniz-structure $(TIME', <', \approx', T')$.

Ockham-validity does not entail Leibniz-validity. With the aim of showing why this is the case, we define the operators Ap and Sp as respectively $Hp \wedge p \wedge Gp$ and $Pp \vee p \vee Fp$. As pointed out in (Nishimura, 1979b), the following formula is Ockham-valid:

$$\neg(p1 \wedge p2)$$

where

$$p1 = A\Box S(Hq \wedge q \wedge G\neg q)$$

and

$$p2 = A\Box A(q \Rightarrow \Diamond Fq)$$

It is straightforward to prove that the formula above is not Leibniz-valid by giving a Leibniz-structure in which it is not true. It should, however, be mentioned that the Leibniz-model agrees with the Ockham-model with respect to the formulae involved in the Master Argument.

The most important difference between the two systems is that in the Leibnizian semantics, the truth-value of a proposition depends on the instant in time only, and not explicitly on the branch, since every instant belongs to one and only one temporal branch. For this reason it can be said that the system comes closer to Ockham's ideas on the future than Prior's Ockhamistic system. The Leibnizian logic seems to have a number of attractive qualities as an interdeterministic temporal logic which allows the notion of plain truth.

THE THIN RED LINE

In an interesting paper, (Belnap and Green, 1994), Nuel Belnap and Mitchell Green have considered another aspect of the Ockhamistic model. They have argued that the model not only has to specify a preferred branch corresponding to the true history (past, present, and future). If one wants to insist on a concept of future which is different from the possible future as well as the necessary future, it must be assumed that there is a preferred branch at every (counterfactual) instant.

Belnap and Green have based their argument on the following example of a statement from ordinary language which any Ockhamistic logic in their opinion should be able to deal with:

> The coin will come up heads. It is possible, though, that it will come up tails, and then *later* (*) it will come up tails again (though at that moment it could come up heads), and then, inevitably, still *later* it will come up tails yet again. ((Belnap and Green, 1994), p. 379)

As Belnap and Green explains, the trouble is that at (*) the example says that tails *will* happen, not merely that it *might*. The point is that (*) is future relative to a counterfactual instant (which is induced by the use of the notion of possibility). Belnap and Green argue that the Ockhamistic model cannot deal with such a situation because only one preferred branch is given. They remedy this deficiency by introducing a function TRL (*T*hin *R*ed *L*ine) which to each instant assigns a preferred branch; the so-called thin red line. Using this new notion, our understanding of the sentence above can be represented by the branching time structure

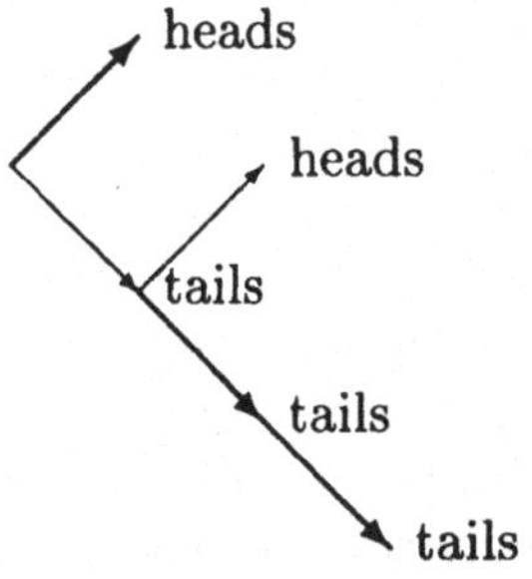

where we have used a thick line to represent the future part of a thin red lines (note that the past part needs no representation as it is uniquely determined).

Now, which conditions should the function TRL satisfy? Belnap and Green have argued that the condition

(TRL1) $t \in TRL(t)$

should hold in general, and that in addition the following

(TRL2) $t < t' \Rightarrow TRL(t) = TRL(t')$

may be considered. However, they show that (TRL2) is inconsistent with branchingness of time; to be precise, the condition implies that the before-relation is forwards linear. So condition (TRL2) seems to be too strong a requirement. Rather than (TRL2), we propose the weaker condition

(TRL2') $(t < t' \wedge t' \in TRL(t)) \Rightarrow TRL(t) = TRL(t')$

which seems to be natural in relation to the Ockhamistic notion of branching time and which is certainly consistent with it.

Belnap and Green have argued that any such TRL-function should give rise to a semantics in which the following formulae are valid:

(T1) $PPq \Rightarrow Pq$

(T2) $FFq \Rightarrow Fq$

(T3) $q \Rightarrow PFq$

No formal semantics is given by Belnap and Green, but they seem to assume that tense operators are interpreted only relative to an instant. This amounts to interpreting tenses using a two-place valuation operator:

$$T(t, Pp) \quad \text{iff} \quad \exists t' < t.\, T(t', p)$$
$$T(t, Fp) \quad \text{iff} \quad \exists t' > t.\, t' \in TRL(t) \wedge T(t', p)$$

Given such a semantics, it is straightforward to check that (T1) is valid without (TRL2'). On the other hand, (T2) is not valid without (TRL2'), but it is if this assumption is made. The formula (T3) is not valid even if (TRL2') is assumed; it is not true in the following structure:

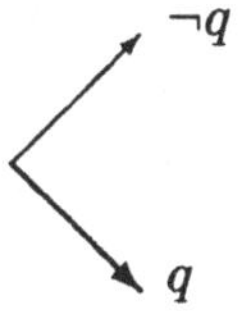

(If somebody would like to accept $q \Rightarrow PFq$ is invalid, one would have to say something like this: The counterfactual assumption of q does not invalidate the truth of the past prediction $PF\neg q$. If I am awake now, it certainly was true yesterday that I was going to be awake after one day. That prediction was true (but of course not necessary) even if I now - while being awake - imagine that I were asleep. For this reason one might say, that the truth of $q \wedge PF\neg q$, where q stands for 'I am asleep', is in fact conceiveable. But this piece of argumentation is somewhat strained.)

However, it is not obvious how to extend the above described semantics of tenses to deal with necessity. The problem is that in an Ockhamistic branching time model, interpretation of the necessity operator involves interpretations relative to counterfactual branches. This amounts to interpreting necessity using a three-place valuation operator:

$$T(t, c, \Box p) \quad \text{iff} \quad \forall c' \in C.\, t \in c' \Rightarrow Ock(t, c', p)$$

There does not seem to be any reasonable way to combine this interpretation of necessity (which depends on interpretations relative to counterfactual branches) with the above mentioned interpretation of tenses (where branches are redundant).

We shall therefore propose another semantics of tenses as well as necessity. As usual, we need a set $TIME$ equipped with a transitive and backwards linear relation $<$ together with a function T which assigns a truth value to each pair consisting of an instant and a propositional letter. Furthermore, adopting Belnap and Green's idea, we assume that there be a function TRL which to each instant assigns a branch such that the conditions (TRL1) and (TRL2') are satisfied. A novel feature of the semantics we give here is the notion of a (counterfactual) branch with the property that at any future instant it coincides with the corresponding thin red line. Given an instant t, the set $C(t)$ of such branches is defined as follows:

$$C(t) = \{c \in C \mid t \in c \wedge \forall t' > t.\ t' \in c \Rightarrow TRL(t') = c\}$$

Note that (TRL1) and (TRL2') say exactly that $TRL(t) \in C(t)$. Also, note that $C(t)$ might contain more branches than just $TRL(t)$. This allows for counterfactuality.

We shall now give a valuation operator where truth is relative to an instant as well as to a branch with the above mentioned property. We define the valuation operator T as follows:

$$\begin{array}{rll} T(t,c,p) & \text{iff} & T(t,p),\ \text{where } p \text{ is a propositional letter} \\ T(t,c,p \wedge q) & \text{iff} & T(t,c,p) \wedge T(t,c,q) \\ T(t,c,\neg p) & \text{iff} & \neg T(t,c,p) \\ T(t,c,Pp) & \text{iff} & \exists t' < t.\ T(t',c,p) \\ T(t,c,Fp) & \text{iff} & \exists t' > t.\ t' \in c \wedge T(t',c,p) \\ T(t,c,\Box p) & \text{iff} & \forall c' \in C(t).\ T(t,c',p) \end{array}$$

A formula p is said to be valid if and only if p is true in any structure $(TIME, <, T, TRL)$, that is, we have $T(t,c,p)$ for any instant t and branch c such that $c \in C(t)$.

The tense operators P and F are interpreted as in the usual Ockhamistic semantics. This makes all of the formulae (T1), (T2) and (T3) valid. On the other hand, the necessity operator is interpreted differently in the sense that fewer (counterfactual) branches are taken into account. This invalidates the formula

$$F\Diamond Fp \Rightarrow \Diamond FFp$$

which is valid in the usual Ockhamistic semantics (note that in the context of dense time, this formula is equivalent with $F\Diamond Fp \Rightarrow \Diamond Fp$). The formula is not true in the following structure:

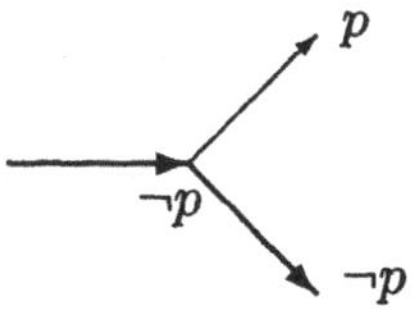

So Ockham-validity in the usual sense does not imply validity in the sense proposed above. It should, however, be mentioned that all S5 axioms are validated.

The invalidity of the above mentioned formula actually reveals a notable feature of this new semantics: it allows for the emergence of possibility in time, that is, it allows for a situation in which the truth of a proposition is possible in the future whereas the future truth of the proposition is impossible.

Further investigation of the new Ockhamistic semantics given above has to be left to future work. It can be said, however, that the idea of a semantical model in which truth is relative to a branch as well as to an instant is problematic from a philosophical point of view since the ontological status of branches is rather debatable. For this reason one might prefer the Leibnizian system presented earlier in this paper. According to this system, the branches have to be viewed as a set of 'parallel lines' on which there is defined a relation corresponding to identity up to a certain moment. In such a model made up of 'parallel lines' the TRL-function will be trivial. On the Leibnizian view, $q \Rightarrow PFq$ holds, whereas $q \Rightarrow P\Box Fq$ does not hold.

Acknowledgements

We wish to thank the staff at the Bodleian Library in the Department for Western Manuscripts, and especially in the Modern Papers Reading Room, for being always helpful when working with Prior's papers there. Thanks must also go to Balliol College for a gracious offer of accomodation while working in Oxford on Prior's papers. We are also indebted to Jack Copeland for answering various questions. Furthermore, thanks to Alberto Zanardo for stimulating discussions at the ICTL '97 conference.

Further information on Prior's life and work can be found at the following web-site:

`http://www.hum.auc.dk/~poe/prior.html`

References

Belnap, N. and M. Green: 1994, 'Indeterminism and the Thin Red Line'. *Philosophical Perspectives* **8:365-88**.

Copeland, J.: 1996, 'Prior's Life and Legacy'. In: J. Copeland (ed.): *Logic and Reality: Essays in the Legacy of Arthur Prior.* Oxford University Press/Clarendon Press, Oxford. pp. 1-40.

Findlay, J. N.: 1941, 'Time: A Treatment of Some Puzzles'. *Australasian Journal of Psychology and Philosophy* **19:216-35**. Reprinted in: A. Flew (ed.): 1951, Essays on Logic and Language. Blackwell, Oxford.

Hasle, P.: 1997, 'The Problem of Predestination - as a prelude to A. N. Prior's tense logic'. To appear in M. Wegener (editor): Time, Creation and World-Order. 22 p.

Kenny, A. J. P.: 1970, 'Arthur Normann Prior (1914-1969)'. In: *Proceedings of the British Academy*, Vol. LVI. pp. 321-49.

Klocker, H. R.: 1968, *God and the Empiricists.* The Bruce Publishing Company.

Łukasiewicz, J.: 1970, 'On Three-Valued Logic'. In: *Jan Łukasiewicz: Selected Works.* Amsterdam. Original dating 1920.

Mates, B.: 1949, 'Diodorean Implication'. *Philosophical Review* **58:234-44**.

Mates, B.: 1953, *Stoic Logic.* University of California Press.

Nishimura, H.: 1979a, 'Is the Semantics of Branching Structures Adequate for Chronological Modal Logics?'. *Journal of Philosohical Logic* **8:469-75**.

Nishimura, H.: 1979b, 'Is the Semantics of Branching Structures Adequate for Non-metric Ockhamist Tense Logics?'. *Journal of Philosohical Logic* **8:477-78**.

Øhrstrøm, P. and P. Hasle: 1993, 'A. N. Prior's Rediscovery of Tense Logic'. *Erkenntnis* **39:23-50**.

Øhrstrøm, P. and P. Hasle: 1995, *Temporal Logic: from Ancient Ideas to Artificial Intelligence.* Kluwer Academic Publishers.

Peirce, C. S.: 1931-58. In: *Collected Papers of Charles Sanders Peirce*, Vol. I-VIII. Harvard University Press.

Prior, A. N.: 1955, 'Diodorean Modalities'. *The Philosophical Quarterly* **5:205-13**.

Prior, A. N.: 1962a, 'The Formalities of Omniscience'. *Philosophy* **37:114-29**.

Prior, A. N.: 1962b, 'Tense-logic and the continuity of Time'. *Studia Logica* **13:133-48**.

Prior, A. N.: 1967, *Past, Present and Future.* Oxford.

Prior, A. N.: 1996, 'Some Free Thinking About Time'. In: J. Copeland (ed.): *Logic and Reality: Essays in the Legacy of Arthur Prior.* Oxford University Press/Clarendon Press, Oxford. With introduction by P. Øhrstrøm in which the original dating is discussed, pp. 43-44 and pp. 47-51.

Prior, A. N.: ca. 1944, 'Determinism in philosophy and theology'. Typed, 4 p. The Bodleian Library, Prior's Papers, Box 6. Unpublished.

Prior, A. N.: ca. 1951, 'The Craft of Formal Logic'. Typed, 806 p. The Bodleian Library, Prior's Papers, Box 22. Unpublished. A part of the manuscript has been published as 'The Doctrine of Propositions and Terms', ed. P. T. Geach and A. J. P. Kenny, Duckworth 1976.

Rescher, N. and A. Urquhart: 1971, *Temporal Logic.* Springer.

William of Ockham: 1983, *Predestination, God's Foreknowledge, and Future Contingents.* Hackett Publishing Company. Translated and with introduction by M. M. Adams and N. Kretzmann.

Modelling Linguistic Events

Miguel Leith (mfl@doc.ic.ac.uk) and Jim Cunningham (rjc@doc.ic.ac.uk)
Department of Computing, Imperial College of Science, Technology and Medicine, 180 Queen's Gate, London, England

Abstract. In this paper we investigate the logical representation of linguistic categories of tense and aspect with an emphasis on ease of computation. We show how temporal readings of simple sentences involving tense, the progressive and the perfect may be expressed concisely using a fragment of Halpern and Shoham's interval logic HS, and describe how to obtain the readings using a compositional semantic approach. By modelling the readings as timelines we show how linguistic entailments may be checked simply and systematically.

1. Introduction

Linguistic categories of tense and aspect allow us to describe the temporal arrangement and structure of events in natural language. Interval relations such as *later than, adjoins, overlaps, begins, ends* and *includes* may be used to analyse and represent some of these phenomena. For example the simple sentence *Max stopped running* might give rise to the informal temporal reading in Figure 1.

Even for quite simple sentences it is hard to find a straightforward formalism for representing temporal readings that may be arrived at by a compositional semantic analysis and that may be modelled intuitively so that entailments may be checked easily. Our work here focuses on a formalism which meets these criteria for a range of simple examples.

Tense and aspect in natural language is the subject of a great deal of theoretical work and has many open issues which will not be discussed here. A thorough discussion may be found in Verkuyl (Verkuyl, 1993). The work described here is related to that of Kent (Kent, 1993) who draws on the work of Moens (Moens, 1987), Lascarides (Lascarides, 1988) and Moens and Steedman (Moens and Steedman, 1988) to develop a formal account of categories of tense and aspect in an interval logic related to *IQ* (Richards et al (Richards et al., 1989)).

We too have adopted a formal semantic approach and chosen an interval logic to represent temporal readings. We analyse sentences using compositional semantic rules into readings expressed using the interval tense logic known as HS, which was described by Halpern and Shoham in (Halpern and Shoham, 1986). We define a fragment of HS which is sufficient for our expressive needs and for which formulae may be mod-

H. Barringer et al. (eds.), Advances in Temporal Logic, 207–222.

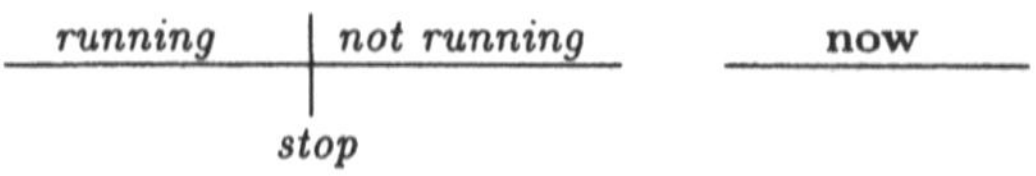

Figure 1. Time periods for the sentence *Max stopped running.*

elled as intuitive data structures called timelines. Pairs of timelines may then be compared as a way of checking linguistic entailments.

The structure of this paper, then, is as follows. In Sect.2 we give a summary of the logic HS, and discuss the modal operators which are of immediate use to us. In Sect.3 we describe our logic fragment HSF1 and our method for constructing timeline models. We then outline an interesting type of timeline that we term a Closest Position Model. In Sect.4 we describe a systematic way of obtaining the HSF1 temporal readings of simple sentences using a compositional semantic approach. This leads us to give examples of the analysis of sentences into their temporal readings and thence into timelines. Examples of the types of sentences we will be looking at are given in Figure 2. In Sect.5 we

1 - *Mike is playing tennis.*
2 - *Max stopped running.*
3 - *Jack has finished eating dinner.*
4 - *Mark will start racing.*
5 - *Fred has been playing the piano.*

Figure 2. Example sentences for which we wish to model temporal readings.

summarise the details of an entailment checker which can carry out comparisons between timelines, and demonstrate its use in modelling linguistic entailments. Examples of the types of entailments we wish to check are given in Figure 3. Finally in Sect.6 we make our conclusions.

1 - *Max stopped running* $\Rightarrow$ *Max was running*
2 - *Jack stopped eating dinner* $\not\Rightarrow$ *Jack finished eating dinner*
3 - *Mike is playing tennis* $\Rightarrow$ *Mike has been playing tennis*

Figure 3. Example entailments to check.

2. The Logic HS

HS is the interval tense logic of Halpern and Shoham, which was first described in (Halpern and Shoham, 1986) and further investigated by Venema in (Venema, 1990). In HS the set of possible worlds is the set of closed intervals in a temporal frame which may be dense or discrete.

current interval   ————————

$\langle\underline{L}\rangle\varphi$	——	1
$\langle\underline{A}\rangle\varphi$	———	2
$\langle\underline{O}\rangle\varphi$	————	3
$\langle\underline{E}\rangle\varphi$	—————	4
$\langle\underline{D}\rangle\varphi$	——————	5
$\langle B\rangle\varphi$	———	6
$\langle\underline{B}\rangle\varphi$	—————	7
$\langle D\rangle\varphi$	——	8
$\langle E\rangle\varphi$	———	9
$\langle O\rangle\varphi$	————	10
$\langle A\rangle\varphi$	———	11
$\langle L\rangle\varphi$	——	12
$[[BP]]\varphi$		13
$[[EP]]\varphi$		14

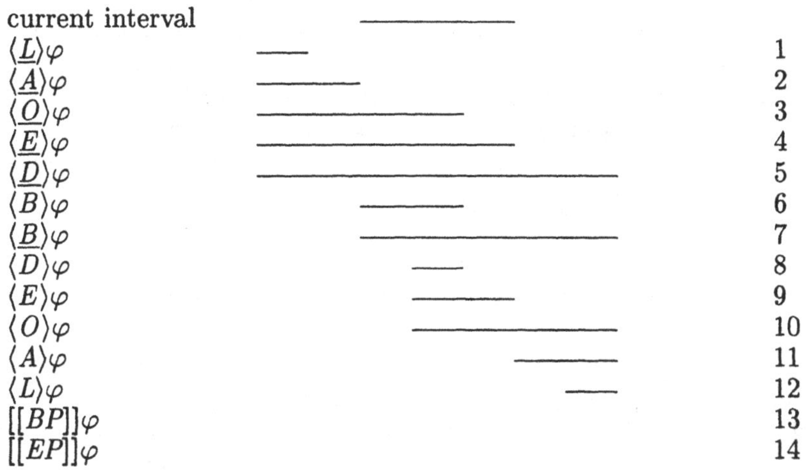

Figure 4. Pictorial illustration of relative interval positions and their HS formulae

Its six basic modal operators and six common derived modal operators correspond to relative interval positions as illustrated by the diagram in Figure 4. We give additionally the derived beginning and end point operators. The corresponding necessitation operators are defined in the usual way.

One may notice a similarity between some of the HS operators and those of Allen's interval logic (see (Allen, 1984)). HS operators in fact express the same interval relations as Allen's operators, the difference being that the latter use explicit interval variables.

Allen	HS
DURING$(i1, i2)$	$\langle D\rangle$ and $\langle\underline{D}\rangle$
STARTS$(i1, i2)$	$\langle B\rangle$ and $\langle\underline{B}\rangle$
FINISHES$(i1, i2)$	$\langle E\rangle$ and $\langle\underline{E}\rangle$
BEFORE$(i1, i2)$	$\langle L\rangle$ and $\langle\underline{L}\rangle$
OVERLAP$(i1, i2)$	$\langle O\rangle$ and $\langle\underline{O}\rangle$
MEETS$(i1, i2)$	$\langle A\rangle$ and $\langle\underline{A}\rangle$

2.1. A SUBSET OF HS OPERATORS

An interesting subset of HS operators is the union of the set of all possibility operators with one of the necessitation operators, $[D]$, which means informally *all intervals during the current interval.* The $[D]$ operator is of particular use here as it may be used to effect *homogeneity*, an important concept in event representation. A proposition is said to be homogeneous if it is true at all sub-intervals of any intervals at which it holds. Homogeneity is used by Kent in (Kent, 1993) as a property on states. If someone was running for five minutes, for instance, then they were running at all periods of time within that five minutes, unless an exception is specified, in which case the proposition may be split into two or more homogeneous periods instead. By having a formula $[D](run(max))$ hold at an interval we would be denoting that Max is running at all intervals/points during that interval.

3. HSF1: A Fragment of HS Logic

We begin this section by defining the fragment HSF1 which we later use to represent temporal readings of sentences. We describe how to model the formulae from the fragment as *timelines*. Lastly we give a full example of a timeline being constructed, and define a type of timeline that is of particular interest.

The syntax of HS is defined using grammar rules as follows.

HSF1 : Exp2 | PossOp HSF1
PossOp : $\langle D\rangle$ | $\langle \underline{D}\rangle$ | $\langle E\rangle$ | $\langle \underline{E}\rangle$ | $\langle B\rangle$ | $\langle \underline{B}\rangle$ | $\langle L\rangle$ | $\langle \underline{L}\rangle$ |
$\langle A\rangle$ | $\langle \underline{A}\rangle$ | $\langle O\rangle$ | $\langle \underline{O}\rangle$ | $[[EP]]$ | $[[BP]]$
Exp2 : ExtendedAtomic | HSF1 $\wedge$ HSF1
ExtendedAtomic : Atomic | $[D]$ Atomic
Atomic : PropConstant | $\neg$ PropConstant
PropConstant : p_1 | p_2 | p_3 ...

We impose a restriction on frames by requiring that the set of points T in a temporal frame is dense, linear and unbounded.

3.1. Modelling HSF1 Formulae

PROPOSITION 1. *Any HSF1 formula may be modelled by a finite number of alternative data structures which we will refer to as timelines.*

DEFINITION 1. *A timeline is a triple* $\langle$*Start Interval, Ordering, Interval Mapping*$\rangle$ *where Start Interval is* **now**, *Ordering is a sequence of point variables in linear dense unbounded time and Interval Mapping is a mapping between intervals and the sets of extended atomic formulae. An interval is represented by a pair of point variables from the ordering e.g.* (t_1, t_2).

Proposition 1 is justified because in the case of the HSF1 fragment, the semantics of an HSF1 formula can be interpreted by a procedure which builds the timelines that may model it. The procedure by which timelines are built is defined informally below. In general there may be more than one possible timeline for a given HSF1 formula although some formulae may give rise to only one.

Timelines are built by an algorithm called the Timeline Construction Algorithm (TCA). The algorithm is a recursive traversal of an HSF1 formula that builds an ordering and an interval mapping. At each stage of the recursive traversal we are at a particular *current interval* denoted by a pair of point variables e.g. (t_1, t_2). Each modal operator (except for $[D]$) is a possibility operator in its positive form (e.g. $\langle E\rangle$, $\langle L\rangle$, $[[EP]]$ etc.) and effectively asserts the existence of an interval in given temporal relation to the current interval. Thus the effect of encountering a possibility operator in the algorithm is to change the current interval and update the ordering to reflect the points of the new current interval in their correct positions. If the algorithm encounters an extended atomic formula it will augment the interval mapping to reflect the formula being true at the current interval. For instance if it encounters a formula $[D]$(p) (or a formula p) at current interval (t_4, t_3) it will add $[D](p)$ (or p) to the set of formulae holding at (t_4, t_3).[1] We henceforth refer to a single entry of an interval mapping $((t_x, t_y) : \{ F_1, F_2 \ldots F_n \})$ as a *binding*.

By the preceding argument a timeline modelling an HSF1 formula of depth n (where we count the starting interval and each possibility modality as adding 1 to the depth) has at most $2n$ time points in its ordering. When we take into account the possible branching effects when new intervals are introduced, a potentially large but finite set of alternative timelines may arise.

3.2. Timeline Construction Example

There follows a timeline construction example for the formula:

$$\langle \underline{L}\rangle([[EP]](\langle \underline{A}\rangle([D](p))))$$

proceeding as described in Figures 5 to 9, where Figure 5 shows the initial state of the algorithm. The ordering is represented as a sequence of point variables marked by a vertical line | with a variable name attached. An asterisk *, positioned between two point variables, marks the start interval. The current interval is marked by a curved line joining the interval's left bound with its right bound, and the previous interval is marked with a dotted curved line in a similar way (initially there is no previous interval).

[1] By the semantics of the operator $[D]$, the formula $[D](p)$ must hold at all internal durative intervals and p must hold at all internal intervals/points. These possible inferences must be reflected in any algorithms that reason using timelines.

Initial timeline state:

Next formula: $\langle \underline{L} \rangle([[EP]](\langle \underline{A} \rangle([D](p))))$

Figure 5. Timeline construction example, part 1

Interpreting: $\langle \underline{L} \rangle$

Next formula: $[[EP]](\langle \underline{A} \rangle([D](p)))$

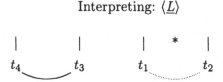

Figure 6. Timeline construction example, part 2

In Figure 6 the $\langle \underline{L} \rangle$ operator, which means *earlier than*, is analysed. The previous interval (t_1, t_2), marked by a dotted curve, is the current interval of Figure 5. Two new point variables are inserted into the ordering in positions earlier than (t_1, t_2), and become the left and right bounds of the new current interval, (t_4, t_3).

In Figure 7 the $[[EP]]$ operator, which means *end point of*, is analysed. The end point of the previous interval, t_3, becomes both the left and right bound of the new current interval, (t_3, t_3). The ordering does not have to be changed as no new points have to be added.

In Figure 8 the $\langle \underline{A} \rangle$ operator, which means *adjoins at the beginning of*, is analysed. The beginning point of the previous interval, t_3, becomes the right bound of the new current interval, but there are many alternative positions where the left bound of the new current interval

Interpreting: $[[EP]]$

Next formula: $\langle \underline{A} \rangle([D](p))$

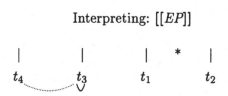

Figure 7. Timeline construction example, part 3

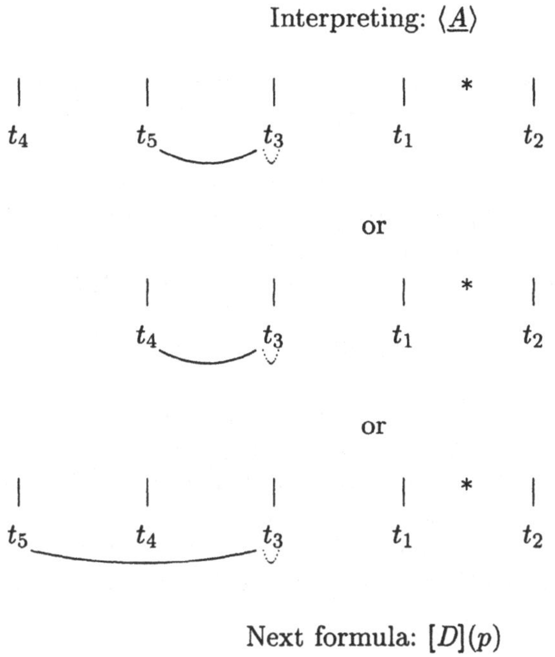

Figure 8. Timeline construction example, part 4

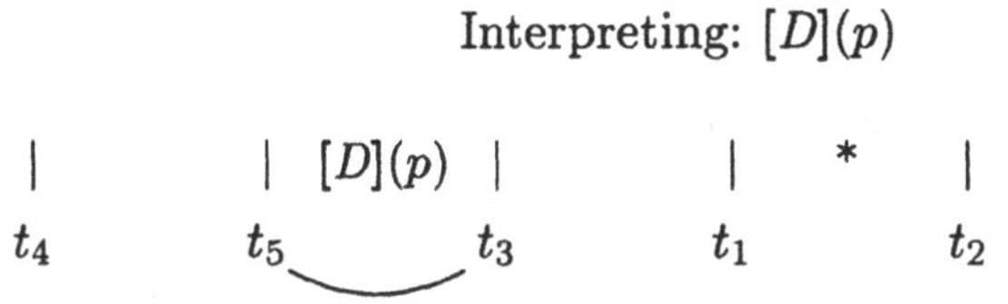

Figure 9. Timeline construction example, part 5

may go, because there is an existing point variable positioned before t_3.

We pick the first alternative from Figure 8 as the state from which to continue our construction example from, noting that the full example would fan out into alternative timelines at this juncture. In Figure 9 the extended atomic formula $[D](p)$ is analysed. The interval mapping is updated, represented in our diagram by the formula $[D](p)$ being positioned at the current interval, (t_5, t_3).

Finally the point variable t_4 is removed because it is superfluous i.e. it is present in the ordering but it is not present in the start interval, focus or interval mapping. The timeline is then complete.

3.3. Closest Position Models

Of the timelines that may model an HSF1 formula, there will be one timeline that we term its *closest position model* (CPM). The idea arises as follows. Each time the TCA interprets the semantics of a possibility operator such as $\langle \underline{A} \rangle$, $\langle \underline{E} \rangle$ etc. it changes the current interval by selecting point variables in appropriate positions in the timeline ordering (see Figures 6, 7 and 8). Whenever a new point variable is to be inserted before/after some point variable t_n the CPM is constructed by selecting a new point variable immediately before/after t_n in the ordering. For example the first alternative in the analysis of the operator $\langle \underline{A} \rangle$ in Figure 8 enables a CPM.

CPMs separate future, past, and present, in preference to alternatives with overlapping intervals. They are required for the simple compositional semantics given in Sect.4. Natural language entailments can be demonstrated without restricting the class of models (see (Leith and Cunningham, 1997)), but psychological interpretation may not consider all logical alternatives (see (Johnson-Laird and Byrne, 1991) for related discussion).

4. Temporal Sentence Semantics

In this section we show how formulae from the HSF1 fragment may be used to represent the temporal readings of the sentences given in Figure 2.

4.1. The Treatment of Start, Stop and Culmination

In our temporal readings of natural language sentences great importance is given to three special types of time point: a state's start point, stop point and culmination point. In this subsection we define each of the three point types in terms of the HSF1 formulae that hold at them and illustrate the principles with diagrams. But first we define what we mean by a state.

DEFINITION 2. *An HSF1 state is a formula of the form* $[D](P)$ *where* P *may be any atomic formula. The state of an atomic proposition* p *is* <u>*on*</u> *at any interval/point at which* $\langle \underline{D} \rangle([D](p))$ *holds. The state of an atomic proposition* p *is* <u>*off*</u> *at any interval/point at which* $\langle \underline{D} \rangle([D](\neg p))$ *holds.*

A start point of a state is a point of transition from the state being off to the state being on. This means that $[D](\neg p)$ must hold at an interval adjoining the point on the left, and $[D](p)$ must hold at an interval adjoining the point on the right.

DEFINITION 3. *A start point of a state* $[D](p)$ *is any point at which the formula* $\langle \underline{A} \rangle ([D](\neg p)) \wedge \langle A \rangle ([D](p))$ *holds.*

$$| \ [D](\neg p) \ | \ \ [D](p) \ \ |$$

Figure 10. Start point of a state

A stop point of a state is a point of transition from the state being on to the state being off. This means that $[D](p)$ must hold at an interval adjoining the stop point on the left, and $[D](\neg p)$ must hold at an interval adjoining the stop point on the right.

DEFINITION 4. *A stop point of a state* $[D](p)$ *is any point at which the formula* $\langle \underline{A} \rangle ([D](p)) \wedge \langle A \rangle ([D](\neg p))$ *holds.*

$$| \ \ [D](p) \ \ | \ \ [D](\neg p) |$$

Figure 11. Stop point of a state

A culmination point of a state is a point at which the state is said to be *finished* or *completed.* Culmination points are critical to our understanding of linguistic expressions such as *Jack finished eating dinner.* An issue for some time in event representation has been the question of what a proposition's truth value is at a stop or start point of a state. We define a culmination of a state as a stop point of the state for which the state's atomic proposition is true. Formally, this is defined below.

DEFINITION 5. *A culmination point of a state* $[D](p)$ *is any stop point of* $[D](p)$ *at which* p *holds.*

$$| \ \ [D](p) \ \ |_p \ [D](\neg p) |$$

Figure 12. Culmination point of a state

4.2. Obtaining Temporal Readings Compositionally

To work out the HSF1 temporal readings of simple sentences like those in Figure 2 we adopt a compositional approach. Semantic operators correspond to syntactic features and are represented as lambda terms which are combined in a prescribed order. The composite lambda term is then reduced to yield the temporal reading of the sentence. We describe the lambda terms next.

PERF: $\lambda w.\langle \underline{A} \rangle(w)$
Feature correspondence(s): auxiliary *have* coupled with perfective form of verb.

PAST: $\lambda x.\langle \underline{L} \rangle(x)$
Feature correspondence(s): past tense form of main verb or auxiliary.

PRES: $\lambda x.x$
Feature correspondence(s): present tense form of main verb or auxiliary.

FUTURE: $\lambda x.\langle L \rangle(x)$
Feature correspondence(s): auxiliary *will* coupled with base form of verb.

BE: $\lambda y.\lambda f.\langle \underline{D} \rangle((f)(y))$
Feature correspondence(s): auxiliary *be.*

CULM: $\lambda y.\lambda f.[[EP]](y \wedge \langle \underline{A} \rangle((f)(y)) \wedge \langle A \rangle((f)(\neg y)))$
Feature correspondence(s): a verb such as *finish* that records a sense of accomplishment in a verb phrase e.g. *finish reading the book.*

START: $\lambda y.\lambda f.[[EP]](\langle \underline{A} \rangle((f)(\neg y) \wedge \langle A \rangle((f)(y)))$
Feature correspondence(s): a verb denoting the starting of a state e.g. *start.*

STOP: $\lambda y.\lambda f.[[EP]](\langle \underline{A} \rangle((f)(y)) \wedge \langle A \rangle((f)(\neg y)))$
Feature correspondences: a verb denoting the stopping of a state e.g. *stop.*

ING: $\lambda z.[D](z)$
Feature correspondence(s): an *ing*-inflection of a main verb denoting a state of some kind, for instance *running, eating.*

PROP: p (A lambda calculus constant, some atomic proposition for a simple sentence e.g. $run(max)$, $eat(jill, apple)$).
Feature correspondence(s): the basic temporally unmodified meaning of a sentence e.g. *Max run, Jill eat apple.*

The lambda terms are combined into a larger term in the following structure:

$$(\text{Pos1})(\text{Pos2})(((\text{Pos3})\text{PROP})\text{ING})$$

where Pos1 may be occupied by PERF, Pos2 may be occupied by PAST, PRES or FUTURE and Pos3 may be occupied by BE, CULM, STOP or START. A composite term with this structure is reduced to obtain the temporal reading of a sentence.

4.3. Example Temporal Readings

We give here our temporal readings of some of the sentences listed in Figure 2.

Sentence Semantics

$(\text{PRES})(((\text{BE})(play(mike, tennis)))(\text{ING})) \equiv$

reduces to

$\langle \underline{D} \rangle([D](play(mike, tennis)))$

Closest Position Model

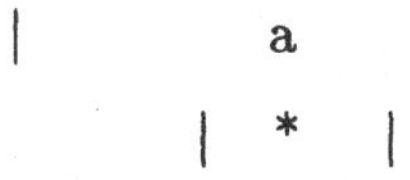

Timeline Index

a: $[D](play(mike, tennis))$ *: *now*

Figure 13. A temporal reading of the sentence *Mike is playing tennis*

The present progressive results in a timeline like the one in Figure 13. The stopped event in Figure 14 is placed in the past.

Perfectives convey a sense of current significance or recency. For instance *Jack has finished eating dinner* appears to denote that the completion of the event took place at a time that is in touch with the current temporal focus i.e. *now*. We attempt to capture this sense by applying the operator $\langle \underline{A} \rangle$ to represent the perfective, meaning that it adjoins the beginning of *now*. The right point of *now* may be seen as the current instant in time.

The start of the racing event in Figure 16 is placed in the future.

Sentence Semantics

$(\text{PAST})(((\text{STOP})(run(max)))(\text{ING})) \equiv$

reduces to

$\langle \underline{L} \rangle([[EP]](\langle \underline{A} \rangle([D](run(max))) \wedge \langle A \rangle([D](\neg run(max)))))$

Closest Position Model

Timeline Index

b: $[D](run(max))$ a: $[D](\neg run(max))$ $*$: now

Figure 14. A temporal reading of the sentence *Max stopped running*

Sentence Semantics

(The proposition $eat(jack, dinner)$ is abbreviated to ejd)

$(\text{PERF})(\text{PRES})(((\text{CULM})(ejd))(\text{ING})) \equiv$

reduces to

$\langle \underline{A} \rangle([[EP]](ejd \wedge \langle \underline{A} \rangle([D](ejd)) \wedge \langle A \rangle([D](\neg ejd))))$

Closest Position Model

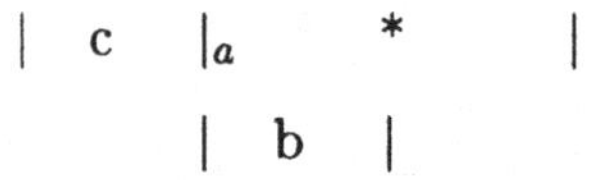

Timeline Index

c: $[D](eat(jack, dinner))$ b: $[D](\neg eat(jack, dinner))$

a: $eat(jack, dinner)$ $*$: now

Figure 15. A temporal reading of the sentence *Jack has finished eating dinner*

5. An Entailment Checker for Timelines

In the general case an entailment between two HSF1 formulae holds if each timeline modelling the left formula entails a timeline modelling the right formula. Therefore if we determine a single timeline for each HSF1 formula there is a one-to-one correspondence between an HSF1 entailment and its timeline entailment. An entailment between a pair of timelines holds if the timeline of the right formula is a *substructure* of the timeline of the left formula, which in informal terms means it will *fit* into it. We refer to the algorithm which compares the timelines

Sentence Semantics

$(\text{FUTURE})(((\text{START})(race(mark)))(\text{ING})) \equiv$

reduces to

$\langle L\rangle([[EP]](\langle \underline{A}\rangle([D](\neg race(mark))) \wedge \langle A\rangle([D](race(mark)))))$

Closest Position Model

| * | | b | a |

Timeline Index

b: $[D](\neg race(mark))$ a: $[D](race(mark))$ *: now

Figure 16. A temporal reading of the sentence *Mark will start racing*

Exact match, atomic case

[$prop$]
[$prop$]

Exact match, homogeneous case

[$[D](prop)$]
[$[D](prop)$]

Right homogeneous match

[$[D](prop)$]
[$[D](prop)$]

Left homogeneous match

[$[D](prop)$]
[$[D](prop)$]

Internal match, homogeneous case

[$[D](prop)$]
[$[D](prop)$]

Internal match, atomic case

[$[D](prop)$]
[$prop$]

Figure 17. How pairs of formulae may be matched by the SFA

as the Substructure Fitting Algorithm (SFA). The algorithm succeeds if a new timeline may be created which represents the match of the two timelines. We will use the terms *substructure timeline*, *superstructure timeline* and *matched timeline* to distinguish between the operative timelines in this algorithm.

The objective of the SFA is to build a new timeline (the matched timeline) in which every formula in every binding of the substructure timeline is matched by a formula from a binding in the superstructure timeline in one of the ways depicted in Figure 17 such that the matched ordering remains consistent with the superstructure ordering and substructure ordering. Full details of the SFA will be given in a forthcoming thesis.

Entailment Check

$\langle \underline{L} \rangle([[EP]](\langle \underline{A} \rangle([D](\neg\, run(max))) \wedge \langle A \rangle([D](run(max))))) \models_{cpm}$
$\langle \underline{L} \rangle(\langle \underline{D} \rangle([D](run(max))))$

Superstructure Timeline (CPM)

| B | A | | * |

Timeline Index

B: $[D](run(max))$ A: $[D](\neg run(max))$

$*$: *now*

Substructure Timeline (CPM)

| a | | * |

Timeline Index

a: $[D](run(max))$ $*$: *now*

Match Timeline

| Ba | A | | * |

Figure 18. Entailment check: *Max stopped running* ⇒ *Max was running*

5.1. Entailment Examples

In Figures 18 and 19 we show checks for the logical entailments corresponding to two of the natural language entailments in Figure 3.

6. Conclusion

In this paper we have demonstrated a route to compositional temporal semantics of natural language events using interval temporal logic. We believe that our representation of states, start points, stop points and in particular culmination points are a contribution towards intuitive formal semantics of natural language events. The restrictions on the fragment of the logic and the models we allow in order to achieve compositional correspondences may be of psychological relevance and a hint for a new logic. Work is currently in progress to refine our analysis of tense and the perfective by linking with the work of Reichenbach ((Reichenbach, 1947)).

Entailment check:
(The proposition $eat(jack, dinner)$ is abbreviated to ejd)
$\langle \underline{L} \rangle([[EP]](\langle \underline{A} \rangle([D](ejd)) \wedge \langle A \rangle([D](\neg ejd)))) \models_{cpm}$
$\langle \underline{L} \rangle([[EP]](ejd \wedge \langle \underline{A} \rangle([D](ejd)) \wedge \langle A \rangle([D](\neg ejd))))$

Superstructure Timeline (CPM)

| B | A | | * |

Timeline Index
A: $[D](\neg eat(jack, dinner))$ B: $[D](eat(jack, dinner))$
*: Now

Substructure Timeline (CPM)

| b $|_c$ a | | * |

Timeline Index
a: $[D](\neg eat(jack, dinner))$ b: $[D](eat(jack, dinner))$
c: $eat(jack, dinner)$ *: Now

Match Timeline:
None.

Figure 19. Entailment check: Jack stopped eating dinner $\not\Rightarrow$ Jack finished eating dinner

References

Allen, J.: 1984, 'Towards a General Theory of Action and Time'. *Artificial Intelligence* **23**(2), 123–154.

Halpern, J. and Y. Shoham: 1986, 'A Propositional Modal Logic of Time Intervals'. In: *Proceedings of Symposium on Logic in Computer Science.* Cambridge, Massachusetts, pp. 279–292.

Johnson-Laird, P. and R. M. Byrne: 1991, *Deduction.* Erlbaum.

Kent, S.: 1993, 'Modelling Events from Natural Language'. Ph.D. thesis, Imperial College, London.

Lascarides, A.: 1988, 'A Formal Semantic Theory of the Progressive'. Ph.D. thesis, University of Edinburgh.

Leith, M. and J. Cunningham: 1997, 'Representing and Reasoning with Events from Natural Language'. In: *First International Joint Conference on Qualitative and Quantitative Practical Reasoning.*

Moens, M.: 1987, 'Tense, Aspect and Temporal Reference'. Ph.D. thesis, University of Edinburgh.

Moens, M. and M. Steedman: 1988, 'Temporal Ontology and Temporal Reference'. *Computational Linguistics 14* pp. 15–28.

Reichenbach, H.: 1947, *Elements of Symbolic Logic.* Macmillan, London.

Richards, B., I. Bethke, J. van der Does, and J. Oberlander: 1989, *Temporal Representation and Inference.* Academic Press.

Venema, Y.: 1990, 'Expressiveness and Completeness of an Interval Tense Logic'. *Notre Dame Journal of Formal Logic* **31**(4), 529–547.
Verkuyl, H.: 1993, *A theory of aspectuality.* Cambridge University Press.

A Dynamic Temporal Logic for Aspectual Phenomena in Natural Language

Ralf Naumann
Seminar für Allgemeine Sprachwissenschaft, University of Düsseldorf

Abstract. It is usually thought that Dowty's decompositional approach to aspect and that of event-semantics are incompatible with each other. In this paper this is shown to be wrong by presenting a theory that combines both approaches. The decompositional component is represented by a two-sorted variant B of TL with corresponding structure **B** introduced by Van Benthem et al. (1994) in which the second sort besides that of formulas is that of procedures. The event-component is represented by an eventuality-structure **E**. By combining both structures one gets a double perspective: whereas at the level of **E** events are structureless objects, they are assigned an internal structure in form of a sequence in **B** according to which events are interpreted as state-transformers. This makes it possible to define various aspectual classes that cannot be defined in event-semantics.

1. Introduction

Any theory of aspect has to account for the data given in (1).

(1) a. John ate an apple in ten minutes/*for ten minutes.
b. John pushed a cart *in ten minutes/for ten minutes.

A sentence with the verb *eat* (in the simple past tense and arguments that denote single (atomic) objects) can be modified with an *in*-adverbial but not with a *for*-adverbial. This property will be called terminativity. A sentence with the verb *push*, on the other hand, admits of modification with a *for*-adverbial but not with an *in*-adverbial. This property will be called durativity. Next consider the data in (2).

(2) a. Bill knocked at the door at three o'clock.
b. John arrived at three o'clock.
c. John was ill at three o'clock/*in ten minutes/for three days.
d. Mary washed her hands in ten minutes/for ten minutes.

The sentence (2a) gives an example of a verb which always admits of modification with an *at*-adverbial if the arguments are atomic. This does not hold for verbs like *push* and *eat* and therefore shows that a verb like *knock* has aspectual properties that differ from those of both *push* and *eat*. The verb *arrive* shows a similar behaviour to *knock* with respect to *at*-adverbials. But both verbs behave differently with respect to the perfect of result (PR).

H. Barringer et al. (eds.), Advances in Temporal Logic, 223–253.

(3) a. * Bill has knocked at the door.

b. John has arrived.

The sentence in (3a) cannot be used to make the claim that there is a past knocking the result of which still holds at speech time (It can only be used as an experiential perfect meaning that Bill has knocked *at least once* in the past at the door). In the case of *arrive* no such restriction exists. In its PR-reading the example in (3b) means that John arrived some time ago and is still here. Consequently, *knock* must be distinguished from *arrive*. A stative verb like *be ill* also admits of modification with an *at*-adverbial, like *knock* and *arrive*, but it can only be modified with a *for*-adverbial (2c). A verb like *wash*, finally, admits of modification both with an *in*-adverbial as well as with a *for*-adverbial (2d). This is different from all five classes considered so far. According to the above discussion, at least six aspectual classes must be distinguished. In (4) they are listed by their standard names which (mostly) go back to Vendler (1967).

(4)	Accomplishments	eat, build, cross
	Activities (Processes)	run, push, walk
	States (stative Verbs)	be ill, hate, love
	Points	knock, hit, hickup, twinkle
	Achievements	arrive, die, reach
	Proc-Acco Verbs	wash, hammer, iron

This variety of aspectual distinctions at the level of verbs poses serious difficulties both for event-semantics (ES) (see (Krifka, 1992)) and the index-approach developed by Verkuyl (1993). In both approaches the class of so-called occurrences (represented by non-stative verbs) is characterized by only one subdivision which is defined by the notions of cumulativity and being quantized (ES) or those of durativity and terminativity (Verkuyl). But then only a distinction between activities and accomplishments can be made. The classes of achievements, points and Proc-Accos remain unaccounted for. Furthermore the class of stative verbs is problematic for both approaches (for further arguments and discussion see Naumann (1995)). A second problem has to do with the relation between tense and aspect. Both dimensions are analyzed independently of each other in the sense that the aspectual properties of the verb and therefore of an untensed proposition have no influence on the interpretation of the temporal dimension. This independence gives rise to problems with respect to the interpretation of the perfect. The perfect has both a temporal and an aspectual dimension. The aspectual dimension is most salient in the so-called perfect of result as discussed with respect to (3b), *John has arrived*, above. For this sentence to be felicitous John must still be here at speech time. In the common analysis which goes back to Reichenbach this condition is not accounted for. There is nothing in the difference between the analysis of the present perfect (E<R=S) that guarantees that John is still here as opposed to the corresponding sentence in the past tense (E=R<S) for which no such implication holds (i.e. John can already have left again). A further problem, already mentioned above, has to do with aspectual restrictions. Not each aspectual class can be used as a perfect of result but only those that characterize the lexical content of a verb as having a well-defined result. Thus *John has arrived* is fine whereas *John has run* is not because a running has no well-defined result state which is reached as the outcome of this activity.

2. The Decompositional Approach of Dowty

In Dowty (1979) Dowty presented a decompositional approach to aspect. This approach is based on the following assumptions. Stative verbs like *be ill* or *hate* are taken as basic, i.e., they do not receive a decompositional analysis. The reason is that they can be evaluated at single time points and can therefore be interpreted as properties of times (or states). All verbs belonging to another aspectual class are semantically decomposed by means of stative predicates and aspectual operators. The difference between accomplishments and activities is explained in terms of the notion of change. Whereas accomplishments express a definite change, activities express an indefinite change. Informally, the difference between a definite and an indefinite change is explained as follows. Each verb is evaluated with respect to an interval I. Furthermore, to each non-stative verb V there corresponds a specific stative predicate P_v. For an accomplishment-verb V_{acco} one then gets $[[V_{\text{acco}}]](\ldots x_i \ldots) = 1$ on an interval I just in case $[[P_{V_{\text{acco}}}]](x_i) = 1$, at the end of I (where x_i is one of the arguments of V_{acco}), $[[P_{V_{\text{acco}}}]](x_i) = 0$ at the beginning of I and there is no proper subinterval I' of I such that the two conditions hold for I'. The last requirement expresses a kind of minimality-condition. It is this condition which characterizes the notion of definite change. For an activity-verb V_{act}, on the other hand, there is no such minimality-condition. Rather their interpretation is determined by the first two conditions: $[[V_{\text{act}}]](\ldots x_i \ldots) = 1$ on an interval I just in case $[[P_{V_{\text{act}}}]](x_i) = 0$ at the beginning of I and $[[P_{V_{\text{act}}}]](x_i) = 1$ at the end of I.

For accomplishment-verbs these informal considerations are made formally precise by means of the BECOME-operator. This operator is a generalization of Von Wright's NEXTTIME-operator to an interval-structure. It is defined in (5b). In (5a) the decomposition for an accomplishment-verb is given.

(5) a. $V_{\text{acco}} \Rightarrow \text{CAUSE}(\phi, \text{BECOME}(P_v(x_1, \ldots, x_n)))$[1]

b. $[[\text{BECOME}(\phi)]]^{M,g}(I) = 1$ iff

(i) $[[\phi]]^{M,g}(J) = 0$ where J is the lower bound of I

(ii) $[[\phi]]^{M,g}(K) = 1$ where K is the upper bound of I

(iii) there is no $I' \subset I$ satisfying conditions (i) and (ii)

From the definition in (5b) it follows that non-stative verbs can only be true on non-minimal (non-singleton) intervals, i.e. intervals that do not correspond to a single time point. If one assumes that time is linear and discrete, intervals can be defined as (convex) subsets of S, i.e. the set of time points. The definition in (5b) can then be rewritten as (6).

(6) $[[\text{BECOME}(\phi)]]^{M,g}(I) = 1$ iff

(i) $[[\phi]]^{M,g}([s_0]) = 0$

(ii) $[[\phi]]^{M,g}([s_n]) = 1$ (where $I = [s_0, s_n]$)

(iii) there is no $I' \subset I$ satisfying conditions (i) and (ii)

An accomplishment-verb is interpreted as a kind of state-transformer: it transforms a state s_0 at which ϕ does not hold to a state where ϕ holds, where s_n is the first state, relative to s_0, where ϕ holds. There is the following serious problem that faces the definition in (6) (or that in (5b)). It allows only changes on intervals $[s, s']$ consisting of exactly two time points. Any longer interval necessarily does not satisfy the minimality condition in (iii). If, on the other hand, the minimality condition is dropped, one again gets results that are intuitively not correct. For instance, suppose that John moved from New York to San Francisco in 1940 and has lived there from then on until today. Then *John moved from New York to San Francisco* is true on the interval [1940, 1997] according to (6). But for that interval one must use the present perfect *John has moved to San Francisco* and not the past tense.

3. A Decompositional Dynamic Temporal Logic

Dowty's mistake can be analyzed as follows. The minimality-condition must not be expressed in terms of the subintervals of I. Rather it must

[1] In the sequel I will write ϕ instead of $P_V(x_1, \ldots, x_n)$.

be expressed in terms of ϕ relative to s, i.e. the first point of I, as it has already been done above when the interpretation of an accomplishment-verb was described as a minimal state-transformer: given a state s in which ϕ does not hold, $[[V_{\text{acc}}]]$ transforms this state into the first state s' (given the linear order) at which ϕ does hold. Applied to the example from the preceding section, John started in New York and continued moving (driving) until he eventually arrived in San Francisco. Let the corresponding interval be $I = [s_0, s_5]$. The temporal structure corresponding to this action is then given in Figure 1. The structure in

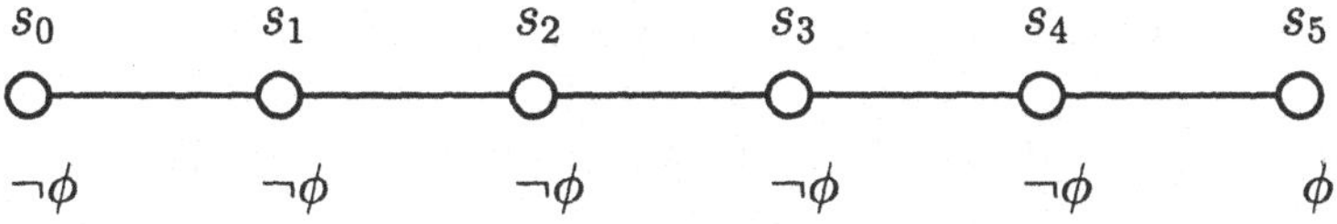

Figure 1.

Figure 1 can be described by the Until-operator from temporal logic (TL).

(7) $M, s_0 \models \neg\phi \wedge \mathbf{Until}(\neg\phi, \mathbf{G}\bot \wedge \phi)$

For an action described by an activity-verb V_{act} one gets the structure in Figure 2. The condition ϕ that has to hold in the last state $s_n (= s_5)$

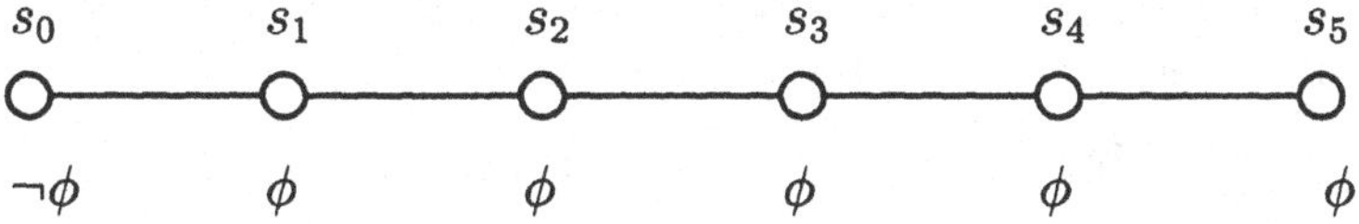

Figure 2.

of the interval holds for all states s_i, except the initial state s_0. For instance, in the case of the activity-verb *run* ϕ can be taken to be the condition that the location of the agent is different from the location he started from (or, alternatively, that the agent has traversed a non-empty path).

The intervals I on which stative and non-stative verbs are evaluated can be decomposed into three different parts.

(8) $]s_0, s_n[$ = Development Portion (DP) (Preparatory Process)
$[s_0]$ = Inception Point (IP)
$[s_n]$ = Culmination (Cul)
$[s_n, s_m]$ = Consequent State (CS)[2]

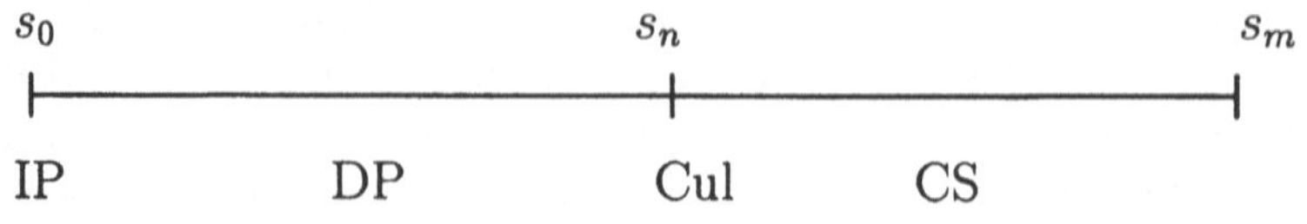

Figure 3.

The terminology is taken from Moens and Steedman (1988)[3]. The quadripartite structure is called a nucleus-structure. Intuitively, it can be interpreted as providing the internal structure of an event (or an action). The inception point denotes the beginning of the event, the culmination its end and the development portion that time where the event occurs.

The difference between accomplishment-verbs and activity-verbs concerns the development-portion. Whereas the negation of ϕ constantly holds on the DP for accomplishment-verbs, ϕ is constantly true on the DP in the case of an activity-verb. When viewed as simple input/output-relations, there is no difference: $\neg\phi$ holds in the input and ϕ in the output. What is needed is a language that makes it possible to directly express properties of these three parts of the nucleus-structure. Ordinary (linear) TL is not well suited for this purpose. As (7) above shows, the Until-operator does not directly admit to express that its second argument be true in the final state of a sequence. This must be separately expressed by means of $\mathbf{G}\bot$. This becomes clear if one looks at the usual interpretation of the Until-operator in TL.

(9) $M, s \models \mathbf{Until}(\psi, \phi)$ iff there is an $s' > s$ s.t. $M, s' \models \phi$ and for all s'' with $s < s'' < s'$, $M, s'' \models \psi$

If the underlying domain corresponds to the evaluation interval $I = [s_0, s_n]$ with $s_0 = s$, (9) does not guarantee that $s' = s_n$. A similar argument holds in the case of sequences characterizing activity-verbs: $\mathbf{G}\phi$ does not distinguish between Cul and DP.

In what follows I will use a variant of the two-sorted language B of temporal logic with composite 'between'-procedures that has been introduced by Van Benthem et al. (1994). The principle operator of this language is the Until-operator. The basic idea is to introduce this operator not as primitive but rather to decompose it into different components which are then interpreted by means of different procedures.

[2] The notion of a consequent state plays a role in the analysis of the perfect of result; see Naumann (1996b).

[3] Except for the notion of an inception point.

The decomposition is based on the fact that Until has both a universal and an existential component, represented by a universal quantification and an existential one in its truth-conditions. The Until-operator can then be defined as a (procedural) boolean combination of the different components. The syntax of B is defined in (10).

(10) a. B-formulas $\phi ::= p \mid \neg\phi \mid \phi \wedge \phi' \mid \mathrm{DO}(\alpha)$
b. B-procedures $\alpha ::= R\phi \mid L\phi \mid\ \sim\alpha \mid \alpha\,;\alpha' \mid \alpha \cap \alpha'$

The semantics is defined in a way that is used in two-dimensional modal logics. The interpretation of formulas ϕ is given with respect to a single state (11), whereas procedures are defined as relations between states by the corresponding satisfaction relation (12).

(11) a. $M, s \models p$ iff $s \in V(p)$ for p atomic
b. $M, s \models \neg\phi$ iff it is not the case that $M, s \models \phi$
c. $M, s \models \phi \wedge \psi$ iff $M, s \models \phi$ and $M, s \models \psi$

(12) a. $M, s, s' \models R\phi$ iff $M, s' \models \phi$
b. $M, s, s' \models L\phi$ iff $M, s \models \phi$
c. $M, s, s' \models \sim\alpha$ iff it is not the case that $M, s, s' \models \alpha$
d. $M, s, s' \models \alpha \cap \alpha'$ iff $M, s, s' \models \alpha$ and $M, s, s' \models \alpha'$
e. $M, s, s' \models \alpha\,; \alpha'$ iff there is some s'' s.t. $s < s'' < s'$ and $M, s, s'' \models \alpha$ and $M, s'', s' \models \alpha'$, where $<$ is an ordering on S, the underlying set of states
f. $M, s, s' \models \nabla$ always (where $\nabla =_{\mathrm{def.}} R\top$, i.e. the universal relation)

The modality ; corresponds to sequencing in dynamic logic (DL). Its definition refers to an intermediate state s'' of the sequence $\sigma = (s, s')$. The modalities L and R can be used to impose conditions on the initial state s and the final state s' of the sequence, respectively. The modalities $\sim$ and $\cap$, finally, are used for making boolean combinations.

The relation between formulas and procedures is established by means of the so-called domain modality DO: $\mathrm{DO}(\alpha)$ is true at a state s just in case s is in the domain of the relation denoted by α:

(13) $M, s \models \mathrm{DO}(\alpha)$ iff there is an s' with $s < s'$ s.t. $M, s, s' \models \alpha$

The idea underlying the decomposition of the Until-operator is to interpret the two clauses of its truth-conditions in TL as two separate procedures by means of the $\cap$-construct, where one procedure corresponds to the $\forall$-part of the truth-conditions of the Until-operator and the other procedure to the $\exists$-part. This leads to (14).

(14) a. $M, s, s' \models \sim[R\neg\psi; \nabla]$ iff for all s'' s.t. $s < s'' < s'$: $M, s'' \models \psi$

b. $M, s, s' \models R\phi$ iff $M, s' \models \phi$

c. $M, s, s' \models \sim(R\neg\psi; \nabla) \cap R\phi$ iff $M, s' \models \phi$ and $\forall s''$ s.t. $s < s'' < s' : M, s'' \models \psi$

In (14c) the procedure corresponding to the Until-operator is given. (14b) corresponds to the existential component. It imposes on the final state s' the condition that ϕ holds there. (14a) corresponds to the universal component of the Until-operator and therefore to the betweenness-condition. Let me explain this part by decomposing it.

(15) a. $M, s, s' \models R\phi; \nabla$ iff there is an s'' with $s < s'' < s'$ and $M, s'' \models \phi$

b. $M, s, s' \models \sim(R\phi; \nabla)$ iff there is no s'' with $s < s'' < s'$ and $M, s'' \models \phi$

iff for all s'' with $s < s'' < s'$ it holds that $M, s'' \models \neg\phi$

For a given ϕ, $R\phi\ ; \nabla$, (15a), is used to make an assertion about an intermediate state of σ, i.e., that ϕ holds somewhere on $]s, s'[$. Analogously, $R\neg\phi\ ; \nabla$ says that $\neg\phi$ holds somewhere in between s and s'. By negating $R\phi; \nabla$ one gets the assertion that there is no state s'' in between s and s' at which ϕ holds or, equivalently, that for all states s'' in between s and s' $\neg\phi$ holds. Thus an expression of the form $\sim(R\phi\ ; \nabla)$ can be used for expressing that the negation of ϕ, $\neg\phi$, holds on all states in between s and s' and therefore to impose a betweenness-condition on $\sigma = [s, s']$. For instance, $\sim(R\neg\phi\ ; \nabla)$ holds on σ just in case ϕ holds for all s'' with $s < s'' < s'$. But this is just the universal part of the Until-operator. Putting this procedure together with the procedure $R\psi$ by means of $\cap$ one arrives at the decomposition in (14c). The fact that the formula **Until**(ψ, ϕ) holds at s is then expressed by $M, s \models \mathrm{DO}(\sim(R\neg\psi\ ; \nabla) \cap R\phi)$.

If the Until-operator is used to describe properties of the nucleus-structure of an accomplishment-verb, $\psi = \neg\phi$ holds, i.e., ψ is the negation of the condition that has to hold in the final state s'. For that reason an accomplishment-verb can be interpreted as a state (or sequence-) transformer which takes a state s and transforms it to the first state s' such that ϕ holds there relative to s. In contrast to the Until-operator the interpretation of an accomplishment-verb requires that ϕ does not hold in the initial state s. In the language B this can be expressed by requiring $L\neg\phi$.

Let $R = [[\phi]]$ and $R' = [[\psi]]$ be the interpretations of ϕ and ψ, respectively. The perspective of the Until-operator as a state (or sequence) transformer can then be expressed as in (16a,b). In (16c) the variant U' that is used to interpret an accomplishment-verb is given.

(16) a. $[[U(\psi,\phi)]] = [[U]]([[\psi]],[[\phi]])$
b. $[[U]](R',R) = \{\sigma \mid \sigma = s_0 \ldots s_n \wedge s_n \in R \wedge \forall j : 0 < j < n, s_j \in R'\}$
c. $[[U']](R) = \{\sigma \mid \sigma = s_0 \ldots s_n \wedge s_n \in R \wedge \forall j : 0 \leq j < n, s_j \notin R\}$

If $[[\phi]]$ is interpreted as a temporal property, (16c) expresses a minimal change with respect to this property. The difference between accomplishment-verbs and activity-verbs concerns the condition that is imposed on the DP. Whereas accomplishments satisfy the requirement $\sim(R\phi\,;\nabla)$, activities satisfy the condition $\sim(R\neg\phi\,;\nabla)$. Thus, one arrives at the decompositions in (17).

(17) a. Accomplishment (e_{acco}): $L\neg\phi \cap \sim(R\phi\,;\nabla) \cap R\phi$
b. Activity (e_{act}): $L\neg\phi \cap \sim(R\neg\phi\,;\nabla) \cap R\phi$

In terms of these decompositions one can define suitable operators as abbreviations.

(18) a. Min-BECϕ $=_{\text{def.}}$ $L\neg\phi \cap \sim(R\phi\,;\nabla) \cap R\phi$
b. BECϕ $=_{\text{def.}}$ $L\neg\phi \cap \sim(R\neg\phi\,;\nabla) \cap R\phi$

For both classes it is possible that there are intermediate states s'' in between s and s'. The difference concerns the fact that in the case of an accomplishment one is interested in the *minimal* change with respect to ϕ (relative to s). From this it follows that ϕ only holds in the final state s' and in no other state. For activity-verbs this does not hold. Rather something like the converse is true. The negation of the condition ϕ only holds in the initial state.

3.1. The Interpretation of Other Aspectual Classes

3.1.1. *Achievement-Verbs*

Accomplishment- and activity-verbs are interpreted on sequences that are at least normally non-minimal and are not part of longer sequences. Achievement-verbs like *die* or *arrive*, on the other hand, will be interpreted on sequences that are part of a longer sequence. For instance *arrive* can be analyzed as denoting suffixes of movements that can be referred to by *go to the station* or *drive from New York to Chicago*. Contrary to the assumption made by Vendler that achievements can only be true on minimal sequences (intervals), I will interpret them on non-minimal sequences σ such that σ is a suffix of another sequence σ'

which will be of type accomplishment[4]. In general, σ cannot be an arbitrary suffix but there will be a constant k determining how long σ can be. For instance, *arrive* will normally only denote short suffixes whereas *die* can denote rather long suffixes of a corresponding process (for details see Naumann (1997c)). Below in Figure 4 the nucleus-structure for an achievement-verb and in (18) its decomposition are given.

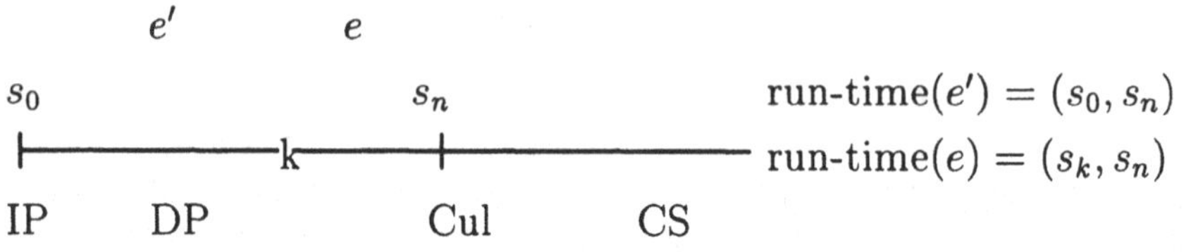

Figure 4. nucleus-structure for an achievement-verb

(18) Achievement (e_{ach}): $L\neg\phi \cap \sim(R\phi\,;\nabla) \cap R\phi$ and each execution sequence σ is a suffix of a (possibly) larger sequence σ' which belongs to e_{acco}

The condition that each sequence must be a suffix of a sequence that belongs to an accomplishment-verb can be expressed by an appropriate axiom. In the theory developed in section (4) below the condition can be expressed in terms of the corresponding events:

$$\forall e[e \in P_{\text{Ach}} \Rightarrow \exists e'[e' \in P_{\text{Acco}} \wedge \beta(e) = \beta(e') \wedge \alpha(e') \leq \alpha(e)]].$$

3.1.2. *Point-Verbs*

The difference between point-verbs and accomplishment- or activity-verbs concerns the existence of a DP. For the latter two classes this part of the nucleus will normally be non-empty, i.e., there are intermediate states s''. For point-verbs, on the other hand, the DP is always empty: there are no states in between s and s'. For that reason the betweenness-condition $\sim(R\phi\,;\nabla)$ holds vacuously. In contrast to achievement-verbs, point verbs are simple culminations, i.e., they denote events such that the corresponding execution sequences σ have length 1, i.e. $\sigma = ss'$. Thus it is possible to decompose them as in (19a).

(19) a. Point(e_{point}): $L\neg\phi \cap \sim(R\phi\,;\nabla) \cap R\phi$ and $\sigma = s_0 s_1$, i.e., each execution sequence consists of just two

[4] One may object that achievement-verbs denote only the culmination point Cul of procedures corresponding to accomplishment-verbs. Nevertheless, there are verbs, e.g. *fade*, for which this interpretation holds.

states

b. Point(e_{point}): $L\neg\phi \cap \sim(R\neg\phi \,;\nabla) \cap R\phi$ and $\sigma = s_0 s_1$, i.e., each execution sequence consists of just two states

According to (19a), point-verbs correspond to 'minimal' accomplishment-verbs in the sense that they are true on minimal sequences. This explains why they can be modified with *in*-adverbials in the sense that they allow a single event-reading. On the other hand, the condition $\sim(R\neg\phi \,;\nabla)$ also holds vacuously. Thus, (19b) is also a possible decomposition of a point-verb. According to this decomposition, a point-verb corresponds to a minimal activity-verb because the condition ϕ holds at all states except the first one. This similarity explains why point-verbs can be modified with a *for*-adverbial (although one gets an iterative interpretation) (for details, see Naumann (1997c)). If both conditions on the DP are imposed together, the requirement that all sequences must have length 1 is already defined by the decomposition because no state in between s and s' can satisfy ϕ and $\neg\phi$. This yields (19c).

(19) c. Point(e_{point}): $L\neg\phi \cap \sim(R\neg\phi \,;\triangle) \cap \sim(R\phi \,;\nabla) \cap R\phi$

The second representation in (19b) turns out to be significant for the interpretation of the perect of result: only those verbs allow a perect of result that satisfy the strong betweenness-condition defined in (20).

(20) (i) $\forall\sigma : M,\sigma \models L\neg\phi \cap \sim(R\phi \,;\nabla) \cap R\phi$
(ii) $\neg\forall\sigma : M,\sigma \models \sim(R\neg\phi \,;\nabla) \cap R\phi$

Expressed in terms of TL the second condition says that $\mathbf{G}\phi$ does not hold for all execution sequences. This condition is satisfied for accomplishments and achievements but not for points[5].

3.1.3. *Stative Verbs*

Stative verbs do not denote changes. In Dynamic Logic this can be expressed by interpreting them as tests, i.e. programs which do not change the state. In the present context they will be interpreted on sequences that are characterized by the condition that ϕ holds at all states of the sequence. The decomposition is given in (21).

[5] There is the following problem. There can be models in which all execution sequences corresponding to an accomplishment-verb are minimal, i.e. consist of just two states. In this case (20ii) is false. A similar argument holds for activity-verbs: there can be models in which all executions are minimal such that (20i) is satisfied. One must therefore either assume that the models are 'rich' enough, i.e., they contain both a non-minimal execution for each accomplishment-verb and for each activity-verb, or one must define the strong betweenness-condition not with respect to a particular model M but with respect to all models M.

(21) stative Verb (e_{state}) $L\phi \cap {\sim}(R\neg\phi\,;\nabla) \cap R\phi$

In (22) the decompositions of the five aspectual classes are summarized by using suitable operators.

(22)

a.	Accomplishment	Min-BECϕ	=	$L\neg\phi \cap {\sim}(R\phi\,;\nabla) \cap R\phi$
b.	Activity	BECϕ	=	$L\neg\phi \cap {\sim}(R\neg\phi\,;\nabla) \cap R\phi$
c.	Point	CHANGEϕ	=	$L\neg\phi \cap {\sim}(R\phi\,;\nabla) \cap {\sim}(R\neg\phi\,;\nabla) \cap R\phi$
d.	Achievement	Min-BEC$_{suf}\phi$	=	$L\neg\phi \cap {\sim}(R\phi\,;\nabla) \cap R\phi$ and each execution sequence σ is a suffix of a (possibly) larger sequence σ' of type e_{acco}
e.	stative Verb	HOLDϕ	=	$L\phi \cap {\sim}(R\neg\phi\,;\nabla) \cap R\phi$

The fact that a sentence S is true in state s is expressed by $M, s \models \text{Do}(\alpha)$ where α is the decomposition of S in terms of procedures from B as given in (22) and the operator Do is defined by $M, s \models \text{Do}(\alpha)$ iff there is an s' with $s \leq s'$ such that $M, s, s' \models \alpha$[6].

To each propositional variable $\phi \in \text{VAR}$ and operator OP defined in (22) corresponds a binary relation $\text{R}_{\text{OP}(\phi)}$ on S defined in (23).

(23) $\text{R}_{\text{OP}(\phi)} = \{(s, s') \mid M, s, s' \models \text{OP}\phi\}$

For all $\phi \in \text{VAR}$ the corresponding relations $\text{R}_{\text{OP}(\phi)}$ agree with respect to the valuation of ϕ on all three parts of the nucleus-structure. For a given operator OP the relation corresponding to all $\phi \in \text{VAR}$ is given in (24a). In (24b) the restriction to a particular ϕ is defined.

(24) a. $R_{\text{OP}} = \bigcup_{\phi\in\text{VAR}} R_{\text{OP}(\phi)}$
b. $R_{\text{OP}(\phi)} = \{(s, s') \mid (s, s') \in R_{\text{OP}} \wedge M, s, s' \models \text{OP}\phi\}$

The relations $R_{\text{OP}(\phi)}$ cannot, at least in general, directly be interpreted as sequences on which verbs (or the corresponding sentences) belonging to a particular aspectual class are true. For instance, John's being not in Manchester was the case long before he moved there. But the whole sequence will not be a sequence on which a sentence like 'John moved/went/drove/ to Manchester' is true because on an initial subsequence there was no movement towards Manchester. But each

[6] The view that sentences are evaluated at states is, of course, too simple. See Naumann (1996b) for a more adequate analysis.

sequence on which a verb (sentence) belonging to a particular class is true is an element of R_{OP} where OP is the operator characterizing this class. R_{OP} can therefore be interpreted as determining the type of sequences on which verbs (sentences) belonging to a particular aspectual class can be true by specifying the way the condition ϕ is valuated in each state of this type of sequence. For that reason the R_{OP} can be interpreted as aspectual classes. In the present context there are five operators corresponding to the different aspectual classes $ac \in \mathrm{AC} = \{$Acco(mplishment), Act(ivity), Ach(ievement), Point, State$\}$. The corresponding relations will be referred to as (25).

(25) $\{R_{\mathrm{ac}}\}\ ac \in \mathrm{AC}$.

The procedures that are used to define the nucleus-structure express at the same time properties of this structure (i.e., the nucleus-structure is defined in terms of its properties). The procedures can therefore be compared to the notions that are used in event-semantics to define aspectual classes, in particular those of a gradual and a constant role (for details, see Krifka (1992) or Naumann (1997a)). In contrast to event-semantics which allows only the distinction of these two properties, the language B admits one to distinguish more properties in terms of the boolean condition ϕ in relation to the three parts of the nucleus-structure.

(i)	ϕ holds in IP and Cul:	yes: stative verbs
		no: non-stative verbs (occurrences)
(ii)	betweenness-condition:	yes: events (accomplishment, achievement, point)
	(20i)	no: activities, stative verbs
(iii)	strong betweenness:	yes: proper events (accomplishment, achievement)
	(20i,ii)	no: points, activities, stative verbs

4. A Decompositional Dynamic Temporal Logic of States and Events

From the perspective of event semantics the approach presented in the previous section suffers from the same defect as that by Dowty. If (non-stative) verbs express changes, then there should be some object which brings about the change, namely an event. But so far there are no entities in the model that are interpreted as bringing about the change from the initial state s of the sequence σ to its final state s'. A structure **B** for the language B from the previous section is a tuple

(26) $\mathbf{B} =< S, <, [[\cdot]], V >$

such that

- S is a (non-empty) domain of states
- $<$ is a discrete linear ordering on S
- $[[\cdot]]$: Proc $\rightarrow 2^{S \times S}$ is a function which interprets the procedural part of the language in accordance with the definition given in (12) above
- V : VAR $\rightarrow 2^{S}$ is a function which assigns to each propositional variable $p \in$ VAR a subset of S.

Whereas there are two sorts of syntactic category, formulas and procedures, such that B is a two-sorted language, there is only one domain in the structure **B**, namely the domain S of states. Whereas this domain can be interpreted as corresponding to the syntactic category of formulas, there is no corresponding domain for the other category, i.e. that of procedures. At the level of the interpretation of verbs this is reflected by the fact that a verb (or its dynamic component) denotes a set of sequences (s, s'), i.e. a binary relation on S. In terms of Dynamic Logic this means that it is interpreted as a program and therefore as an input/output relation.

A possible reply to this objection is that there is no need for such a separate domain. The decompositional analysis in the previous section has shown that aspectual distinctions can be explained without the notion of event. Although this reply is correct, the argument has two other aspects, one which is more philosophical in nature and another one which is more specificly linguistic. First, above in section (3) (the interpretation of) a verb was informally described as a state transformer: it transforms a given input-state into an output-state. Although this transformation (or change) corresponds to a sequence of states, it cannot be identified with it. What is missing is that the transformation (change) is brought about by something, and this is an event. It is this intuition which is not captured in the model if verbs are interpreted as denoting sequences of states and are therefore equated with input/output relations. Second, there are other linguistic phenomena outside the domain of aspect for which the notion of event is used. Let me only give two examples: constructions involving verbs of perception and anaphoric relations. Consider the following data.

(27) a. John saw Mary cross the street.
b. Yesterday John lost his purse. That made him very sad.

The sentence in (27a) can be analyzed as follows. The verb of perception is interpreted as a relation between an individual and an event (or, more generally, a situation). What John saw is a particular event which was a crossing the agent of which was Mary and the theme of which was a particular, contextually determined street. If one accepts the notion of an event at all, there is no problem in assuming that individuals can see events (or situations). A similar argument can be adduced in the case of anaphoric relations. The example in (27b) shows that it is possible to refer back to something that is not one of the (syntactic) arguments of the verb. Using the notion of event, this relationship can be explained by assuming that in (27b) *that* is anaphorically related to the event that is introduced in the first sentence: it is the event of loosing the purse which caused John's sadness.

This perspective can be modeled by an eventuality structure **E**. For the present purpose an eventuality structure **E** of signature $<\mathrm{Verb}, \mathrm{AC}>$ can be taken to be a tuple

$$\mathbf{E} =< E, \{E_{\mathrm{ac}}\}ac \in \mathrm{AC}, \{P_v\}\ v \in \mathrm{Verb} >,$$

where E is a domain of event-occurrences, P_v $v \in$ Verb are unary relations on E that correspond to particular event-types like eating, running or pushing. They can be thought of as the (atomic) event-predicates used in event-semantics. Therefore Verb can be taken to consist of (a subset of) the verbs of English: Verb = { eat, walk, drink, drive, love, hit ...}. For instance, P_{eat} is the set of all eating events (in the structure **E**). Each E_{ac} $ac \in$ AC is interpreted as an aspectual class. In the present context AC will be the set AC = { Accomplishment, Activity, Point, Achievement, State } from section (3). Each E_{ac} corresponds to a subset of $\{P_v\}$, namely all those verbs which belong to the aspectual class *ac*.

The crucial question that has to be answered is what properties distinguish elements from one aspectual class *ac* from elements of another, different class *ac'*. So far each aspectual class E_{ac} is simply taken as primitive. One way of answering this question is to combine the structure **E** with an interval structure $\mathbf{I} =< I, <, \sqsubseteq>$[7] to a complex structure **E-I** such that different aspectual classes can be defined in terms of the properties the run-times of the events belonging to this class have, where the run-time is an interval from **I**. This strategy is chosen by Blackburn et al. (1994). For instance, states are distinguished from other aspectual classes by having run-times that are unbounded intervals as opposed to bounded intervals in the case of a non-stative

[7] I is a domain of intervals and $<$ and $\sqsubseteq$ are binary relations on I, the precedence and subinterval relation, respectively.

aspectual class. The problem that this approach faces is that accomplishments and activities are characterized by the same axiom, given in (28). (In (28) z is a function from E to I assigning to each event e its run-time $z(e)$)

(28) $\forall e[E_{\text{Act}}(e) \vee E_{\text{Acco}}(e) \rightarrow z(e)$ is an non atomic bounded interval]

In the informal discussion Blackburn et al. use a further property which distinguishes accomplishments from activities: whereas the former are culminating (+culm), the latter are non culminating (-culm). But this property is not formally defined and it seems impossible to define it in terms of properties the run-times of events (i.e. intervals) have. Blackburn et al. try to solve this problem by assuming a function Compl(etes) which is a partial function from E_{Acco} to E_{Ach}. Accomplishments are then distinguished from activities by the fact that only the former class but not the latter stands in the Compl-relation to achievements. But this function is taken as primitive and is not defined in terms of more basic notions[8].

What is needed is a structure that admits both perspectives. How can this be achieved? Let me begin by comparing the structures **B** and **E** with a labeled transition system (LTS). An **LTS** of signature $< L >$ is a tuple

$$\textbf{LTS} =< S, T, \alpha, \beta, \lambda >$$

where S is a set of states, T a set of transitions such that each $t \in T$ is indivisible in the sense that it cannot be decomposed into a sequence $t_1 \dots t_n$ of elements from T, α and β are functions from T to S assigning to each $t \in T$ its source-state $\alpha(t)$ and target-state $\beta(t)$, respectively, and $\lambda : T \rightarrow L$ is a labeling-function that assigns to each transition t its label $\lambda(t)$. The label $\lambda(t)$ can intuitively be interpreted as indicating the type of the transition. Complex transitions can be modeled by finite paths of transitions, $c = t_1 \dots t_n$, for some n such that for $\forall i : 1 \leq i < n$, $\beta(t_i) = \alpha(t_{i+1})$. The set of all finite paths will be T^+. The mappings α and β are extended to elements from T^+ by defining $\alpha(t_1 \dots t_n) = \alpha(t_1)$ and $\beta(t_1 \dots t_n) = \beta(t_n)$. To each path $c \in T^+$ of length n there uniquely corresponds a sequence of states $\alpha(t_1)\beta(t_1)\beta(t_2)\dots\beta(t_n)$. The labeling function λ can also be extended to T^+ such that $\lambda(c)$ assigns to c its label from L[9]. **LTS** and the structures **E** and **B** can be compared to

[8] Note furthermore that Compl cannot be interpreted as a formal representation of the informal feature +culm because not only accomplishments but achievements too are +culm.

[9] It may be assumed that the extension of λ to T^+ yields only a partial function. For instance, a path c may be assigned a label only if all of its (atomic) subtransitions t_i have the same label $\lambda(t_i)$.

each other in the following way. Let c be a finite path $c = t_1 \ldots t_n$ with sequence $\alpha(t_1)\beta(t_1)\beta(t_2)\ldots\beta(t_n)$ in the **LTS**, then

(i) c corresponds to an event e from E

(ii) The sequence $\alpha(t_1)\beta(t_1)\beta(t_2)\ldots\beta(t_n)$ corresponds to a sequence (s, s') in **B**

(iii) The label $\lambda(c)$ indicates the type of the event e, i.e., if viewed from **E**, L, the set of labels, can be taken to be identical to Verb: $L =$ Verb[10].

Whereas (i) cannot be expressed in **B**, (ii) cannot be expressed in **E**. If, on the other hand, **E** and **B** are combined, one can say that to each event e from E in **E** there uniquely corresponds a particular sequence of states which can be thought of as its execution sequence and the label $l \in L$ indicates the type of the event. The combined structure will relate the domains E and B in exactly the same way as the two domains of an LTS are related to each other[11]. The crucial observations are the following.

(i) Each event $e \in E$ is related to a pair of states (s, s') which will be called its source-state and target-state, respectively, following the use in an **LTS**. This relation will be defined by means of two functions $\alpha : E \rightarrow S$ and $\beta : E \rightarrow S$, similarly to the two functions in the structure **LTS**[12].

(ii) Each event $e \in E$ is (at least) of one type P_v where P_v is interpreted as in the eventuality structure **E** defined above.

(iii) From (i) and (ii) it follows that each event type P_v corresponds to a subset R_v of $S \times S$ which is defined as follows.

$$R_v = \{(s, s') \mid \alpha(e) = s \text{ and } \beta(e) = s' \text{ for some } e \in P_v\}.$$

Thus, by means of α and β it is possible to recover (or simulate) the relational view: each event $e \in E$ corresponds to a particular sequence $(s, s') \in S \times S$ which is its execution sequence. The product mapping $< a, \beta > (= \tau)$ corresponding to an event type

[10] Below it will be assumed that an event $e \in E$ can belong to different types of verbs such that λ will in general not be functional.

[11] Note that there is the following difference between an **LTS** and **B** with respect to the domain S: whereas in **B** an ordering is defined on S, no such ordering is assumed in the **LTS**.

[12] There will be at least the following restriction on α and β: it must respect the ordering on S, that is, $\forall e, s, s'[\alpha(e) = s \wedge \beta(e) = s' \rightarrow s \leq s']$.

P_v, i.e. R_v, is comparable to the binary relations R_π that are used in the standard translation ST of PDL into FOL: $\mathrm{ST}([\pi]\phi) = \forall y[R_\pi(x,y) \rightarrow [x/y]\ \mathrm{ST}(\phi)]$. In the present context where events (transitions) are taken as primitive one gets instead a ternary relation $R(s,s',e)$ which can be defined by α and β: $R(s,s',e)$ iff $\alpha(e) = s \wedge \beta(e) = s'$.

Note that not all events $e \in P_v$ for some $v \in$ Verb are supposed to satisfy ϕ_v in their target state, i.e., $\beta(e) \in V(\phi)$ need not always hold where ϕ_v is the property of states that can be brought about by events of type v. Thus, events of a particular type need not be (totally) correct. They can 'terminate' without it being the case that ϕ_v holds in the target-state, i.e., their execution sequence need not satisfy the aspectual restriction imposed by P_v (see below for details). From this it follows that R_v is not necessarily a subset of one of the R_{ac} defined in (25) above.

(iv) Each verb $v \in$ Verb is characterized by a particular property which gets changed by the execution of an event of this type. For instance, in the case of 'eat' the property can be taken to be the mass of the object denoted by the internal argument, say the apple in 'eat an apple'. In the target-state of each event of this type a condition ϕ on the property must hold, expressing the result which the event has brought about. This condition will be assumed to be atomic, i.e., to be an atomic predicate, and can therefore be expressed by some propositional variable $\phi \in$ VAR. The relationship will be expressed by a function δ : Verb $\rightarrow$ VAR which assigns to each verb the property of states $\delta(v)$ which must hold in the target-state of a ('terminating') event of this type[13].

(v) So far the structures **E** and **B** are linked by the functions α, β and δ. A fourth link will relate the unary relations P_v $v \in$ Verb to particular sequences in **B**, i.e. binary relations on S. In the structure **B** the different aspectual classes were defined by subsets of $S \times S$ which are characterized by valuating the propositional variables ϕ in the same way, i.e. the R_{ac} $ac \in$ AC defined in (24,25) above in section (3). Each verb type v will be assigned that subset of one of the R_{ac} which corresponds to $\delta(v)$, i.e., the property of states that can be brought about by that

[13] Instead of defining δ in terms of labels corresponding to the P_v and the unary relations interpreting the elements of VAR, one can define δ directly at the semantic level: δ is then a partial function $2^E \rightarrow 2^S$ which is defined for exactly those subsets of E which correspond to one of the P_v and the value of δ for P_v will be $Q_\phi (= V(\phi))$ such that $\delta(v) = \phi$ if δ is interpreted as given in the text, i.e., as a function from Verb into VAR. Thus, one gets $\delta : \{P_v\}\ v \in Verb \rightarrow \{Q_\phi\}\ \phi \in \mathrm{VAR}$.

type. At the formal level this will be done by a partial function $\gamma : 2^E \rightarrow 2^{S \times S}$ which is defined for each P_v from $\{P_v\} v \in$ Verb and which assigns to this subset of E one of the relations $R_{\text{OP}(\phi)}$ where it is required that $\phi = \delta(v)$[14]. Therefore, one has in effect $\gamma : \{P_v\}\ v \in \text{Verb} \rightarrow \{R_{\text{OP}(\phi)}\}\ \phi \in \text{VAR}$, OP $\in$ { Min-BEC, BEC, Min-BEC$_{suf}$, CHANGE, HOLD}. For instance, if $v =$ eat, one gets $\gamma(P_{\text{eat}}) = R_{\text{Min-BEC}(\delta(\text{eat}))}$ which means that 'eat' is of type accomplishment.

The language underlying the combined structure is defined in (29).

(29) a. A formula ϕ is of one of the following four forms:

(i) $p \in \text{VAR}$ (i.e., ϕ is a propositional variable)

(ii) $\neg p$ (i.e., ϕ is the negation of a propositional variable)

(iii) $\top$ (i.e., $\phi = \top$, where $\top$ holds in each state: $M, s \models \top$ always)

(iv) $(e_v)\text{OP}\phi$, where OPϕ is one of the operators defined in (22) in section (2) above

b. A procedure α is of one of the following three forms:

(i) $L(\neg)\phi$ or $R(\neg)\phi$, where ϕ is a propositional variable

(ii) $R\top$

(iii) if α and α' are procedures, then $\sim\alpha$, $\alpha\,;\alpha'$ and $\alpha \cap \alpha'$ are also procedures

A sorted dynamic-temporal eventuality structure **B-E** is defined as follows.

(30) A sorted dynamic-temporal event-structure **B-E** of signature $<\text{VAR}, \text{Verb}>$ is a tuple $<\mathbf{B}, \alpha, \beta, \delta, \gamma, \mathbf{E}>$ such that

- **B** is a dynamic-temporal structure defined in (26).
- $\mathbf{E} = <E, P_v\ v \in \text{Verb}>$ where E is a domain of event-occurrences and the $P_v\ v \in$ Verb are unary relations on E corresponding to eatings, runnings etc. Thus, Verb corresponds to (a subset of) the verbs in English, i.e., Verb =

[14] It is possible to dispense with δ by defining γ in such a way that it directly assigns both an appropriate operator OP and an appropriate property ϕ. This will not be done in order to keep the two functions distinct. A third alternative is to define γ as a (partial) two-place function which assigns to an event-type P_v and some $\phi \in$ VAR an element from $\{R_{\text{OP}(\phi)}\}$. γ will only be defined if $\phi = \delta(v)$. Thus one gets:

$$\gamma(P_v, \delta(v)) = R_{\text{OP}}(\delta(v)) \text{ for some OP.}$$

{eat, run, arrive, knock, hate, ...}. The elements from Verb will be written e_v or simply v, respectively, and are interpreted by the corresponding P_v.

- $\alpha : E \to S$ and $\beta : E \to S$ are functions which assign to each element of e its source-state $\alpha(e)$ and target-state $\beta(e)$, respectively.
- δ : Verb $\to$ VAR is a function which assigns to each element $v \in$ Verb an element $\delta(v)$ from VAR.
- $\gamma : 2^E \to 2^{S \times S}$ is a partial function which assigns to each P_v from $\{P_v\}$ $v \in$ Verb a binary relation $R_{\mathrm{OP}(\phi)}$ with $\phi = \delta(v)$ which determines the aspectual class of the verb type v.

Instead of formulas Do(α) in **B** one now has formulas $(e_v)\mathrm{OP}\phi$. Their interpretation is defined in (31a). In (31b) a particular example is given.

(31) a. **B-E**,$s \models (e_v)\mathrm{OP}\phi$ iff there are e, s' s.t. (i) $s \leq s'$[15] and $e \in P_v$, (ii) $\alpha(e) = s$ and $\beta(e) = s'$, (iii) $\delta(v) = \phi$ and $(s, s') \in \gamma(P_v)$ and (iv) **B-E**,$s, s' \models \mathrm{OP}\phi$

b. **B-E**,$s \models (e_{\mathrm{eat}})\mathrm{Min\text{-}BEC}\phi$ iff there are e, s' s.t. (i) $s \leq s'$ and $e \in P_{\mathrm{eat}}$, (ii) $\alpha(e) = s$ and $\beta(e) = s'$, (iii) $\delta(\mathrm{eat}) = \phi$ and $(s, s') \in R_{\mathrm{Min-BEC}(\phi)}$ and (iv) **B-E**,$s, s' \models \mathrm{Min\text{-}BEC}\phi$

According to (31a,b), '$(e_v)\mathrm{OP}\phi$' is true at state s in **B-E** just in case in s an event e of type v is executed such that ϕ, the property related to v by δ, holds in the target-state $\beta(e)(= s')$ of e and ϕ is brought about by e according to the way that is characteristic of type v. The latter condition means that ϕ must be brought about in the way determined by the operator OP corresponding to the aspectual class $\gamma(P_v)$. The condition $\delta(v)$ guarantees that only those formulas '$(e_v)\mathrm{OP}\phi$' can be true where ϕ (or e_v) is of the correct type, i.e., if an event of type v can bring about ϕ. This corresponds to the intuitive interpretation of '$(e_v)\mathrm{OP}\phi$' as 'an event e of type v brings about ϕ'. If the condition $\delta(v) = \phi$ were dropped, it could happen that '$(e_v)\mathrm{OP}\phi$' is true for some ϕ which an event of type v cannot bring about. For instance, one can have '$(e_{\mathrm{eat}})\mathrm{Min\text{-}BEC}\phi_{\mathrm{drink}}$' such that this formula is true at a state s because there is an event of eating which brings about ϕ_{eat} in some state s' such that $(s, s') \in \gamma(P_{\mathrm{eat}})$ and there is an event e' of type 'drink' which brings about ϕ_{drink} on the same sequence (i.e., $(s, s') \in \gamma(P_{\mathrm{drink}})$). In this situation '$(e_{\mathrm{eat}})\mathrm{Min} - \mathrm{BEC}\phi_{\mathrm{drink}}$' would be true in s but could not be interpreted in the sense that it was an event

[15] If it is assumed that $\alpha(e) \leq \beta(e)$ for all $e \in E$, this condition is redundant because it already follows from clause (ii).

e of eating which brought about ϕ_{drink}. It may seem that the condition $(s, s') \in \gamma(P_v)$ is redundant, in particular if one looks at the example of the accomplishment-verb 'eat' given in (31b), because it is already accounted for by the fourth clause. But **B-E**, $s, s' \models \mathrm{OP}\phi$ only requires that ϕ was brought about *in some way.* This does not guarantee that it was brought about in the correct way, i.e. the way required by the verb type v. For instance, it is possible to have '$(e_{\text{eat}})\mathrm{BEC}\phi$'. This should in general be false because a sequence evaluating $\delta(\text{eat})(= \phi)$ as true in all non-initial states is no eating of, say, an apple. There are at least two reasons of why ϕ can already hold at intermediate states of an execution of eating or, more generally, of an event of type accomplishment. Recall that so far it has been assumed that a formula of the form '$(e_v)\mathrm{OP}\phi$' corresponds to a sentence where the arguments are singular, i.e. denote atomic objects, like an apple in 'eat an apple'. The property ϕ, then, expresses the condition, say, that the apple has vanished. But if instead of one apple two apples were eaten, ϕ already holds after the first apple has been eaten[16]. These other cases like, e.g., eating two apples, can be captured by defining δ not as a function which assigns to each $v \in$ Verb an element from VAR but rather a subset of VAR. Each element from $\delta(v)$ expresses a particular change with respect to the property that can be changed by an event of type v, corresponding to eat half an apple/an apple/two apples and so forth. The function γ assigns then to P_v a set of binary relations $R_{\mathrm{OP}(\phi)}$, one for each $\phi \in \delta(v)$ (see section (5) for details).

The second possible case is given, e.g., by a situation in which in effect one apple is eaten but the target-state of the event falls into the consequent state. Such cases should be excluded because there cannot be any event of eating without there being a corresponding change. This can be achieved by requiring that if $\tau(e) = (s, s')$ and e is of type v (i.e., $e \in P_v$), then for each pair (s_i, s_{i+1}) of successive states of the sequence (s, s') there must be a change with respect to an element from $\delta(v)$.

The condition $(s, s') \in \gamma(P_v)$ therefore functions as a test with respect to the operator OP: is OP the appropriate operator for events of type v? Thus, whereas the condition $\delta(v) = \phi$ checks whether ϕ is the correct property for events of type v, the test with respect to γ checks whether the change was brought about in the appropriate way[17]. These considerations suggest the following alternative representation. Instead

---

[16] Thus, it is assumed that ϕ is preserved on the sequence if the eating continues.

[17] Alternatively, one could drop the conditions expressed by δ and γ in (31a) and define a predicate 'brings-about(e_v, ϕ)' ('an event of type v brings about ϕ'): $[[\text{brings-about}(e_v, \phi)]](s) = 1$ iff **B-E**,$s \models (e_v)\mathrm{OP}\phi$, $\gamma(P_v) = R_{\mathrm{OP}(\phi)}$ and $\delta(v) = \phi$.

of formulas of the form '$(e_v)\mathrm{OP}\phi$' one has formulas of the form '$(e_v)R\phi$' the interpretation of which is exactly the same as that for '$(e_v)\mathrm{OP}\phi$'.

(31) c. $\mathbf{B\text{-}E},s \models (e_v)R\phi$ iff there are s', e s.t. (i) $s \leq s'$ and $e \in P_v$, (ii) $\alpha(e) = s$ and $\beta(e) = s'$, (iii) $\delta(v) = \phi$ ($\phi \in \delta(v)$) and $(s, s') \in \gamma(P_v)$ ($(s, s') \in \gamma_\phi(P_v)$)[18], (iv) $\mathbf{B\text{-}E},s,s' \models R\phi$.

Clause (iv) now only requires that ϕ was brought about on the sequence (s,s') (in the case of a non-stative verb). It is clause (iii) which alone tests whether the change was brought about in the appropriate way with respect to the appropriate property (or an appropriate property).

In (32a) the predicate 'accomplishment(e, v)' is defined.

(32) a. $\text{accomplishment}(e, v) =_{\text{def.}} e \in P_v \wedge \gamma(P_v) \subseteq R_{\text{Acco}}$

b. $\text{accomplishment}(e) =_{\text{def.}} \exists v.\text{accomplishment}(e, v)$

According to (32a), an event cannot be said to be an accomplishment in an absolute way but only relative to some verb type v (or relative to some unary relation P_v on E of which it is an element). For instance, an event e can be an accomplishment relative to 'cross the street' but be an activity with respect to 'walk', i.e., it is a walking across the street. For the other aspectual classes the definitions are given in an analogous way.

In (33) the revised decompositional schemes are given.

(33)

Accomplishment:	$(e_{\text{Acco}})\text{Min-BEC}\phi_{\text{Acco}}$
Activity:	$(e_{\text{Act}})\text{BEC}\phi_{\text{Act}}$
Achievement:	$(e_{\text{Ach}})\text{Min-BEC}_{suf}\phi_{\text{Ach}}$
Point:	$(e_{\text{Point}})\text{CHANGE}\phi_{\text{Point}}$
stative Verb:	$(e_{\text{State}})\text{HOLD}\phi_{\text{State}}$

As was said above, the function of Do in **B** is taken over by (e_v) for some $v \in$ Verb in **B-E**. Alternatively, one can evaluate an expression of the form '$(e_v)\mathrm{OP}\phi$' on sequences of states[19]. This yields (34).

(34) $\mathbf{B\text{-}E},s,s' \models (e_v)\mathrm{OP}\phi$ iff there is an e s.t. (i) $e \in P_v$, (ii) $\alpha(e) = s$ and $\beta(e) = s'$, (iii) $\delta(v) = \phi$ and $(s, s') \in \gamma(P_v)$ and (iv) $\mathbf{B\text{-}E},s,s' \models \mathrm{OP}\phi$

[18] $\gamma_\phi(P_v)$ is that element of $\gamma(P_v)$ that corresponds to ϕ.

[19] This assumes that the language is changed accordingly because '$(e_v)\alpha$' is then of category procedure.

This format is more suitable if temporal phenomena are taken into account because a sentence will then be evaluated on an interval (which is taken to be a sequence of states). If one wishes, it is possible to reintroduce the Do-operator which can then be applied to the procedure expression '$(e_v)\mathrm{OP}\phi$' to yield an expression of category formula.

Next the relation between the two structures **B** and **E** will be analyzed in more detail. As was said above, in the eventuality structure **E** the elements of E are structureless points. For that reason the aspectual classes E_{ac} can only be taken as primitive. This is insufficient in order to analyze the various aspectual classes that can be found in natural language. Conversely, in the dynamic temporal structure **B** verbs are interpreted by a decompositional analysis according to which they are state-transformers (or run-transformers) and each aspectual class corresponds to a particular type of state-transformer. The disadvantage is that on this perspective verbs denote only 'second-class objects', namely sequences of states that are defined in terms of the only domain S. Verbs do not denote objects of a separate domain, as e.g. in **E**, where the elements of this domain are interpreted as bringing about changes.

On the present perspective a change has two different components. First, there is the event which brings about the change and second there is the result which is brought about by the event. The event is interpreted as an element from E whereas the result is interpreted as a property of states that holds upon termination of the event, that is at the target-state of e. Event semantics focuses on the event, i.e. the object that brings about the change, but leaves out the result of the change brought about by the event. Dowty's decompositional approach, on the other hand, stresses the result of the change but does not represent the event.

Both components are independent of each other. For instance, a change from being in London to being in Manchester can be brought about by different types of events, e.g. a walking or a running. On the other hand, an event e can be described as bringing about a change from being in London to being in Manchester but also simply as bringing about a change of location. Thus, an event of walking to Manchester satisfies ϕ =be in Manchester in its target state but also the (weaker) condition that the location in the target state is different from that in the source-state. The first case can be described by the sentence 'John walked to Manchester' whereas the second corresponds to the sentence 'John walked'. The distinction is aspectually relevant because the first sentence will be classified as an accomplishment whereas the second as an activity.

In the approach presented above both components are explicitly represented. The event component corresponds to the eventuality structure **E** and the result (state) component to the dynamic-temporal structure **B**. Aspectually, the structure **B** is the more important one because aspectual classes are defined in terms of binary relations on the domain S underlying **B**. The fact that a change can be decomposed into two components, namely the event which brings about the change and the result that is brought about by the event, is reflected in a formula of the form '$(e_v)\mathrm{OP}\phi$' in the following way. The part '(e_v)' corresponds to that component which brings about the change and the second part '$\mathrm{OP}\phi$' expresses the result component (as well as the way the result was brought about).

The combined structure **B-E** can be interpreted as a structure of events (actions) and states. Let me explain this in more detail by comparing the present approach with Dynamic Logic (DL). In DL, the main type of formula is of the form $(\pi)\phi$ where $(\cdot)$ is one of the modalities used in DL, i.e., either $< \cdot >$ or $\Box$. In DL 'π' is interpreted as a binary relation on S, the underlying set of states. Programs (actions) are therefore only derived objects and no 'first class-citizens' like states. DL is therefore only a logic of states and derived actions. In the structure **B-E**, on the other hand, there is a second domain E besides that of states. Thus, programs (actions), now interpreted as events, have become first class-citizens. At the formal level this is reflected by the interpretation of π and e_v. Whereas the former is interpreted as a binary relation on S, the latter is a unary relation on E[20]. From the perspective of **B** the domain E can be seen as a kind of reification of the second-class objects. This corresponds to a view on which the programs-letters in PDL are not interpreted as relations between states, as it is normally done, but as elements of a separate domain of objects (for this perspective see Naumann (1997a) and Van Benthem and Bergstra (1995)).

A second difference concerns the boolean part ϕ. DL focuses on the result of a program (action). The way this result was brought about, i.e., what holds at intermediate states of an execution, is not taken into account: ϕ must hold in the output-state s' of the program. This reflects the pure input/output interpretation of a program. In contrast, on the present account $\mathrm{OP}\phi$ expresses a condition on each state of an execution with respect to ϕ and not only with respect to the output (target)-state. Formulating this part in this way has two advantages.

[20] Due to this parallel, a formula of the form '$(e_v)\phi$' could be interpreted in an appropriately modified DL, DL*, as given in (*).

(*) $M^{DL*}, s \models (e_v)\phi$ iff there is an $e \in P_v$ such that $\alpha(e) = s$ and $\beta(e) \in V(\phi)$ for atomic ϕ.

First, interpreting the 'boolean' part on a sequence stresses the relation that holds between an event (an event type) and a pair of states (s, s') (a set of pair of states). Second, this interpretation links the present approach to (some variants of) Temporal Logic (TL), where formulas are interpreted on sequences of states instead of at a single state. A sequence is then interpreted as the execution (sequence) of a program where the program is left implicit[21]. A program itself is identified with a set of execution sequences. This corresponds exactly to the interpretation of verbs given in section (3).

TL can therefore be taken as a logic which focuses on the result of a change as well as the way this result was brought about. What the present approach adds to TL is an explicit representation of the first component of a change, i.e. the object that brings about the change. Applying the perspective taken on a program in TL to the present approach, the '(e_v)'-part represents the change as unstructured, which is reflected by the fact that an event is a basic object in the model, whereas 'OPϕ' represents the change as having an internal structure, that is, it represents it as the evolution from one state s to another state s'. The advantage of the temporal logical perspective is that it rather easily admits to make assertions about intermediate states of a computation whereas in PDL one first has to define special operators. The structure **B** can be seen as a combination of both perspectives: only the states are represented but the sequence is subdivided into three different parts about which assertions can be made by means of different procedures. In terms of minimal sequences the operator L corresponds to making an assertion about the input-state and R to making an assertion about the output-state. A combination of ;, $\sim$ and R is used to make assertions about intermediate states of an execution in case the program is non-atomic. It is just this perspective which admits the comparison of different types of programs (actions, events) with respect to the quadripartite structure (the nucleus-structure from Figure (3)) of their executions.

Taking into consideration not only the 'pure' result but also the way the result was achieved is aspectually important. If non-stative verbs were interpreted simply as expressing changes, there would be no difference between them because they all express changes. Aspectual distinctions between non-stative verbs must be explained as differences with respect to the types of change expressed by the verbs.

In terms of Kripke-structures that are used for PDL a sequence σ can be seen as the execution sequence of a program π. Sequences of

[21] TL therefore is an endogeneous formalism, i.e., there are no explicit constructs that are interpreted as programs.

length 1 correspond to atomic programs and tests whereas sequences of length > 1 correspond to non-atomic programs[22]. As is well known, the execution of a complex program can be represented as the (relational) composition (or sequencing) of atomic programs and tests. As each atomic program and test corresponds to a simple input/output relation between two states s and s', to each complex program corresponds a sequence of states consisting of the input/output states of the atomic programs (and tests) from which it is built up. It is this latter perspective that is adopted in TL and in the dynamic temporal structure **B**. Each sequence σ is understood as an execution sequence of one global program that is left implicit. On the other hand, each execution sequence can be characterized by the properties it has. One way of specifying these properties is by using (linear) TL as a specification language.

(35) a. A property **P** is any subset of S^*: $\mathbf{P} \subseteq S^*$, where S^* is the collection of all finite sequences of states of the domain S of the underlying structure (respecting the ordering on S).

b. A property **P** is specified by the TL formula ϕ iff $\sigma \in \mathbf{P}$ iff $\sigma \models \phi$

c. A program π has property **P** iff each execution σ of π satisfies ϕ, where a sequence σ satisfies a formula ϕ iff the first state of σ satisfies ϕ ($\sigma \models \phi$ iff $s_0 \models \phi$)

In the present approach instead of TL the dynamic two-sorted variant **B** is used to express the properties. In terms of programs one gets the relationship in (36a): Each aspectual class determines a type of program π which is determined by a set of properties which can be expressed by formulas ϕ which each regular execution sequence of π must satisfy. In the combined structure **B-E** the programs are reified and interpreted as events. This yields (36b).

(36) a. aspectual class AC $\rightarrow$ program $\pi \leftarrow$ formulas ϕ expressing properties of π

b. aspectual class AC $\rightarrow$ event-type $e_{ac} \leftarrow$ procedures α expressing properties of the regular execution sequences of e_{ac}.

Regular sequences are those sequences $(s, s') \in R_v$ such that $\mathbf{B\text{-}E}, s, s' \models \mathrm{OP}\delta(v)$ where OP is the operator characterizing the aspectual class to

[22] This view of programs as denoting (finite) sequences of states instead of relations between states is the one adopted in Process Logic (PL). These two views can be proved to be equivalent as long as neither infinite nor aborting sequences are admitted and atomic programs are interpreted as (indivisible) relations between states; for details see (Wilm, 1988).

which verbs of type v belong. This subset of R_v will be denoted by $R_{\text{regular}(v)}$. This relativization is necessary because not all sequences $(s, s') \in R_v$ need be regular as shown above. This can be the case because $\delta(v)$ does not hold in s' or, e.g. in the case of an accomplishment-verb, $\delta(v)$ already holds at intermediate states.

In the structure **B-E** events that correspond to complex transitions, i.e. paths in an LTS or non-atomic programs in DL, are not represented as being composed out of atomic events. This corresponds again to the view adopted in TL, which is an endogeneous formalism. But in effect, there is no need for such an explicit representation because the atomic programs can be thought of to be of the same type v as the complex event. For instance, they are just minimal eating- or pushing-events. The complex event can be interpreted as the n-fold (relational) composition of such minimal events (where n is the length of the sequence) (see Naumann (1997a), where this view is made explicit).

The problem which the approach by Blackburn et al. faces can be solved in the present approach in the following way. A verb of a particular type v expresses a change that is described as having a culmination just in case all regular execution sequences (s, s') have the property expressed by the set of sequences given in (37) where $\phi = \delta(v)$.

(37) $R_{\text{culm}(\phi)} = \{(s, s') \mid \mathbf{B\text{-}E}, s, s' \models L\neg\phi \cap {\sim}(R\phi\,;\nabla) \cap R\phi\}$

Regular sequences corresponding to accomplishment-, achievement- and point-verbs have this property, whereas regular sequences corresponding to stative-verbs and activity-verbs do not all have this property. Sequences corresponding to point-verbs satisfy this condition vacuously because the development portion is always empty as there are no states in between s and s'. The sequences therefore also have the property expressed by $R_\phi = \{(s, s') \mid \mathbf{B\text{-}E}, s, s' \models {\sim}(R\neg\phi\,;\nabla) \cap R\phi\}$. A verb the regular sequences of which have both properties will be called *weakly culminative* whereas verbs all of whose corresponding regular sequences belong to $R_{\text{culm}(\phi)}$ but not all belong to R_ϕ will be called *strongly culminative*. The property of being strongly culminative is exactly the property that was defined as *strong betweenness* in (20) above in section (2), which is repeated here for convenience.

(20) (i) $\forall\sigma : \mathbf{B\text{-}E}, \sigma \models L\neg\phi \cap {\sim}(R\phi; \nabla) \cap R\phi$

(ii) $\neg\forall\sigma : \mathbf{B\text{-}E}, \sigma \models {\sim}(R\neg\phi\,;\nabla) \cap R\phi$

From this it follows that if a verb is strongly culminative, this means that there is the possibility of intermediate states at which the postcondition ϕ does not (yet) hold. If, on the other hand, the verb is only weakly culminative, all regular execution sequences corresponding to events of this type have length 1, i.e., consist of two states.

The distinction between two kinds of culminativity plays a role in the interpretation of modified VPs of the form 'VP in n time' (see Naumann (1997c) for details) as well as the perfect of result (see Naumann (1997b) for details). The predicates 's-culm' and 'w-culm' are defined in (38)[23].

(38) a. $\text{w-culm}(v) =_{\text{def.}} \forall\sigma[\sigma \in R_{\text{regular}(v)} \Rightarrow \sigma \in R_{\text{culm}(\delta(v))}]$

b. $\text{s-culm}(v) =_{\text{def.}} \text{w-culm}(v) \wedge \neg\forall\sigma[\sigma \in R_{\text{regular}(v)} \Rightarrow \sigma \in R_{\delta(v)}]$

A corresponding predicate 'culm(e,v)' for events is defined in (39) where 'culm' is either 'w-culm' or 's-culm'.

(39) $\text{culm}(e, v)$ iff $e \in P_v \wedge (\alpha(e), \beta(e)) \in R_{\text{regular}(v)} \wedge \text{culm}(v)$

'Culm(e, v)' is to be read as 'event e culminates when described as being of type v'. According to this definition, an event e cannot be said to culminate or to not culminate in an absolute way, but only with respect to a particular verb v and therefore a particular event type P_v to which it belongs. This relativization of the culm-predicate to a particular verb v (or the corresponding event type) is necessary because, for instance, an event e can be said to culminate if it is described as a walking to the station (accomplishment) whereas it does not culminate if it is described simply as a walking (activity). This is a consequence of the fact that a sequence (s, s') can belong to different R_{ac} for different properties ϕ. Thus, for one property ϕ it can belong to $R_{\text{Acco}(\phi)}$ but for another property ϕ' it is an element of $R_{\text{Act}(\phi')}$. The properties ϕ and ϕ' will belong to different verbphrases v and v' via the function δ, i.e., $\delta(v) = \phi$ and $\delta(v') = \phi'$. It is therefore possible to have an event e with $\alpha(e) = s$, $\beta(e) = s'$, $e \in P_v$ and $e \in P_{v'}$ such that the sequence $(s, s') \in R_{\text{Acco}(\phi)}$ and at the same time $(s, s') \in R_{\text{Act}(\phi')}$. As all sequences $(s, s') \in R_{\text{Acco}(\phi)}$ satisfy (38b) with respect to ϕ, e culminates with respect to its being an element of the event type P_v. But it does not culminate with respect to its being an element of the event type $P_{v'}$ because not all elements from $R_{\text{Act}(\phi')}$ satisfy (38b) (nor do they all satisfy (38a)).

5. The Plural Level

In this final section I will sketch how the influence of plural NPs at the VP-level can be explained (for details, see Naumann (1997d)). Consider the data in (40).

[23] To the definition applies the same qualification as to the definition of the property of strong betweenness. See footnote (5) above.

(40) a. John ate five apples in one hour/*for one hour.

b. John ate apples *in one hour/for one hour.

c. Bill pushed five carts one after the other in one hour/together for one hour.

d. Bill pushed carts *in one hour/for one hour.

These data show that a particular verb type, e.g. *eat*, can correspond to different types of sequences, where the type of sequence is determined by the different aspectual criteria. The relation between a verb v and a particular type of sequence (or a particular temporal operator) is therefore not one-one but rather one-many. At the VP-level, the types of sequences that correspond to a particular verb v are determined by two different types of information.

(i) the information determined by the verb

(ii) the cardinality information specified by the determiner inside the (internal) argument NP

The information determined by the verb concerns the property that is changed by that particular type of verb. So far this property has been determined by δ. If the determiner specifies the cardinality 1, the type of sequence is determined in accordance with δ alone, i.e., the type corresponds to the value of γ . From this it follows that one can distinguish two different cases: the base case, (ii'), and the plural case, (ii').

(i') the determiner specifies the cardinality 1

(ii') the determiner specifies a cardinality > 1 or is a bare plural

To the base case corresponds a function $\delta_{bc} = \delta$. In the plural case a function δ_{PL} is assumed which assigns to each verb v a set of plural properties corresponding to the cardinality information of the different plural determiners.

For a given verb v to each element of $\delta' = \delta_{bc} \cup \delta_{PL}$ there corresponds a particular type of sequence. Formally, this can be expressed by defining a two-place function γ' which assigns to each event type P_v and a property from $\delta'(v)$ a type of sequence. The crucial question is whether it is possible to determine the value of $\gamma'(P_v)$ for elements from δ_{PL} from that of γ in the base case, i.e., from $\gamma'(P_v, \delta_{\mathrm{bc}}(v))$. This is possible (for details, see Naumann (1997d)). In (41) the case for an accomplishment-verb is given.

(41) a. If the ES of a verb type v, e.g. *eat*, belong to R_{ACCO} for the basic property $\delta_{\text{bc}}(v)$, then the ES for non-bare elements of $\delta_{\text{PL}}(v)$ belong to R_{ACCO} too[24].

b. The ES of a bare element from $\delta_{\text{PL}}(v)$ of a verb type v belong to R_{ACT}.

c. $\gamma'(P_v, \delta_{\text{bc}}(v)) \subseteq R_{\text{ACCO}}(\delta_{\text{bc}}(v)) \Rightarrow \forall\phi \in \delta_{\text{nb-PL}}(v)\ \gamma'(P_v, \phi) \subseteq R_{\text{ACCO}}(\phi)$

In (41c) $\delta_{\text{nb-PL}}$ denotes for a given v the set of all properties corresponding to non-bare plural determiners.

Already from this sketch one can conclude that each verb v has a particular 'aspectual' potential which is determined by the functions δ and γ from section (4), i.e. by the base case. Given this base case, the aspectual properties (and therefore the corresponding types of sequences) which result at higher levels (VP, sentence) due to the influence of nominal arguments and other constituents (see Naumann (1997d)) can be calculated.

References

Blackburn, P. et al. (1994): Back and forth through Time and Events, in: Gabbay, D./Ohlbach, H. (eds.): *Temporal Logic*, Berlin, Springer, 1994, pp. 225-237

Dowty, D. (1979): *Word Meaning and Montague Grammar*, Dordrecht, Reidel, 1979

Krifka, M. (1992): Thematic Relations as Links between Nominal Reference and Temporal Constitution, in: Sag, I. et al. (eds.): *Lexical Matters*, CSLI Lecture Notes, no.24, CSLI, Stanford University, pp. 29-54, 1992

Moens, M./Steedman, M. (1988): Temporal Ontology and Temporal Reference, in: *Computational Linguistics*, vol.14, pp. 15-28, 1988

Naumann, R. (1995): *Aspectual Composition and Dynamic Logic*, Habilitationsschrift, University of Düsseldorf, 1995

Naumann, R. (1996a): Aspectual Composition in Dynamic Logic, in: Dekker, P./Stokhof, M. (eds.): Proc. 10th Amsterdam Colloquium, ITLI, University of Amsterdam, 1996, pp. 567-586

Naumann, R. (1996b): Hybrid Languages for Tense and Aspect, manuscript, University of Düsseldorf

Naumann, R. (1997a): A Dynamic Decompositional Event Semantics, manuscript University of Düsseldorf, submitted to *Linguistics and Philosophy*

Naumann, R. (1997b): The Present Perfect in Dynamic Semantics, paper presented at the Conference Logical Aspects of Computational Linguistics, Nancy, France, 23-25.9.96

Naumann, R. (1997c): The Aspectual Analysis of English Verbs in a Dynamic Decom positional Event Semantics, manuscript, University of Düsseldorf

[24] This is too simple in view of determiners like *at most n* or *at least n*; for details see Naumann (1997d).

Naumann, R. (1997d): An Analysis of Aspectual Phenomena at the Plural Level in a Dynamic Temporal Logic, manuscript, University of Düsseldorf

Van Benthem, J. et al. (1994): Modal Logic, Transition Systems and Processes, *Journal of Logic and Computation*, vol.4, pp. 811-855, 1994

Van Benthem, J./Bergstra, J. (1995): Logic of Transition Systems, in: JoLLI, vol.3, 1995, pp. 247-283

Vendler, Z. (1967): Verbs and Times, in: *Linguistics in Philosophy*, Ithaca, Cornell University Press, 1967

Verkuyl, H. (1993): *A Theory of Aspectuality*, Cambridge, Cambridge University Press, 1993

Wilm, A. (1988): *Aussagenlogische Dynamische Logik als Programmlogik*, dissertation, University of Kiel (Germany), 1988

A Decidable Temporal Logic for Temporal Prepositions

Ian Pratt (ipratt@cs.man.ac.uk)
Department of Computer Science, Manchester University, Manchester M13 9PL, UK

Nissim Francez *
Department of Computer Science, Technion-IIL, Haifa, Israel

Abstract. This paper investigates the connection between temporal logic and the semantics of some temporal expressions in English. Specifically, we show how the meanings of English temporal preposition-phrases can be expressed as modal operators in an interval-based temporal logic which we call ETL0. We show how cascaded temporal preposition phrases and temporal preposition phrases with complex complements can be translated into ETL0. Finally, we present a decision-procedure for ETL0 and establish its correctness. The main contribution of the paper is to provide a characterization, in terms familiar to modal logicians, of the expressive resources made available by an important class of temporal expressions in English.

1. Introduction

This paper aims to characterize the expressive resources provided by a subset of the temporal preposition-phrases (TPPs) in English. We show, informally, how sentences involving these phrases can be translated into an interval-based, modal temporal logic. The emphasis throughout is on the tailoring of a temporal logic to the expressive power of temporal English, rather than on the formal semantics of English itself. The resulting logic, called ETL0, is sufficient to model some of the more common prepositional constructions in English. As the '0' in its name suggests, it is conceived of as a minimal system: extensions of this logic have been devised to capture larger temporal fragments of English.

There is now a great deal of interest in the development of logics specialized for (various aspects of) natural language (see, for example (Ali and Shapiro, 1993),(Franconi, 1993),(Hwang and Schubert, 1993); for an earlier reference, see (Suppes, 1991)), and the present paper falls into this tradition. The interval-based temporal logic which we use to account for the semantics of temporal preposition phrases uses operators familiar from dynamic logic (see (Harel, 1984)) and description logics (see (Schild, 1991) for an overview of the relationship between dynamic

* The authors would like to thank Mr. Dominik Schoop for his comments on the manuscript. The authors also gratefully acknowledge the support of the EPSRC, grant number GR/L/07529.

H. Barringer et al. (eds.), Advances in Temporal Logic, 255–278.

logic and description logics) and, more generally, from relation-algebras (see (Ladkin and Maddux, 1994)). The closest well-known temporal logic to the one developed here is probably Halpern and Shoham's interval logic HS (Venema, 1990). One important feature of ETL0 is its decidability. For results on the decidability of related systems, see, e.g., Henkin *et al.* (Henkin and Tarski, 1985) and Nemeti (Nemeti, 1991).

2. Temporal preposition phrases as modal operators

Consider the sentences:

(1) Mary kissed John

(2) Mary kissed John during every meeting

(3) Mary kissed John during every meeting one day

(4) Mary kissed John during every meeting one day in January.

The semantics of tensed expressions are notoriously complicated (references are too numerous to cite; for a recent survey, see (Steedman, 1996)), but, simplifying somewhat, we may take sentence (1) to state that Mary kissed John at some time within some (contextually determined) period, which we call the *time of interest* (*toi*). We take the *toi* to depend on several factors, most saliently, verb-tense and -aspect, but we shall not concern ourselves in this paper with how it is established. Rather, we are interested in how temporal preposition phrases are employed to quantify over the *toi*. For example, sentence (2) states that for every sub-interval within the *toi* over which a meeting occurred, Mary kissed John some time within that meeting-interval. Sentence (3) asserts that, for some day within the *toi*, for every sub-interval of that day over which a meeting occurred, Mary kissed John some time within that meeting-interval. Finally, sentence (4) asserts that there is a unique January in the *toi*, and that, for some day within that January, for every meeting on that day, Mary kissed John some time within that meeting-interval.[1]

Thus, in sentences (1)–(4), we see that each additional temporal preposition phrase functions so as to restrict the temporal quantification in the previous sentence to sub-intervals of a particular interval or collection of intervals. The most natural way to model this effect is to think of the meaning of a sentence as a function from temporal intervals to truth-values. To see how this works in detail, we avail ourselves of the the following three relations:

[1] In this paper, presuppositions will be assimilated to truth-conditions in the obvious (e.g. Russellian) way. This strategy avoids complications irrelevant to the treatment of temporal prepositions.

(5) $\lambda P \lambda Q[\mathbf{every}(P, Q)] =_{\mathrm{def}} \lambda P \lambda Q[\forall x(P(x) \rightarrow Q(x))]$

(6) $\lambda P \lambda Q[\mathbf{a}(P, Q)] =_{\mathrm{def}} \lambda P \lambda Q[\exists x(P(x) \wedge Q(x))]$

(7) $\lambda P \lambda Q[\mathbf{the}(P, Q)] =_{\mathrm{def}} \lambda P \lambda Q[\exists ! x(P(x)) \wedge \exists x(P(x) \wedge Q(x))]$.

Then, taking the variables I_i and J_i to range over time-intervals, we can write:

(8) $[\![\text{Mary kissed John}]\!] = \lambda I_0[\exists J_0[\textsc{kiss}(\textsc{mary}, \textsc{john})(J_0) \wedge J_0 \subseteq I_0]]$

(9) $[\![\text{Mary kissed John during every meeting}]\!] =$
$\lambda I_1[\mathbf{every}(\lambda J_1[\textsc{meeting}(J_1) \wedge J_1 \subseteq I_1],$
$\lambda I_0[\exists J_0[\textsc{kiss}(\textsc{mary}, \textsc{john})(J_0) \wedge J_0 \subseteq I_0]])]$

(10) $[\![\text{Mary kissed John during every meeting one day}]\!] =$
$\lambda I_2[\mathbf{a}(\lambda J_2[\textsc{day}(J_2) \wedge J_2 \subseteq I_2],$
$\lambda I_1[\mathbf{every}(\lambda J_1[\textsc{meeting}(J_1) \wedge J_1 \subseteq I_1],$
$\lambda I_0[\exists J_0[\textsc{kiss}(\textsc{mary}, \textsc{john})(J_0) \wedge J_0 \subseteq I_0]])])]$

(11) $[\![\text{Mary kissed John during every meeting one day in January}]\!] =$
$\lambda I_3[\mathbf{the}(\lambda J_3[\textsc{January}(J_3) \wedge J_3 \subseteq I_3],$
$\lambda I_2[\mathbf{a}(\lambda J_2[\textsc{day}(J_2) \wedge J_2 \subseteq I_2],$
$\lambda I_1[\mathbf{every}(\lambda J_1[\textsc{meeting}(J_1) \wedge J_1 \subseteq I_1],$
$\lambda I_0[\exists J_0[\textsc{kiss}(\textsc{mary}, \textsc{john})(J_0) \wedge$
$J_0 \subseteq I_0]])])])]$.

An assertion of any of these sentences can then be seen as an application of the relevant function to the *toi*. We follow a widely-adopted policy (e.g. (Allen, 1984),(Herweg, 1991)) and take events to occur at time-*intervals*, rather than time points. We assume intervals to be closed and bounded, but we admit single points as intervals.

There is another way to view such functions, of course: namely, as formulae in an interval-based temporal logic. On this view, the temporal preposition phrases are modal operators in this logic. As we shall be using sentence (1) throughout this paper, let us abbreviate its meaning, given in (8), by ϕ. Then we can think of the meanings (9)–(11) as having the forms:

(12) $[\![$ Mary kissed John $[_{\text{TPP}}$ during every meeting$]$ $]\!] =$
$\text{Op}_{\text{during every meeting}}(\phi)$

(13) $[\![$ Mary kissed John $[_{\text{TPP}}$ during every meeting$]$ $[_{\text{TPP}}$ one day$]$ $]\!] =$
$\text{Op}_{\text{one day}}(\text{Op}_{\text{during every meeting}}(\phi))$

(14) $[\![$Mary kissed John $[_{\text{TPP}}$ during every meeting$][_{\text{TPP}}$ one day $]$
$[_{\text{TPP}}$ in January $]]\!] =$
$\text{Op}_{\text{in January}}(\text{Op}_{\text{one day}}(\text{Op}_{\text{during every meeting}}(\phi)))$.

Our task now is to explain how these operators can be constructed, and how they relate to the components of the temporal preposition phrases whose meanings they express.

As a first step, we define the modal operators $\langle\alpha\rangle$, $[\alpha]$ and $\{\alpha\}$, where α is a *binary relation* on intervals (note that we write $\models_{I,J} \alpha$ to mean that (I, J) stand in the relation α):

(15) $\models_I [\alpha]\phi$ iff for all J, $\models_{I,J} \alpha$ implies $\models_J \phi$

(16) $\models_I \langle\alpha\rangle\phi$ iff there exists a J such that $\models_{I,J} \alpha$ and $\models_J \phi$

(17) $\models_I \{\alpha\}\phi$ iff there exists an unique J such that $\models_{I,J} \alpha$ and there exists a J such that $\models_{I,J} \alpha$ and $\models_J \phi$.

These definitions will be familiar from dynamic logic and description logic. They allow a binary relation α to combine with a formula ϕ to produce another formula ϕ'.

Next, we isolate the semantic contribution of the *content-words* in the complements of the temporal preposition-phrases occurring in sentences (2)–(4). Our proposal is to take the meanings of meeting, day and January to be the following *binary relations*:

(18) $[\![\text{meeting}]\!] = \lambda I \lambda J[\text{MEETING}(J) \wedge J \subseteq I]$

(19) $[\![\text{day}]\!] = \lambda I \lambda J[\text{DAY}(J) \wedge J \subseteq I]$

(20) $[\![\text{January}]\!] = \lambda I \lambda J[\text{JANUARY}(J) \wedge J \subseteq I]$,

where MEETING, DAY and JANUARY are suitable predicates of intervals. These meanings allow us to give the meanings of the TPPs in sentences (2)–(4) using the operators (16)–(17), thus:

(21) $[\![\text{during every meeting}]\!] = [[\![\text{MEETING}]\!]]$

(22) $[\![\text{one day}]\!] = \langle[\![\text{DAY}]\!]\rangle$

(23) $[\![\text{in January}]\!] = \{[\![\text{JANUARY}]\!]\}$.

The full advantages for taking these meanings to be binary relations rather than unary predicates will become clear only as we proceed.

Thus, on this account, the sentence-meanings (8)–(11) have the following representations:

(24) $[\![\text{Mary kissed John}]\!] = \phi$

(25) $[\![\text{Mary kissed John during every meeting}]\!] = [[\![\text{MEETING}]\!]]\phi$

(26) $[\![\text{Mary kissed John during every meeting one day}]\!] = \langle[\![\text{DAY}]\!]\rangle[[\![\text{MEETING}]\!]]\phi$

(27) $[\![\text{Mary kissed John during every meeting one day in January}]\!] = \{[\![\text{JANUARY}]\!]\}\langle[\![\text{DAY}]\!]\rangle[[\![\text{MEETING}]\!]]\phi$.

3. Temporal prepositions and relation composition

Consider the sentence:

(28) John kissed Mary during every meeting on a Wednesday.

At what times must John have kissed Mary for this sentence to be true? There are two possible answers: (i) at some time during every meeting falling on *some particular Wednesday* in the *toi*, and (ii) at some time during every meeting falling on *any Wednesday* in the *toi*. Let us call these two readings of (28) the 'weak' and 'strong' truth-conditions, respectively, since the latter (assuming there is a Wednesday in the *toi*) implies the former. They may be expressed, respectively, as follows:

(29) $[\![$John kissed Mary during every meeting on a Wednesday$]\!]_1 =$
$$\lambda I_2[\mathbf{a}(\lambda J_2[\textsc{wednesday}(J_2) \wedge J_2 \subseteq I_2], \lambda I_1[\mathbf{every}(\lambda J_1[\textsc{meeting}(J_1) \wedge J_1 \subseteq I_1], \lambda I_0[\exists J_0[\textsc{kiss}(\textsc{john},\textsc{mary})(J_0) \wedge J_0 \subseteq I_0]])])]$$

(30) $[\![$John kissed Mary during every meeting on a Wednesday$]\!]_2 =$
$$\lambda I_2[\mathbf{every}(\lambda J_1[\mathbf{a}(\lambda J_2[\textsc{wednesday}(J_2) \wedge J_2 \subseteq I_2], \lambda I_1[\textsc{meeting}(J_1) \wedge J_1 \subseteq I_1])], \lambda I_0[\exists J_0[\textsc{kiss}(\textsc{john},\textsc{mary})(J_0) \wedge J_0 \subseteq I_0]]).$$

It is possible to argue that the ambiguity of (28) is structural—specifically, that the weak truth-condition corresponds to the phrase-structure

(31) John kissed Mary $[_{\text{TPP}}$ during every meeting $]$ $[_{\text{TPP}}$ on a Wednesday$]$

and the strong truth-condition to the phrase-structure

(32) John kissed Mary
$[_{\text{TPP}}$ during every $[_{\text{N}'}$ meeting $[_{\text{TPP}}$ on a Wednesday $]]]$.

Here, the category N′ denotes a noun-phrase before the application of a determiner. This notation suggests that meeting on a Wednesday has the same type of meaning as, e.g. meeting in sentence (2).

The phrase-structure (31) is unproblematic. Here we have two TPPs modifying an underlying sentence, and, according to the above proposals, the resulting truth-condition is given by the formula

(33) $\langle[\![\textsc{wednesday}]\!]\rangle[[\![\textsc{meeting}]\!]]\phi$.

The phrase-structure (32), by contrast, is more difficult. Here, meeting on a Wednesday acts as a semantic, as well as a syntactic, unit. We take its meaning to be

(34) $[\![$meeting on a Wednesday$]\!] =$
$$\lambda I_2 \lambda J_1[\mathbf{a}(\lambda J_2[\textsc{wednesday}(J_2) \wedge J_2 \subseteq I_2], \lambda I_1[\textsc{meeting}(J_1) \wedge J_1 \subseteq I_1])],$$

for then, we can simply write, by analogy with (25):

(35) ⟦Mary kissed John during every meeting on a Wednesday⟧ = [⟦MEETING ON A WEDNESDAY⟧]ϕ

which can be seen to be equivalent to the suggested truth-condition (30).

It remains for us only to construct the meaning of ⟦meeting on a Wednesday⟧ as given in (34). This meaning can be seen as a composite of the meanings

(36) ⟦ Wednesday ⟧ $= \lambda I_2 \lambda J_2[\text{WEDNESDAY}(J_2) \wedge J_2 \subseteq I_2]$

(37) ⟦ meeting ⟧ $= \lambda I_1 \lambda J_1[\text{MEETING}(J_1) \wedge J_1 \subseteq I_1]$

in which the λJ_2 of (36) and the λI_1 of (37) are 'bound' by the second-order relation **a**. This situation is reminiscent of role composition in description logic and relation logic, or of program-sequencing in dynamic logic. Thus, defining (existential) relation composition in the normal way as

(38) $\models_{I,J} \alpha;\beta$ iff there exists a K such that $\models_{I,K} \alpha$ and $\models_{K,J} \beta$,

we can re-write the meaning-assignment (34) quite simply as

(39) ⟦ meeting on a Wednesday ⟧ = ⟦ Wednesday ⟧; ⟦ meeting ⟧,

whence the truth-condition in (30) can be re-written

(40) ⟦John kissed Mary during every meeting on a Wednesday ⟧ = [⟦Wednesday⟧; ⟦meeting⟧]ϕ.

Of course, by varying the temporal preposition and the determiner in the phrase meeting on a Wednesday, we vary the manner in which the two roles are composed. Thus, to express the meaning

(41) ⟦ meeting on Wednesday ⟧ =
$\lambda I_2 \lambda J_1[\mathbf{the}(\lambda J_2[\text{WEDNESDAY}(J_2) \wedge J_2 \subseteq I_2],$
$\lambda I_1[\text{MEETING}(J_1) \wedge J_1 \subseteq I_1])]$

(where on Wednesday is assumed to have the force of on *the* Wednesday), we propose the relation-combining operator

(42) $\models_{I,J} \alpha!\beta$ iff there exists an unique K such that $\models_{I,K} \alpha$ and there exists a K such that $\models_{I,K} \alpha$ and $\models_{K,J} \beta$,

whence

(43) ⟦John kissed Mary during every meeting on Wednesday ⟧ = [⟦Wednesday⟧!⟦meeting⟧]ϕ.

We shall see below, however, that, with only minimal concessions in respect of the ability to express English senences, we can do without these more complicated forms of relation composition.

One final point about relation composition. There is certainly no point, in the present context, in introducing a universal relation composition operator (i.e. the dual of ";"):

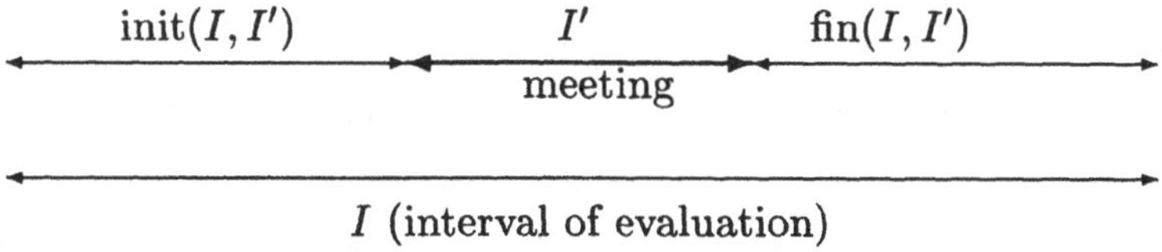

Figure 1. Representation of before, during and after.

(44) $\models_{I,J} \alpha\forall\beta$ iff $\models_{I,K} \alpha$ implies $\models_{K,J} \beta$ for all K,

for such a relation will be satisfied only in trivial and degenerate cases, given the atomic relations to which it is likely to be applied in natural language. Thus, for example, the concept

(45) $[\![$meeting every Wednesday $]\!] =$
$$\lambda I_2 \lambda J_1[\mathbf{every}(\lambda J_2[\text{WEDNESDAY}(J_2) \wedge J_2 \subseteq I_2], \lambda I_1[\text{MEETING}(J_1) \wedge J_1 \subseteq I_1])]$$

finds no application except in the degenerate case where there is only one Wednesday in the *toi*, since Wednesdays are disjoint, and no meeting can take place on all of them. Therefore, we do not need universal relation composition in this temporal logic.

4. Expressing temporal order

Whereas the TPP during the meeting serves to locate events in the sub-interval of the interval of evaluation *occupied by* the meeting, so the TPP before the meeting serves to locate them in the sub-interval of the interval of evaluation *leading up* to the interval occupied by the meeting (symmetrically for the TPP after the meeting). These relationships are depicted in fig. 1. The partial functions init and fin are functions defined as follows:

(46) $\text{init}([a, b], [c, d]) =_{\text{Def}} [a, c]$ if $a \leq c \leq d \leq b$
$\text{fin}([a, b], [c, d]) =_{\text{Def}} [d, b]$ if $a \leq c \leq d \leq b$.

Intuitively, the role of these functions is to nudge the interval of subsequent evaluation either forwards or backwards. Accordingly, we call them *temporal warp functions.* Now we define the modal operators:

(47) $\models_I \{\alpha\}_{>}\phi$ iff there exists an unique J such that $\models_{I,J} \alpha$ and there exists a J such that $\models_{I,J} \alpha$ and $\models_{\text{init}(I,J)} \phi$.

(48) $\models_I \{\alpha\}_{<}\phi$ iff there exists an unique J such that $\models_{I,J} \alpha$ and there exists a J such that $\models_{I,J} \alpha$ and $\models_{\text{fin}(I,J)} \phi$.

Then we have:

(49) $[\![$Mary kissed John before the meeting$]\!] = \{[\![\text{MEETING}]\!]\}_{\geq}\phi$

(50) $[\![$Mary kissed John after Wednesday$]\!] = \{[\![\text{WEDNESDAY}]\!]\}_{\leq}\phi$,

which give intuitively correct truth-conditions.

Of course, the meaning proposed here for before assumes the sense *any time before* rather than the equally possible *just before*. However, by choosing a different warp-function, for example,

(51) just-before$([a, b], [c, d]) =_{\mathrm{Def}} [c - \epsilon, c]$ if $a \leq c - \epsilon$ and $c \leq d \leq b$

where $\epsilon > 0$ is a contextually determined parameter, we could define an appropriate modal operator, say $\{\alpha\}_{\geq\epsilon}$ to model this meaning as well. Similar remarks apply to more complex prepositional constructions such as 5 minutes before etc. Other temporal propositions, such as until, since and by can also be handled by the framework introduced here. For example, we have

(52) $[\![$John kissed Mary every Wednesday until Christmas$]\!] =$
$\{[\![\text{CHRISTMAS}]\!]\}_{\geq}[[\![\text{WEDNESDAY}]\!]]\phi$.

However, since these prepositions introduce linguistic technicalities well beyond the scope of this paper, and do not change the nature of the logic required, we shall not discuss them further.

It is important to note that using before in the sense of *any time before* limits the available choices of the determiner in the embedded noun-phrase. Thus, if we consider the sentences:

(53) Mary kissed John before a meeting

(54) Mary kissed John before every meeting

before has to mean something like *just before*. For, on the *any time before* interpretation, and given commonsense assumptions about the distribution of meetings, (53) would mean that Mary kissed John before the *last* meeting in the *toi*, and (54) would mean that he did so before the *first* meeting in the *toi*. Yet these meanings do not seem available. Presumably, this is because of the semantic redundancy involved: if you mean that Mary kissed John before the first meeting, then you should say so explicitly. Corresponding remarks apply to after.

Can temporal prepositions such as before and after modify preposition complements and so introduce new kinds of relation-combination? The answer is that they can. In the sentence

(55) Mary kissed John during the $[_{N'}$ meeting before the dinner $]$

the meaning of $[\![$meeting before the dinner$]\!]$ can be seen as a combination of the temporal relations $[\![$meeting$]\!]$ and $[\![$ dinner$]\!]$. Thus, we might write

(56) $\models_{I,J} \alpha !_{\geq} \beta$ iff there exists an unique K such that $\models_{I,K} \alpha$ and there exists a K such that $\models_{I,K} \alpha$ and $\models_{\text{init}(I,K),J} \beta$.

whence

(57) $[\![$Mary kissed John during the $[_{N'}$ meeting before the dinner $]]\!] =$ $\{[\![\text{dinner}]\!]!_{\geq}[\![\text{meeting}]\!]\}\phi$.

The basic framework just introduced can in principle cope with preposition phrases nested in this way to any depth. In practice, however, a point is quickly reached where the multiplying structural ambiguities and sheer difficulty of understanding such phrases render further investigation of this matter pointless. Consider, for example, the possible meanings of the preposition phrase(s) before the meeting after the dinner on Wednesday.

5. Introducing some simplifications

We have shown how a range of English temporal preposition-phrases can be viewed as modal operators in an interval-based temporal logic, where binary relations can be combined with formulae. Before we show how this logic can be equipped with a decision procedure, however, some simplifications must be introduced, which make the logic easier to understand without significantly reducing the coverage of English sentences.

The most important simplification is the restriction of relation composition to the existential combinator, ";". Thus, we lose the ability to model temporal preposition-phrases such as during the meeting before the dinner (which involved the combinator "$!_{\geq}$"). However, a fully adequate treatment of the semantics of these phrases would in any case take us beyond the logic investigated here, and is better left for future work. More seriously, we *appear* to lose the ability to model temporal preposition-phrases such as *before the meeting on Wednesday* (taking Wednesday to be implicitly definitely quantified). However, iterated use of the the operator $\{\alpha\}$ gives us the ability to assert the existence of a unique meeting on the unique Wednesday in the *toi* if necessary, say using the formula

(58) $\{[\![\text{Wednesday}]\!]\}\{[\![\text{meeting}]\!]\}\top$

where $\top$ holds at all intervals. So complex temporal preposition complements involving embedded definite quantification can be accounted for anyway.

A second simplification is possible. We saw in connection with sentences (53) and (54) that, *in English*, before (in the sense of *any time before*) cannot be combined with an indefinitely quantified complement.

However, it turns out to be simpler to do precisely that. Thus, we employ, instead of the $\{\alpha\}_{\geq}$ and $\{\alpha\}_{\leq}$ as defined in (47) and (48), the operators

(59) $\models_I \langle\alpha\rangle_{\geq}\phi$ iff there exists a J such that $\models_{I,J} \alpha$ and $\models_{\mathsf{init}(I,J)} \phi$

(60) $\models_I \langle\alpha\rangle_{\leq}\phi$ iff there exists a J such that $\models_{I,J} \alpha$ and $\models_{\mathsf{fin}(I,J)} \phi$.

For we can replace $\{\alpha\}_{\geq}\phi$ with $\langle\alpha\rangle_{\geq}\phi$ and $\{\alpha\}\top$.[2] Corresponding remarks apply to $\{\alpha\}_{\leq}$.

Thus, to model a wide range of English temporal preposition-phrases, it suffices to employ an interval-based temporal logic with the modal operators $\langle\alpha\rangle$, $[\alpha]$, $\{\alpha\}$, $\{\alpha\}_{\geq}$ and $\{\alpha\}_{\leq}$, and the relation-combinator ";". With this logic we can represent the meanings of all the example sentences mentioned in this paper except for sentence (55).

6. Syntax and semantics of ETL0

In the remainder of this paper, we describe a logic, ETL0, based on the modal operators introduced above—in effect, a logic for the temporal preposition system of English. After defining the syntax and semantics of ETL0, we present a tableau-based decision procedure and prove its correctness. The one feature of the formal language which we have not linked to English is a negation operator (with restricted syntax). However, it should be clear, in principle, how the account developed here could be extended to include a range of negative English sentences, even if the details are somewhat complicated.

An *event-atom* is a member of the infinite set $E = \{e_1, e_2, \ldots\}$. Intuitively, the e_i should be thought of as standing for those parts of sentence-meanings that we have not decomposed in the foregoing sections, for example, $\lambda J[\textsc{kiss}(\textsc{mary}, \textsc{john})(J)]$, $\lambda J[\textsc{meeting}(J)]$ or $\lambda J[\textsc{monday}(J)]$.

The *temporal roles* of ETL0 are the smallest set of expressions obeying the rules:

1. If p is an event atom, then p is a temporal role

2. If ρ, σ are temporal roles, so is $\rho\sigma$.

[2] Actually, we need a conjunction operator in the logic to handle embedded instances of this translation; however, the addition of such an operator is completely routine.

Intuitively, the temporal roles are the binary relations appearing as the (undetermined) contents of temporal preposition phrases, for example $[\![\text{MEETING}]\!] = \lambda I \lambda J[\text{MEETING}(J) \wedge J \subseteq I]$. Notice that, in taking event-atoms to *be* temporal roles, we are implicitly turning them into relations by "inserting" $I \subseteq J$.

The *formulae* of ETL0 are the smallest set of expressions obeying the rules:

1. If p is an event atom, then $\Diamond p$ and $\neg\Diamond p$ are formulae

2. If ϕ is a formula and τ is a temporal role, then $\langle\tau\rangle\phi$, $[\tau]\phi$, $\{\tau\}\phi$, $\langle\tau\rangle_{\geq}\phi$ and $\langle\tau\rangle_{\leq}\phi$ are formulae.

Note the restriction of $\neg$ to the inner-most formulae.

Event atoms in E are interpreted as true or false at closed, bounded intervals of the real line. We denote the set of such intervals by Ω. Thus, each element of E denotes a type of event which may be instantiated at any number of intervals in Ω. In practice, however, the kinds of events reported in natural language are not arbitrarily distributed, but obey strong constraints on when they can occur. For example, except in very artificial cases, it is reasonable to assume that no single event-type in E is instantiated infinitely often in a bounded interval.[3]

More interestingly, it is difficult for most event-types to overlap in time. For example, if John's closing the door takes place over an interval I, that same type of event cannot ordinarily take place over an interval I' overlapping I. It may be objected that several meetings can take place in overlapping intervals if they occur, for example, in different rooms. But in that case, we reply, we have not one event-type, but several (one for each room). Our event-atoms are *purely* functions of time, so that, for example, simultaneous events described by the same event-atom are identical. And in that case, we maintain, it is sensible to assume that event-atoms are never instantiated at overlapping intervals.

Now, we relax this assumption slightly, and insist that no event-atom is ever instantiated at intervals I and I' such that $I \subseteq I'$. The reason for this relaxation is purely technical: since we only need the weaker assumption for our decision procedure to work, that is the one we make. Building in the stronger assumption requires only a straightforward change to that decision procedure.

Definition: Let E and Ω be as above. An *interpretation* is a set $M \subset E \times \Omega$ obeying the condition that, if $\langle p, I_1\rangle \in M$ and $\langle p, I_2\rangle \in M$

[3] Such an artificial example might be the event-type of the temperature's becoming a rational number in a room that is steadily getting hotter.

then $I_1 \subseteq I_2$ implies $I_1 = I_2$.

Thus, properly nested intervals of the same type are banned in interpretations. We refer to this ban as the *ontological assumption.*

Definition: Let $I = [i_1, i_2], J = [j_1, j_2] \in \Omega$ with $J \subseteq I$. Define $\text{init}(I, J) = [i_1, j_1], \text{fin}(I, J) = [j_2, i_2]$.

Let M be an interpretation. The following semantic rules define the notion of truth in M for ETL0. Throughout, we let the variables I, J and K range over Ω, p over E, ρ, σ and τ over temporal roles, and ϕ over formulae.

1. $M \models_{I,J} p$ if $\langle p, J\rangle \in M$ and $J \subseteq I$.
2. $M \models_{I,J} \rho\sigma$ if there exists a K such that $M \models_{I,K} \sigma$ and $M \models_{K,J} \rho$
3. $M \models_I \Diamond p$ if there exists a $J \subseteq I$ such that $\langle p, J\rangle \in M$.
4. $M \models_I \neg\Diamond p$ if p there exists no $J \subseteq I$ such that $\langle p, J\rangle \in M$.
5. $M \models_I \langle\tau\rangle\phi$ if there exists a J such that $M \models_{I,J} \tau$ and $M \models_J \phi$.
6. $M \models_I [\tau]\phi$ if $M \models_{I,J} \tau$ implies $M \models_J \phi$ for all J.
7. $M \models_I \{\tau\}\phi$ if there is an unique J such that $M \models_{I,J} \tau$, and there is a J such that $M \models_{I,J} \tau$ and $M \models_J \phi$.
8. $M \models_I \langle\tau\rangle_{\geq}\phi$ if there is a J such that $M \models_{I,J} \tau$ and $M \models_{\text{init}(I,J)} \phi$.
9. $M \models_I \langle\tau\rangle_{\leq}\phi$ if there is a J such that $M \models_{I,J} \tau$ and $M \models_{\text{fin}(I,J)} \phi$.

As usual, if Φ is a set of formulae and M an interpretation such that, for some $I \in \Omega$, $M \models_I \phi$ for all $\phi \in \Phi$, we say that M *models* Φ (at I) and write $M \models_I \Phi$.

Given these semantics, we introduce the following notational conventions. First, all temporal roles may be written as $p_1 \ldots p_n$, without parentheses. Second, since $\langle\rho\sigma\rangle\phi$ is equivalent to $\langle\rho\rangle\langle\sigma\rangle\phi$ and $[\rho\sigma]\phi$ to $[\rho][\sigma]\phi$, we assume that the operators $\langle\tau\rangle$ and $[\tau]$ occur only where τ is an event atom.

7. Decision procedure for ETL0

We now describe a procedure for determining whether a finite set of formulae Φ in ETL0 is consistent—that is, has a model.

The first stage is to set up a collection of *nodes* for the *node-processing rules* to operate on. There are three sorts of nodes: t-nodes, u-nodes and f-nodes. A t-node is a pair $\langle R, C\rangle$ consisting of an *interval relation* R and a *conditions-list* C, for example:

(61) $\langle \bar{I}_2 = \bar{I}_2, \{\bar{I}_1 \subseteq \bar{I}_3, \bar{I}_2 \subseteq \bar{I}_4\}\rangle$.

The interval relation is one of the forms $\bar{I} \subseteq \bar{J}$, $\bar{I} = \bar{J}$ or $\perp$, and the conditions-list is a finite set of such interval relations. The arguments to these interval relations are *interval labels*, which we may think of as names for intervals in Ω. We use the variables $\bar{I}$, $\bar{I}_1$, etc. to range over interval labels. Intuitively, $\langle R, C\rangle$ states that, if the interval relations in C all hold, then so does R. (Here, $\perp$ denotes the empty relation.) Interval labels are either atoms or compounds of the forms $\text{init}(\bar{I}, \bar{J})$ and $\text{fin}(\bar{I}, \bar{J})$ where $\bar{I}$ and $\bar{J}$ are interval labels.

A u-node $\langle \tau, \bar{I}, \bar{J}, C\rangle$ consists of four elements: a temporal role τ, two interval labels $\bar{I}$ and $\bar{J}$ and a conditions-list C, for example:

(62) $\langle p_1 \ldots p_n, \bar{I}_1, \bar{I}_2, \{\bar{I}_2 \subseteq \bar{I}_3\}\rangle$.

Intuitively, a u-node asserts that, if the conditions in C all hold, then $\bar{J}$ names the unique interval J such that the relation τ holds between the interval named by $\bar{I}$ and J.

An f-node $\langle \phi, \bar{I}, C, A\rangle$ consists of four elements: an *assertion* ϕ, an interval label $\bar{I}$, a conditions-list C and an *A-list* A, for example:

(63) $\langle \{p_1 \ldots p_n\}\langle p_{n+1}\rangle\neg\Diamond p_{n+2}, \bar{I}_1, \{\bar{I}_1 \subseteq \bar{I}_2\}, \emptyset\rangle$.

The assertion is either a formula of the logic or an event atom. The interval label and conditions-list are as explained above. The A-list is a technical device concerned with the termination of the algorithm. Intuitively, the f-node $\langle \phi, \bar{I}, C, A\rangle$ states that, if the conditions in C all hold, then ϕ is true at the interval named by $\bar{I}$. A *node set* is just a set of (t-, u- and f-) nodes.

The notion of an interpretation's modelling a set of formulae Φ can be extended in a natural way to that of an interpretation's modelling a node set S. The important point here is that, since interpretations make events true at *intervals* while nodes involve interval *labels*, we need some way of associating the latter with the former. We define:

Definition: Let μ be a function mapping the interval labels in a node set S to intervals in Ω. We consider only those mappings obeying the constraint that, if $\mu(\bar{I}) = [a, b]$ and $\mu(\bar{J}) = [c, d]$, then $\mu(\text{init}(\bar{I}, \bar{J})) = [a, c]$ and $\mu(\text{fin}(\bar{I}, \bar{J})) = [d, b]$. We say that the interval relation $\bar{I} \subseteq \bar{J}$ *holds under* μ if $\mu(\bar{I}) \subseteq \mu(\bar{J})$. Similarly for $\bar{I} = \bar{J}$. The relation $\perp$ never holds under μ.

We say that S is *suitable under* μ if, for all t-nodes $\langle R, C\rangle \in S$, if all of the temporal relations in C hold under μ, then so does R.

We say that S is *suitable* if, for some μ, S is suitable under μ.
We say that M *models* S under μ, written $M \models_\mu S$, if (i) S is suitable under μ, (ii) for all u-nodes $\langle \tau, \bar{I}_1, \bar{I}_2, C \rangle$, if the conditions C hold under μ, then $\mu(\bar{I}_2)$ is the only $J \in \Omega$ such that $J \subseteq \mu(\bar{I}_1)$ and $M \models_{\mu(\bar{I}_1),J} \tau$, and (iii) for all f-nodes $\langle \phi, \bar{I}, C, A \rangle$, if C hold under μ, then $M \models_{\mu(I)} \phi$.

For brevity, in the sequel, we shall sometimes refer to "the interval $\bar{I}$" when, strictly, we mean the interval *named by* $\bar{I}$, under some mapping μ. We now set up an initial set of nodes corresponding to the the formulae in Φ. We define:

Definition: Let Φ be a set of formulae and $\bar{I}_0$ an interval label. We say that the *initial node set constructed from* Φ *at* $\bar{I}_0$ is the set of f-nodes: $S_{\text{init}} = \{\langle \phi, \bar{I}_0, \emptyset, \emptyset, \rangle | \phi \in \Phi\}$.

In other words, the initial node set constructed from Φ is just a collection of f-nodes stating that the formulae in Φ are unconditionally true at some interval named by $\bar{I}_0$. Note that, if Φ has a model, then there is some interval I_0 at which all the formulae in Φ are made true by that model. Therefore, according to the above definition, setting $\mu(\bar{I}_0) = I_0$ gives us $M \models_\mu S_{\text{init}}$.

The decision procedure is as follows. Let Φ be a finite set of formulae of ETL0. Let $S = S_{\text{init}}$ be the initial node set constructed from Φ at $\bar{I}_0$. We shall add to the nodes in S using our *node-processing rules.* Here is one such rule:

(64) $\{\langle \langle p \rangle \phi, \bar{I}, C, A \rangle\} \Longrightarrow \{\langle p, \bar{J}, C, A \rangle, \langle \phi, \bar{J}, C, A \cup \{p\} \rangle, \langle \bar{J} \subseteq \bar{I}, \emptyset \rangle\}$.

Intuitively, the rule says: if you have a proposition of the form $\langle p \rangle \phi$ true at interval $\bar{I}$ under conditions C, then the proposition ϕ and the event atom p will be true at an interval $\bar{J}$ ($\bar{J}$ is a new interval-label) under conditions C, where $\bar{J}$ is (without conditions) a sub-interval of $\bar{I}$. Accordingly, we stipulate that this rule can be fired for a node set S if S contains a member matching the left-hand side; in firing the rule, we simply add the nodes on the right-hand side to S.

Here is another node-processing rule:

(65) $\{\langle [p]\phi, \bar{I}_1, C_1, A_1 \rangle, \langle p, \bar{I}_2, C_2, A_2 \rangle\} \Longrightarrow$
$\{\langle \phi, \bar{I}_2, C_1 \cup C_2 \cup \{\bar{I}_2 \subseteq \bar{I}_1\}, A_1 \cup A_2 \cup \{p\} \rangle\}$ (if $p \notin A_2$).

Intuitively, the rule says: if you have a formula of the form $[p]\phi$ true at interval $\bar{I}_1$ under conditions C_1, and an event atom p true at interval $\bar{I}_2$ under conditions C_2, then the proposition ϕ is true at $\bar{I}_2$ under conditions C_1 together with C_2 together with the condition that $\bar{I}_2$ is a subset of $\bar{I}_1$.

The full set of node-processing rules we require is as follows. We adopt the convention that the variables $\bar{J}$ and $\bar{J}_i$ are used only on the right hand sides of rules, and indicate that *new* interval labels must be selected. Moreover, we assume that no rule may be applied more than once to the very same node (or pair of nodes).

1. $\{\langle \Diamond p, \bar{I}, C, A\rangle\} \Longrightarrow \{\langle p, \bar{J}, C, A\rangle, \langle \bar{J} \subseteq \bar{I}, \emptyset\rangle\}$

2. $\{\langle\langle p\rangle\phi, \bar{I}, C, A\rangle\} \Longrightarrow \{\langle p, \bar{J}, C, A\rangle, \langle \phi, \bar{J}, C, A \cup \{p\}\rangle, \langle \bar{J} \subseteq \bar{I}, \emptyset\rangle\}$

3. $\{\langle\{p_1 \ldots p_n\}\phi, \bar{I}, C, A\rangle\} \Longrightarrow$
 $\{\langle p_i, J_i, C, A_i\rangle, \langle \bar{J}_i \subseteq \bar{J}_{i-1}, \emptyset\rangle | 1 \leq i \leq n\} \cup$
 $\{\langle \phi, \bar{J}_n, C, A_n \cup \{p_n\}\rangle, \langle p_1 \ldots p_n, \bar{I}, \bar{J}_n, C\rangle\}$ where $\bar{J}_0 = \bar{I}$, $A_1 = A$ and $A_i = A_{i-1} \cup \{p_{i-1}\}$ for all i $(2 \leq i \leq n)$

4. $\{\langle\langle p_1 \ldots p_n\rangle_{>}\phi, \bar{I}, C, A\rangle\} \Longrightarrow$
 $\{\langle p_i, J_i, C, A_i\rangle, \langle \bar{J}_i \subseteq \bar{J}_{i-1}, \emptyset\rangle | 1 \leq i \leq n\} \cup \{\langle \phi, \text{init}(\bar{I}, \bar{J}_n), C, A\rangle\}$
 where $\bar{J}_0 = \bar{I}$, $A_1 = A$ and $A_i = A_{i-1} \cup \{p_{i-1}\}$ for all i $(2 \leq i \leq n)$

5. $\{\langle\langle p_1 \ldots p_n\rangle_{<}\phi, \bar{I}, C, A\rangle\} \Longrightarrow$
 $\{\langle p_i, J_i, C, A_i\rangle, \langle \bar{J}_i \subseteq \bar{J}_{i-1}, \emptyset\rangle | 1 \leq i \leq n\} \cup \{\langle \phi, \text{fin}(\bar{I}, \bar{J}_n), C, A\rangle\}$
 where $\bar{J}_0 = \bar{I}$, $A_1 = A$ and $A_i = A_{i-1} \cup \{p_{i-1}\}$ for all i $(2 \leq i \leq n)$

6. $\{\langle [p]\phi, \bar{I}_1, C_1, A_1\rangle, \langle p, \bar{I}_2, C_2, A_2\rangle\} \Longrightarrow$
 $\{\langle \phi, \bar{I}_2, C_1 \cup C_2 \cup \{\bar{I}_2 \subseteq \bar{I}_1\}, A_1 \cup A_2 \cup \{p\}\rangle\}$ (if $p \notin A_2$)

7. $\{\langle \neg\Diamond p, \bar{I}_1, C_1, A_1\rangle, \langle p, \bar{I}_2, C_2, A_2\rangle\} \Longrightarrow \{\langle \bot, C_1 \cup C_2 \cup \{\bar{I}_2 \subseteq \bar{I}_1\}\rangle\}$

8. $\{\langle p_1 \ldots p_n, \bar{I}_0, \bar{I}', C\rangle\} \cup \{\langle p_i, \bar{I}_i, C_i, A_i\rangle | 1 \leq i \leq n\} \Longrightarrow$
 $\{\langle \bar{I}_n = \bar{I}', C_1 \cup \ldots \cup C_n \cup C \cup \{\bar{I}_i \subseteq \bar{I}_{i-1} | 1 \leq i \leq n\}\rangle\}$

9. $\{\langle p, \bar{I}_1, C_1, A_1\rangle, \langle p, \bar{I}_2, C_2, A_2\rangle\} \Longrightarrow \{\langle \bar{I}_1 = \bar{I}_2, C_1 \cup C_2 \cup \{\bar{I}_1 \subseteq \bar{I}_2\}\rangle\}$

We show below that, if we apply these rules repeatedly to S, a point will be reached at which no new nodes can be added. We will further show that Φ has a model if and only if this final node set is suitable. So we have only to address the problem of determining whether a node set is suitable, and our decision procedure is complete.

To determine whether a node set is suitable, we first translate any t-node $\langle R, C\rangle$, where $C = \{R_1, \ldots, R_n\}$, into a horn-clause involving interval relations:

(66) $R_1 \wedge \ldots \wedge R_n \rightarrow R.$

We call such a clause an *interval clause*. From the definition, S is suitable if and only if the interval clauses generated from all its t-nodes are simultaneously satisfiable. From the node-processing rules above, we know that all of the R_i will be of the form $\bar{J}_i \subseteq \bar{I}_i$, and R will be of the forms $\bar{J}_i \subseteq \bar{I}_i$, $\bar{J}_i = \bar{I}_i$ or $\bot$, where $\bar{I}_i$ and $\bar{J}_i$ are interval labels.

We can then translate, in the obvious way (and taking account of labels involving init and fin), the set of these interval clauses into a set of *order-clauses* of the two forms

$$x_1 \leq y_1 \wedge \cdots \wedge x_n \leq y_n \quad \rightarrow \quad x_{n+1} \leq y_{n+1}$$
$$x_1 \leq y_1 \wedge \cdots \wedge x_n \leq y_n \quad \rightarrow \quad \bot$$

where the x_i and y_i are the start- and end-points of the various interval-labels in S. We assume that we have an algorithm `order_test` to determine the satisfiability of such order-clauses, returning "consistent" or "inconsistent" in response to a set of order-clauses of the above forms.

Thus we have the following decision-procedure for ETL0:

```
begin consis_test(Φ)
    let S be the initial node set constructed
        from Φ with interval label Ī₀
    until S stabilizes do
        let S be the result of applying any
            node-processing rule to S
    end until
    forall t-nodes ⟨R, {R₁, ..., Rₙ}⟩ in S do
        form the set of corresponding order clauses
    let O be the union of all these sets
    return order_test(O)
end consis_test
```

7.1. Correctness

We now state the main results guaranteeing that our decision procedure will terminate with the correct answer.

Lemma (termination): Suppose the node-processing rules for ETL0 are applied repeatedly to a finite node set S. After a finite number of rule applications, a point will be reached where no rule application adds any new nodes.

Proof: It suffices to show that S bounds the length of the derivation of any f-node.

At any point in such a derivation, if a node N is produced by rule 6, we call the nodes $\langle [p]\phi, \bar{I}_1, C_1, A_1 \rangle$ and $\langle p, \bar{I}_2, C_2, A_2 \rangle$ on the left hand side of the rule the *mother* and *father* of N, respectively. If N is produced by some other rule, we again call the single node on the left hand side of the rule the mother of N (so that N has no father). By examining each rule, we see that the assertion (formula) in N's mother is strictly more complex than the assertion in N, and that the A-list in N's mother is no larger than the A-list in N. Since the maximum complexity of any assertion is bounded by the complexity of the formulas mentioned in S, this imposes a bound on the length of any purely female ancestral lines. By examining rule 6, we note that the A-list of N's father (if any) must be strictly smaller than the A-list of N. Since the size of any A-list is bounded by the number of event-atoms mentioned in S, the number of N's male ancestors is also bounded. □

The following lemma will be used by the completeness lemma below. Its function is to establish that the restriction on node-processing rule 6, that $p \notin A_2$, does not block any valid inferences. The critical assumption required to establish this result is the ontological assumption, that events of the same type do not nest one inside the other. It is also at this point that the restriction of quantification to subintervals of the current interval of evaluation is used.

Lemma (technical): Suppose $\langle \phi, \bar{I}, C, A \rangle$ is a node in a node set S derived from some initial node set S_{init} (with ϕ a formula or an event atom) such that C all hold under some interval-label mapping μ under which S is suitable. Then, for any event atom $p' \in A$, there exists a node $\langle p', \bar{I}', C', A' \rangle$ in S such that $p' \notin A'$, $C' \subseteq C$ and $\mu(\bar{I}) \subseteq \mu(\bar{I}')$.

Proof: By induction on the length of the derivation of $\langle \phi, \bar{I}, C, A \rangle$.

1. If the node is in S_{init}, then $A = \emptyset$ so there is nothing to prove.

2. Suppose the node is derived by means of the one of the rules 1–5 . We take rule 2 as an example:

$$\{\langle \langle p \rangle \phi, \bar{I}, C, A \rangle\} \Longrightarrow \{\langle p, \bar{J}, C, A \rangle, \langle \phi, \bar{J}, C, A \cup \{p\} \rangle, \langle \bar{J} \subseteq \bar{I}, \emptyset \rangle\}.$$

   We concentrate on the second of the two nodes created; the result for the first node then follows immediately. We must prove that, if $p' \in A \cup \{p\}$, there exists a node $\langle p', \bar{I}', C', A' \rangle$ in S such that $p' \notin A'$, $C' \subseteq C$ and $\mu(\bar{J}) \subseteq \mu(\bar{I}')$.

   If $p \notin A$, then $p = p'$, so the node $\langle p, \bar{J}, C, A \rangle$ will do. If, on the other hand, $p' \in A$, we can apply the inductive hypothesis to the parent node $\langle \langle p \rangle \phi, \bar{I}, C, A \rangle$ from which it follows that there exists a node $\langle p', \bar{I}', C', A' \rangle$ in S such that $p' \notin A'$, $C' \subseteq C$ and $\mu(\bar{I}) \subseteq \mu(\bar{I}')$. And from the presence of the t-node $\langle \bar{J} \subseteq \bar{I}, \emptyset \rangle$ and the fact that S is suitable for μ, it follows that $\mu(\bar{J}) \subseteq \mu(\bar{I}')$.

3. Suppose the node is derived by means of the rule 6:

$$\{\langle [p]\phi, \bar{I}_1, C_1, A_1\rangle, \langle p, \bar{I}_2, C_2, A_2\rangle\} \Longrightarrow$$
$$\{\langle \phi, \bar{I}_2, C_1 \cup C_2 \cup \{\bar{I}_2 \subseteq \bar{I}_1\}, A_1 \cup A_2 \cup \{p\}\rangle\}$$

We must show that, if $p' \in A_1 \cup A_2 \cup \{p\}$, and the $C_1 \cup C_2 \cup \{\bar{I}_2 \subseteq \bar{I}_1\}$ all hold under μ, then there exists a node $\langle p', \bar{I}', C', A'\rangle$ in S such that $p' \notin A'$, $C' \subseteq C_1 \cup C_2 \cup \{\bar{I}_2 \subseteq \bar{I}_1\}$ and $\mu(\bar{I}_2) \subseteq \mu(\bar{I}')$.

If $p' \notin A_1 \cup A_2$ then $p' = p$, so the node $\langle p, \bar{I}_2, C_2, A_2\rangle$ will do. If $p' \in A_1$, then, by inductive hypothesis, there exists a node $\langle p', \bar{I}', C', A'\rangle$ in S such that $p' \notin A'$, $C' \subseteq C_1$ and $\mu(\bar{I}_1) \subseteq \mu(\bar{I}')$. Then since $C_1 \subseteq C_1 \cup C_2 \cup \{\bar{I}_2 \subseteq \bar{I}_1\}$, the result follows immediately. If $p' \in A_2$, then an even more direct argument applies.

4. Rules 7–9 yield no nodes containing formulae, and no nodes from which other nodes can be derived. □

Finally, we address the soundness and completeness of the node processing rules. The following lemma guarantees that, if the final node set S' yields an inconsistent set of interval relations, the original set Φ of formulae cannot have a model.

Lemma (soundness): Suppose S is a node set and S' is the result of applying one of the node-processing rules to S. Then, if $M \models_\mu S$, there is some μ' extending μ such that $M \models_{\mu'} S'$. (Note that S' may contain some interval labels not mentioned in S.)

Proof: Taking the node-processing rules one by one, we extend μ to μ' (by defining μ for labels mentioned in S' but not in S) such that: (i) for any new t-node, $\langle R, C\rangle$, if C all hold under μ', then so does R, (ii) for any new u-node $\langle\langle p_1 \ldots p_n\rangle, \bar{I}, \bar{I}', C\rangle$, if C all hold under μ', then $\mu'(\bar{I}')$ is the only $J \subseteq \mu'(\bar{I})$ such that $M \models_{\mu(\bar{I}),J} p_1 \ldots p_n$. (iii) for any new f-node $\langle \phi, \bar{I}, C, A\rangle$ if C all hold under μ', then $M \models_{\mu'(\bar{I})} \phi$ (where ϕ is a formula), or $\langle p, \mu'(\bar{I})\rangle$ (where ϕ is an event-atom).

1. Consider the rule:

$$\{\langle \Diamond p, \bar{I}, C, A\rangle\} \Longrightarrow \{\langle p, \bar{J}, C, A\rangle, \langle \bar{J} \subseteq \bar{I}, \emptyset\rangle\}.$$

The node $\langle \Diamond p, \bar{I}, C, A\rangle$ is already in S. If the C do not all hold under μ, then we let $\mu(\bar{J})$ be any subinterval of $\mu(\bar{I})$, and we have $M \models_\mu S'$. If all the C hold under μ, then $M \models_{\mu(\bar{I})} \Diamond p$. By semantic rule 3, choose any $J \subseteq \mu(\bar{I})$ such that $\langle p, J\rangle \in M$ and set $\mu(\bar{J}) = J$. Then we have $M \models_\mu S'$.

2. Consider the rule:

$$\{\langle\neg\Diamond p, \bar{I}_1, C_1, A_1\rangle, \langle p, \bar{I}_2, C_2, A_2\rangle\} \Longrightarrow \{\langle\bot, C_1 \cup C_2 \cup \{\bar{I}_2 \subseteq \bar{I}_1\}\rangle\}.$$

The nodes $\langle\neg\Diamond p, \bar{I}_1, C_1, A_1\rangle$ and $\langle p, \bar{I}_2, C_2, A_2\rangle$ are already in S. It follows that $C_1 \cup C_2 \cup \{\bar{I}_2 \subseteq \bar{I}_1\}$ do not all hold under μ. For otherwise, since $M \models_\mu S$, we would have $M \models_{\mu(\bar{I}_1)} \neg\Diamond p$, $\langle p, \mu(\bar{I}_2)\rangle \in M$, and $\mu(\bar{I}_2) \subseteq \mu(\bar{I}_1)$, which is impossible by semantic rule 4. Therefore, $M \models_\mu S'$.

3. Consider the rule:

$$\{\langle\langle p\rangle\phi, \bar{I}, C, A\rangle\} \Longrightarrow \{\langle p, \bar{J}, C, A\rangle, \langle\phi, \bar{J}, C, A \cup \{p\}\rangle, \langle\bar{J} \subseteq \bar{I}, \emptyset\rangle\}.$$

The node $\langle\langle p\rangle\phi, \bar{I}, C, A\rangle$ is already in S. If the C do not all hold under μ, then we let $\mu(\bar{J})$ be any subinterval of $\mu(\bar{I})$, and we have $M \models_\mu S'$. If all the C hold under μ, then $M \models_{\mu(\bar{I})} \langle p\rangle\phi$. By semantic rule 5, choose any $J \subseteq \mu(\bar{I})$ such that $M \models_{\mu(\bar{I}),J} p$ and $M \models_J \phi$ and set $\mu(\bar{J}) = J$. Then we have $M \models_\mu S'$.

4. Consider the rule:

$$\{\langle[p]\phi, \bar{I}_1, C_1, A_1\rangle, \langle p, \bar{I}_2, C_2, A_2\rangle\} \Longrightarrow$$
$$\{\langle\phi, \bar{I}_2, C_1 \cup C_2 \cup \{\bar{I}_2 \subseteq \bar{I}_1\}, A_1 \cup A_2 \cup \{p\}\rangle\} \text{ (if } p \notin A_1 \cup A_2)$$

The nodes $\langle[p]\phi, \bar{I}_1, C_1, A_1\rangle$ and $\langle p, \bar{I}_2, C_2, A_2\rangle$ are already in S. If $C_1 \cup C_2 \cup \{\bar{I}_2 \subseteq \bar{I}_1\}$ do not all hold under μ, then, there is nothing to prove. Otherwise, since $M \models_\mu S$, we have $M \models_{\mu(\bar{I}_1)} [p]\phi$, $\langle p, \mu(\bar{I}_2)\rangle \in M$ and $\mu(\bar{I}_2) \subseteq \mu(\bar{I}_1)$. By the semantic rule 6, we must have $M \models_{\mu(\bar{I}_2)} \phi$, and so $M \models_\mu S'$.

5. Consider the rule:

$$\{\langle\{p_1 \dots p_n\}\phi, \bar{I}, C, A\rangle\} \Rightarrow \{\langle p_i, J_i, C, A_i\rangle, \langle\bar{J}_i \subseteq \bar{J}_{i-1}, \emptyset\rangle | 1 \leq i \leq n\}$$
$$\cup\{\langle\phi, \bar{J}_n, C, A_n \cup \{p_n\}\rangle, \langle p_1 \dots p_n, \bar{I}, \bar{J}, C, \emptyset\rangle\}$$

The node $\langle\{p_1 \dots p_n\}\phi, \bar{I}, C, A\rangle$ is already in S. If the C do not all hold under μ, then we let $\mu(\bar{J}_i)$ be any subinterval of $\mu(\bar{J}_{i-1})$ for all i ($1 \leq i \leq n$), with $\bar{J}_0 = \bar{I}$, and we have $M \models_\mu S'$. If all the C hold under μ, we have $M \models_{\mu(\bar{I})} \{p_1 \dots p_n\}\phi$. By semantic rule 7, choose the unique $J_n \subseteq \mu(\bar{I})$ such that $M \models_{I,J} p_1 \dots p_n$. And let $J_1, \dots, J_{n-1}$ be any intervals such that $\langle p_i, J_i\rangle \in M$ and $J_i \subseteq J_{i-1}$ for all i ($1 \leq i \leq n$) with $J_0 = I$. Setting $\mu(\bar{J}_i) = J_i$ for all i ($1 \leq i \leq n$), we have $M \models_\mu S'$.

6. Consider the rule:

$$\{\langle\langle p_1 \ldots p_n\rangle_{\geq}\phi, \bar{I}, C, A\rangle\} \Longrightarrow$$
$$\{\langle p_i, J_i, C, \bar{A_i}\rangle, \langle \bar{J_i} \subseteq \bar{J}_{i-1}, \emptyset\rangle | 1 \leq i \leq n\} \cup \{\langle \phi, \text{init}(\bar{I}, \bar{J}_n), C, A\rangle\}$$

The node $\langle\langle p_1 \ldots p_n\rangle_{\geq}\phi, \bar{I}, C, A\rangle$ is already in S. If the C do not all hold under μ, then we let $\mu(\bar{J}_i)$ be any subinterval of $\mu(\bar{J}_{i-1})$ for all i $(1 \leq i \leq n)$, with $\bar{J}_0 = \bar{I}$, and we have $M \models_\mu S'$. If all the C hold under μ, we have $M \models_{\mu(\bar{I})} \langle p_1 \ldots p_n\rangle_{\geq}\phi$. By semantic rules 2 and 8, choose $J_1, \ldots, J_n$ such that $\langle p_i, J_i\rangle \in M$ and $J_i \subseteq J_{i-1}$ for all i $(1 \leq i \leq n)$ with $J_0 = I$ and such that $M \models_{\text{init}(\mu(\bar{I}), J_n)} \phi$. Setting $\mu(\bar{J}_i) = J_i$ for all i $(1 \leq i \leq n)$, and noting that $\mu(\text{init}(\bar{I}, \bar{J}_n)) = \text{init}(\mu(\bar{I}), \mu(\bar{J}_n))$, we have $M \models_\mu S'$.

7. The rule

$$\{\langle\langle p_1 \ldots p_n\rangle_{<}\phi, \bar{I}, C, A\rangle\} \Longrightarrow$$
$$\{\langle p_i, J_i, C, \bar{A_i}\rangle, \langle \bar{J_i} \subseteq \bar{J}_{i-1}, \emptyset\rangle | 1 \leq i \leq n\} \cup \{\langle \phi, \text{fin}(\bar{I}, \bar{J}_n), C, A\rangle\}$$

is similar to the previous case.

8. Consider the rule:

$$\{\langle p_1 \ldots p_n, \bar{I}_0, \bar{I}', C\rangle\} \cup \{\langle p_i, \bar{I}_i, C_i, A_i\rangle | 1 \leq i \leq n\} \Longrightarrow$$
$$\{\langle \bar{I}_n = \bar{I}', C_1 \cup \ldots \cup C_n \cup C \cup \{\bar{I}_i \subseteq \bar{I}_{i-1} | 1 \leq i \leq n\}\rangle\}.$$

The nodes $\{\langle p_1 \ldots p_n, \bar{I}_0, \bar{I}', C_1\rangle\} \cup \{\langle p_i, \bar{I}_i, C_i, A_i\rangle | 1 \leq i \leq n\}$ are already in S. If $C_1 \cup \ldots \cup C_n \cup \{\bar{I}_i \subseteq \bar{I}_{i-1} | 1 \leq i \leq n\}$ do not all hold under μ, then there is nothing to prove. Otherwise, since $M \models_\mu S$, it follows that: (i) $\mu(\bar{I}')$ is the only subinterval $J \subseteq \mu(\bar{I}_0)$ such that $M \models_{\mu(\bar{I}_0), J} p_1 \ldots p_n$, (ii) $M \models_{\mu(\bar{I}_0), \mu(\bar{I}_n)} p_1 \ldots p_n$ by semantic rules 1 and 2, and (iii) $\mu(\bar{I}_n) \subseteq \mu(\bar{I}_1)$. Whence $\mu(\bar{I}_n) = \mu(\bar{I}')$ and $M \models_\mu S'$.

9. Consider the rule:

$$\{\langle p, \bar{I}_1, C_1, A_1\rangle, \langle p, \bar{I}_2, C_2, A_2\rangle\} \Longrightarrow \{\langle \bar{I}_1 = \bar{I}_3, C_1 \cup C_2 \cup \{\bar{I}_1 \subseteq \bar{I}_2\}\rangle\}.$$

The nodes $\langle p, \bar{I}_1, C_1, A_1\rangle$ and $\langle p, \bar{I}_2, C_2, A_2\rangle$ are already in S. If $C_1 \cup C_2$ do not all hold under μ, then there is nothing more to prove. Otherwise, we have $\langle p, \mu(\bar{I}_1)\rangle \in M$ and $\langle p, \mu(\bar{I}_2)\rangle \in M$. By the ontological assumption, $\mu(\bar{I}_1) \subseteq \mu(\bar{I}_2)$ implies $\mu(\bar{I}_1) = \mu(\bar{I}_2)$, so $M \models_\mu S'$. □

The following lemma guarantees that, if the final node set yields a consistent set of interval relations, then the original set Φ of formulae must have a model.

Lemma (completeness): Suppose S is a final node set (i.e., closed under the application of the node-processing rules). If S is suitable for μ, then, for some interpretation M, $M \models_\mu S$.

Proof: Consider the following interpretation:

$$M = \{\langle p, \mu(\bar{I})\rangle | p \text{ is en event atom and there is a node } \langle p, \bar{I}, C, A\rangle \in S \text{ such that } C \text{ all hold under } \mu\}.$$

We must check that M obeys the ontological assumption, namely, that if $\langle p, I_1\rangle \in M$, $\langle p, I_2\rangle \in M$ and $I_1 \subseteq I_2$, then $I_1 = I_2$. But, by the construction of M, $\langle p, I_1\rangle \in M$ and $\langle p, I_2\rangle \in M$ only if there are nodes $\langle p, \bar{I}_1, C_1, A_1\rangle$, $\langle p, \bar{I}_2, C_2, A_2\rangle$ in S with $\mu(\bar{I}_1) = I_1$, $\mu(\bar{I}_2) = I_2$ and $C_1 \cup C_2$ all holding under μ. Then, by the finality of S and node-processing rule 9, the node $\langle \bar{I}_1 = \bar{I}_2, C_1 \cup C_2 \cup \{\bar{I}_1 \subseteq \bar{I}_2\}\rangle$ is in S, so that, by the suitability of S under μ, if $I_1 \subseteq I_2$, then $I_1 = I_2$.

We note in addition that, if $\langle p_1 \ldots p_n, \bar{I}, \bar{J}, C\rangle$ is any u-node in S with the C all holding under μ, then $\mu(\bar{J})$ is the only $J \subseteq \mu(\bar{I}_1)$ such that $M \models_{\mu(\bar{I}),J} p_1 \ldots p_n$. For, by the construction of M, we can have a $J \subseteq \mu(\bar{I})$ with $M \models_{\mu(\bar{I}),J} p_1 \ldots p_n$ only if there are nodes in S of the form $\langle p_i, \bar{J}_i, C_i, A_i\rangle$ with $C_i \cup \{\bar{J}_i \subseteq \bar{J}_{i-1}\}$ holding under μ for all i $(1 \leq i \leq n)$ with $\bar{J}_0 = \bar{I}$ and $\mu(\bar{J}_n) = J$. But then, by the finality of S and node-processing rule 8 we must have a node $\langle \bar{J}_n = \bar{J}, C^*\rangle$, in S, where $\mu(C^*)$ all hold under μ, so that, by the suitability of M for μ, $J = \mu(\bar{J})$ as required.
If the f-node $\langle p, \bar{I}, C, A\rangle$ is in S, where p is an event atom, and C all hold under μ, then $\langle p, \mu(\bar{I})\rangle \in M$, by the construction of M. We now show that, if the f-node $\langle \psi, \bar{I}, C, A\rangle$ is in S, where ψ is a formula, and the C all hold under μ, then $M \models_{\mu(\bar{I})} \psi$. We proceed by structural induction on ψ, according to the syntax rules for ETL0. We note in passing the technical lemma and the ontological assumption are used in clause 4; this move is crucial in securing the decidability of ETL0.

1. If ψ is $\Diamond p$, by the finality of S and node-processing rule 1, we have in S the nodes $\{\langle p, \bar{J}, C, A\rangle, \langle \bar{J} \subseteq \bar{I}, C\rangle\}$ so that, by construction of M and suitability of S under μ, we have $\langle p, \mu(\bar{J})\rangle \in M$ and $\mu(\bar{J}) \subseteq \mu(\bar{I})$. Then, by the semantic rule 3, we have $M \models_{\mu(\bar{I})} \psi$.

2. If ψ is $\neg\Diamond p$, then, by the construction of M, we can have a $J \subseteq \mu(\bar{I})$ with $\langle p, J\rangle \in M$ only if there is a node in S of the form $\langle p, \bar{I}', C', A'\rangle$ with $C' \cup \{\bar{I}' \subseteq \bar{I}\}$ all holding under μ. By the finality of S and node-processing rule 7, we must have the node $\langle \bot, C \cup C' \cup \{\bar{I}' \subseteq \bar{I}\}\rangle$, in S, contradicting the suitability of S under μ. Therefore, there is no $J \subseteq \mu(\bar{I})$ with $\langle p, J\rangle \in M$, so, by semantic rule 4 , we have $M \models_{\mu(\bar{I})} \psi$.

3. If ψ is $\langle p\rangle\phi$, by the finality of S and node-processing rule 2, we have in S the nodes $\langle p, \bar{J}, C, A\rangle$ and $\langle \phi, \bar{J}, C, A\rangle$, $\langle \bar{J} \subseteq \bar{I}, C\rangle$ so that, by the construction of M and suitability of S under μ, we have $\langle p, \mu(\bar{J})\rangle \in M$, and $\mu(\bar{J}) \subseteq \mu(\bar{I})$. By inductive hypothesis, $M \models_{\mu(\bar{J})} \phi$. Then, by the semantic rules 1 and 5, we have $M \models_{\mu(\bar{I})} \psi$.

4. If ψ is $[p]\phi$, then, by the construction of M, we can have a $J \subseteq \mu(\bar{I})$ with $\langle p, J\rangle \in M$ and $J \subseteq \mu(\bar{I}_1)$ only if $J = \mu(\bar{I}_2)$ for some node in S of the form $\langle p, \bar{I}_2, C_2, A_2\rangle$ with $C_2 \cup \{\bar{I}_2 \subseteq \bar{I}\}$ all holding under μ. Suppose first that $p \notin A_2$. By the finality of S and node-processing rule 6 , we must have the node $\langle \phi, \bar{I}_2, C \cup C_2 \cup \{\bar{I}_2 \subseteq \bar{I}\}, A \cup A_2 \cup \{p\}\rangle$ in S, so that, by inductive hypothesis, $M \models_{\mu(\bar{I}_2)} \phi$. But $\mu(\bar{I}_2) = J$. Suppose, on the other hand, that $p \in A_2$. By the technical lemma, S contains a node $\langle p, \bar{I}', C', A'\rangle$ with $p \notin A'$, $C' \subseteq C_2$ and $\mu(\bar{I}_2) \subseteq \mu(\bar{I}')$. By construction of M, we must have $\langle p, \mu(\bar{I}')\rangle$, and by assumption, $\langle p, \mu(\bar{I}_2)\rangle \in M$. Since $\mu(\bar{I}_2) \subseteq \mu(\bar{I}')$, we have $\mu(\bar{I}_2) = \mu(\bar{I}')$, by the ontological assumption. Again, by the finality of S and node-processing rule 6 , we must have the node $\langle \phi, \bar{I}', C \cup C' \cup \{\bar{I}' \subseteq \bar{I}\}, A \cup A' \cup \{p\}\rangle$, so that, by inductive hypothesis, $M \models_{\mu(\bar{I}')} \phi$. But $\mu(\bar{I}') = \mu(\bar{I}_2) = J$. Thus, for any interval J with $M \models_{\mu(\bar{I}), J} p$, we have $M \models_J \phi$. Therefore, by semantic rule 6, we have $M \models_{\mu(\bar{I})} \psi$.

5. If ψ is $\{p_1 \ldots p_n\}\phi$, by the finality of S and node-processing rule 3, we have in S nodes $\langle p_i, \bar{J}_i, C, A_i\rangle$ for all i $(1 \leq i \leq n)$ with $\bar{J}_0 = \bar{I}_0$, as well as $\langle \phi, \bar{J}_n, C, A_n \cup \{p_n\}\rangle$ and $\langle p_1 \ldots p_n, \bar{I}, \bar{J}, C\rangle$. By construction of M, the suitability of S under μ and semantic rules 1 and 2, and since C all hold under μ, we have $M \models_{\mu(\bar{I})\mu(\bar{J}_n)} p_1 \ldots p_n$. By inductive hypothesis, we have $M \models_{\mu(\bar{J}_n)} \phi$. Moreover, by the presence of the node $\langle p_1 \ldots p_n, I, J_n, C\rangle$, $\mu(\bar{J}_n)$ must be the only $J \subseteq \bar{I}$ such that $M \models_{\mu(\bar{I}), J} p_1 \ldots p_n$. Therefore, by semantic rule 7, we have $M \models_{\mu(\bar{I})} \psi$.

6. If ψ is $\langle p_1 \ldots p_n\rangle_{\geq}\phi$, by the finality of S and node-processing rule 4, we have in S nodes $\langle p_i, \bar{J}_i, C, A_i\rangle$ and $\langle \bar{J}_i \subseteq \bar{J}_{i-1}\rangle$ for all i $(1 \leq i \leq n)$ with $\bar{J}_0 = \bar{I}_0$ with $\bar{J}_0 = \bar{I}$, as well as $\langle \phi, \text{init}(\bar{I}, \bar{J}_n), C, A\rangle$. By construction of M, the suitability of S under μ and semantic rules 1 and 2, and since C all hold under μ, we have $M \models_{\mu(\bar{I})\mu(\bar{J}_n)} p_1 \ldots p_n$. By inductive hypothesis, $M \models_{\mu(\text{init}(\bar{I}, \bar{J}_n))} \phi$. Since $\mu(\text{init}(\bar{I}, \bar{J}_n)) = \text{init}(\mu(\bar{I}), \mu(\bar{J}_n))$, we have, by semantic rule 8, $M \models_{\mu(\bar{I})} \psi$.

7. The final case is similar to the previous one.□

Theorem (correctness): The procedure `consis_test`(Φ) using the node-
processing rules given above is a correct procedure for determining whether a set of formulae Φ of the language ETL0 is consistent.

Proof: We assume that the procedure `order_test`(C), when passed a set C of order clauses (i.e. a set of disjunctions of (in)equalities), terminates with the result "consistent" if C is consistent with the theory of linear order and "inconsistent" otherwise.
Let Φ be a set of formulae of ETL0. S starts off as the initial node set formed from Φ for interval label I_0; moreover, by the termination lemma, S eventually stabilizes and the **until**-loop terminates. Since `order_test`(C) terminates with either "consistent" or "inconsistent", so will `consis_test`(Φ).
If M is a structure and $I \in \Omega$ such that $M \models_I \Phi$, by setting $\mu(\bar{I}_0) = I$, we have $M \models_\mu S_{\mathrm{init}}$. By soundness, there is an extension μ' of μ such that $M \models_{\mu'} S$ is an invariant of the **until**-loop, so that S is always suitable for μ' by the definition of $M \models_{\mu'} S$. Therefore, the order clauses C will be consistent, `order_test` will return "consistent", and `consis_test`(Φ) will return "consistent."
Conversely, if `consis_test`(Φ) returns "consistent", the order-clauses C passed to it will be satisfiable, so that the final value of S is suitable for some μ. By completeness, there is some M such that $M \models_\mu S$. Since S simply gets added to throughout the **until**-loop, $\models_{\mu(\bar{I}_0)} \Phi$. Therefore Φ is consistent. □

8. Conclusion

We have shown how a decidable, interval-based, temporal logic can be used to give the semantics of a range of temporal preposition phrases in English. The emphasis throughout has been on tailoring the logic closely to the expressive resources of English, deviating from the patterns of quantification found in everyday language only for simplicity. We showed how the modal operators $\langle\alpha\rangle$, $[\alpha]$, $\{\alpha\}$, $\langle\alpha\rangle_{\geq}$ and $\langle\alpha\rangle_{\leq}$, together with (existential) role-composition was sufficient to capture a wide range of temporal prepositional constructions in English, including cascades of temporal preposition phrases and temporal preposition phrases with prepositionally modified complements. Thus, we have made a first step towards characterizing the expressive power of the temporal preposition system in English. Further steps will be made as the minimal logic ETL0 is extended to cover a wider range of constructions.

References

Ali, S. S. and S. Shapiro: 1993, 'Natural language processing using a propositional semantic network'. *Minds and Machines* **3**(4), 421–452.

Allen, J. F.: 1984, 'Towards a general theory of action and time'. *Artificial Intelligence* **23**, 123–154.

Franconi, E.: 1993, 'A treatment of plurals and plural quantifications based on a theory of collections'. *Minds and Machines* **3**(4), 453–474.

Harel, D.: 1984, 'Dynamic Logic'. In: D. Gabbay and F. Guenthner (eds.): *Handbook of Philosophical logic, Vol. II.* D. Reidel, Dordrecht, pp. 89 – 133.

Henkin, L., J. M. and A. Tarski: 1985, *Cylindric Algebras*, Vol. II. Amsterdam: North Holland.

Herweg, M.: 1991, 'A critical examination of two approaches to aspect'. *Journal of Semantics* **8**.

Hwang, C. H. and L. K. Schubert: 1993, 'Episodic Logic: A comprehensive, Natural Representation for Language Understanding'. *Minds and Machines* **3**(4), 381–420.

Ladkin, P. and R. Maddux: 1994, 'On Binary Constraint Problems'. *Journal of the Association for Computing Machinery* **41**(3), 435–469.

Nemeti, I.: 1991, 'Algebraizations of quantifier logics: and introductory overview'. *Studia Logica* **50**(3–4), 485–569.

Schild, K.: 1991, 'A correspondence theory for terminological logics: preliminary report'. In: *Proceedings, IJCAI.* pp. 466–471.

Steedman, M.: 1996, 'Temporality'. In: J. van Benthem and A. ter Meulen (eds.): *Handbook of Logic and Language.* Elsevier.

Suppes, P.: 1991, *Languages for Humans and Robots.* Blackwell.

Venema, Y.: 1990, 'Expressiveness and Completeness of an Interval Tense Logic'. *Notre Dame Journal of Formal Logic* **31**(4).

Transitions in Continuous Time, with an Application to Qualitative Changes in Spatial Relations

Antony Galton (antony@dcs.exeter.ac.uk)
Department of Computer Science, University of Exeter, Exeter EX4 4PT, UK

Abstract. We investigate the definition of transition events within a formal framework for representing time, states, and events. Continuity considerations lead to the introduction of eight distinct kinds of transition. This proliferation is justified in the subsequent application to the qualitative description of changes in relation between spatial regions, where the fineness of discrimination provided by the inventory of transition-types is needed in order to adequately cover all the possibilities.

1. Introduction

The formal representation of knowledge with a temporal dimension has received extensive coverage in the research literature on logic, linguistics, computer science, and artificial intelligence. As a result of this coverage, a number of issues have been singled out as fundamental, issues over which there is still no clear consensus amongst scholars pursuing research in this area. Amongst the most prominent of these issues are the following:

1. Should time be defined in terms of change, or change in terms of time?
2. Should we distinguish at a fundamental level between states, events, and maybe other categories such as processes?
3. Should states be defined in terms of events, or events in terms of states?
4. Should intervals be defined in terms of instants, or instants in terms of intervals?
5. Should the ordering of the instants be modelled as discrete or dense, and if dense, should it be continuous?
6. Should changes be modelled as continuous or discontinuous, or should both types be allowed?

It has become clear that there is no single definitive answer to any of these questions. In each case, the answer will be dependent on the context in which the question is asked. Because our concern is with modelling, the appropriateness of any particular answer has to be judged in relation to what it implies for the representational power of the model and its inferential properties, and not on the basis of philosophical or metaphysical questions as to what the world is 'really like'.

For all that, it is only natural that any detailed discussion of temporal modelling has to make some basic assumptions in order to lay down

H. Barringer et al. (eds.), Advances in Temporal Logic, 279–297.

a framework within which the technicalities can be intelligibly pursued. The present paper is no exception to this, and in what follows we shall be taking a stance on all of the issues enumerated above. Our framework, then, is established by means of the following assumptions:

1. We shall assume that a temporal framework is given prior to, and thus independently of, any changes, states, or events.
2. We shall regard states and events as fundamentally distinct kinds of entity, requiring distinct syntactical resources for their expression.
3. Although we shall not rule out the possibility of defining some states in terms of events, we shall assume that in general states are more fundamental, and that events are to be defined in terms of them.
4. We shall take the view that since instants and intervals are mutually interdefinable, it does not matter in principle which are taken as conceptually prior in our model, but for the sake of convenience we shall base our temporal model on instants, and define intervals in terms of them.
5. We shall assume a continuous order for the instants; specifically, we shall assume that the temporal ordering on the instants is of the same type as the 'less/greater' ordering on the real numbers.
6. We shall assume that at some level of description all changes are continuous; this leaves open, however, the possibility that at other levels of description some changes may appear discontinuous.

These points will all be further elaborated in what follows. The meaning of most of them will become clear in the next two sections, where we lay out the formal basis for our representations.

2. A temporal framework

The formal model we shall use is as follows. Time itself is represented by means of a set T of instants ordered by a continuous linear order $\prec$, such that $(T, \prec)$ has the same order type as $(\mathbb{R}, <)$. Having decided what the order type of the set of instants is, it is not necessary for us to write down specific axioms; but it should be noted that first-order logic is not sufficient to characterise this order-type completely, since the first-order properties of the relation $<$ on the real numbers are identical to the first-order properties of $<$ on the rationals.

For any two instants t_1 and t_2 such that $t_1 \prec t_2$, we can specify the interval between them by means of the ordered pair $\langle t_1, t_2 \rangle$. As is well known, there are five possible qualitative relations in which an instant t can stand to an interval $\langle t_1, t_2 \rangle$. To facilitate their definitions, we introduce the notations $\inf(i)$ and $\sup(i)$ for the instants which respectively begin and end interval i; thus $i = \langle \inf(i), \sup(i) \rangle$. We now define the five instant-interval relations as follows:

t is before i: $t < i \triangleq t \prec \inf(i)$.

t begins i: $begins(t, i) \triangleq t = \inf(i)$.

t is in i: $t \mathsf{E} i \triangleq \inf(i) \prec t \prec \sup(i)$.

t ends i: $ends(t, i) \triangleq t = \sup(i)$.

t is after i: $t > i \triangleq \sup(i) \prec t$.

Note that we do not identify an interval i with the set $\{t \mid t \mathsf{E} i\}$, or indeed with any set of instants, nor do we regard the interval as being in any sense made up out of or composed of the instants which fall within it. Intervals are regarded purely as *logical* constructions out of instants; they are not 'constructed' from intervals in any other way. Philosophically it might be regarded as more reasonable to regard intervals as more fundamental, since they possess the basic temporal property of *duration*, which instants lack (suggesting that intervals are not aggregates of instants); but from a technical point of view it is more convenient to take instants as our starting point.

The fundamental relation on intervals is *immediate succession*, defined by

$$i \| j \triangleq \sup(i) = \inf(j).$$

When $i \| j$, we say that i *meets* j. In particular, if $\sup(i) = \inf(j) = t$, we shall say that i meets j at t, and write this as $i \|_t j$.

As was pointed out by Hamblin (1971) and Allen (1984), if time is totally-ordered then there are thirteen exhaustive and mutually exclusive basic relations in which one interval can stand to another, which can be defined in terms of $\|$. We shall not list them here, as they are well-known, and we shall not need to use any of the basic thirteen, apart from 'meets' itself, in what follows. In fact, the only additional relation on intervals that we shall need is the *subinterval* relation, which may be defined as follows[1]:

$$j \sqsubseteq i \triangleq \inf(i) \preceq \inf(j) \wedge \sup(j) \preceq \sup(i).$$

3. Introducing states

We shall specify what happens in time in terms of a set of properties called *states*. The state-based view is based on the idea that the world

[1] Note that the definition of this relation given by Allen and Ferguson (1994) is defective in that it excludes the cases where i is either an initial or final subinterval of j.

can be in any of a (possibly infinite) number of different states, and that the history of the world is specified by determining which state it is in at each instant. If W is the set of possible world-states, then a possible world-history is a function $h : T \to W$.

The world-states $w \in W$ are supposed to be complete 'snapshots' of the world at a time; they are the *situations* of McCarthy and Hayes (1969). In practice, we never specify world-states in full detail—indeed, we would be quite unable to even if we wanted to. Instead, again following McCarthy and Hayes, we use *fluents*, which are time-variable properties of the world, to pick out those aspects of the total world-state we are interested in describing. An example is the fluent 'The temperature in Exeter'; in each world-state this fluent has a determinate value (e.g., 18°C), so we can model the fluent as a mapping from world-states to values. Different fluents may have different value-ranges associated with them.

A particularly important class consists of the *Boolean fluents* or *(partial) states*, whose values are taken from the set $\{\top, \bot\}$ (where $\top$ may be read as 'true' and $\bot$ as 'false'). Each Boolean fluent divides the set of world-states into those in which the fluent is true and those in which it is not. An example is 'It is raining in Exeter', which serves to discriminate the set of all world-states in which it is raining in Exeter from the set of all world-states in which it is not raining in Exeter. Any fluent gives rise to Boolean fluents; if $f : W \to V$ takes values from some set V, then for each $v \in V$ there is a Boolean fluent $[f = v]$ defined by the rule

$$[f = v](w) = \begin{cases} \top & \text{if } f(w) = v \\ \bot & \text{if } f(w) \neq v \end{cases}$$

Again, for each $U \subset V$, there is a Boolean fluent $[f \in U]$ defined by the rule

$$[f \in U](w) = \begin{cases} \top & \text{if } f(w) \in U \\ \bot & \text{if } f(w) \notin U \end{cases}$$

If f is the fluent 'The temperature in Exeter', then this enables us to treat statements such as 'The temperature in Exeter is 18°C' and 'The temperature in Exeter is lower than 18°C' as Boolean fluents.

There is a three-way relationship amongst world-states, fluents, and times. The history function h assigns a world-state $h(t)$ to each time t; a fluent f assigns a value $f(w)$ to each world state w; hence the history function indirectly assigns the value $f(h(t))$ to the fluent f at time t. For a Boolean fluent s it is customary, following Allen (1984), to write $Holds(s, t)$ to mean that $s(h(t)) = \top$. Note that h is not specified in the predicate $Holds$; it is tacitly assumed that we know what history we are talking about (with luck, the true history of the world).

It is useful to distinguish the case in which a state holds at an instant from the case in which it holds over an interval. We shall use $Holds_{\mathbf{T}}$ for the former and $Holds_{\mathbf{I}}$ for the latter. We can straightforwardly[2] define $Holds_{\mathbf{T}}$ by

$$Holds_{\mathbf{T}}(s,t) \triangleq s(h(t)) = \top,$$

and then define $Holds_{\mathbf{I}}$ in terms of $Holds_{\mathbf{T}}$ by

$$Holds_{\mathbf{I}}(s,i) \triangleq \forall t \in T(t \sqsubseteq i \rightarrow Holds_{\mathbf{T}}(s,t)).$$

An immediate consequence of this definition is that states are *homogeneous* over intervals, that is, if a state holds over an interval then it must also hold over all subintervals of that interval:

HOM $\forall i,j(j \sqsubseteq i \rightarrow (Holds_{\mathbf{I}}(s,i) \rightarrow Holds_{\mathbf{I}}(s,j)))$

In many ways Boolean fluents (i.e., states) resemble propositions, the main difference being that whereas philosophers have generally regarded propositions as timelessly true or false, states may hold at some times but not at others. This resemblance to propositions prompts the natural question whether states can be combined in the same way as propositions, e.g., using the truth-functional connectives of the Propositional Calculus. This proves to be straightforward so long as we are considering the assignment of statements to instants, but becomes more complicated when intervals are brought in as well.

The only utterly unproblematic case is conjunction. We can define the conjunction $s.s'$ of two states s and s' by the rule

$$s.s'(w) = \begin{cases} \top & \text{if } s(w) = s'(w) = \top \\ \bot & \text{otherwise} \end{cases}$$

Then we have, for an instant t,

$$Holds_{\mathbf{T}}(s.s',t) \leftrightarrow Holds_{\mathbf{T}}(s,t) \wedge Holds_{\mathbf{T}}(s',t),$$

and for an interval i,

$$Holds_{\mathbf{I}}(s.s',i) \leftrightarrow Holds_{\mathbf{I}}(s,i) \wedge Holds_{\mathbf{I}}(s',i).$$

Turning now to negation, we can define the negation $-s$ by the rule

$$-s(w) = \begin{cases} \top & \text{if } s(w) = \bot \\ \bot & \text{if } s(w) = \top \end{cases}$$

[2] Although this may seem straightforward, many people have been unhappy about $Holds_{\mathbf{T}}$. Hamblin (1971) adduced a number of reasons why we might want to disallow talk about states holding at instants, and Allen, in setting up his Interval Calculus (Allen, 1984), was motivated by a desire to do away with instants altogether.

giving us

$$Holds_{\mathbf{T}}(-s,t) \leftrightarrow \neg Holds_{\mathbf{T}}(s,t).$$

For an interval i we have

$$Holds_{\mathbf{I}}(-s,i) \leftrightarrow \neg\exists t \sqsubset i(Holds_{\mathbf{T}}(s,t)),$$

which is called the *strong* or *internal* negation[3], as opposed to the *weak* or *external* negation seen in

$$\neg Holds_{\mathbf{I}}(s,i) \leftrightarrow \exists t \sqsubset i \neg Holds_{\mathbf{T}}(s,t).$$

4. Introducing events

Events differ from states in a number of fundamental ways: in fact, describing what goes on in terms of events constitutes a radically different approach to the description of the world. The language we use reflects this. Whereas we speak of states *holding* or *obtaining* at times, we speak of events *happening* or *occurring*; the logic of these terms is quite different.

We must distinguish between generic *event-types* and individual *event-tokens*. An event-token is a particular piece of history, occurring at a definite and unique time. By its nature it must happen once and only once. An event-type, on the other hand, is a general description under which more, or fewer, than one event-token can fall. The event-tokens falling under a given event type are its *occurrences*; a given event-type may occur on many different occasions.

Historically, the logic of event-types and the logic of event-tokens have generally been studied separately rather than together. The systems of McDermott (1982), Galton (1984) and Allen (1984), albeit in rather different ways, use formalisms involving reference to event-types but not to event-tokens. By contrast, the influential work of Davidson (1967) on the logical analysis of 'action sentences', as well as the more recent Event Calculus of Kowalski and Sergot (1986) involve reference to event-tokens but not (or only derivatively) to event-types. The system used in this paper resembles the former group rather than the latter in this respect.

---

[3] Hamblin and Allen, having rejected the assignment of states to instants, use a different definition of $-s$:

$$Holds_{\mathbf{I}}(-s,i) \;\triangleq\; \forall j \sqsubseteq i \neg Holds_{\mathbf{I}}(s,j).$$

This is only equivalent to our definition of $-s$ on condition that a state cannot hold at an isolated instant, i.e., $\forall t \in T(Holds_{\mathbf{T}}(s,t) \rightarrow \exists i((begins(t,i) \vee ends(t,i)) \wedge Holds_{\mathbf{I}}(s,i)))$. Equivalently, states always persist over intervals.

We introduce two predicates $Occurs_{\mathbf{T}}$ and $Occurs_{\mathbf{I}}$, relating event-types to times. To say that type E has an occurrence which takes up exactly the interval i, we write $Occurs_{\mathbf{I}}(E, i)$. To say that it has an instantaneous occurrence at the instant t, we write $Occurs_{\mathbf{T}}(E, t)$. It is necessary to recognise instantaneous events, even if one does not allow states to hold at isolated instants (Galton, 1994).

5. Constructing events from states

Many event-types can be specified in terms of changes of state. In general, to define a new event-type we must specify its *occurrence conditions*, which are necessary and sufficient conditions for there to be an occurrence of the type on a given interval or at a given instant. We shall call event-types defined in this way *derived event-types*. In this paper we shall concentrate on one particular kind of derived event-type, called a *transition* (Gooday, 1994; Gooday and Galton, 1997). As we shall see, the detailed specification of transitions in our framework is rather complex. Later we shall discuss the meaning and implications of this complexity.

To specify a transition we must pick an *initial state* s and a *final state* s'. A transition from s to s' is an event-type which occurs whenever a time at which s holds is followed, not necessarily immediately, by a time at which s' holds. There are several possibilities here, according to whether the times involved are instants or intervals, and whether the initial and final states are separated by an interval, an instant, or not at all. If they are separated by an interval, then the transition is durative; otherwise it is instantaneous. A durative transition from s to s' will be denoted $Trans_{\mathbf{I}}(s, s')$; an instantaneous one is denoted $Trans_{\mathbf{T}}(s, s')$. We cannot give single occurrence conditions to either of these; instead we must split them up into cases according to the nature of the holding times of the initial and final states. To a considerable extent, these variations are determined in any particular case by the nature of the states involved.

In order to give precise occurrence conditions we shall define eight different transition operators, four durative and four instantaneous. Our notation for them uses the basic forms $Trans_{\mathbf{I}}$ and $Trans_{\mathbf{T}}$ decorated with superscripts suggestive of the corresponding pattern of temporal incidence. The eight kinds of transition are illustrated schematically in Figure 1, in which the labels correspond to those in the following enumeration:

DURATIVE TRANSITIONS

(1a) $Trans_{\mathbf{I}}^{iii}(s,s')$ occurs when an interval over which s holds meets an interval over which neither s nor s' holds, which in turn meets an interval over which s' holds:

$$Occurs_{\mathbf{I}}(Trans_{\mathbf{I}}^{iii}(s,s'),i) \triangleq \exists j,k(j\|i\|k \wedge Holds_{\mathbf{I}}(s,j) \wedge Holds_{\mathbf{I}}((-s).(-s'),i) \wedge Holds_{\mathbf{I}}(s',k)).$$

The purpose of the penultimate conjunct here is to ensure that i is a minimal interval within which the transition takes place, which can then be designated as the time of a particular occurrence (token) of $Trans_{\mathbf{I}}(s,s')$.

(1b) $Trans_{\mathbf{I}}^{iit}(s,s')$ occurs when an interval over which s holds meets an interval over which neither s nor s' holds, which is ended by an instant at which s' holds:

$$Occurs_{\mathbf{I}}(Trans_{\mathbf{I}}^{iit}(s,s'),i) \triangleq \exists j(j\|i \wedge Holds_{\mathbf{I}}(s,j) \wedge Holds_{\mathbf{I}}((-s).(-s'),i) \wedge Holds_{\mathbf{T}}(s',\mathrm{sup}(i))).$$

(1c) $Trans_{\mathbf{I}}^{tii}(s,s')$ occurs when an instant at which s holds begins an interval over which neither s nor s' holds, which meets an interval over which s' holds:

$$Occurs_{\mathbf{I}}(Trans_{\mathbf{I}}^{tii}(s,s'),i) \triangleq \exists k(i\|k \wedge Holds_{\mathbf{T}}(s,\mathrm{inf}(i)) \wedge Holds_{\mathbf{I}}((-s).(-s'),i) \wedge Holds_{\mathbf{I}}(s',k)).$$

(1d) $Trans_{\mathbf{I}}^{tit}(s,s')$ occurs when an instant at which s holds begins an interval over which neither s nor s' holds, which is ended by an instant at which s' holds:

$$Occurs_{\mathbf{I}}(Trans_{\mathbf{I}}^{tit}(s,s'),i) \triangleq Holds_{\mathbf{T}}(s,\mathrm{inf}(i)) \wedge Holds_{\mathbf{I}}((-s).(-s'),i) \wedge Holds_{\mathbf{T}}(s',\mathrm{sup}(i)).$$

INSTANTANEOUS TRANSITIONS

(2a) $Trans_{\mathbf{T}}^{ii}(s,s')$ occurs when an interval over which s holds meets an interval over which s' holds:

$$Occurs_{\mathbf{T}}(Trans_{\mathbf{T}}^{ii}(s,s'),t) \triangleq \exists i,j(i\|_t j \wedge Holds_{\mathbf{I}}(s,i) \wedge Holds_{\mathbf{I}}(s',j)).$$

(2b) $Trans_{\mathbf{T}}^{iti}(s,s')$ occurs when an interval over which s holds meets an interval over which s' holds at an instant at which neither s nor s'

holds:

$$Occurs_{\mathbf{T}}(Trans^{iti}_{\mathbf{T}}(s,s'),t) \triangleq \exists i,j(i\|_t j \wedge Holds_{\mathbf{I}}(s,i) \wedge Holds_{\mathbf{T}}((-s).(-s'),t) \wedge Holds_{\mathbf{I}}(s',j)).$$

(2c) $Trans^{it}_{\mathbf{T}}(s,s')$ occurs when an interval over which s holds is ended by an instant at which s' holds:

$$Occurs_{\mathbf{T}}(Trans^{it}_{\mathbf{T}}(s,s'),t) \triangleq \exists i(ends(t,i) \wedge Holds_{\mathbf{I}}(s,i) \wedge Holds_{\mathbf{T}}(s',t)).$$

(2d) $Trans^{ti}_{\mathbf{T}}(s,s')$ occurs when an instant at which s holds begins an interval over which s' holds:

$$Occurs_{\mathbf{T}}(Trans^{ti}_{\mathbf{T}}(s,s'),t) \triangleq \exists j(begins(t,j) \wedge Holds_{\mathbf{T}}(s,t) \wedge Holds_{\mathbf{I}}(s',i))$$

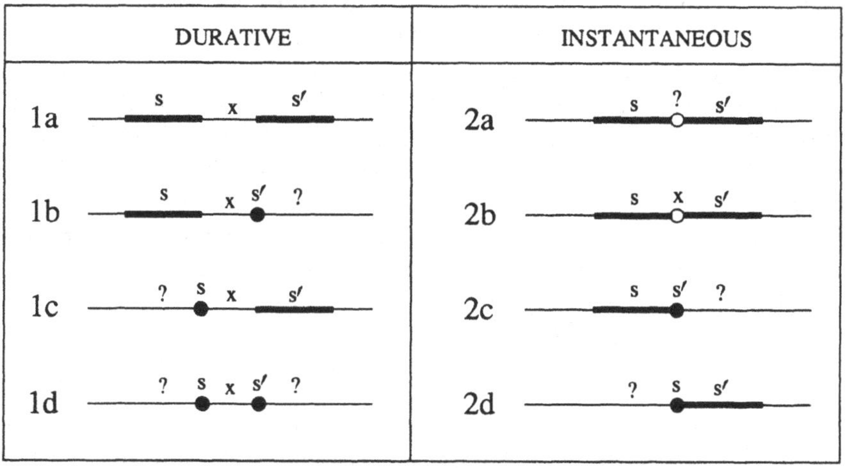

Figure 1. Eight kinds of transition from state s to s′. ('x' denotes the state (−s).(−s′), '?' indicates that any state is allowed.)

Note that any occurrence of $Trans^{ii}_{\mathbf{T}}(s,s')$ must also be an occurrence of one of $Trans^{iti}_{\mathbf{T}}(s,s')$, $Trans^{it}_{\mathbf{T}}(s,s')$, and $Trans^{ti}_{\mathbf{T}}(s,s')$; which of these it is will depend on the nature of the states s and s' involved in it. If one wishes to follow Allen and Hamblin in denying that a state can hold at an isolated instant, then the only two transition types available are $Trans^{iii}_{\mathbf{I}}$ and $Trans^{ii}_{\mathbf{T}}$. These could then be unambiguously denoted by $Trans_{\mathbf{I}}$ and $Trans_{\mathbf{T}}$ respectively. The resulting simplicity is attractive, but as argued by Galton (1990), it has the disadvantage of not being able to handle continuous change satisfactorily.

The reader may be wondering why we need to distinguish between so many different kinds of transition. The full answer to this will appear later, in §6.2, where we use these different transition operators to model various events of interest. For the present we shall just give a simple (perhaps not wholly convincing) example of why we must distinguish between $Trans_{\mathbf{I}}^{iii}$ and $Trans_{\mathbf{I}}^{tit}$; the reader will be able to supply analogous examples for the other cases. Consider, then, a car which is initially at rest over an interval i ending at t_0, then accelerates over an interval j to reach a speed of 100 km/h, which is then maintained over a further interval k, where $i \| j \| k$. Thus there is an occurrence of the event $Trans_{\mathbf{I}}^{iii}(speed(car) = 0, speed(car) = 100)$ taking up exactly the interval j. Assume now that over the interval j the speed increases strictly monotonically. Then there must be instants t_1 and t_2 in j at which the speed of the car is 20km/h and 60km/h respectively. We can then say that over the interval $\langle t_1, t_2 \rangle$ there is an occurrence of the event $Trans_{\mathbf{I}}^{tit}(speed(car) = 20, speed(car) = 60)$, but no occurrence of $Trans_{\mathbf{I}}^{iii}(speed(car) = 20, speed(car) = 60)$, since neither of these speeds is maintained over any interval, however short, contained in j.

A special case of a transition is when the initial state is the negation of the final state. If we exclude cases of "intermingling" (Galton, 1996), in which the state changes value infinitely often over a finite interval, then any such transition must be of one of the forms $Trans_{\mathbf{T}}^{it}(-s, s)$ or $Trans_{\mathbf{T}}^{ti}(-s, s)$; some, but not all, occurrences of these will also be occurrences of $Trans_{\mathbf{T}}^{ii}(-s, s)$. In any other case the state $(--s).(-s)$ would have to hold at the time of the transition itself, and this is impossible. The occurrence condition for $Trans_{\mathbf{T}}^{ii}(-s, s)$ reduces to

$$\begin{aligned} &Occurs_{\mathbf{T}}(Trans_{\mathbf{T}}^{ii}(-s, s), t) \leftrightarrow \\ &\qquad \exists i, j (i \|_t j \wedge Holds_{\mathbf{T}}(-s, i) \wedge Holds_{\mathbf{T}}(s, j)). \end{aligned}$$

A final note regarding the expressiveness of our formalism. We do not, in this paper, need to handle transitions of the kind expressible in ordinary language by "the value of x increases by δ", where x is a fluent taking numerical values. Such transitions are readily expressed in languages with an assignment operator, as for example $x \leftarrow x + \delta$. We can express any particular instance of this transition, for example from the state $x = a$ to the state $x = a + \delta$, as an occurrence of $Trans_{\mathbf{I}}(x = a, x = a + \delta)$. Using this representation, we can state that the general transition occurs over the interval i by the formula

$$\exists x_0 Occurs_{\mathbf{I}}(Trans_{\mathbf{I}}(x = x_0, x = x_0 + \delta), i),$$

but no subformula of this represents the event-type itself. The obvious ploy is to introduce a general notion of a "quantified event-type", defined by the rule

$$Occurs_{\mathbf{I}}(\exists x E(x), i) \quad \triangleq \quad \exists x Occurs_{\mathbf{I}}(E(x), i).$$

Definitions of this kind require care, particularly in relation to the interaction with other connectives such as negation; we shall not pursue this further here.

6. Case study: qualitative change in spatial relations

We shall illustrate several of the varieties of transition in the context of analysing varieties of change in spatial relationships. We shall use as the formal basis for describing spatial relationships a system developed in the context of Artificial Intelligence by a group of researchers led by Cohn (Randell et al., 1992), from whose work the material in §6.1 is derived. This approach to space is often called the 'RCC theory', where 'RCC' can be taken as standing for either Randell, Cui and Cohn, the authors of the original paper in which it was presented, or Regional Connection Calculus[4].

6.1. The RCC theory

The RCC theory makes use of a single primitive relation C, where the intended meaning of $C(x, y)$ is that the spatial region x is *connected* to the spatial region y. This may, if desired, be expressed in terms of point-set topology as the condition that the closures of the two regions are non-disjoint, i.e. $cl(X) \cap cl(Y) \neq \emptyset$. The relation C is stipulated to be both reflexive and symmetric. We can then define a part relation P in terms of C as follows:

$$P(x, y) \quad \triangleq \quad \forall z(C(z, x) \rightarrow C(z, y)).$$

In other words, x is a part of y just so long as anything connected to x is connected to y. As Cohn himself has pointed out, the definition of P leads to a problem if space is discrete, since any region connected to an atomic region is connected to the complement of that region, which would make an atomic region part of its complement. Thus there is a presumption here that space is not discrete.

In terms of P we can define the notions of *proper part* (PP) and *overlap* (O) as follows:

$$PP(x, y) \quad \triangleq \quad P(x, y) \wedge \neg P(y, x).$$
$$O(x, y) \quad \triangleq \quad \exists z(P(z, x) \wedge P(z, y)).$$

[4] The first use of the designation 'RCC' is, I think, in (Galton, 1993).

Using connection as a primitive, the RCC theory defines a set of eight relations such that any two regions of the same dimension and co-dimension must stand in exactly one of the eight relations. These eight relations are the analogue for spatial regions of the the 13 interval-interval relations of Allen's interval calculus. Six of them are

- x is disconnected from y: $DC(x,y) \triangleq \neg C(x,y)$.
- x is externally connected to y: $EC(x,y) \triangleq C(x,y) \wedge \neg O(x,y)$.
- x partially overlaps y: $PO(x,y) \triangleq O(x,y) \wedge \neg P(x,y) \wedge \neg P(y,x)$.
- x coincides with (or equals) y: $EQ(x,y) \triangleq P(x,y) \wedge P(y,x)$.
- x is a tangential proper part of y:

$$TPP(x,y) \triangleq PP(x,y) \wedge \exists z(EC(z,x) \wedge EC(z,y)).$$

- x is a non-tangential proper part of y:

$$NTPP(x,y) \triangleq PP(x,y) \wedge \neg TPP(x,y).$$

The remaining two relations are the inverses of TPP and $NTPP$. All eight relations are illustrated using circular regions in Figure 2. This diagram is a *conceptual neighbourhood* diagram, in the sense of Freksa (1992); there is a link in the diagram between relations R_1 and R_2 if and only if R_1 can be continuously transformed into R_2, without any third relation holding at any time during the transformation[5]. Note that since the RCC relations form a finite, and therefore discrete set, any transition between them must be discontinuous at that level of description; but the low-level changes underlying these apparent discontinuities are continuous.

6.2. Change in RCC relations

Using the RCC system, we can readily describe several kinds of movement, giving precise occurrence conditions in each case. We illustrate this with a number of examples, adapted from (Galton, 1997). The moving object is a in each case; its movement may be specified with

[5] The use of continuity to establish the conceptual neighbourhood diagram presupposes that the regions themselves are in some sense well-behaved. "Pathological" examples such as regions with fractal boundaries may lead to anomalies in this respect. We shall not pursue this matter further here.

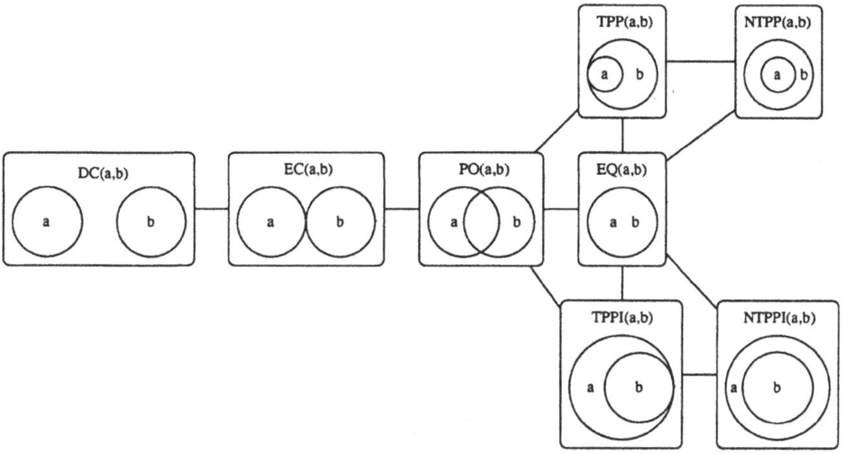

Figure 2. The eight irreducible region-region relations.

respect to another object b or to one or more regions r, r_1, etc. We shall make use of the various kinds of transitions defined above.

Since the RCC relations are defined as relations between regions, the strictly correct way to describe one object a being, say, wholly separate from another object b is $DC(pos(a), pos(b))$, where the relation DC is said to hold between the positions of the two objects rather than between the objects themselves. For simplicity, we shall from now on abbreviate expressions of the form '$R(pos(a), pos(b))$', where R is an RCC relation other than EQ, as '$R(a, b)$'. That is, we shall describe an *object* as bearing a certain spatial relation to another, when what we should say, within the strict terms of our formalism, is that the *position* of the first object bears that relation to the position of the second. Having agreed to abuse our notation in this way, we might as well also allow mixed expressions of the form $R(a, r)$, in which a is an object and r is a region, when what is really meant is $R(pos(a), r)$. In addition, we shall use the notation $at(a, r)$ for $pos(a) = r$ (or equivalently $EQ(pos(a), r)$).

6.2.1. *Entering a region*

We shall express the event 'a enters region r'. The object a starts off outside r and ends up inside r. Three different ways in which this can happen are illustrated in Figure 3. In each case the entering event occurs over the interval $\langle t_1, t_2 \rangle$. It is clearly a transition of some kind, and our task is to determine precisely what transition it is. To this end, we list below all the transitions of which each of the events illustrated is an occurrence.

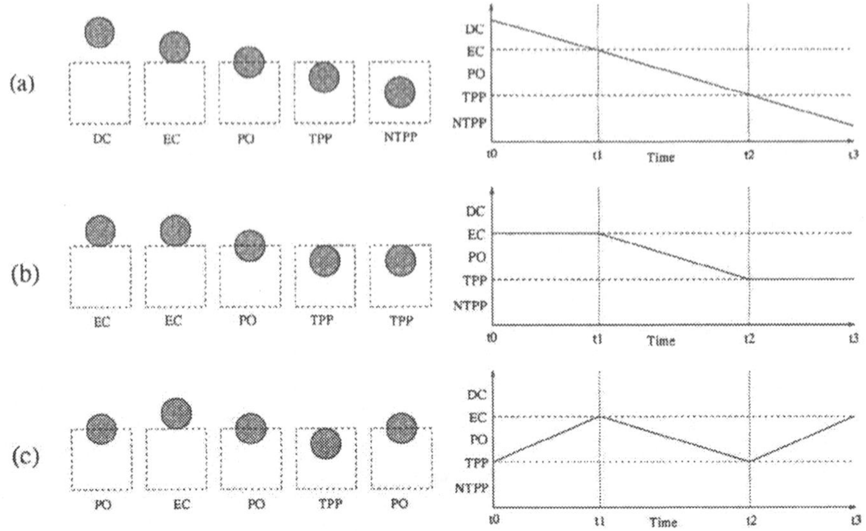

Figure 3. Possible sequences of RCC relations when an object (circle) enters a region (square).

In (a), object a moves towards r from a distance, then crosses the boundary of r and moves away from the boundary into the interior of r. One way of describing this is to say that an interval on which $DC(a, r)$ holds meets an interval on which $PO(a, r)$ holds, which meets an interval on which $NTPP(a, r)$ holds. Thus the event is an occurrence of the type $Trans_{\mathrm{I}}^{iii}(DC(a, r), NTPP(a, r))$. This is not the only way of describing it, however, since we also know that $EC(a, b)$ holds at the instant t_1 which marks the beginning of the interval over which the entering occurs, and $TPP(a, b)$ holds at the instant t_2 which marks its end. Thus the event in this case is also an occurrence of the types $Trans_{\mathrm{I}}^{tit}(EC(a, r), TPP(a, r))$, $Trans_{\mathrm{I}}^{tii}(EC(a, r), NTPP(a, r))$, and $Trans_{\mathrm{I}}^{iit}(DC(a, r), TPP(a, r))$.

In (b), our object a rests a while just outside r, touching its boundary, then moves inside r, where it rests a while just inside, again touching the boundary. From considerations of continuity, it is evident that as in the previous case we have $EC(a, b)$ holding at t_1 and $TPP(a, b)$ holding at t_2. Thus the transition is an occurrence of each of the types $Trans_{\mathrm{I}}^{x}(EC(a, b), TPP(a, b))$, where $x \in \{iii, iit, tii, tit\}$.

In (c), our object starts off inside the region, then comes outside it, only remaining outside (in the EC position) for an instant before entering it again. (The extreme positions, at t_0 and t_3, are not illustrated, though the graph on the right indicates them). Having entered the region it only remains there (in the TPP position) for an instant before once more leaving r. What happens between the instantaneous holding of EC and the instantaneous holding of TPP is quite naturally described as a's entering r; indeed, there seemed nothing unnatural about my so describing it in the previous two sentences. This may not be a paradigmatic case of entering, but I think we must admit that it *is* a case of entering nonetheless. It differs from the preceding cases in that there is only one type of transition we can regard it as an occurrence of, namely $Trans_{\mathbf{I}}^{tit}(EC(a,b), TPP(a,b))$[6].

Reviewing our examples, we see that what they all have in common is that they are occurrences of the type $Trans_{\mathbf{I}}^{tit}(EC(a,b), TPP(a,b))$, and it seems reasonable therefore to take this as our definition of the entering event:

$$Enter(a,r) \quad \triangleq \quad Trans_{\mathbf{I}}^{tit}(EC(a,b), TPP(a,b)).$$

We do not in fact have to specify the type so narrowly. If we were to put

$$Enter(a,r) \quad \triangleq \quad Trans_{\mathbf{I}}(EC(a,b), TPP(a,b))$$

instead, that would do fine, since we can infer from continuity that any occurrence of the latter type must be an occurrence of the former, whatever other types it may also be an occurrence of. However, the advantage of using the more narrowly defined type is that we have an explicit occurrence condition for events of this type, which in this instance gives us

$$\begin{aligned} Occurs_{\mathbf{I}}(Enter(a,r), i) \quad \leftrightarrow \quad & Holds_{\mathbf{T}}(EC(a,r), \inf(i)) \wedge \\ & Holds_{\mathbf{I}}(-EC(a,r). -TPP(a,r), i) \wedge \\ & Holds_{\mathbf{T}}(TPP(a,r), \sup(i)). \end{aligned}$$

This occurrence condition implies that throughout the time of any occurrence of $Enter(a,r)$, the position of a must be neither EC nor TPP. Assuming that the motion of a is continuous, we can read off from the conceptual neighbourhood diagram for the RCC relations (Figure 2) that throughout that period, the position of a must be PO to r. This is a welcome conclusion, for it simply says that when an object is in the process of entering a region, part of it must already be inside the region, and part of it must still be outside.

6.2.2. *Coming into contact*

For two objects to come into contact is for there to be a transition from their being separated in space to their being in contact. There are essentially three ways in which this can happen, as illustrated in Figure 4.

[6] Note that it is not an occurrence of, for example, $Trans_{\mathbf{I}}^{iit}(PO(a,b), TPP(a,b))$, because $PO(a,b)$ holds during the course of the transition.

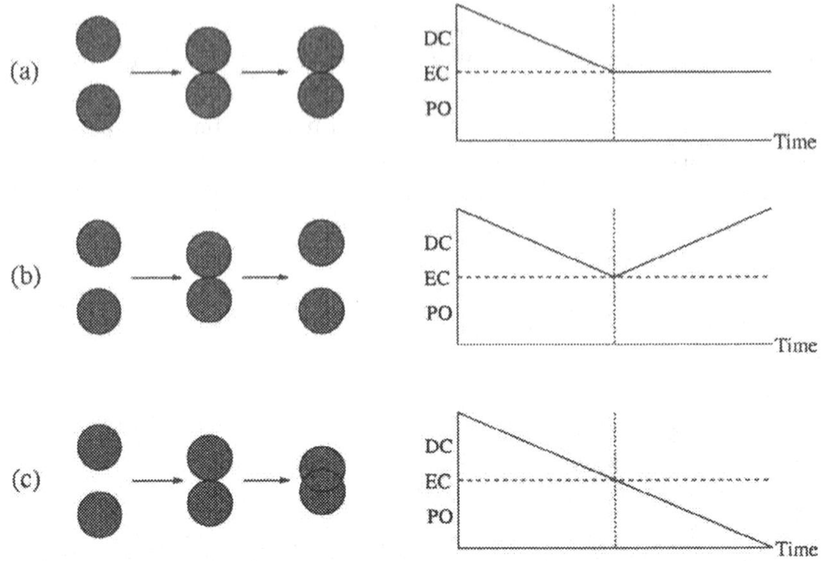

Figure 4. Possible sequences of RCC relations when two objects come into contact.

In (a), the two objects approach one another and when they have made contact they remain in contact over an interval:

$$i \| j \wedge Holds_{\mathbf{I}}(DC(a,b), i) \wedge Holds_{\mathbf{I}}(EC(a,b), j).$$

This is a clear case of $Trans^{ii}_{\mathbf{T}}(DC(a,b), EC(a,b))$, occurring at the instant at which i meets j. Notice, though, that if the motion is continuous, then the final state $EC(a,b)$ must already hold at the instant of the transition, which means that more specifically the transition is an instance of $Trans^{it}_{\mathbf{T}}(DC(a,b), EC(a,b))$.

In (b), the objects are only in contact for an instant, moving away again as soon as they have met. We can describe this as follows:

$$i \|_t j \wedge Holds_{\mathbf{I}}(DC(a,b), i) \wedge Holds_{\mathbf{T}}(EC(a,b), t) \wedge Holds_{\mathbf{I}}(DC(a,b), j).$$

This means that we have an occurrence of $Trans^{it}_{\mathbf{T}}(DC(a,b), EC(a,b))$ at t; note that symmetrically we also have, at the same instant, an occurrence of $Trans^{ti}_{\mathbf{T}}(EC(a,b), DC(a,b))$: the bodies make and break contact simultaneously. Whether such an event could happen in real life is debatable, but it is a useful idealisation for describing events such as a pair of billiard balls bouncing off one another, where the moment of contact is effectively an instant at the granularity of everyday observation.

In (c) we have interpenetrable objects which begin to overlap as soon as they have made contact. The making contact in this case may also be described using $Trans^{it}_{\mathrm{T}}(DC(a,b), EC(a,b))$, although other equally valid descriptions in this case are $Trans^{ii}_{\mathrm{T}}(DC(a,b), PO(a,b))$ and $Trans^{iti}_{\mathrm{T}}(DC(a,b), PO(a,b))$.

The upshot of these observations is that every case of making contact can be described using the transition type $Trans^{it}_{\mathrm{T}}$, enabling us to define:

$$connect(a,b) \triangleq Trans^{it}_{\mathrm{T}}(DC(a,b), EC(a,b)).$$

Note the implied asymmetry between DC and EC here; one might wonder why we could not instead use $Trans^{ti}_{\mathrm{T}}(DC(a,b), EC(a,b))$, but a little reflection will show that this would violate continuity: if two bodies are separated at t, then an interval must elapse during which the distance of separation smoothly diminishes to zero before they can make contact.

6.2.3. *Moving from one region to another*

Our last example is 'a moves from r_1 to r_2'. This is to be distinguished from the case of a moving from one position to another, which can be represented as $Trans_{\mathrm{I}}(at(a,p), at(a,q))$ or, more exactly, as $Trans^{tit}_{\mathrm{I}}(at(a,p), at(a,q))$. In that case p and q are possible positions for a, i.e., regions of space geometrically congruent to a. Here we attempt to approach more closely to the kind of locutions we use in everyday life, as in 'Ann went from the kitchen to the bathroom', or 'Bob moved from the desk to the wall'. Here Ann is congruent to neither the kitchen nor the bathroom, and Bob is congruent to neither the desk nor the wall.

These two cases are quite different, in that in Ann's case the movement is described as starting inside the first region and ending inside the second, whereas in Bob's case the starting point is only adjacent to the first region and the end point is adjacent to the second (taking adjacency as a first approximation to the relation Bob bears first to the desk and later to the wall).

We need not labour the details of these cases. Considerations similar to those of the preceding cases will lead us to define two kinds of *Move* event as follows:

$$Move_1(a, r_1, r_2) \triangleq Trans^{tit}_{\mathrm{I}}(TPP(a, r_1), TPP(a, r_2))$$
$$Move_2(a, r_1, r_2) \triangleq Trans^{tit}_{\mathrm{I}}(EC(a, r_1), EC(a, r_2))$$

If need be, one could define two other cases to represent movement from *TPP* to *EC* and vice versa.

7. Conclusion

We have set up a formal framework for representing time, states, and events, and have investigated the definition of transition events within this framework. Because of the requirement of continuity, we found that we had to define eight distinct kinds of transition; this proliferation was justified in the subsequent application to the qualitative description of changes in relation between spatial regions, where we needed the fineness of discrimination provided by our inventory of transition-types in order to adequately cover all the possibilities.

One is left with a feeling of dissatisfaction: surely there must be some way of handling these distinctions without the introduction of such a complex formalism? If one is unwilling to accept the complexity, then the only way out, I believe, is to relinquish one or more of our fundamental assumptions. One possibility is to deny that there are genuinely continuous changes; this is implied by the assumption that states can only hold over intervals, not at isolated instants. This assumption goes hand in hand with the determination to banish instants from the formalism altogether (although this cannot be done completely: we still need instants to be the times of occurrence of instantaneous transitions, though one might perhaps get away without referring to them explicitly).

Another possibility, which has many of the same consequences, is to insist that our temporal model is discrete, the fundamental units being atomic intervals which have no proper subintervals. Then the closest we can approach to continuity is the quasi-continuity expressed by the phrase 'one step at a time', by which at neighbouring atomic intervals, neighbouring or identical states must hold. This in turn implies that the set of world-states is discrete.

Under different circumstances these various assumptions can be justified; the purpose of this paper is not so much to advocate any particular set of assumptions as to examine in detail certain consequences of one natural set of assumptions that is in line with the dominant tradition in the mathematical modelling of physical processes.

References

Allen, J.: 1984, 'Towards a general theory of action and time'. *Artificial Intelligence* **23**, 123–154.

Allen, J. and G. Ferguson: 1994, 'Actions and Events in Interval Temporal Logic'. *Journal of Logic and Computation* **4**, 531–579.

Davidson, D.: 1967, 'The logical form of action sentences'. In: N. Rescher (ed.): *The Logic of Decision and Action.* University of Pittsburgh Press. Reprinted in D. Davidson, *Essays on Actions and Events*, Oxford, Clarendon Press, 1980.

Freksa, C.: 1992, 'Temporal reasoning based on semi-intervals'. *Artificial Inteliigence* **54**, 199–227.

Galton, A. P.: 1984, *The Logic of Aspect: an Axiomatic Approach.* Oxford: Clarendon Press.

Galton, A. P.: 1990, 'A critical examination of Allen's theory of action and time'. *Artificial Intelligence* **42**, 159–188.

Galton, A. P.: 1993, 'Towards an integrated logic of space, time, and motion'. In: R. Bajcsy (ed.): *Proceedings of the Thirteenth International Joint Conference on Artificial Intelligence.*

Galton, A. P.: 1994, 'Instantaneous Events'. In: H. J. Ohlbach (ed.): *Temporal Logic: Proceedings of the ICTL Workshop.* Saarbrücken.

Galton, A. P.: 1996, 'An investigation of 'non-intermingling' principles in temporal logic'. *Journal of Logic and Computation* **6, 2**, 271–294.

Galton, A. P.: 1997, 'Space, Time and Movement'. In: O. Stock (ed.): *Spatial and Temporal Reasoning.* Kluwer Academic Publishers.

Gooday, J. M.: 1994, 'A Transition-Based Approach to Reasoning about Action and Change'. Ph.D. thesis, University of Exeter.

Gooday, J. M. and A. P. Galton: 1997, 'The Transition Calculus: a high-level formalism for reasoning about action and change'. *Journal of Experimental and Theoretical Artificial Intelligence* **9**, 51–66.

Hamblin, C. L.: 1971, 'Instants and intervals'. *Studium Generale* **24**, 127–134.

Kowalski, R. A. and M. J. Sergot: 1986, 'A logic-based calculus of events'. *New Generation Computing* **4**, 67–95.

McCarthy, J. and P. J. Hayes: 1969, 'Some philosophical problems from the standpoint of artificial intelligence'. In: B. Melzer and D. Michie (eds.): *Machine Intelligence 4.* Edinburgh University Press.

McDermott, D.: 1982, 'A temporal logic for reasoning about processes and plans'. *Cognitive Science* **6**, 101–155.

Randell, D. A., Z. Cui, and A. G. Cohn: 1992, 'A spatial logic based on regions and connection'. In: *Proc. 3rd Int. Conf. on Knowledge Representation and Reasoning.* pp. 165–176. Cambridge MA, Oct. 1992.

A Modal Logic of Durative Actions
*

Isabel Nunes (in@di.fc.ul.pt) and Jose Luiz Fiadeiro (llf@di.fc.ul.pt)
Department of Informatics, Faculty of Sciences, University of Lisbon, Campo Grande, 1700 Lisboa, Portugal

Wladyslaw M. Turski (wmt@mimuw.edu.pl)
Institute of Informatics, Warsaw University, ul. Banacha 2,02-097 Warsaw, Poland

Abstract. Having in mind the development of formal support for programming in transactional and multiprocessor systems, a modal logic is presented which is based on durative actions, i.e. actions for which the state in which they finish executing is not necessarily the same in which they started. For this purpose, dyadic action modalities are introduced which allow us to relate three states: the launching, the accepting and the resulting states of every action execution. The characterisation of the new modalities is given both through an extension of the traditional notion of transition system and an axiomatisation that is proved to be complete for the class of extended transition systems. Finally, we show how the proposed logic can be used to express and verify properties of specifications in the new behavioural paradigm introduced by W.Turski.

1. Introduction

Traditionally, modal logics have been based on actions that are atomic in two senses: (1) they execute without any interference from the environment, i.e. the results that are produced by the execution of an action are a function of the starting state; (2) they have no duration, i.e. the state of the system does not change while an action is executing. These two kinds of atomicity have led to models of system behaviour based on transition systems where actions are modelled as functions (or relations in the case of non-deterministic actions) on a set of worlds, and global system behaviour is modelled through sequences of worlds (runs).

In this paper, we extend this model by relaxing the second assumption of atomicity. That is to say, while assuming that actions compute their effects without interference from the environment by executing on a local copy of the state, we develop a logic in which actions can have

* This work was partially supported by JNICT through contracts PRAXIS XXI 2/2.1/MAT/46/94 (ESCOLA), PCSH/OGE/1038/95 (MAGO) and PRAXIS XXI 2/2.1/TIT/1662/95 (SARA).

H. Barringer et al. (eds.), Advances in Temporal Logic, 299–317.

duration, i.e. we admit that the global state of the system can change while an action is executing. Notice that we use the term durative not to express the real-time duration of action execution but, instead, the fact that the system state can change while actions execute.

The motivation for this generalisation is twofold. On the one hand, several application domains, namely in Artificial Intelligence (e.g. for agent-based programming) and Databases (e.g. for transaction management), require logical support for actions that are durative.

On the other hand, the continuous development of multiprocessor and distributed systems give us the means for truly parallel and concurrent process execution, which prompts the need for new methods and supporting logics.

A new paradigm of programming that is representative of such extended use of multiprocessing capabilities has been recently proposed by W. Turski. A behaviour specification (Turski, 1992) is a collection of actions which describe the reactions to the possible states of the system and such that:

- associated with each action there are two conditions – the pre-guard (desirability condition) and the post-guard (acceptability condition) – which define the system states in which an action should be undertaken and accepted, respectively;
- actions are performed by agents, of which there is an unspecified number, that are equipped with memories capable of storing a copy of the system state. Agents can copy a state into memory – loading – and copy values from memory to system objects – discharging;
- associated with each action there is a loading list which consists of the objects whose values are to be copied into the agent's memory when the action is launched, and a discharging list which consists of the objects of the system which are to be updated at a discharge with the values of the identically named objects in the agent's memory;
- loading and discharging are instantaneous; action execution is not;
- when the system is in a state that satisfies the pre-guard of an action a:
    - this action is assigned to an unoccupied agent, which loads to its memory the values of the objects in the loading list of a and starts execution;
    - the agent performs action a in its memory, without interference from the environment;

- when the execution of action a terminates and if the post-guard of a is satisfied by the system state, then the agent discharges the values of the objects referred in the discharging list of a into the system, updating it. Testing the post-guard and discharging the local state is performed atomically (*test and set*).

As an example the following is the specification of the "dining philosophers" (c.f. (Turski, 1990)) as seen from an individual philosopher.

$$
\begin{array}{rcl}
(\neg H, \neg H) & \rightarrow & [\ |\ think \mid H] \\
(H \wedge PR \wedge PL, H \wedge PR \wedge PL) & \rightarrow & [\ |\ eat \mid H, PR, ROT, PL, LOT] \\
(H \wedge PL \wedge \neg PR \wedge \neg ROT, \neg ROT) & \rightarrow & [\ |\ rel\_L \mid PL, LOT] \\
(H \wedge PR \wedge \neg PL \wedge \neg LOT, \neg LOT) & \rightarrow & [\ |\ rel\_R \mid PR, ROT] \\
(H \wedge LOT, H \wedge LOT) & \rightarrow & [\ |\ take\_L \mid PL, LOT] \\
(H \wedge ROT, H \wedge ROT) & \rightarrow & [\ |\ take\_R \mid PR, ROT]
\end{array}
$$

where

H	–	the philosopher is hungry
LOT	–	the left fork is on the table
ROT	–	the right fork is on the table
PL	–	the philosopher possesses the left fork
PR	–	the philosopher possesses the right fork

Each *dga* takes the form

(pre-guard, postguard) $\rightarrow$ [loading | program | discharging]

In this example all loading lists are empty. This means that the computations performed locally by each action do not depend on the state in which the action is launched. The actual transformations performed by the actions are specified separately.

The intended semantics of the rel_R action specification, for example, is the following. When a philosopher is hungry (H), possesses the right fork (PR) and does not possess the left fork ($\neg PL$), and the left fork is not on the table ($\neg LOT$), he will release the right fork (the action is launched because the pre-guard is true). This is so specified in order to prevent deadlock. If, while our philosopher is in the process of releasing its right fork, the philosopher at his left releases his right fork (our philosopher's left fork), then our philosopher should not bring about the action. This is expressed through the post-guard: if the left fork is on the table (LOT) when the philosopher finishes executing the release, the action is rejected.

Programming in this new paradigm requires logical support that cannot be adequately given using current modal logics. Whereas action execution in the behavioural paradigm involves three states – the launching, accepting and resulting states – the action modalities that are currently available allow us to relate only two states – the launching and the resulting state. Indeed, in the absence of state changes during action execution, the launching and accepting states are the same, thus reducing the characterisation of action execution to two states.

Hence, the main challenge raised by durative actions, and one of the main contributions of the paper, is the characterisation of action modalities that relate three states. This characterisation is given both through an extension of the traditional notion of transition system and an axiomatisation that is proved to be complete for the class of extended transition systems. Due to the limitations on the size of the paper, we shall concentrate on the proof-theoretic aspects of the logic, namely the completeness proof and the way the new action modalities characterise the behavioural paradigm. Details on the underlying computational model and model-theory can be found in (Turski, 1992; Turski, 1990) and (Nunes et al., 1996; Nunes et al., 1997), respectively.

2. Syntax and semantics of DTL

Labelled transition systems are usually defined (e.g. as in (Manna and Pnueli, 1991; Stirling, 1992)) in terms of pairs $(\Sigma, \{\xrightarrow{a} \mid a \in \mathcal{A}\})$ where Σ is a non-empty set (of states), $\mathcal{A}$ is a non-empty set (of actions/labels) and, for each $a \in \mathcal{A}$, $\xrightarrow{a} \subseteq \Sigma \times \Sigma$ is the transition relation. Given $\sigma, \sigma' \in \Sigma$, $\sigma \xrightarrow{a} \sigma'$ means that action a takes the system from state σ to state σ'. If we are modelling systems in which actions are deterministic, each $\xrightarrow{a}$ is a partial function.

In this context, each transition within the system is completely defined by an action and two states, the launching and the resulting ones. When we consider durative actions, we are faced with the possibility of state changes occurring during action execution due to the effects of other agents executing concurrently. Hence, the traditional notion of transition system is no longer adequate.

In order to provide a model that takes action duration into account, we present what we call durative transition systems.

DEFINITION 1. *A durative transition system is a pair* $(\Sigma, \{\xrightarrow{a} \mid a \in \mathcal{A}\})$ *where* Σ *is a non-empty set (of states),* A *is a non-empty set (of actions) and, for each* $a \in \mathcal{A}$, $\xrightarrow{a} : \Sigma \times \Sigma \times \Sigma$ *is a ternary relation.*

We adopt the notation $\sigma \xrightarrow{a} (\sigma' \rightarrow \sigma'')$ which reflects the idea that action a is launched in state σ, takes its "time" executing, is accepted in state σ' and results in state σ''. The use of the "long" arrow is meant to suggest that the transition from σ to σ' is durative and the "short" arrow is meant to suggest that the transition from σ' to σ'' is atomic.

Durative transition systems will be used as models for the logic that we have developed for the behavioural paradigm – Durative Transition Logic (DTL). DTL is based on dyadic operators that are expressive enough to allow us to refer to the three different moments, or system states, that characterise the execution of actions.

DEFINITION 2. *The language $\mathcal{L}(\mathcal{P}, \mathcal{A})$ of the propositional modal logic DTL defined over a set $\mathcal{P}$ of atomic sentences and a set $\mathcal{A}$ of actions is defined, in abstract syntax form, by*

$$A ::= p \mid \neg A \mid A_1 \supset A_2 \mid A_1\,[a]\,A_2 \mid A_1\,\overline{[a]}\,A_2$$

where $p \in \mathcal{P}$ and $a \in \mathcal{A}$.

The language $\mathcal{L}(\mathcal{P}, \mathcal{A})$ is interpreted on durative transition systems as follows:

DEFINITION 3. *An interpretation structure I for a language $\mathcal{L}(\mathcal{P}, \mathcal{A})$ is a pair $(\mathcal{D}, \mathcal{V})$ where $\mathcal{D} = (\Sigma, \{\xrightarrow{a} \mid a \in \mathcal{A}\})$ is a durative transition system and $\mathcal{V} : \Sigma \to 2^{\mathcal{P}}$, the valuation function, maps each state $\sigma \in \Sigma$ to a subset of $\mathcal{P}$ (the subset that consists of the atomic sentences that are defined to be true at σ). We say that I is an interpretation structure based on $\mathcal{D}$.*

DEFINITION 4. *Given an interpretation structure $I = (\mathcal{D}, \mathcal{V})$ for a language $\mathcal{L}(\mathcal{P}, \mathcal{A})$ we define the satisfaction relation $\models_I \subseteq \Sigma \times \mathcal{L}(\mathcal{P}, \mathcal{A})$ as follows (where $\sigma \in \Sigma$ and $p \in \mathcal{P}$):*

$$\begin{array}{lll}
\sigma \models_I p & \textit{iff} & p \in \mathcal{V}(\sigma) \\
\sigma \models_I \neg A & \textit{iff} & \sigma \not\models_I A \\
\sigma \models_I A_1 \supset A_2 & \textit{iff} & \sigma \models_I A_1 \textit{ implies } \sigma \models_I A_2 \\
\sigma \models_I A_1\,[a]\,A_2 & \textit{iff} & \textit{for every } \sigma', \sigma'' \in \Sigma, \\
& & \sigma \xrightarrow{a} (\sigma' \to \sigma'') \textit{ and } \sigma' \not\models_I A_1 \textit{ implies } \sigma'' \models_I A_2 \\
\sigma \models_I A_1\,\overline{[a]}\,A_2 & \textit{iff} & \textit{for every } \sigma', \sigma'' \in \Sigma, \\
& & \sigma' \xrightarrow{a} (\sigma \to \sigma'') \textit{ and } \sigma' \not\models_I A_1 \textit{ implies } \sigma'' \models_I A_2
\end{array}$$

We further say that A is true in I, $(\models_I A)$, iff $\sigma \models_I A$ for all states $\sigma \in \Sigma$.

We will use the traditional propositional connectives as abbreviations, together with the "possibility" operators $\langle a \rangle$ and $\overline{\langle a \rangle}$ defined as follows:

$A_1 \langle a \rangle A_2$ is an abbreviation for $\neg(A_1\, [a]\, \neg A_2)$
$A_1 \overline{\langle a \rangle} A_2$ is an abbreviation for $\neg(A_1\, \overline{[a]}\, \neg A_2)$

It can easily be checked that

$$\sigma \models_I A_1 \langle a \rangle A_2 \text{ iff}$$
$$\text{there is } \sigma', \sigma'' \in \Sigma, \sigma \stackrel{a}{\longrightarrow} (\sigma' \to \sigma'') \text{ and } \sigma' \not\models_I A_1 \text{ and } \sigma'' \models_I A_2$$
$$\sigma \models_I A_1 \overline{\langle a \rangle} A_2 \text{ iff}$$
$$\text{there is } \sigma', \sigma'' \in \Sigma, \sigma' \stackrel{a}{\longrightarrow} (\sigma \to \sigma'') \text{ and } \sigma' \not\models_I A_1 \text{ and } \sigma'' \models_I A_2$$

Notice how the new modal operators allow us to refer to the three critical moments of the life of a given action a: the sentence $(A_1 \supset A_2\, [a]\, A_3)$ is true if whenever a is launched in a state satisfying A_1 and accepted in a state *refusing* A_2, the resulting state satisfies A_3.

In particular, the sentence $(\mathrm{ff}\, \langle a \rangle\, \mathrm{tt})$ allows us to identify *launching* states for action a. Hence, for instance, the fact that all states satisfying A are launching states for action a corresponds to the truth of $(A \supset \mathrm{ff}\, \langle a \rangle\, \mathrm{tt})$.

The first argument of the modal operator is negated for technical reasons: it simplifies the presentation of the axiomatisation. From an intuitive point of view, it corresponds to a condition for the *rejection* of the effects of the action.

The operator $\overline{[a]}$ is a kind of reverse of $[a]$ in the sense that it points "backwards" in the transition relation (Stirling, 1992): the accepting state is the state of evaluation of the formula $(A_1 \supset A_2\, \overline{[a]}\, A_3)$ and the launching state is the state where A_2 is evaluated.

For instance, the sentence $(\mathrm{ff}\, \overline{\langle a \rangle}\, \mathrm{tt})$ allows us to identify accepting states for action a. Hence, the fact that all accepting states for a satisfy property A corresponds to the truth of $(\mathrm{ff}\, \overline{\langle a \rangle}\, \mathrm{tt} \supset A)$.

In section 4, we shall analyse other kind of properties that are of interest in the context of the behavioural paradigm.

DEFINITION 5. *Given a set $\mathcal{M}$ of interpretation structures, an interpretation I, a sentence $A \in \mathcal{L}(\mathcal{P}, \mathcal{A})$ and a set Γ of sentences, we say that :*

- *A is true in $\mathcal{M}$, $\models_{\mathcal{M}} A$, iff for all $I \in M$, $\models_I A$;*
- *Γ entails A in I, $(\Gamma \models_I A)$ iff for all states $\sigma \in \Sigma$, if $\sigma \models_I \Gamma$ then $\sigma \models_I A$;*

– *Γ entails A in $\mathcal{M}$, ($\Gamma \models_{\mathcal{M}} A$) iff for all $I \in M$ and for all $\sigma \in \Sigma$, if $\sigma \models_I \Gamma$ then $\sigma \models_I A$.*

Notice that we have defined a *local* semantic consequence relation (Stirling, 1992).

3. A complete and decidable axiomatisation

In this section, we present an axiomatisation for the proposed operators and generalise the typical canonical model style of completeness proof for unary operators to the binary case.

DEFINITION 6. *The logic DTL over the language $\mathcal{L}(\mathcal{P}, \mathcal{A})$ is given by the following proof system:*

Axioms:

GA1.	*Any instance of a propositional tautology*
GKR.	$A_1\,\lfloor a \rfloor\,(A_2 \supset A_3) \supset (A_1\,\lfloor a \rfloor\,A_2 \supset A_1\,\lfloor a \rfloor\,A_3)$
G$\overline{\text{K}}$R.	$A_1\,\lceil a \rceil\,(A_2 \supset A_3) \supset (A_1\,\lceil a \rceil\,A_2 \supset A_1\,\lceil a \rceil\,A_3)$
GKL.	$(A_1 \supset A_2)\,\lfloor a \rfloor\,A_3 \supset (A_1\,\lfloor a \rfloor\,A_3 \supset A_2\,\lfloor a \rfloor\,A_3)$
G$\overline{\text{K}}$L.	$(A_1 \supset A_2)\,\lceil a \rceil\,A_3 \supset (A_1\,\lceil a \rceil\,A_3 \supset A_2\,\lceil a \rceil\,A_3)$

Rules:

MP.	*if* $\vdash A$ *and* $\vdash A \supset B$ *then* $\vdash B$
GNecR.	*if* $\vdash A$ *then* $\vdash B\,\lfloor a \rfloor\,A$
GNecL.	*if* $\vdash A$ *then* $\vdash A\,\lfloor a \rfloor\,B$
GR1.	$\vdash A_1 \supset (A_2\,\lfloor a \rfloor\,A_3)$ *iff* $\vdash \neg A_2 \supset (\neg A_1\,\lceil a \rceil\,A_3)$

The axiom schemes express a principle of distributivity of the modal operators with respect to the conditional. They are a generalisation of the axiom K of normal modal logics. The first two, GKR and G$\overline{\text{K}}$R, express distributivity in what concerns post-conditions. Axiom GKL gives distributivity over rejection conditions, while G$\overline{\text{K}}$L gives distributivity over pre-conditions.

The rule of necessitation "at the right", GNecR, states that a theorem is necessarily true at all resulting states. The rule of necessitation "at the left", GNecL, states that a theorem used as a rejection condition for an action, leads to no accepting state and, hence, no transition. Rule GR1 establishes the relationship between each modal operator and its "reverse": it exchanges between the roles of the launching and accepting states.

We are able to derive within DTL several rules and theorems that are expected to hold when we are dealing with modal operators, such as:

$(A_1\,[a]\,A_2 \vee A_1\,[a]\,A_3) \supset A_1\,[a]\,(A_2 \vee A_3)$
$(A_1\,[a]\,A_3 \vee A_2\,[a]\,A_3) \supset (A_1 \vee A_2)\,[a]\,A_3$
$(A_1\,[a]\,A_2 \wedge A_1\,[a]\,A_3) \equiv A_1\,[a]\,(A_2 \wedge A_3)$
$(A_1\,[a]\,A_3 \wedge A_2\,[a]\,A_3) \equiv (A_1 \wedge A_2)\,[a]\,A_3$
$(A_1\,[a]\,A_2 \wedge A_1\,\langle a\rangle\,A_3) \supset A_1\,\langle a\rangle\,(A_2 \wedge A_3)$
if $\vdash (A_1 \wedge \ldots \wedge A_n) \supset A$ then $\vdash (B\,[a]\,A_1 \wedge \ldots \wedge B\,[a]\,A_n) \supset B\,[a]\,A$ (GRkR)
if $\vdash (A_1 \wedge \ldots \wedge A_n) \supset A$ then $\vdash (A_1\,[a]\,B \wedge \ldots \wedge A_n\,[a]\,B) \supset A\,[a]\,B$ (GRkL)

and also the corresponding ones for the operator $\overline{[a]}$.

We will now establish soundness and completeness of the axiomatic system just given with respect to the class $\mathcal{M}$ of interpretation structures as given in definition 3.

PROPOSITION 1. *The axiomatic system given in definition 6 is sound with respect to the class $\mathcal{M}$ of interpretation structures, that is, if $\Gamma \subseteq \mathcal{L}(\mathcal{P}, \mathcal{A})$ then*

$$\text{If } \Gamma \vdash A \text{ then } \Gamma \vDash_{\mathcal{M}} A$$

Completeness is proved using canonical models. For simplicity, we consider the case of the "forward" operators $[a]$ only.

The definition used for canonical models extends the one that is typically found in the literature for unary modal operators.

DEFINITION 7. *The canonical model for DTL is the structure*

$$I_{DTL} = (\Sigma_{DTL}, \{\xrightarrow{a} \mid a \in \mathcal{A}\}, \mathcal{V}_{DTL})$$

where:

a) $\Sigma_{DTL} = \{\sigma \subseteq \mathcal{L}(\mathcal{P}, \mathcal{A}) \mid \sigma \textit{ is maximal DTL-consistent}\}$

b) $\sigma \xrightarrow{a} (\sigma' \to \sigma'')$ *iff, for every* $B \notin \sigma', \{A \mid (B\,[a]\,A) \in \sigma\} \subseteq \sigma''$

c) $\mathcal{V}_{DTL}(\sigma) = \{p \in \mathcal{P} \mid p \in \sigma\}$, *for* $\sigma \in \Sigma_{DTL}$

Notice that this is a good definition because, DTL being sound, Σ_{DTL} is not empty.

PROPOSITION 2. *Durative transitions in a canonical model satisfy,*

$$\sigma \xrightarrow{a} (\sigma' \to \sigma'') \textit{ iff, for every } B \notin \sigma',\ \{(B\,\langle g\rangle\,A) \mid A \in \sigma''\} \subseteq \sigma$$

THEOREM 1. *The axiomatization given in definition 6 is complete with respect to class M of interpretation structures, that is,*

$$\textit{if} \models_{\mathcal{M}} A \textit{ then} \vdash A$$

It is also strongly complete, that is, if $\Gamma \subseteq \mathcal{L}(\mathcal{P}, \mathcal{A})$ *then*

$$\textit{if } \Gamma \models_{\mathcal{M}} A \textit{ then } \Gamma \vdash A$$

Proof. The proof relies on a few lemmas. Let $A, B \in \mathcal{L}(\mathcal{P}, \mathcal{A})$ and $\sigma \in \Sigma_{DTL}$.

P1. $A\,[a]\,B \in \sigma$ iff for every $\sigma', \sigma'' \in \Sigma_{DTL}, \sigma \xrightarrow{a} (\sigma' \to \sigma'')$ and $A \notin \sigma'$ implies $B \in \sigma''$

P2. $\sigma \models_{I_{DTL}} A$ iff $A \in \sigma$

P3. $\models_{I_{DTL}} A$ implies $\vdash A$

P4. $\Gamma \models_{I_{DTL}} A$ implies $\Gamma \vdash A$

We prove completeness of DTL, as usual, by using P3: $\models_{\mathcal{M}} A$ implies $\models_{I_{DTL}} A$ which, by P3, implies $\vdash A$.

For proving strong completeness we can argue in a similar way using P4.

The only lemma that is really interesting, because it is new in that it has to deal with the dyadic operators, is P1. Hence, we will sketch its proof:

LEMMA 1. *Given* $A, B \in \mathcal{L}(\mathcal{P}, \mathcal{A})$ *and* $\sigma \in \Sigma_{DTL}$, $A\,[a]\,B \in \sigma$ *iff, for every* $\sigma', \sigma'' \in \Sigma_{DTL}$, $\sigma \xrightarrow{a} (\sigma' \to \sigma'')$ *and* $A \notin \sigma'$ *implies* $B \in \sigma''$.

Proof:

$\boxed{\Rightarrow}$ Suppose $A\,[a]\,B \in \sigma$ and take arbitrary $\sigma', \sigma'' \in \Sigma_{DTL}$. Suppose also that $\sigma \xrightarrow{a} (\sigma' \to \sigma'')$ and $A \notin \sigma'$. By the definition of canonical model we have

$\sigma \xrightarrow{a} (\sigma' \to \sigma'')$ iff, for every $D \notin \sigma'$, $\{C \mid (D\,[a]\,C) \in \sigma\} \subseteq \sigma''$

Because $A \notin \sigma'$ and $A\,[a]\,B \in \sigma$, it follows that $B \in \sigma''$

$\boxed{\Leftarrow}$ We have to prove that

$\forall \sigma', \sigma'' \in \Sigma_{DTL}.(\sigma \xrightarrow{a} (\sigma' \to \sigma'') \wedge A \notin \sigma' \Rightarrow B \in \sigma'') \Rightarrow A\,[a]\,B \in \sigma$

That is to say

$\forall \sigma'' \in \Sigma_{DTL}.(\exists \sigma' \in \Sigma_{DTL}.(\sigma \xrightarrow{a} (\sigma' \to \sigma'') \wedge A \notin \sigma') \Rightarrow B \in \sigma'') \Rightarrow A\,[a]\,B \in \sigma$

Suppose then, that

$\forall \sigma'' \in \Sigma_{DTL}.(\exists \sigma' \in \Sigma_{DTL}.(\sigma \xrightarrow{a} (\sigma' \to \sigma'') \wedge A \notin \sigma') \Rightarrow B \in \sigma'')$

By lemma 2 (in the Appendix),

$\forall \sigma'' \in \Sigma_{DTL}.(\{C \mid (A\,[a]\,C) \in \sigma\} \subseteq \sigma'' \Rightarrow \exists \sigma' \in \Sigma_{DTL}.(\sigma \xrightarrow{a} (\sigma' \to \sigma'') \wedge A \notin \sigma'))$

Then, we have that

$\forall \sigma'' \in \Sigma_{DTL}.(\{C \mid (A\,[a]\,C) \in \sigma\} \subseteq \sigma'' \Rightarrow B \in \sigma'')$

which is equivalent to

$\{C \mid (A\,[a]\,C) \in \sigma\} \vdash B$

and which implies that there are $A_1, \ldots, Am \in \{C \mid (A\,[a]\,C) \in \sigma\}$ s.t. $\vdash (A_1 \wedge \ldots \wedge Am) \supset B$.

If $m = 0$, then B is a theorem of DTL which, through GNecR, implies that $A\,[a]\,B$ is a theorem. Because σ is maximal DTL-consistent, $A\,[a]\,B \in \sigma$.

If $m \neq 0$, then we can infer $\vdash (A\,[a]\,A_1 \wedge \ldots \wedge A\,[a]\,A_m) \supset A\,[a]\,B$.

Because $A_1, \ldots, A_m \in \{C \mid (A\,[a]\,C) \in \sigma\}$, we can conclude that $A\,[a]\,A_1, \ldots, A\,[a]\,A_m \in \sigma$ which, because σ is maximal DTL-consistent, implies $A\,[a]\,B \in \sigma$.

We can further prove:

PROPOSITION 3. *The DTL-system is decidable.*

This result is proved by showing that DTL has the finite model property: if $\nvdash A$ then there is a finite model I s.t. $\nvDash_I A$, that is, every non-theorem of DTL is false in some finite model of DTL. The finite model property itself is proved by a straightforward adaptation of filtration techniques.

4. Application to the behavioural paradigm

In this section, we show how the proposed logic can be put to use in expressing and verifying properties of behavioural specifications.

4.1. Locality and frame restrictions

As motivated in the introduction, actions in the behavioural paradigm have two associated sets of propositional symbols, which account for the "attributes" or "system variables" that the actions need to read and write, respectively. The first set consists of the attributes that are loaded to the local memory of any agent that engages in executing an action – loading list. The second consists of the attributes that are discharged from the local state of the agent into the global state of the system – discharging list. The discharging list is particularly important in determining the effect of an action in the state of the system: attributes of the discharging list get the values computed locally by the executing agent (locality condition); attributes not in the discharging list remain unchanged, i.e. get the same values that they had in the accepting state (frame condition).

These restrictions can be brought into our logic by assigning, to every action $a \in \mathcal{A}$, two sets of propositional symbols – L_a and D_a – accounting for the loading and discharging lists, respectively.

In order to formalise the locality and frame restrictions over durative transition systems, we need the following auxiliary definitions:

DEFINITION 8. *Given a set $\Omega \subseteq \mathcal{P}$ of propositional symbols and an interpretation structure $I = (\Sigma, \{\xrightarrow{a} \mid a \in \mathcal{A}\}, V)$, we say that two states σ and σ' are Ω-equivalent – written $\sigma \underset{\Omega}{\equiv} \sigma'$ – iff the values of all symbols in Ω are the same in both states:*

$$\sigma \underset{\Omega}{\equiv} \sigma' \textit{ iff } \mathcal{V}(\sigma) \cap \Omega = \mathcal{V}(\sigma') \cap \Omega$$

DEFINITION 9. *Let $\Omega \subseteq P$ and $\Gamma, \Gamma' \subseteq \mathcal{L}(\mathcal{P}, \mathcal{A})$. We say that Γ Ω-implies Γ' – written $\Gamma \underset{\Omega}{\supset} \Gamma'$ – iff all states where Γ holds have an Ω-equivalent state where Γ' holds,*

$\Gamma \underset{\Omega}{\supset} \Gamma'$ *iff for every I and σ, $\sigma \vDash_I \Gamma$ implies that there is σ' s.t.*

$$\sigma \underset{\Omega}{\equiv} \sigma' \textit{ and } \sigma' \vDash_I \Gamma'$$

DEFINITION 10. *An interpretation structure is said to be* behavioural *iff:*

a) For every state $\sigma_1, \sigma_2, \sigma_1', \sigma_2', \sigma_1'', \sigma_2''$ and action a,

$\sigma_1 \xrightarrow{a} (\sigma_1' \to \sigma_1'')$ *and* $\sigma_2 \xrightarrow{a} (\sigma_2' \to \sigma_2'')$ *and* $\sigma_1 \underset{L_a}{\equiv} \sigma_2$ *imply*

$$\sigma_1'' \underset{D_a}{\equiv} \sigma_2''$$

b) For every $\sigma, \sigma', \sigma''$, and action a, $\sigma \xrightarrow{a} (\sigma' \to \sigma'')$ implies

$$\sigma' \underset{P \setminus D_a}{\equiv} \sigma''$$

Restriction a) says that the values of attributes in the discharging list (D_a) in a resulting state of a transition only depend on the values that the attributes in the loading list (the ones in L_a) had in the corresponding launching state. This restriction reflects the notion of local execution of actions. We are modelling deterministic actions, which is the model adopted in the behavioural paradigm.

Restriction b) says that the values for attributes other than the ones in the discharging list only depend on the accepting state σ', not on the launching state. This restriction reflects the fact that actions are durative. It is called the frame condition because it states that attributes outside the discharging list remain unchanged.

Notice that we are modelling deterministic actions, which is the model adopted in the behavioural paradigm: a durative transition is a function because, given a launching and an accepting states, the resulting state is unique, and it is partial because not all states are necessarily launching states, and not every state is necessarily an accepting state for a given action a.

DEFINITION 11. *The logic* DTL^{++} *is obtained from DTL through the addition of the axioms*

GSy. $\neg P\,[a]\,P$ *where* P *is* p *or* $\neg p$ *for* $p \notin D_a$

and the rules

GLo. $A \vdash \mathrm{ff}\,\langle a\rangle\,P$ *implies* $A' \vdash \mathrm{ff}\,[a]\,P$ *where* $A' \underset{L_a}{\supset} A$ *and* P *is* p *or* $\neg p$ *for* $p \in D_a$

PROPOSITION 4. DTL^{++} *is sound and complete with respect to the class of behavioural interpretation structures.*

Proof:

We will only prove completeness here. We need to prove that the canonical model for DTL^{++} is behavioural, that is, that it satisfies a) and b) of definition 10. We will only prove that it satisfies a) here.

Consider states $\sigma_1, \sigma_2, \sigma_1', \sigma_2', \sigma_1'', \sigma_2'' \in \Sigma_{DTL}$ such that

1. $\sigma_1 \xrightarrow{a} (\sigma_1' \rightarrow \sigma_1'')$
2. $\sigma_2 \xrightarrow{a} (\sigma_2' \rightarrow \sigma_2'')$
3. $\sigma_1 \underset{L_a}{\equiv} \sigma_2$

So, we want to prove that $\sigma_1'' \underset{D_a}{\equiv} \sigma_2''$, that is, $\mathcal{V}_{DTL}(\sigma_1'') \cap D_a = \mathcal{V}_{DTL}(\sigma_2'') \cap D_a$. Because σ_1'', σ_2'' are maximal DTL-consistent, then we have $\sigma_1'' \models_{I_{DTL}} A$ iff $A \in \sigma_1''$ and $\sigma_2'' \models_{I_{DTL}} A$ iff $A \in \sigma_2''$. By definition, $\sigma \models_{I_{DTL}} p$ iff $p \in \mathcal{V}_{DTL}(\sigma)$. So, we want to prove $\sigma_1'' \cap D_a = \sigma_2'' \cap D_a$.

Now let,

4. $P \in \sigma_1''$ where P is p or $\neg p$ for $p \in D_a$

We want to prove $P \in \sigma_2''$.

5. for every $B \notin \sigma_2' \{A \mid (B\,[g]\,A) \in \sigma_2\} \subseteq \sigma_2''$ — From 2 and Def. 7
6. for every $B \notin \sigma_1' \{(B\,\langle g\rangle\,A) \mid A \in \sigma_1''\} \subseteq \sigma_1$ — From 1 and Prop. 2
7. $(\mathrm{ff}\,\langle g\rangle\,P) \in \sigma_1$ — From 4 and 6
8. $\sigma_1 \models_{I_{DTL}} \mathrm{ff}\,\langle g\rangle\,P$ — From 7 and maximal DTL-consistency of σ_1
9. $\sigma_1 \vdash \mathrm{ff}\,\langle g\rangle\,P$ — By strong completeness
10. $\sigma_2 \models \sigma_2$ and $\sigma_1 \models \sigma_1$
11. $\sigma_2 \underset{L_a}{\supset} \sigma_1$ — From 3, 10 and Def. 9
12. $\sigma_2 \vdash \mathrm{ff}\,[g]\,P$ — From 9, 11 and rule Glo
13. $\sigma_2 \models_{I_{DTL}} \mathrm{ff}\,[g]\,P$ — By strong soundness
14. $\sigma_2'' \models_{I_{DTL}} P$ — From 2 and 13
15. $P \in \sigma_2''$ — Because σ_2'' is maximal DTL-consistent

4.2. Properties of Behavioural Specifications

A behavioural specification can be identified with a collection of DTL-sentences for a choice of propositional symbols (program "variables" or actions). Indeed, we have seen in section 2 how the new modalities allow us to express properties of the three states that characterise action execution – launching, accepting and resulting states.

We are now interested in typical properties of system behaviour in the new paradigm.

DEFINITION 12. *Given a set Γ of sentences corresponding to a behavioural specification, we denote by $\vdash_\Gamma$ the extension of* DTL^{++} *with the sentences of Γ as axioms.*

DEFINITION 13. *Given a behavioural specification Γ, a sentence A expresses a stability property of the system iff*

$$\vdash_\Gamma \bigwedge_{a\in\mathcal{A}} (\neg A\,[a]\,A)$$

That is to say, all actions accepted in A states, preserve A.

Stability can be seen, as in Unity(Chandy and Misra, 1988), as a particular case of *unless* properties:

DEFINITION 14. *Given a behavioural specification Γ and sentences A and B, the system satisfies the property A* unless *B iff*

$$\vdash_\Gamma \bigwedge_{a\in\mathcal{A}} ((A \supset B)\,[a]\,(A \vee B))$$

That is to say, all actions accepted in states that satisfy A but not B, either preserve A or establish B.

Notice that, because action execution is not atomic, relating the launching state with the resulting state as in Unity, would not work. That is, defining $A \,\mathrm{unless}\, B$ through

$$\bigwedge_{a \in \mathcal{A}} ((A \supset B) \overline{[a]} (A \vee B))$$

would not ensure that, once A is established, either B never holds and A continues to hold forever, or B holds eventually and A continues to hold at least until B holds. This has to do with the possible existence of pending actions launched in states not satisfying A and not B. These pending actions could establish resulting states where neither A nor B hold. If, as proposed, we ensure that all actions accepted when A is true and B is false either preserve A or establish B, then, because the "transition" from the accepting to the resulting state is atomic, we have the desired effect.

EXAMPLE: A typical safety property expressed through unless properties is, for the dining philosophers presented in section 1, the property $\neg PL \,\mathrm{unless}\, H$, that is, a philosopher cannot possess his left fork unless he is hungry:

We have to prove that $\bigwedge_{a \in \mathcal{A}} ((\neg PL \supset H) [a] (\neg PL \vee H))$ is a DTL consequence of the behavioural specification Γ. We will only prove for the rel_L and take_R actions here.

Assume that, from the behavioural specification Γ, we can infer

1. $H \wedge PL \wedge \neg PR \wedge \neg ROT \equiv \mathrm{ff} \langle rel\_L \rangle \,\mathrm{tt}$ and
2. $H \wedge PL \wedge \neg PR \wedge \neg ROT \vdash_\Gamma \mathrm{ff} \langle rel\_L > \rangle \neg PL$

So,

3. $H \wedge PL \wedge \neg PR \wedge \neg ROT \vdash_\Gamma \mathrm{ff} [rel\_L] \neg PL$
   from 2 and rule GLo ($PL \in D_{rel\_L}$)
4. $H \wedge PL \wedge \neg PR \wedge \neg ROT \supset \mathrm{ff} [rel\_L] \neg PL$ — From 3
5. $\mathrm{ff} \langle rel\_L \rangle \,\mathrm{tt} \supset \mathrm{ff} [rel\_L] \neg PL$ — From 1, 4
6. $\neg(\mathrm{ff} \langle rel\_L \rangle \,\mathrm{tt}) \vee \mathrm{ff} [rel\_L] \neg PL$ — From 5
7. $\mathrm{ff} [rel\_L] \mathrm{ff} \vee \mathrm{ff} [rel\_L] \neg PL$ — From 6
8. $(\mathrm{ff} [rel\_L] \mathrm{ff} \vee \mathrm{ff} [rel\_L] \neg PL) \supset \mathrm{ff} [rel\_L] (\mathrm{ff} \vee \neg PL)$ — Theorem
9. $\mathrm{ff} [rel\_L] (\mathrm{ff} \vee \neg PL)$ — From 7, 8, MP
10. $\mathrm{ff} [rel\_L] \neg PL$ — From 9
11. $(\mathrm{ff} \supset (\neg PL \supset H)) [rel\_L] \neg PL$ — GNecL
12. $((\mathrm{ff} \supset (\neg PL \supset H)) [rel\_L] \neg PL) \supset ((\mathrm{ff} [rel\_L] \neg PL) \supset (\neg PL \supset H) [rel\_L] \neg PL)$ — GKL
13. $((\mathrm{ff} [rel\_L] \neg PL) \supset (\neg PL \supset H) [rel\_L] \neg PL)$ — 11, 12, MP
14. $(\neg PL \supset H) [rel\_L] \neg PL$ — 10, 13, MP
15. $(\neg PL \supset H) [rel\_L] (\neg PL \vee H)$ — From 14

In what concerns action take_R:

1.	$\neg\neg PL\,[take\_R]\,\neg PL$	Gsy $(PL \notin D_{take\_R})$
2.	$(\neg\neg PL \vee H)\,[take\_R]\,\neg PL$	From 1
3.	$(\neg PL \supset H)\,[take\_R]\,\neg PL$	From 2
4.	$(\neg PL \supset H)\,[take\_R]\,(\neg PL \vee H)$	From 3

In (Nunes et al., 1996) we present behavioural specifications where the locality conditions are expressed by rules that allow us to infer DTL-formulas from formulas written in the language of *local* logics. These local formulas express local behaviour of actions, therefore allowing us to separate computation from coordination concerns.

Stability guarantees that a system never visits states outside a certain set (captured by the sentence). An extreme case is the one in which a system never leaves a certain state. Such states are called *stationary*. We can classify stationary states in two classes:

Alive stationary – states where no action is accepted.

Dead stationary – states where no action is either accepted or desired.

An *Alive* stationary state is so called because actions can continue to be launched although they cannot be accepted. In terms of program termination we cannot say that a program terminates when it reaches such a state, because it still "lives", but we can guarantee that, for all purposes, it will not evolve because no action can be accepted to take it to another state. On the other hand, a *Dead* stationary state corresponds to complete termination of activity.

DEFINITION 15. *The following predicates capture the notions of stationarity just discussed:*

$$Alive \equiv \bigwedge_{a\in\mathcal{A}}(\neg(\mathrm{ff}\,\overline{\langle a\rangle}\,\mathrm{tt}))$$
$$Dead \equiv \bigwedge_{a\in\mathcal{A}}(\neg(\mathrm{ff}\,\overline{\langle a\rangle}\,\mathrm{tt} \wedge \neg(\mathrm{ff}\,\langle a\rangle\,\mathrm{tt})))$$

Notice the use, as discussed in section 2, of $\mathrm{ff}\,\overline{\langle a\rangle}\,\mathrm{tt}$ to identify accepting states, and of $\mathrm{ff}\,\langle a\rangle\,\mathrm{tt}$ to identify launching states of action a.

When we think about the classical notion of deadlock, we see that *Alive* stationary states can correspond to deadlock situations: actions that are desired (and, hence, launched) cannot be accepted. As an example, the rel_R action specification presented in section 1, together with its left version, rel_L, which are included in the specification given in (Turski, 1990) for the "dining philosophers" in order to prevent deadlock, cause the predicate *Alive* to be false, that is, the system never deadlocks because in all states there is always an action that can be accepted.

Another class of properties is concerned with the possibility of reaching certain states.

DEFINITION 16. *Given a behaviour specification* Γ *and sentences* A *and* B, *the system satisfies the property* A *reachable from* B *iff*

$$\vdash_\Gamma B \supset \bigvee_{a \in \mathcal{A}} (\mathrm{ff}\,\langle a \rangle\, A \vee \mathrm{ff}\,\overline{\langle a \rangle}\, A))$$

That is to say, A is reachable from B if any state satisfying B is either a launching state of some action which causes an A-resulting state, or the accepting state of some action that causes an A-resulting state.

Reachability can be used to express properties similar to *ensures* and *leads-to* properties of Unity(Chandy and Misra, 1988).

5. Concluding remarks

We presented a modal logic supporting the behavioural paradigm introduced in (Turski, 1990) in which actions execute atomically but are durative, i.e. the global state of the system can change while they are executing. The proposed logic is based on an extended notion of transition system in which action execution is characterised not by two but three states – the state in which actions are launched, the state in which they are considered for acceptance, and the resulting state. The corresponding modal operator is, accordingly, dyadic. An axiomatisation was proposed for the new class of operators. This axiomatisation was proved to be complete by adapting from the usual proofs for unary operators based on canonical models. Finally, the logic was applied to the behavioural paradigm by restricting the class of models to those that satisfy locality of execution and a frame condition on attributes outside the discharging list of actions. The expressive power of the logic was illustrated by showing how properties like stability, stationarity and reachability, in the context of a given behavioural specification, can be formalised.

Other developments of this formalism exist that, because of space limitations, could not be presented, namely bisimilarity notions for durative transition systems and corresponding equivalences between specifications. More details on the semantics of the behavioural paradigm can be found in (Nunes et al., 1996; Nunes et al., 1997). Methodological aspects of the paradigm and worked examples can also be found in (Turski, 1992; Turski, 1990).

As future work, we intend to study the relationship between DTL and interval-based temporal logics, e.g. (Venema, 1991; Kutty et al.,

1994) (where we also find a completeness proof for an axiom system that involves operators based on ternary relations) as well as the real-time aspects that can be associated with durative actions (Chaochen et al., 1991).

References

Chandy, K. and J. Misra: 1988, *Parallel Program Design.* Addison-Wesley.

Chaochen, Z., C. Hoare, and A. Ravn: 1991, 'A Calculus of Durations'. *Information Processing Letters* **40**, 269–276.

Kutty, G., L. Moser, P. Melliar-Smith, L. Dillon, and Y. Ramakrishna: 1994, 'First-Order Future Interval Logic'. In: D. Gabbay and H. Ohlbach (eds.): *Temporal Logic,*, No. 827 in LNAI. Springer-Verlag, pp. 195–209.

Manna, Z. and A. Pnueli: 1991, *The Temporal Logic of Reactive and Concurrent Systems.* Springer-Verlag.

Nunes, I., J. Fiadeiro, and W. Turski: 1996, 'Semantics of Behavioural Specifications'. Research report, Department of Informatics, Faculty of Sciences, University of Lisbon.

Nunes, I., J. Fiadeiro, and W. Turski: 1997, 'Coordinating Durative Actions'. In: D. Garlan and D. L. Métayer (eds.): *Coordination Languages and Models*, Vol. 1282 of *LNCS.* pp. 115–130.

Stirling, C.: 1992, 'Modal and Temporal Logics'. In: S. Abramsky, D. Gabbay, and T. Maibaum (eds.): *Handbook of Logic in Computer Science*, Vol. 2. Oxford: Clarendon Press, pp. 477–563.

Turski, W.: 1990, 'On Specification of Multiprocessor Computing'. *Acta Informatica* **27**, 685–696.

Turski, W.: 1992, 'Extending the Computing Paradigm'. *Structured Programming* **13**, 1–9.

Venema, Y.: 1991, 'A Modal Logic for Chopping Intervals'. *Journal of Logic and Computation* **1**(4), 453–476.

Appendix

LEMMA 2. $\forall \sigma'' \in \Sigma_{DTL}.$

$$(\{C \mid (A\,[a]\,C) \in \sigma\} \subseteq \sigma'' \Rightarrow \\ \exists \sigma' \in \Sigma_{DTL}.(\sigma \xrightarrow{a} (\sigma' \to \sigma'') \wedge A \notin \sigma')).$$

Proof:
By definition of canonical model we know that:
$\sigma \xrightarrow{a} (\sigma' \to \sigma'')$ iff $\forall B \notin \sigma'.\{D \mid (B\,[a]\,D) \in \sigma\} \subseteq \sigma''$
So, we have to show that, for every σ'',

$$\forall C.((A\,[a]\,C) \in \sigma \Rightarrow C \in \sigma'') \Rightarrow$$
$$\exists \sigma'.(\forall D.\forall B \notin \sigma'.((B\,[a]\,D) \in \sigma \Rightarrow D \in \sigma'') \wedge A \notin \sigma')$$

Assume given an arbitrary σ'' for which $\forall C.((A\,[a]\,C) \in \sigma \Rightarrow C \in \sigma'')$ holds.
Let $B_0, B_1, \ldots, B_n, \ldots$ be an enumeration of $\mathcal{L}(\mathcal{P}, \mathcal{A})$. Let

$$\sigma'_0 = \{\neg A\}$$
$$\sigma'_{n+1} = \begin{cases} \sigma'_n \cup \{B_n\} & \text{if } \sigma'_n \vdash B_n \\ \sigma'_n \cup \{\neg B_n\} & \text{otherwise} \end{cases}$$
$$\sigma' = \bigcup_{n \geq 0} \sigma'_n$$

We know that σ' is a maximal DTL-consistent set. Moreover, $A \notin \sigma'$ because $\neg A \in \sigma'$. We still have to prove that, for this specific σ',

$$\forall D.\forall B \notin \sigma'.((B\,[a]\,D) \in \sigma \Rightarrow D \in \sigma'')$$

Assume that, for every $\sigma' i$,

$$\forall D.((((\bigvee_{B \in \sigma'_i} \neg B)\,[a]\,D) \in \sigma) \Rightarrow D \in \sigma'')(*)$$

We then have,

$$\forall D.((\bigvee_{B \in \sigma'_i} (\neg B\,[a]\,D) \in \sigma) \Rightarrow D \in \sigma'')$$

and also,

$$\forall D.(\bigvee_{B \in \sigma'_i} ((\neg B\,[a]\,D) \in \sigma) \Rightarrow D \in \sigma'')$$

that is,

$$\forall D.(\bigwedge_{B \in \sigma'_i} ((\neg B\,[a]\,D) \in \sigma \Rightarrow D \in \sigma''))$$

So, for all formula D and for all $\sigma' i$ we have, for every $B \in \sigma' i$, $((\neg B\,[a]\,D) \in \sigma \Rightarrow D \in \sigma'')$
We know that σ' is maximal DTL-consistent. So, by definition, for all formula X, either $X \in \sigma'$ or $\neg X \in \sigma'$. Hence we have that:

$$\forall B \notin \sigma'.((B\,[a]\,D) \in \sigma \Rightarrow D \in \sigma'')$$

That is, a maximal DTL-consistent set σ' exists such that

$$\forall D.\forall B \notin \sigma'.((B\,[a]\,D) \in \sigma \Rightarrow D \in \sigma'') \wedge A \notin \sigma'$$

It remains to prove (*), i.e. for every σ'_i,

$$\forall D.((((\bigvee_{B\in\sigma'_i} \neg B)\,[a]\,D) \in \sigma) \Rightarrow D \in \sigma'')$$

The proof is done by induction.
– induction base:
$\sigma'_0 = \{\neg A\}$ so (*) reduces to $\forall C((A\,[a]\,C) \in \sigma \Rightarrow C \in \sigma'')$ which was assumed to be true.
– induction step: assume that (*) holds for $\sigma' i$;
Let us abbreviate $(\bigvee_{B\in\sigma' i} \neg B)$ by $\neg\sigma'_i$ instead. Because $\neg\sigma'_{i+1} \Leftrightarrow \neg\sigma'_i \vee \neg B'_i$, where B'_i is B_i or $\neg B_i$, we have that $\forall D.(((\neg\sigma'_{i+1})\,[a]\,D) \in \sigma \Rightarrow D \in \sigma'')$ is equivalent to $\forall D.(((\neg\sigma'_i \vee \neg B'_i)\,[a]\,D) \in \sigma \Rightarrow D \in \sigma'')$. So, this is what we are going to prove now.
We distinguish three cases:

i) if $\sigma'_i \vdash B'_i$ then $\neg\sigma'_{i+1} \Leftrightarrow \neg\sigma'_i \vee \neg B'_i \Leftrightarrow \neg\sigma'_i$
The result then follows from the induction hypothesis.

ii) if $\sigma'_i \nvdash B'_i$ and $\sigma'_i \vdash \neg B'_i$ the reasoning is similar to i).

iii) in the case where ($\sigma'_i \nvdash B'_i$ and $\sigma'_i \nvdash \neg B'_i$) we show that, for our purposes, adding B'_i to σ'_i or adding $\neg B'_i$ is equally correct. If not, suppose we have:

$$((\neg\sigma'_i \vee \neg B'_i))\,[a]\,D_1) \in \sigma \text{ and } D_1 \notin \sigma'' \text{ and}$$
$$((\neg\sigma'_i \vee B'_i)\,[a]\,D_2) \in \sigma \text{ and } D_2 \notin \sigma''$$

Because σ'' is maximal DTL-consistent, if $D_1 \notin \sigma''$ and $D_2 \notin \sigma''$ then $\neg D1 \in \sigma''$ and $\neg D2 \in \sigma''$, and so $(\neg D1 \wedge \neg D2) \in \sigma''$, hence $(D_1 \vee D_2) \notin \sigma''$. Moreover, $((\neg\sigma'_i \vee \neg B'_i)\,[a]\,(D_1 \vee D_2)) \in \sigma$ and $((\neg\sigma'_i \vee B'_i)\,[a]\,(D_1 \vee D_2)) \in \sigma$ and $(D_1 \vee D_2) \notin \sigma''$ which implies $(((\neg\sigma'_i \vee \neg B'_i) \wedge (\neg\sigma'_i \vee B'_i))\,[a]\,(D_1 \vee D_2)) \in \sigma$ and $(D_1 \vee D_2) \notin \sigma''$, that is, $\neg\sigma'_i\,[a]\,(D_1 \vee D_2)) \in \sigma$ and $(D_1 \vee D_2) \notin \sigma''$.

By the induction hypothesis, we reach a contradiction.

About Real Time, Calendar Systems and Temporal Notions

Hans Jürgen Ohlbach (h.ohlbach@doc.ic.ac.uk) *
Department of Computing, Imperial College of Science, Technology and Medicine, 180 Queen's Gate, London SW7 2BZ

Abstract. A specification language is presented for describing every day temporal notions, like weekends, holidays, office hours etc. The time model underlying this language exploits in a particular way the well investigated algorithms for calendar systems currently in use. All peculiarities and irregularities of the calendar systems, like time zones, leap years, daylight savings time are respected. The specification language and the underlying temporal model allows us to convert between arbitrary calendar systems and to decide a number of questions, for example whether a certain point in time is within the time intervals specified by a term in this language. Besides neglecting the phenomena of relativity theory, there are no idealizing assumptions. The language is the basis for a temporal logic with quantifiers over real time temporal notions.

1. Introduction

It would not be the first time that I tried to phone a colleague in another country, maybe on another continent, I look up into my database, get his or her office-phone number, dial it, and get no answer. Maybe I forgot the time shift between different time zones, and it is actually in the middle of the night over there, or in this country there is just a public holiday, or it is lunch time, or they moved to daylight savings time, and everything is shifted by one hour compared to last week, when I phoned him or her at the same time of the day. A clever database would know about all this, and if I asked for the phone number for Mr. X, it would give me the actually valid number, or at least warn me that it is not very likely to get somebody at the phone at this time of the day or the year.

One can implement of course a notion like 'office-hour' with a special algorithm that takes into account all the phenomena of real calendar systems (c.f. Dershowitz and Reingold's excellent book on calendrical calculations (Dershowitz and Reingold, 1997)). Many commercial software products offer some kind of calendar manipulations, but they are in general limited to one calendar system and don't convert between different systems. Much more convenient and flexible, however, would be a simple abstract specification language for these kind of temporal

* This work was supported by EPSRC Research Grant GR/K57282.

H. Barringer et al. (eds.), Advances in Temporal Logic, 319–338.

notions, which reduces the actual computation tasks to some standard algorithms. Such a specification language must be based on a mathematical model of calendar systems with all their specialties.

Let's start with the western calendar system, the Gregorian calendar. We have years, months, weeks, days, hours, minutes, seconds, milliseconds etc. Although at first glance, the system looks quite simple, there are a lot of phenomena which make a formal model difficult. First of all, there are different time zones. A year in Europe is not the same as for example a year in America. Not all years have the same length. In leap years a year is a day longer. Not all days have the same length. In many countries one day in the spring is an hour shorter and one day in the autumn is an hour longer than usual. Even not all minutes might have the same length if some 'time authorities' decide to insert some seconds into a minute to re-calibrate their reference clocks, for example to compensate for earth's decreasing angular velocity.

Some time measures are exactly in phase with each other. For example a new year starts exactly at the same time point when a new month, day, hour, minute, second starts. But it does not start at the same time point as a new week. Nevertheless, weeks are quite often counted as week 1 in a year, week 2 etc. Many businesses allocate their tasks to particular week numbers in a year. If you go into a furniture shop and buy a new sofa, they might well tell you that the sofa will be delivered in week 25, and you have to look into your calendar to find out, when the hell, is week 25.

Weeks are not in phase with months and years, but they are in phase with hours and minutes and seconds, which themselves are in phase with months and years. Quite confusing, isn't it? Things would be much easier if calendar systems would be decimal systems, with one fixed unit of time as basis, and all other units as fractions or multiples of this basis. Since this is not the case, we need to model the time units in a different way. Therefore in the next section we begin with the definition of time units as independent partitionings of a universal time axis. Based on this, a specification language for temporal notions denoting sets of time points will be presented. Its semantics is such that we obtain algorithms for deciding whether a given point in time is within the set of time points denoted by a given time term or not (for example whether it is currently an office hour, say, in Recife, Brazil).

2. Reference Time and Time Units

There is no simple way of choosing a popular unit such as a year as the basis for a mathematical model, and define everything else relative to this unit. Years have different length, and they depend on the time zone. Fortunately there is a well established time standard, which is available on most computer systems and in most programming languages. The origin of time in this standard is 1.1.1970, Greenwich Mean Time (GMT). There are functions, which give you the number of seconds (or even milliseconds) elapsed from the origin of time until right now. Other functions can map this number back to a common date format (year, month, day, hour, minute, second), either in GMT or in local time.

In our reference time standard we therefore assume a linear time axis and a time structure which is isomorphic to the set of real numbers. In a particular implementation we are free to move point 0 in this time axis to some point in history, 1.1.1970 in Unix systems, or 1.1.0 relative to the Gregorian calendar, or even at the time of the Big Bang. This is only a linear transposition by a fixed number.

Definition 2.1. (Reference Time Line) Let $\mathcal{T}$ be an isomorphic copy of the set of real numbers. $\mathcal{T}$ is the reference time line. The time point 0 is called the *reference origin of time.* ◁

Unfortunately it is not the case that the common time units, minutes, hours, days, weeks, months, years can be defined as multiples of a smallest unit. Nevertheless, all of them define a partitioning of the time axis into sequences of time intervals. This means, we can count for example years as 'year 0 after the origin, year 1 after origin' etc. The only difficulty is that different years may have different length (measured in the reference system). To specify a time unit U, for example 'GMTyear' as a sequence of intervals of different length, we need three things. First of all, these time units have their own coordinate system which is isomorphic to the integers. That means we can identify each time interval measured by the given unit by a particular integer. Moreover, we know, if n is the coordinate of a given interval, then $n+1$ is the coordinate of the next interval in this sequence.

Secondly we need for each time unit U a function $U_{\mathcal{N}}$ that maps reference time points to the time unit's own coordinates. For example if the reference time axis is the GMT time measured in seconds from 1.1.1970, then for $U =$ GMTday, $U_{\mathcal{N}}$ might map all the points in the half open interval[1] $[0, 86400[$ to the GMTday coordinate 0 (if we count

[1] The notation $[b, e[$ for half open intervals denotes the set of all points t with $b \leq t < e$. We could have chosen half open intervals $]b, e]$. This does not matter. We can't choose closed intervals because in this case subsequent intervals are not disjoint.

the 1.1.1970 as day 0). In another time zone Z with, say, 1 hour difference to GMT (earlier), we would then have a unit $U' = \text{Zday}$ and a mapping $U'_{\mathcal{N}}$ which maps the interval $[-3600, 82800[$ to day 0.

Finally we need for each unit U a mapping $U_{[[}$ which maps the unit's own coordinates back to the reference time axis. $U_{[[}(n)$ actually computes the beginning and the end of U's time interval n. For $U =$ GMTday, we would for example have $U_{[[}(0) = [0, 86400[$ and for U' we would have $U'_{[[}(0) = [-3600, 82800[$.

It is important to notice, that at this stage of our model, we assume a separate definition of $U_{[[}$ and $U_{\mathcal{N}}$ for each time unit we are interested in. For the standard time units like years, months, days, hours, minutes, seconds, these functions are usually available in one form or another in a programming language, both for the GMT time zone, and for the local time zone, and, via Internet, for all other time zones. All the information about the particular calendar system, leap years, leap seconds, daylight savings time etc. must be encoded in $U_{[[}$ and $U_{\mathcal{N}}$. Therefore these algorithms are usually quite complex.

The correlations between different time units so far are indirect. From a coordinate n in U's system, we can use $U_{[[}$ to go back to the reference system, and from there with $U'_{\mathcal{N}}$ to the coordinate system of some other unit U'. But this way, we can correlate all units with each other, regardless of its calendar system, its time zone, or its granularity. For example we can map the Gregorian calendar GMT-minute with coordinate n to, say the corresponding hour in the Chinese calendar system, provided the two functions GMT-minute$_{[[}$ and Chinese-hour$_{\mathcal{N}}$ for the two time units are available.

Definition 2.2. (Time Units) We define $\mathcal{U}$ to be a set of *time unit symbols*.

If $U \in \mathcal{U}$ is a time unit symbol, we define an *interpretation* $\Im_T(U)$ to be the triple $(\mathcal{N}_U, U_{\mathcal{N}}, U_{[[})$ where

- $\mathcal{N}_U$ is a U-coloured isomorphic copy of the set $\mathbb{Z}$ of integers.[2] $\mathcal{N}_U$ is the time unit's own coordinate system.

- $U_{\mathcal{N}} : \mathcal{T} \mapsto \mathcal{N}_U$ is a function mapping the reference time axis to the U-coordinates.

- $U_{[[} : \mathcal{N}_U \mapsto Int(\mathcal{T})$ is a function mapping the elements n of the U-coordinates to the half open interval $[b, e[$ in the reference time line which corresponds to the U-coordinate n.

[2] A formal definition of $\mathcal{N}_U$ would be $\mathcal{N}_U = \{(i, U) \mid i \in \mathbb{Z}\}$ with the usual structure of integer numbers imposed on $\mathcal{N}_U$.

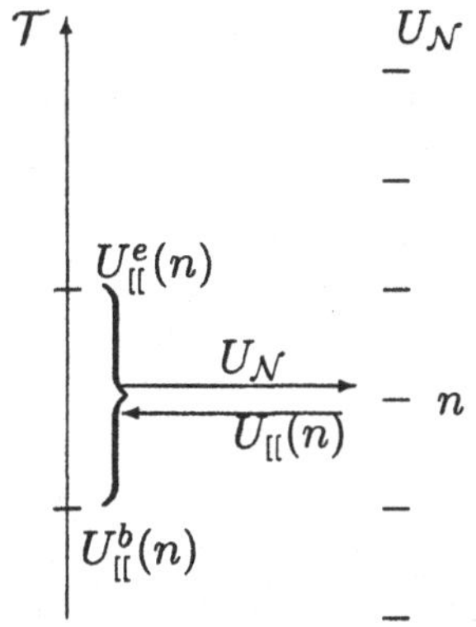

Figure 1. **Relation between the reference time system and a time unit U**

If $U_{[[}(n) = [b, e[$ we denote with $U^b_{[[}(n) = b$ the beginning of the interval and with $U^e_{[[}(n) = e$ the end of the interval.

We require:

- $U^e_{[[}(n) = U^b_{[[}(n+1)$ for all $n \in \mathcal{N}_U$
  (there are no gaps between U-coordinates);
- $\forall t \in \mathcal{T} :\ t \in U_{[[}(U_\mathcal{N}(t))$ and $\forall n \in \mathcal{N}_U :\ \forall t \in U_{[[}(n) :\ U_\mathcal{N}(t) = n$
  ($U_\mathcal{N}$ and $U_{[[}$ are inverse to each other). ◁

It is important to notice that sets of half open intervals which are open all at the same side, in our case the upper side, are closed under union and intersection.

3. Temporal Notions

With the functions $U_\mathcal{N}$ and $U_{[[}$ we can convert back and forth between a reference time system and an arbitrary time unit U. Unfortunately, time units are not the only irregular structures in a calendar where a fully fledged programming language is necessary to compute the details. Some of the Christian holidays also have a quite complex temporal cycle. The date for Easter, for example, is defined to be the first weekend after the first full moon in spring. Since in our context it makes no sense to define a language powerful enough to compute the moon cycle (this should be done in C or any other suitable programming language), we assume that certain functions, for example *easter-time* are available. *easter-time*, for example, would be a function that takes a year-coordinate and gives back a day-coordinate (Easter Sunday).

Given $U_{\mathcal{N}}$ and $U_{[[}$ and special functions like *easter-time*, we need only a few further constructs to specify quite complex temporal notions and to do the corresponding computations.

3.1. The *U_within_V* Constructor

Let us start with something very familiar, the notion of the month 'February'. 'February' denotes a set of time points. There are, however, different possibilities. If I say 'February', do I mean a particular February, or maybe even the set of all Februaries in the history of mankind, or more sophisticated: a function, that gives me for a particular year the particular February in this year? The last interpretation is the most general one, because with this we can reconstruct the other two interpretations. Moreover, it is a quite natural one, because if you ask somebody to define 'February', he would say something like 'February is the second month in a year'. What he has in mind is therefore a notion of 'i^{th} month in a year', which is a function taking a year y and an integer i and yielding a month. If $i = 2$, we get a February, it $i = 3$, we get a March etc.

Both 'month' and 'year' are time units. Therefore we can generalize the notion 'i^{th} month in a year' from the two time units 'month' and 'year' to two arbitrary time units U and V. For each pair of time units U and V we introduce a function *U_within_V* with two arguments, a V-coordinate and an integer. The value of $U\_within\_V(v, i)$ would be a U-coordinate.

Let's go back to the 'February' example. We have U = month and V = year. Stating

$$February(y) \stackrel{\text{def}}{=} \text{month\_}within\text{\_year}(y, 2) \tag{1}$$

we get a definition of a function *February* that takes a year-coordinate y and yields a month-coordinate (as the second month in a year), provided we have a suitable definition of the function month_*within*_year. In this case, we can use the *February*-function to determine whether a given point t in the temporal reference system is in a February. We compute $y = \text{year}_{\mathcal{N}}(t)$ and get the year-coordinate. $m = February(y)$ gives us the corresponding month-coordinate of the February in the year y. Now we check whether t lies in the interval $\text{month}_{[[}(m)$. If this is so, then t is in a February, if not, it is not in a February. Since all the regular and irregular things about the time units 'month' and 'year' are built into the functions $\text{year}_{\mathcal{N}}$ and $\text{month}_{[[}$, and, by the definition of *U_within_V* below, also into this function, $February(y)$ gives us a precise definition of the notion of 'February'.

Before we come to a general definition of the *U_within_V* function, lets see whether there are other useful examples of the application of this function. The date 'February, 20th' denotes a particular day in a year. Following the same ideas as above, we would like to encode this as a function that maps the coordinate of a year to the coordinate of a day, the 20th of February. We have got the function *February* which maps year-coordinates to month-coordinates. In a very similar fashion, 20th-of denotes a function which maps a month-coordinate m to a day-coordinate. Therefore we can define something like:

$$20^{th}\textit{-in-month}(m) \stackrel{\text{def}}{=} \text{day\_}\textit{within}\text{\_month}(m, 20).$$

Composing this with the *February*-function, we get:

$$20^{th}\textit{-in-February}(y) = \text{day\_}\textit{within}\text{\_month}(\textit{February}(y), 20)$$

which maps a year-coordinate y to a day-coordinate. The usual date notation year:month:day:hour:minute:second can in fact be encoded by using the *U_within_V* function for the corresponding time units, and function composition.

As another example,

$$\textit{Monday}(w) \stackrel{\text{def}}{=} \text{day\_}\textit{within}\text{\_week}(w, 1)$$

defines Monday to be the first day in a week, by mapping a week-coordinate w to a day-coordinate. Although we have not yet given a formal definition of the *U_within_V*-function, its meaning seems to be quite obvious. In all cases, where the two units U and V are in phase, i.e. the beginning of a V unit (i.e. year) coincides with the beginning of a U unit (i.e. month), it is in fact obvious.

But sometimes there is a subtlety, we have to consider. For example the beginning of a year does not coincide with the beginning of a week (years and weeks are not in phase). It is therefore not clear whether week_*within*_year should count the first week overlapping with a new year as week number 1 (inclusive interpretation), or only the first week which lies completely within the new year (exclusive interpretation). In this paper we avoid making a choice on this, by simply providing both versions: $U\_within^e\_V$ is the exclusive reading and $U\_within^i\_V$ is the inclusive reading. For all those pairs of time units which are in phase with each other, both versions coincide.

Let us consider the definition of *U_within_V* for these cases first. month_*within*_year(y, i) is an example. In order to get the month coordinate for the year y and integer i, we first need to compute with $k \stackrel{\text{def}}{=} \text{year}^b_{[[}(y)$ the reference time of the beginning of the year. Since years and months are in phase, $\text{month}_{\mathcal{N}}(k)$ yields the month-coordinate of the corresponding January in the year y. If this is counted as the first month, then we need to add $i-1$ to get the month-coordinate of the i-th month in the year y. The definition of $U\_within\_V(n,i)$ in the general case is therefore just

$$U\_within\_V(v,i) \stackrel{\text{def}}{=} U_{\mathcal{N}}(V^b_{[[}(v)) + i - 1.$$

If the two time units are not in phase, we either add $i-1$ or i, in the exclusive reading, depending whether there is an overlap at the beginning of the interval or not. These definitions compute answers even for terms like month_*within*_year(1980, 100) (the 100th month in the year 1980) or year_*within*_month(200000, 100) (the 100th year in the month 200000.) It depends on the application, whether this might be useful or not.

3.2. THE *begin_U* AND *end_U* CONSTRUCTORS

Suppose we have a built-in function *springbegin*(y) which maps a year-coordinate to a day-coordinate (giving the first day in spring). If we want to define 'the Monday in the first week in spring', we cannot use the 'Monday'-function above, because this needs as an input a week coordinate. But given a day-coordinate d, we can of course get the week coordinate w of the week containing the given day. This is just $\text{week}_{\mathcal{N}}(\text{day}^b_{[[}(d))$, i.e. from d we compute the reference time of the beginning of the day d, and from there the week-coordinate corresponding to this reference time.

In this example, we went from the finer day-coordinate to the coarser week-coordinate. We might also want to go the other way round, say from a coarser year-coordinate to the finer day-coordinate, either of the beginning of the year (which could be achieved with the *U_within_V* constructor), or to the end of the year, which could only be achieved with the *U_within_V* constructor if the number of days within a year were fixed.

To do this kind of type casting, we introduce the *begin_U* and *end_U* constructors. For a given time unit U and a given V-coordinate v of another time unit V, *begin_U*(v) yields the U-coordinate of the reference time corresponding to the *beginning* of the v-interval, and *end_U*(v) yields the U-coordinate of the reference time corresponding to the *end* of the v-interval. Both yield the same result if V is coarser than U. A useful example for the application of the *end_U* constructor is

$$\textit{new-years-eve}(y) \stackrel{\text{def}}{=} \textit{end\_}\text{day}(y).$$

This function maps a year coordinate y to the day-coordinate of its last day, new year's eve.

3.3. Set Constructors

With the constructors, we have got so far, we can select single coordinate units within bigger time units. This is still too limited for many purposes. For example, a weekend usually consists of two days, which can only be specified by joining the two days into a set. Therefore we introduce the constructors for manipulating sets of coordinates.

First of all, we introduce the familiar set constructors $\{\ldots\}, \cup, \cap, \setminus$. $\{n\}$ turns the single element n into a singleton set $\{n\}$. $\cup, \cap, \setminus$ denote union, intersection and difference of sets. Now we can define for example

$$weekend(w) \stackrel{\text{def}}{=} \{Saturday(w)\} \cup \{Sunday(w)\}$$

as a function mapping week-coordinates to the union of the two day coordinates corresponding to the Saturday and Sunday in a week. In order to check whether a given reference time t is actually in a weekend, we compute $\{d_1, d_2\} = weekend(\text{week}_{\mathcal{N}}(t))$ and check whether t is within the interval $\text{week}_{[[}(d_1)$ or $\text{week}_{[[}(d_2)$.

Some precautions must be taken with the binary set constructors in order not to mix coordinates of different time units. Although something like $weekend(w) \cup \text{hour\_}within\text{\_week}(w, 36)$ which computes the union of days and hours makes sense set theoretically, it does not make sense in our setting. Therefore in the actual language we need a type mechanism to prevent the mixing of coordinates of different time units.

3.4. The Decomposition Constructor *U_s*

Besides the singleton set constructor $\{\ldots\}$, there is another constructor, which is quite useful for turning individual coordinates into sets. This is the U_s constructor. If, for example, w is a week-coordinate, then $\text{day\_s}(w)$ denotes the set of all day-coordinates within the week w. Now, for example, we can define

$$working\text{-}days(w) \stackrel{\text{def}}{=} \text{day\_s}(w) \setminus weekend(w)$$

as a function mapping week coordinates w to the set of day coordinates corresponding to the complement of the weekend coordinates.

For time units which are not in phase with each other, for example weeks and years, we need a refined U_s constructor. For computing all the weeks within a year, we have to decide whether to include the weeks which overlap partially with the year (inclusive reading) or only the weeks which are completely included in the year (exclusive reading). Therefore here again we introduce two versions of the U_s constructor, the U^e_s (exclusive) and the U^i_s (inclusive) version. The U_s constructor is particularly useful for converting between different time zones or calendar systems. If we have for example a time unit A-day (day in time zone A) and B-hour (hour in time zone B) and a A-day coordinate d then B-hour_s$^e(d)$ yields the set of B-hours contained completely within the A-day d, and B-hour_s$^i(d)$ yields the set of B-hours overlapping with the A-day d.

3.5. Individuals and Sets

Each function ϕ producing a set of U-coordinates for a time unit U can become in a canonical way an argument of a function $\psi(n)$ accepting a single U-coordinate. We just define $\psi(\phi) \stackrel{\text{def}}{=} \{\psi(n) \mid n \in \phi\}$ For example

$$\textit{all-Mondays}(y) \stackrel{\text{def}}{=} \textit{Monday}(\text{week\_s}(y))$$

denotes the set of all Mondays in the year y. week_s(y) yields the set of all week coordinates for the year y, and the definition of *Monday* is applied to all of them separately. The resulting day coordinates are collected in a set.

We finish this section with a more complicated example. We want a realistic definition of the notion of 'office-hour'. There are different ways to define this. We might start with the definition of lunch-time per day:

$$\textit{lunch-time}(d) \stackrel{\text{def}}{=} \text{hour\_}\textit{within}\text{\_day}(d, 12)$$

(Lunch time is between 12 and 1 o'clock.) Next we define *ohd*(d) (office-hours per day):

$$\textit{ohd}(d) \stackrel{\text{def}}{=} \text{hour\_}\textit{within}\text{\_day}(d, [9..17]) \setminus \textit{lunch-time}(d)$$

(hour_*within*_day(d, [9..17])). abbreviates $\{\text{hour\_}\textit{within}\text{\_day}(d, 9)\} \cup \ldots \cup \{\text{hour\_}\textit{within}\text{\_day}(d, 17)\}$. In the next step we define office-hours per week *ohw*(w) and take the weekends out.

$$\textit{ohw}(w) \stackrel{\text{def}}{=} \textit{ohd}(\text{day\_s}(w)) \setminus \text{hour\_s}(\textit{weekend}(w))$$

day_s(w) gives us all day-coordinates of the week w. *ohd*(days_s(w)) is applied to each day-coordinate and the union of the resulting sets is formed. *weekend*(w) yields the two day-coordinates for Saturday and Sunday. The corresponding hours are hour_s(*weekend*(w)), and these are subtracted from the office hours. Let's further assume we have a definition of *holidays*(y) which computes the set of all holidays in the year y. Now we can finally define office-hours per year

$$\textit{ohy}(y) \stackrel{\text{def}}{=} \textit{ohw}(\text{week\_s}(y)) \setminus \text{hour\_s}(\textit{holiday}(y)).$$

In an actual implementation, one would not of course compute these coordinates as sets, but maybe represent them as sets of intervals, and do the corresponding set operations on the intervals.

3.6. Periods of Time

Time units are not only coordinates for fixed intervals in the reference time line. They are also used for specifying other periods of time. If somebody says for example 'for the next two years ...', he usually means 'for a period of time beginning now and lasting for two years'. Unfortunately notions like 'for two years' are not as precise as they seem to be. If next year is a leap year, does 'for two years' mean 365 + 365 days or 366 + 365 days? When does this period start and end? Does it start exactly at this moment and end at some point at some day in two years time, or does it include today as a whole day and the last day after two years time as a whole day?

I propose a flexible encoding of time periods where the user can choose the precise meaning. The constructor is *U_period_V*. For example day_*period*_year$(d, 2)$ denotes the set of day coordinates from the day d up to a day after 2 years time. This includes the begin and the end day. If the 2 years time period are to be located more precisely, say, at a resolution of hours, one can choose hour_*period*_year$(h, 2)$ which yields the set of hour coordinates from hour h up to an hour after 2 years time. (One can of course also define an exclusive reading where the coordinates at the beginning and at the end are not included.)

The problem that a period of 2 years time itself is also not precisely defined, can be solved, or at least be approximated in the following way: Consider again day_*period*_year$(d, 2)$. Day d defines an interval $[d^b, d^e[$ in the reference time line. We compute the middle $t_d = (d^b + d^e)/2$ of this interval, and from this the current year coordinate y. y itself corresponds to an interval $[y^b, y^e[$, and $t_d \in [y^b, y^e[$. t_d therefore divides $[y^b, y^e[$ into a fraction $f = (t_d - y_b)/(y^e - y^b)$. f is the fraction of the year y elapsed since y_b. Now we go 2 years further to $y' = y + 2$ and get the time point $t'_d = y'^b + f \cdot (y'^e - y'^b)$ which corresponds to the same fraction f. From t'_d we get the day coordinate of the corresponding day in two years time.

3.7. Arithmetic

The results of all these functions, we defined so far, are either single coordinates or sets of coordinates, and a coordinate is nothing but an integer. Therefore nothing stops us from embedding the time language into an integer arithmetic language. This way, we can for example define

$$tomorrow(d) \stackrel{\text{def}}{=} d + 1$$

as a function that returns for a given day coordinate d the coordinate of the following day. Even conditional terms with arithmetic comparisons are possible. Borrowing the $c?a : b$ construct from the programming language C (if c holds then a, otherwise b) we can for example define the 'first Monday after the beginning of springtime' by

$$\begin{aligned}&Monday(begin\_\mathrm{week}(springbegin(y))) < springbegin(y)\ ?\\ &\quad Monday(begin\_\mathrm{week}(springbegin(y)) + 1)\ :\\ &\quad Monday(begin\_\mathrm{week}(springbegin(y)))\end{aligned}$$

(if the beginning of springtime lies within a week such that the Monday is actually before the beginning of springtime, then we choose the Monday of the next week, otherwise we take this Monday.)

4. The Time Term Language

We now turn these informal ideas into a formal specification language. Its semantics is purely functional and can be defined in terms of a kind of interpreter which can actually execute the computations indicated above. A specification S of temporal notions is a list of equations $f(u_1, \ldots, u_n) \stackrel{\mathrm{def}}{=} \varphi[u_1, \ldots, u_n]$ where on the left hand side a new function symbol f is introduced, and defined by φ. S should be such that all defined function symbols can be eliminated in a finite sequence of rewrite steps.

In order not to mix up coordinates of different time units (all of them are essentially integers), we need a sorted (typed) language where for each function the sort of the arguments and the sort of the resulting values are specified (Schmidt-Schauß, 1989). The sort structure consists of two parts. The 'single value part' contains the sort symbol $\mathbb{Z}$ (for the integers) and the time unit symbols $\mathcal{U}$ used as further sort symbols denoting the time units' own coordinates. In the 'set value part' we have for each time unit sort symbol $U \in \mathcal{U}$ a symbol U^* denoting sets of U-coordinates. In addition there is a sort *Bool* for arithmetic comparisons.

Most of the function symbols can have 'overloaded' sort declarations where the sort of the function value depends on the sort of its arguments.

Definition 4.1. (The Time Term Language)
We define a first-order (many-sorted) term language $\mathcal{L}$ as follows:

- The set of (element) sort symbols consists of the (finite) set $\mathcal{U}$, the special symbol $\mathbb{Z}$ for integers, and the sort *Bool* for Boolean values. In addition we have for each $U \in \mathcal{U}$ a (set) sort symbol U^* (denoting sets of U-elements).

– For each $U \in \mathcal{U}$, each $u \in \mathcal{N}_U$ is a *constant symbol* of sort U. Each integer is a constant of sort $\mathbb{Z}$[3].

– For each sort $U \in \mathcal{U}$ we have an unlimited supply of variable symbols of that sort (there are no 'set variables', no integer variables and no Boolean variables).

– We can have an arbitrary number of built-in functions of sort $U_1 \times \ldots \times U_n \rightarrow U$. (*eastertime*(*year*) would be a typical built-in function.)[4]

– Depending on the set $\mathcal{U}$ of time units, there is a certain set of derived function symbols.

- For $U, V \in \mathcal{U}$ we have function symbols $U\_within^e\_V$ and $U\_within^i\_V$, both of sort $V \times \mathbb{Z} \rightarrow U$ and $V^* \times \mathbb{Z} \rightarrow U^*$ (a single V-coordinate as input yields a single U-coordinate as output, whereas a set of V–coordinates yields a set of U-coordinates as output.)
- For each $U \in \mathcal{U}$ we have the 'begin' and 'end' constructors *begin_U* and *end_U*, both of sort $V \rightarrow U$ and $V^* \rightarrow U^*$ for all sorts $V \in \mathcal{U}$.
- For each $U \in \mathcal{U}$ we have the set constructor U_s of sort $V \rightarrow U^*$ and $V^* \rightarrow U^*$ for all sorts $V \in \mathcal{U}$.
- For $U, V \in \mathcal{U}$ we have function symbols *U_period_V* and both of sort $U \times V^*$ and $U^* \times V^*$.
- The 'singleton set' operator $\{\ldots\}$ of sort $U \rightarrow U^*$ for each sort $U \in \mathcal{U}$ maps single elements to singleton sets.
- The set connectives $\cap, \cup, \setminus$, all of sort $U^* \times U^* \rightarrow U^*$, for all $U \in \mathcal{U}$ denote the usual set operations intersection, union and set difference.
- For each time unit $U \in \mathcal{U}$ there are two 'decomposition function' symbols U^e_s and U^i_s with sort declarations $V \rightarrow U^*$ for each sort (time unit) $V \in \mathcal{U}$.

– We include the standard integer arithmetic symbols $+, -, *$, all of sort $U \times U \rightarrow U$, $U^* \times U \rightarrow U$ and $U \times U^* \rightarrow U$ for all $U \in \mathcal{U}$.

– There is the special Boolean conditional operator $\ldots ? \ldots : \ldots$ of sort $Bool \times U \times U \rightarrow U$ for all $U \in \mathcal{U}$.

[3] With these constants we have a name for each coordinate. For example the constant 1997 of sort 'year' (actually the pair (1997,year)) denotes the year 1997.

[4] In the sort declarations $U_1 \times \ldots \times U_n \rightarrow U$ for function symbols, the U_k are the *domain-sorts*, and the U is the *range-sort*.

– The standard integer comparison operators $=, <, >, \leq, \geq$ of sort $U \times U \to Bool$ for all $U \in \mathcal{U}$ can be used in the conditional part of the conditional operator.

– Other function symbols can only be used as abbreviations if they have suitable non-recursive definitions of the form $f(u_1, \ldots, u_n) \stackrel{\text{def}}{=} \psi[u_1, \ldots, u_n]$.[5] This way one can always get rid of these function symbols by using their defining equation as rewrite rule (this is actually the main requirement on these definitions). ◁

The sort declarations for the function symbols guarantee that they don't compute coordinates of mixed time units (sorts).

Proposition 4.2. (Unique Range-Sort) Since the terms of our language $\mathcal{L}$ are standard first-order terms with overloaded sort declarations, given the sorts of the free variables in the term they always have a unique range-sort. ◁

We define the semantics of $\mathcal{L}$ terms by giving the definitions of all the built-in functions. From a logical point of view, the semantics of the language $\mathcal{L}$ is based on a single model only.

Definition 4.3. (Semantics of the Time Term Language) Given the interpretation $\Im_T$ for the time unit symbols $\mathcal{U}$ (Def. 2.2) we define an interpretation $\Im = (\Im_T, \mathcal{S}, \mathcal{V}, \mathcal{F})$ for the time term language $\mathcal{L}$:

– $\mathcal{S}$ maps (element) sort symbols U to $\mathcal{N}_U$ and (set) sort symbols U^* to the set of all subsets of $\mathcal{N}_U$.

– $\mathcal{V}$ is the variable assignment. For each variable u of sort U we have $\mathcal{V}(u) \in \mathcal{N}_U$.

– The function interpretation $\mathcal{F}$ maps constant symbols to themselves. It also contains the interpretation of the built-in functions (such as *eastertime*).

– The other functions are interpreted by $\mathcal{F}$ as follows:

- if $v \in \mathcal{N}_V$ and $i \in \mathbb{Z}$ then $\mathcal{F}(U\_within^i\_V)(v, i) = U_{\mathcal{N}}(V^b_{[[}(v)) + i - 1$.
- $\mathcal{F}(U\_within^e\_V)(v, i) = \begin{cases} U_{\mathcal{N}}(V^b_{[[}(v)) + i - 1 & \text{if } U^b_{[[}(U_{\mathcal{N}}(V^b_{[[}(v))) < V^b_{[[}(v) \\ U_{\mathcal{N}}(V^b_{[[}(v)) + i & \text{otherwise} \end{cases}$.

[5] The notation $\psi[u_1, \ldots, u_n]$ means that ψ is a term containing the variables $u_1, \ldots, u_n$ at some places. $u_1, \ldots, u_n$ are ψ's *free variables*.

- if $v \in \mathcal{N}_V$ then
  $\mathcal{F}(U^i\_s)(v) = \{u \in \mathcal{N}_U \mid V_{\mathcal{N}}(U^b_{[[}(u)) = v \text{ or } V_{\mathcal{N}}(U^e_{[[}(u)) = v\}$
  $\mathcal{F}(U^e\_s)(v) = \{u \in \mathcal{N}_U \mid V_{\mathcal{N}}(U^b_{[[}(u)) = v \text{ and } V_{\mathcal{N}}(U^e_{[[}(u)) = v\}$
- $\mathcal{F}(\mathit{begin\_}\mathrm{U})(v) = U_{\mathcal{N}}(V^b_{[[}(v))$ if v is of sort $V \in \mathcal{U}$.
  $\mathcal{F}(\mathit{end\_}\mathrm{U})(v) = U_{\mathcal{N}}(V^e_{[[}(v))$ if v is of sort $V \in \mathcal{U}$.
- $\mathcal{F}(U\_period\_V)(u, n) = \{u, \ldots, w\}$ where w is determined as follows:

Let $t \stackrel{\text{def}}{=} (U^e_{[[}(u) + U^b_{[[}(u))/2$	the middle of the u interval.
Let $v \stackrel{\text{def}}{=} V_{\mathcal{N}}(t)$	the V-coordinate of the middle.
Let $f \stackrel{\text{def}}{=} \frac{t - V^b_{[[}(v)}{V^e_{[[}(v) - V^b_{[[}(v)}$	the fraction elapsed so far.
Let $v' \stackrel{\text{def}}{=} v + n$	the new V-coordinate.
Let $t' \stackrel{\text{def}}{=} V^b_{[[}(v') + f \cdot (V^e_{[[}(v') - V^b_{[[}(v'))$	the end time point.
Let $w \stackrel{\text{def}}{=} U_{\mathcal{N}}(t')$	the end U-coordinate.

- if the argument to one of the functions is a set instead of a single coordinate then the function is applied to each set and the results are collected in a set.
- the conditional ...?... : ..., the set operators and the arithmetic terms are interpreted in the standard way. If one argument to one of the arithmetic functions is a set, then the function is applied to each element and the results are collected in a set.

– Arbitrary time terms $\psi[u_1, \ldots, u_m]$ with free variables $\{u_1, \ldots, u_n\}$ with $sort(u_i) = U_i$ for $1 \leq i \leq n$, and range-sort U can now be interpreted in two ways. The first way is the standard homomorphic extension of the interpretation $\Im$. If $\psi = u$ is a variable then $\Im(u) \stackrel{\text{def}}{=} \mathcal{V}(u)$. If $\psi = f(t_1 \ldots, t_n)$ is a complex term then $\Im(\psi) \stackrel{\text{def}}{=} \mathcal{F}(f)(\Im(t_1), \ldots, \Im(t_n))$. The binding of the variables u_i to some U_i-coordinates is contained in the variable assignment $\mathcal{V}$. Therefore $\Im(\psi)$ is well defined and yields a set of U-coordinates.

– Much more interesting and useful is the interpretation of time terms as functions mapping a reference time point t to some U-coordinates. If for a variable v of sort V, and a V-coordinate n we define $\Im[v/n]$ to be the interpretation which is like $\Im$, but the variable assignment maps v to n, then we can define an $\Im(\psi)(t)$ as a function mapping a reference time point t to some U-coordinates.

If $\psi = u$ is a variable, then

$$\Im(u)(t) \stackrel{\text{def}}{=} \Im[u/U_{\mathcal{N}}(t)](u).$$

If $\psi[u_1, \ldots, u_m]$ is a complex term, then

$$\Im(\psi)(t) \stackrel{\text{def}}{=} \Im[u_1/U_{1\mathcal{N}}(t), \ldots, u_m/U_{m\mathcal{N}}(t)](\psi).$$

Furthermore let

$$\Im'(\psi)(t) \stackrel{\text{def}}{=} \bigcup_{u \in \Im(\psi(t))} U_{[[}(u).$$

◁

$\Im(\psi[u_1, \ldots, u_m])(t)$ computes the value of $\psi[u_1, \ldots, u_m]$ at a given reference time point t by first computing the corresponding U_i coordinates $U_{i\mathcal{N}}(t)$, binding the variables u_i to these coordinates and then evaluating them in the usual way. The result is either a single U-coordinate or a set of U-coordinates.

The $\Im'$ function goes one step further and turns the computed U-coordinates back into intervals in the reference time system.

Using the interpreter $\Im$ we can now check the *instance* relation between a given reference time point and a $\mathcal{L}$ term (for example whether the current moment in time lies in the office hours of my colleague in Brazil).

Definition 4.4. (Instance Relation) A time point t in the reference system is an instance of the term ψ with range-sort U (we write $t \in \psi$), iff $t \in \Im'(\psi[u_1, \ldots, u_m])(t)$ or, which is equivalent,
iff $U_{\mathcal{N}}(t) \in \Im(\psi[u_1, \ldots, u_m])(t)$. ◁

According to this definition, we compute the U-coordinate corresponding to t and check whether this is one of the coordinates computed from ψ. In general this will still be quite inefficient, and better algorithms have to be worked out.

A more complicated relation is the subsumption relation. An example might be the problem to figure out whether my colleague's morning office hours in Brazil lie always within my local afternoon. (Then I could always phone him in the afternoon without first checking it in the database.) Unfortunately this general subsumption relation is usually not decidable This is due to the fact that we treat the functions $U_{\mathcal{N}}$ and $U_{[[}$ as a kind of black boxes where the most peculiar things may be encoded, for example a transition to a completely new calendar system in the year 2100.

For temporal relationships from everyday life, this general subsumption relationship, however, is really too general. I certainly won't be interested to know whether even in the year 2100 my colleague's morning office hours in Brazil lie within my local afternoon. For me it is completely sufficient to know this until his or my own time of retirement.

Definition 4.5. (Subsumption Relation) A term ψ is subsumed by a term φ ($\psi \subseteq \varphi$) iff for all time points t in the reference system $\Im'(\psi)(t) \subseteq \Im'(\varphi)(t)$.

A *restricted subsumption* relation ($\psi \subseteq_i \varphi$) for the time interval $i = [t_1, t_2[$ holds between ψ and φ iff for all time points $t_1 \leq t < t_2$ $\Im'(\psi)(t) \subseteq \Im'(\varphi)(t)$. ◁

Since the restricted subsumption relation still quantifies over an infinite number of time points, this definition is not yet a suitable basis for an algorithm. In order to decide restricted subsumption, we exploit that the time terms depend on the coordinates of the time units, and they are therefore constant over the period in the reference time line corresponding to the same coordinates of the time unit. For example the time term *February*(y) (Def. (1)) depends on the year coordinate. It is therefore constant over a whole year, i.e. over all the points t in the reference time line with $\text{year}_{\mathcal{N}}(t) = y$.

Exploiting this observation, each time term $\psi[u_1, \ldots u_n]$ can be used to partition each finite interval in the reference time line into the finitely many sub-intervals where $\psi[u_1, \ldots u_n]$ represents a constant function. This partitioning can then be used to trigger finite case analysis for checking whether ψ subsumes φ over an interval i. i is partitioned into the set $\{i_1, \ldots, i_n\}$ of sub-intervals over which $\psi \cup \varphi$ is constant, and then for $1 \leq k \leq n$, $\Im(\psi)(t_k) \subseteq \Im(\varphi)(t_k)$ is checked for some arbitrary chosen $t_k \in i_k$

Definition 4.6. (Partitioning Algorithm) For a non-empty interval $i = [i^b, i^e[\subseteq \mathcal{T}$ and a time term $\psi[u_1, \ldots, u_n]$ with $U_k \stackrel{\text{def}}{=} sort(u_k)$, $k = 1, \ldots, n$, we define the *partitioning of i with respect to ψ* as

$$\delta(i, \psi) = \{[i^b, t[\} \cup \delta([t, i^e[, \psi)$$

where

$$t \stackrel{\text{def}}{=} \begin{cases} min(\{U_k{}^b_{[[}(U_{k\mathcal{N}}(i^b) + 1)) \in i \mid 1 \leq k \leq n\}) \\ i^e \text{ if this set is empty.} \end{cases}$$

◁

The partitioning algorithm decomposes i by computing the U_k-coordinate $U_{k\mathcal{N}}(i^b)$ corresponding to the beginning of the interval i and then checking whether the beginning of the next interval corresponding to the next U_k-coordinate still lies in i or not. If it still lies in i, this is a candidate for the lower border of the next sub-interval. The real border is determined by the time unit U_k which gives the smallest such border. (A properly implemented algorithm would of course try the most fine grained time unit U_k first, e.g. seconds before minutes before hours etc.)

Notice that the partitioning function only depends on the time units corresponding to ψ's free variables, not on ψ itself.

Proposition 4.7. (Soundness of the Partitioning Algorithm)
For each interval $j \in \delta(i,\psi)$ (Def. 4.6) and for each $\{t_1,t_2\} \subseteq j$: $\Im'(\psi)(t_1) = \Im'(\psi)(t_2)$.

Proof: by induction on the number of intervals in $\delta(i,\psi)$. In the base case, $\delta(i,\psi) = i$, i.e. i itself is the only component, and for $1 \leq k \leq n$, $U_{k[[}^{b}(U_{k\mathcal{N}}(i^b)+1)) > i^e$. Therefore for all $t \in i$, $U_{k\mathcal{N}}(t)$ yields the same U_k coordinate. Since $\psi[u_1,\ldots,u_n]$ depends on the U_k-coordinates only, $\Im'(\psi)(t_1) = \Im'(\psi)(t_2)$ for all $\{t_1,t_2\} \subseteq i$.
In the induction step, $\delta(i,\psi)$ consists of more than one interval and $[i^b, s[$ is the left most sub-interval of the decomposition. According to the definition of δ, t is the smallest value such that $t = U_{k[[}^{b}(U_{k\mathcal{N}}(i^b)+1)$ lies still in i. That means for all $k \in \{1,\ldots,n\}$: $U_{k\mathcal{N}}(t_1) = U_{k\mathcal{N}}(t_2)$ for all $t_1 \in [i^b,t[$ and $t_2 \in [i^b,t[$. Thus, $\Im'(\psi)(t_1) = \Im'(\psi)(t_2)$ for the first interval in $\delta(i,\psi)$. The induction hypothesis directly applies to the remaining intervals. $\triangleleft$

Proposition 4.8. (Decidability)

- The instance relation is decidable. $\Im'(\psi)(t)$ yields only finitely many intervals. Therefore just check $t \in \Im'(\psi)(t)$, which amounts to finding some $[i^b,i^e[\in \Im'(\psi)(t)$ with $i^b \leq t < i^i$.

- Without special assumptions about the functions $U_{\mathcal{N}}$ and $U_{[[}$, the subsumption relation is in general not decidable.

- The restricted subsumption relation $\psi \subseteq_i \varphi$ over an interval $i = [t_1,t_2[$ is decidable. Using the partitioning algorithm of Def. 4.6, partition this interval into the finitely many sub-intervals where the two time terms are constant functions, choose for each sub-interval j some point $t \in j$, and check the subsumption relation $\Im'(\psi)(t) \subseteq \Im'(\varphi)(t)$ for this point. To do this, check for each $[i^b,i^e[\in \Im'(\psi)(t)$ whether there is some $[j^b,j^e[\in \Im'(\psi)(t)$ with $[i^b,i^e[\subseteq [j^b,j^e[$.[6] $\triangleleft$

4.1. Top-Down Interpretation

The interpretation $\Im(\psi)$ of a time term ψ works bottom-up in the usual recursive style. There is a recursive descent into the term down to the

[6] $\Im'(\psi)(t)$ should be computed in such a way that the sequences of intervals without gaps are comprised into one single interval.

variable level, and then the actual computation is done on the way back to the top-level of the term. This can be quite expensive if large sets of coordinates are generated. For example the interpretation of the term second_s(y) where y is a year-variable causes the generation of the set of all second-coordinates within the year y (a set with about 31 million elements).

For the instance check $t \in$ second_s(y) (Def. 4.4) this would be much too expensive. A *top-down interpretation* is much more efficient in this case. How could this work? For checking $t \in$ second_s(y), we exploit that the term second_s(y) has range-sort 'second'. Therefore one can compute the second-coordinate $s = \mathit{second}_{\mathcal{N}}(t)$, and then compute the set of all year coordinates y such that $s =$ second_s(y). There is only one y in this case, and this can be computed quite easily, using the semantics of the second_s-constructor. If $t \in \mathit{year}_{[[}(y)$ the instance relation holds, otherwise not.

To do this kind of top-down interpretation, one needs for each constructor $f(u_1, \ldots, u_n)$ in the language an algorithm for computing for a given u the set of all $u_1, \ldots, u_n$-tuples such that $u = \mathcal{F}(f)(u_1, \ldots, u_n)$. That means one must compute for a wanted result of a computation the set of all arguments which actually produce this result. A top-down interpreter can then descend into a time term and pass the potential arguments as wanted results to the functions at the deeper level of the term. This has not yet been investigated in detail, but for the constructors in the time term language this seems to be quite feasible.

There are, however, examples, where this strategy also causes the computation of large sets of coordinates. Therefore a mixed bottom-up, top-down strategy seems to be the most efficient way to decide the instance relation.

5. Summary

In this paper I proposed a formal representation of real time and real calendar systems. With corresponding conversion functions the calendar system's time units, years, months, weeks, days, hours, minutes and seconds are individually linked to a reference time system. Each time unit has its own coordinate axis (as integers) which represents an interval in the reference time system. This way one can convert between arbitrary calendar systems and arbitrary time zones.

In the second part a specification language $\mathcal{L}$ for temporal notions has been proposed. It allows us to specify complex temporal notions in a very simple and intuitive way. Since the reference time system represents a single model for this language, and we have the conversion functions, the terms of the language $\mathcal{L}$ can be interpreted and questions like 'is a given point in time within the time intervals specified by a term ψ?' can be decided.

Possible applications of the presented system are temporal databases with automatic conversion between different time systems and user defined notions. Another potential application is the World-Wide-Web, where the author of a WWW-page can use his own temporal notions, and the WWW-browser can convert it into the reader's favourite system.

In (Ohlbach and Gabbay, 1997) we incorporated the time term language into *Calendar Logic*, a propositional temporal logic with operators 'sometimes within ψ' and 'always within ψ' where ψ is a time term. A simple statement which can be expressed in this language is 'yesterday I worked for eight hours with one hour lunch break at noon'. Calendar Logic is decidable, and we provided a tableau based decision procedure.

References

Dershowitz, N. and E. M. Reingold: 1997, *Calendrical Calculations.* Cambridge University Press.

Ohlbach, H. J. and D. Gabbay: 1997, 'Calendar logic'. Technical report, Imperial Colleg.

Schmidt-Schauß, M.: 1989, 'Computational Aspects of an Order-Sorted Logic with Term Declarations'. Vol. 395 of *Lecture Notes in Artificial Intelligence.* Springer-Verlag.

A Model Checking Algorithm for π-Calculus Agents
*

Stefania Gnesi
Istituto di Elaborazione dell'Informazione - C.N.R., Pisa

Gioia Ristori
Dipartimento di Informatica, Università di Pisa

Abstract. This paper presents π-logic, an action-based logic for π-calculus. A model checker is built for this logic, following an automata-based approach. This is made possible by a result which allows finite state Labelled Transition Systems to be associated with a wide class of π-calculus agents by preserving a notion of bisimulation equivalence. The model checker was thus built reusing an efficient model checker for the action based logic ACTL, after a sound translation from π-logic into ACTL has been defined.

1. Introduction

Modal and temporal logics are equipped with verification techniques to show when systems have, or fail to have, certain properties expressed in a given logic. In the case of finite state systems, a classical approach is to define an algorithm called model checker (Clarke et al., 1986) which checks the satisfiability of a formula on the finite state model of a system. Using this approach, the complexity of the model checking problem is a function of the size of the model (number of states and transitions of its corresponding model) and of the size of the formula. For the branching time temporal logic CTL (Emerson and Halpern, 1986), and its action based version ACTL (Nicola and Vaandrager, 1990), there are straightforward linear (in the size of the system and of the formula) time model checking algorithms. As more complex temporal logics are used, e. g. μ-calculus (Kozen, 1983), more complex algorithms are developed. In general, these algorithms are exponential in the alternation depth of the formula (Emerson and Lei, 1986).

In the case of infinite state systems, model checking techniques are not suitable for checking the satisfiability of system properties. In this case, several decision procedures have been defined that rely on local methods (Stirling and Walker, 1989), which are often presented using tableaux.

* Work partially funded by CNR Integrated Project *Metodi e Strumenti per la Progettazione e la Verifica di Sistemi Eterogenei Connessi mediante Reti di Comunicazione.*

H. Barringer et al. (eds.), Advances in Temporal Logic, 339–357.

π-calculus (Milner et al., 1992a) is a formalism for modelling the behaviour of systems where the communication topology among processes can dynamically evolve as the computation progresses. Its primitives are simple but expressive: channel names can be created, communicated (so that process acquaintances can be dynamically reconfigured) and are subject to sophisticated scoping rules. Moreover, names can be used to model objects (Walker, 1991) and higher order communications (Sangiorgi, 1992).
π-calculus has greater expressive power than ordinary process calculi, but also a much more complicated theory. In particular, the usual operational models are infinite state and infinite branching.

Logics have been proposed (Milner et al., 1992b; Dam, 1993) to express properties of π-calculus agents. These logics are extensions, with π-calculus actions and name quantifications and parameterizations, of classical action-based logics (Hennessy and Milner, 1985; Kozen, 1983).

The most popular tool for the verification of behavioural and logical properties for π-calculus is the *Mobility Workbench* (Victor and Moller, 1994) (MWB). The logic verification functionalities offered by the MWB are based on the implementation of the tableau-based proof system for Propositional μ-calculus with name-passing (Dam, 1993). Inference rules are applied to establish whether or not an agent satisfies a given logic formula.

In this paper we present a logic, called π-logic, which is based on the logic formalism ACTL. The *next* operators of ACTL are extended to deal with π-calculus actions, thus π-logic is suitable for expressing properties of π-calculus agents. For this logic it is possible to provide a classical model checking algorithm to verify the satisfiability of the π-logic formulae on π-calculus agents. This is achieved by exploiting the results in (Montanari and Pistore, 1995) which permit, given a finitary π-calculus agent, a finite state LTS to be derived. The model checker for the π-logic is then built using this finite representation and re-using the existing model checker for the ACTL logic (Ferro, 1994; Gnesi, 1997) by means of a sound translation function from π-logic into ACTL.

Our decision not to directly implement the model checker, but to reuse AMC, was because we wished to rapidly build a model checker for π-logic to experiment with the descriptive power of the new logic, while relying on an efficient and continuously upgraded tool.

The paper is organized as follows. Section 2 present some preliminary notions on Labelled Transition Systems and on the logic ACTL. Section 3 briefly describes π-calculus syntax and semantics, and presents π-logic. Section 4 outlines the methodology that allows a π-calculus agent to be translated into a finite state Labelled Transition Systems. Section 5 describes the translation of the π-logic operators into ACTL.

In Section 6 an example is shown of the use of the model checker for the verification of the logical properties of a dynamic protocol. Section 7 concludes the paper.

2. Preliminaries

This section describes Labelled Transition Systems, which are a semantic model for concurrent systems, and ACTL logic (Nicola and Vaandrager, 1990) (the action based version of CTL temporal logic (Emerson and Halpern, 1986)) which is interpreted over Labelled Transition Systems.

DEFINITION 1. (Labelled Transition System). *A Labelled Transition System (LTS) is a 4-tuple* $\mathcal{A} = (Q, q_0, Act \cup \{\mathtt{tau}\}, R)$, *where:*

- Q *is a finite set of states;*
- q_0 *is the initial state;*
- *Act is a finite set of observable actions and* `tau` *is the unobservable action;*
- $R \subseteq Q \times (Act \cup \{\mathtt{tau}\}) \times Q$ *is the transition relation. Whenever* $(q, \alpha, q') \in R$ *we will write* $q \xrightarrow{\alpha} q'$.

Two LTSs are said to be *strongly equivalent*, or strongly bisimilar, if there exists a strong bisimulation relation between their initial states (Milner, 1989).

ACTL is a branching time temporal logic that is suitable for describing the behaviour of systems that perform actions during their working time. In fact, ACTL embeds the idea of "evolution in time by actions". The original definition of ACTL includes an auxiliary logic of action formulae.

DEFINITION 2. (Action formulae). *Given a set of observable actions Act, the language* $\mathcal{AF}(Act)$ *of the action formulae on Act is defined as follows:*

$$\chi ::= \mathtt{true} \mid b \mid \neg\chi \mid \chi \,\&\, \chi$$

where b ranges over Act.
ACTL *is a branching time temporal logic of state formulae (denoted by* ϕ*), in which a path quantifier prefixes an arbitrary path formula (denoted by* π*).*

DEFINITION 3. (ACTL syntax). *The syntax of the* ACTL *formulae is given by the grammar below:*
$\phi ::= \texttt{true} \mid \phi \,\&\, \phi \mid \sim \phi \mid E\pi \mid A\pi$
$\pi ::= X\{\chi\}\phi \mid X\{\texttt{tau}\}\phi \mid [\phi\{\chi\}U\phi] \mid [\phi\{\chi\}U\{\chi'\}\phi]$
where χ, χ' *range over action formulae,* E *and* A *are path quantifiers, and* X *and* U *are the next and the until operators respectively.*

In order to present the ACTL semantics, we need to introduce the notion of paths over an LTS.

DEFINITION 4. (Paths). *Let* $\mathcal{A} = (Q, q_0, Act \cup \{\texttt{tau}\}, R)$ *be an LTS.*

– σ *is a path from* $r_0 \in Q$ *if either* $\sigma = r_0$ *(the empty path from* r_0*) or* σ *is a (possibly infinite) sequence* $(r_0, \alpha_1, r_1)(r_1, \alpha_2, r_2)\ldots$ *such that* $(r_i, \alpha_{i+1}, r_{i+1}) \in R$.

– *The concatenation of paths is denoted by juxtaposition. The concatenation* $\sigma_1\sigma_2$ *is a partial operation: it is defined only if* σ_1 *is finite and its last state coincides with the initial state of* σ_2*. The concatenation of paths is associative and has identities. Actually,* $\sigma_1(\sigma_2\sigma_3) = (\sigma_1\sigma_2)\sigma_3$*, and if* r_0 *is the first state of* σ *and* r_n *is its last state, then we have* $r_0\sigma = \sigma r_n = \sigma$.

– *A path* σ *is called maximal if either it is infinite or it is finite and its last state has no successor states. The set of the maximal paths from* r_0 *will be denoted by* $\Pi(r_0)$.

– *If* σ *is infinite, then* $|\sigma| = \omega$.
If $\sigma = r_0$*, then* $|\sigma| = 0$.
If $\sigma = (r_0, \alpha_1, r_1)(r_1, \alpha_2, r_2)\ldots(r_n, \alpha_{n+1}, r_{n+1})$*,* $n \geq 0$*, then* $|\sigma| = n + 1$*. Moreover, we will denote the* i^{th} *state in the sequence, i.e.* r_i*, by* $\sigma(i)$.

DEFINITION 5. (Action formulae semantics). *The satisfaction relation* $\models$ *for action formulae is defined as follows:*

$a \models \texttt{true}$		*always*
$a \models b$	*iff*	$a = b$
$a \models \sim \chi$	*iff*	*not* $a \models \chi$
$a \models \chi \,\&\, \chi'$	*iff*	$a \models \chi$ *and* $a \models \chi'$

As usual, **false** abbreviates $\sim$ **true** and $\chi \vee \chi'$ abbreviates $\sim (\sim \chi \,\&\! \sim \chi')$.

DEFINITION 6. (ACTL semantics). *Let* $\mathcal{A} = (Q, q_0, Act \cup \{\mathtt{tau}\}, R)$ *be an LTS. Let* $s \in Q$ *and* σ *be a path. The satisfaction relation for* ACTL *formulae is defined in the following way:*

- $s \models \mathtt{true}$ *always*
- $s \models \phi \,\&\, \phi'$ *iff* $s \models \phi$ *and* $s \models \phi'$
- $s \models \sim \phi$ *iff not* $s \models \phi$
- $s \models E\pi$ *iff there exists* $\sigma \in \Pi(s)$ *such that* $\sigma \models \pi$
- $s \models A\pi$ *iff for all* $\sigma \in \Pi(s)$, $\sigma \models \pi$
- $\sigma \models X\{\chi\}\phi$ *iff* $\sigma = (\sigma(0), \alpha_1, \sigma(1))\sigma'$, *and* $\alpha_1 \models \chi$, *and* $\sigma(1) \models \phi$
- $\sigma \models X\{\mathtt{tau}\}\phi$ *iff* $\sigma = (\sigma(0), \mathtt{tau}, \sigma(1))\sigma'$, *and* $\sigma(1) \models \phi$
- $\sigma \models [\phi\{\chi\}U\phi']$ *iff there exists* $i \geq 0$ *such that* $\sigma(i) \models \phi'$, *and for all* $0 \leq j < i$: $\sigma = \sigma'(\sigma(j), \alpha_{j+1}, \sigma(j+1))\sigma''$ *implies* $\sigma(j) \models \phi$, *and* $\alpha_{j+1} = \mathtt{tau}$ *or* $\alpha_{j+1} \models \chi$
- $\sigma \models [\phi\{\chi\}U\{\chi'\}\phi']$ *iff there exists* $i \geq 1$ *such that* $\sigma = \sigma'(\sigma(i-1), \alpha_i, \sigma(i))\sigma''$, *and* $\sigma(i) \models \phi'$, *and* $\sigma(i-1) \models \phi$, *and* $\alpha_i \models \chi'$, *and for all* $0 < j < i$: $\sigma = \sigma'_j(\sigma(j-1), \alpha_j, \sigma(j))\sigma''_j$ *implies* $\sigma(j-1) \models \phi$ *and* $\alpha_j = \mathtt{tau}$ *or* $\alpha_j \models \chi$

As usual, **false** abbreviates $\sim$ **true** and $\phi \vee \phi'$ abbreviates $\sim (\sim \phi \,\&\! \sim \phi')$. Moreover, we define the following derived operators:

- $EF\phi$ stands for $E[\mathtt{true}\{\mathtt{true}\}U\phi]$.
- $AG\phi$ stands for $\sim EF \sim \phi$.
- $< a > \phi$ stands for $E[\mathtt{true}\{\mathtt{false}\}U\{a\}\phi]$.
- $< \mathtt{tau} > \phi$ stands for $E[\mathtt{true}\{\mathtt{false}\}U\phi]$.

ACTL logic can be used to define *liveness* (something good eventually happens) and *safety* (nothing bad can happen) properties of concurrent systems. Moreover, ACTL logic is *adequate* with respect to strong bisimulation equivalence on LTSs (Nicola and Vaandrager, 1990). Adequacy means that two LTSs $\mathcal{A}_1$ and $\mathcal{A}_2$ are strongly bisimilar if and only if $F_1 = F_2$, where $F_i = \{\psi \in \text{ ACTL} : \mathcal{A}_i \text{ satisfies } \psi\}$, $i = 1, 2$.

3. A logic for π-calculus

This Section presents the logic formalism, called π-logic, which we define to express properties of π-calculus agents. We start with a brief description of the syntax and the operational semantics of π-calculus (Milner et al., 1992a), on which the logic formulae are interpreted.

DEFINITION 7. (π-calculus syntax). *Given a denumerable infinite set $\mathcal{N}$ of* names *(denoted by $a, b, \ldots$), π-calculus* agents *over $\mathcal{N}$ are defined as follows*[1]*:*

$$P ::= \texttt{nil} \mid \alpha.P \mid P_1 \| P_2 \mid P_1 + P_2 \mid (x)P \mid [x = y]P \mid A(x_1, \ldots, x_{r(A)})$$

$$\alpha ::= \texttt{tau} \mid x!y \mid x?(y),$$

where $r(A)$ is the range of the agent identifier A.

The occurrences of y in $x?(y).P$ and $(y)P$ are bound; *free names* are defined as usual, and we indicate the set of free names of agent P with $\texttt{fn}(P)$. For each identifier A there is a definition $A(y_1, \ldots, y_{r(A)}) := P_A$ (with y_i all distinct and $\texttt{fn}(P_A) \subseteq \{y_1 \ldots y_{r(A)}\}$), and we assume that each identifier in P_A is in the scope of a prefix (guarded recursion).

The *actions* that agents can perform are defined by the following syntax:

$$\mu ::= \texttt{tau} \mid x!y \mid x!(z) \mid x?y;$$

where x and y are free names of μ ($\texttt{fn}(\mu)$), whereas z is a bound name ($\texttt{bn}(\mu)$); finally $\texttt{n}(\mu) = \texttt{fn}(\mu) \cup \texttt{bn}(\mu)$. In the actions we can distinguish a subject, x, and an object, y and z. The structural rules for the *early operational semantics* are defined in Table I.

The strong early bisimulation (Milner et al., 1992b) is given by the following definition:

DEFINITION 8. *A binary relation B over a set of agents is a strong early bisimulation if it is symmetric and, whenever $(P, Q) \in B$, we have that:*

- *if $P \xrightarrow{\mu} P'$ and $\texttt{fn}(P, Q) \cap \texttt{bn}(\mu) = \emptyset$, then there exists Q' such that $Q \xrightarrow{\mu} Q'$ and $(P', Q') \in B$.*

Two terms are said to be *strong early bisimilar*, written $P \simeq Q$, if there exists a bisimulation B such that $(P, Q) \in B$.

[1] For convenience, we adopt the syntax of the agents that we use to input agents in the environment. We use $(x)P$ for the restriction, $x?(y).P$ for input prefixes, and $x!y.P$ for output prefixes. The syntax of the other operators is standard.

Table I. Early operational semantics.

TAU $\mathtt{tau}.P \xrightarrow{\mathtt{tau}} P$ OUT $x!y.P \xrightarrow{x!y} P$	IN $x?(y).P \xrightarrow{x?z} P\{z/y\}$
SUM $\dfrac{P_1 \xrightarrow{\mu} P'}{P_1 + P_2 \xrightarrow{\mu} P'}$	PAR $\dfrac{P_1 \xrightarrow{\mu} P_1'}{P_1 \Vert P_2 \xrightarrow{\mu} P_1' \Vert P_2}$ if $\mathtt{bn}(\mu) \cap \mathtt{fn}(P_2) = \emptyset$
COM $\dfrac{P_1 \xrightarrow{x!y} P_1' \quad P_2 \xrightarrow{x?y} P_2'}{P_1 \Vert P_2 \xrightarrow{\mathtt{tau}} P_1' \Vert P_2'}$	CLOSE $\dfrac{P_1 \xrightarrow{x!(y)} P_1' \quad P_2 \xrightarrow{x?y} P_2'}{P_1 \Vert P_2 \xrightarrow{\mathtt{tau}} (y)(P_1' \Vert P_2')}$ if $y \notin \mathtt{fn}(P_2)$
RES $\dfrac{P \xrightarrow{\mu} P'}{(x)P \xrightarrow{\mu} (x)P'}$ if $x \notin \mathtt{n}(\mu)$	OPEN $\dfrac{P \xrightarrow{x!y} P'}{(y)P \xrightarrow{x!(z)} P'\{z/y\}}$ if $x \neq y, z \notin \mathtt{fn}((y)P')$
MATCH $\dfrac{P \xrightarrow{\mu} P'}{[x = x]P \xrightarrow{\mu} P'}$	IDE $\dfrac{P_A\{y_1/x_1, \ldots, y_{r(A)}/x_{r(A)}\} \xrightarrow{\mu} P'}{A(y_1, \ldots, y_{r(A)}) \xrightarrow{\mu} P'}$

In order to express properties of π-calculus agents, we introduce a temporal logic, hereafter π-logic. π-logic is based on the logic formalism presented in Section 2, where the modal operators are extended to deal with π-calculus actions. Along with the *strong next* modality, π-logic also includes a *weak next* modality, whose meaning is that a number of tau can be executed before the occurrence of an action. The notation <_> is generally used in modal logics, such as HML (Hennessy and Milner, 1985), to denote the strong next modality, while ≪_≫ is used for the weak next modality. Here we have adopted the ACTL-like notation, where the strong next is denoted by EX and the weak next by <_>. This notation is more convenient in this case, since π-logic formulae will be mapped into ACTL formulae.

DEFINITION 9. (π-logic syntax and semantics). *The syntax of the π-logic is given by:*

$$\phi ::= \mathtt{true} \mid \sim\phi \mid \phi \,\&\, \phi' \mid EX\{\mu\}\phi \mid <\mu>\phi \mid EF\phi.$$

We extend the notion of bound and free names to formulae. The occurrences of y in $EX\{x!(y)\}\phi$ and $<x!(y)>\phi$ are bound. The set of names occurring free in ϕ is denoted as $fn(\phi)$.

The interpretation of the logic formulae defined by the above syntax, is the following:

- $P \models \mathtt{true}$ *always holds;*
- $P \models \sim\phi$ *iff not* $P \models \phi$*;*
- $P \models \phi \,\&\, \phi'$ *iff* $P \models \phi$ *and* $P \models \phi'$*;*

– $P \models EX\{\mu\}\phi$ *iff there exists* P' *such that* $P \xrightarrow{\mu} P'$ *and* $P' \models \phi$, $\mu = \mathtt{tau}, x!y, x?y$;

– $P \models EX\{x!(y)\}\phi$ *iff there exist* P' *and* $w \notin fn(\phi) - \{y\}$ *such that* $P \xrightarrow{x!(w)} P'$ *and* $P' \models \phi\{w/y\}$;

– $P \models <\mu>\phi$ *iff there exist* $P_0, \ldots, P_n$, $n \geq 1$, *such that* $P = P_0 \xrightarrow{\mathtt{tau}} P_1 \ldots \xrightarrow{\mathtt{tau}} P_{n-1} \xrightarrow{\mu} P_n$ *and* $P_n \models \phi$, $\mu = \mathtt{tau}, x!y, x?y$;

– $P \models <x!(y)>\phi$ *iff there exist* $P_0, \ldots, P_n$, $n \geq 1$, *and* $w \notin fn(\phi) - \{y\}$ *such that* $P = P_0 \xrightarrow{\mathtt{tau}} P_1 \ldots \xrightarrow{\mathtt{tau}} P_{n-1} \xrightarrow{x!(w)} P_n$ *and* $P_n \models \phi\{w/y\}$;

– $P \models EF\,\phi$ *iff there exist* $P_0, \ldots, P_n$ *and* $\mu_1, \ldots, \mu_n$, *with* $n \geq 0$, *such that* $P = P_0 \xrightarrow{\mu_1} P_1 \ldots \xrightarrow{\mu_n} P_n$ *and* $P_n \models \phi$.

Starting from the above syntax some derived operators can be defined:

– `false` stands for $\sim$`true`;

– $\phi \vee \phi'$ stands for $\sim(\sim\phi \mathbin{\&} \sim\phi')$;

– $[\mu]\phi$ stands for $\sim<\mu>\sim\phi$. This is the dual version of the weak next operator;

– $AG\,\phi$ stands for $\sim EF \sim\phi$. This is the *always* operator, whose meaning is that ϕ is true now and will always be so in the future.

Note that π-logic is an extension, with the *eventually* operator $EF\phi$ and the weak next operator, of the logic defined in (Milner et al., 1992b) to deal with π-calculus agents. Moreover, π-logic is a subset of the Propositional μ-calculus with name-passing (Dam, 1993). Propositional μ-calculus with name-passing permits a broad class of properties of systems described by π-calculus agents to be expressed, using fixed point operators, name parameterizations and quantifications over the communication objects.

π-logic, with its simple syntax, can however cover an interesting class of properties of mobile systems without using fixed point operators. Moreover, as we will show later, an efficient model checking algorithm can be given for it.

THEOREM 1. *π-logic is adequate with respect to strong early bisimulation equivalence.*

Proof: In (Milner et al., 1992b) it was shown that the subset of the logic given by:

$$\phi ::= \texttt{true} \mid \sim\phi \mid \phi \,\&\, \phi' \mid EX\{\mu\}\phi$$

is adequate with respect to strong early bisimulation equivalence. It is easy to see that strongly equivalent agents satisfy the same set of weak logic formulae. Moreover, the eventually operator simply states that something will happen in the future, hence strongly equivalent agents also satisfy the same set of *eventually* logic formulae. Consequently, the introduction of the weak next modalities and of the eventually operator only adds expressive power to the logic, without changing its adequacy with respect to a strong early bisimulation equivalence.

4. From π-calculus agents to Labelled Transition Systems

In (Montanari and Pistore, 1995) it was proved that finite state LTSs can be built for the class of *finitary* agents: an agent is finitary if there is a bound to the number of parallel components of all the agents that can be reached from it. In particular, all the *finite control* agents, i.e. the agents without parallel composition inside recursion, are finitary.

In order to associate a finite state LTS with a π-calculus agent P, a suitable finite set of object names is considered for representing all the bound output and input transitions of the agent. This set is defined by taking into account the set $an(P)$ of the semantically *active* names, or simply active names, of the agent. The set $an(P)$ is a subset of the free names of the agent. Formally, a name x is active in P, that is $x \in an(P)$, if and only if $(x)P \not\sim P$. Active names have the useful property that bisimilar agents have the same set of active names. Note that this is not true for free names, as Example 1 shows. Informally, the names belonging to $an(P)$ are the free names that are used as subjects of the actions, or as objects of the free output actions.

Assuming that there is a total order among the names in $\mathcal{N}$, only one transition is used, in the LTS associated with the agent P, to represent the infinite set of transitions corresponding to a bound output $x!(y)$. The string chosen as label of the transition is $x!(a)$, where a is the minimal element of the set of names given by $\mathcal{N} - an(P)$. In fact, just suppose that P and Q are bisimilar agents, and that both P and Q can perform a bound output action with subject x. Since bisimilar agents have the same set of active names, the bound output transitions of both P and Q are mapped into the same transition, labelled $x!(a)$, of the corresponding LTSs.

The representation of the infinite set of transitions corresponding to an input action $x?y$, performed by P, is done in a similar way: in the LTS there is a finite set of transitions labelled by strings such as $x?u$, for each active name u of P, along with a transition labelled with $x?(a)$, where a is minimal element of $\mathcal{N} - an(P)$, which represents the input of non active names.

The names chosen to represent, in the LTS of P, non active names, will be hereafter called the *fresh names* of the LTS.

EXAMPLE 1. *Consider the bisimilar agents* $P = x?(y).\mathtt{nil}$ *and* $Q = x?(y).\mathtt{nil} + (w)w!z.\mathtt{nil}$. *The two agents have different sets of free names,* $\{x\}$ *and* $\{x,z\}$ *respectively. However, they have the same set of active names* $\{x\}$. *Actually, the name* z *is a free name for* Q *but it is not an active name in* Q. *In fact, we have that* $(z)Q \simeq Q$.

Although it is generally undecidable whether $(x)P \simeq P$, in (Montanari and Pistore, 1995) algorithms for computing the active names and the finite state LTSs of finitary agents without matching[2] are given. These algorithms ensure that equivalent (i.e. strong or weak bisimilar) π-calculus agents are mapped into equivalent (i.e. strong or weak bisimilar) LTSs. The construction of the state space of the π-calculus agent has a worst case complexity that is exponential in the syntactical size of the agent (Montanari and Pistore, 1995).

We do not give here a complete account of the algorithms. Instead, we give a description of the finite state LTS associated with an agent. Hereafter $\Downarrow P$ denotes the restriction $(y_1)\dots(y_n)P$, where $\{y_1,\dots,y_n\} = fn(P) - an(P)$. Note that $fn(\Downarrow P)$ coincides with $an(P)$.

DEFINITION 10. *Let* P_0 *be a* π*-calculus agent. The LTS associated with* P_0 *is* $(Q, \Downarrow P_0, Act \cup \{\mathtt{tau}\}, \mapsto)$, *where* Q, Act, $\mapsto$ *are defined by the following rules. For* $P \in Q$:

- *if* $P \xrightarrow{\mathtt{tau}} P'$ *then* $\Downarrow P' \in Q$ *and* $\Downarrow P \overset{\mathtt{tau}}{\mapsto} \Downarrow P'$;

- *if* $P \xrightarrow{x!y} P'$ *then* $\Downarrow P' \in Q$, $x,y! \in Act$ *and* $\Downarrow P \overset{x!y}{\mapsto} \Downarrow P'$;

- *if* $P \xrightarrow{x!(y)} P'$ *and* $y = min(\mathcal{N} - fn(P))$ *then* $\Downarrow P' \in Q$, $x!(y) \in Act$ *and* $\Downarrow P \overset{x!(y)}{\mapsto} \Downarrow P'$;

- *if* $P \xrightarrow{x?y} P'$ *and* $y \in fn(P)$ *then* $\Downarrow P' \in Q$, $x?y \in Act$ *and* $\Downarrow P \overset{x?y}{\mapsto} \Downarrow P'$;

[2] The same algorithm also works for π-calculus with a restricted form of matching (Montanari and Pistor, 1997; Ferrari et al., 1997) that is general enough for most practical applications.

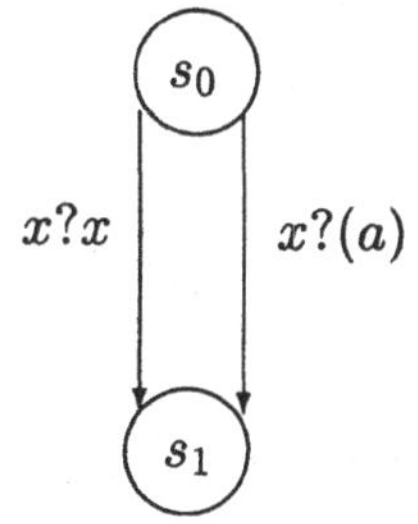

Figure 1. The LTS of P and Q.

– *if* $P \xrightarrow{x?y} P'$ *and* $y = min(\mathcal{N} - fn(P))$ *then* $\Downarrow P' \in Q$, $x?(y) \in$ *Act and* $\Downarrow P \overset{x?(y)}{\mapsto} \Downarrow P'$.

Note that the set of the fresh names of the LTS can easily be recognized, since the fresh names are those that appear in parentheses in the transition labels of the LTS.

EXAMPLE 2. *Consider again agents P and Q in Example 1. They have associated the same LTS (see Figure 1), with two states, say* s_0 *and* s_1*, and two transitions* $s_0 \overset{x?x}{\mapsto} s_1$ *and* $s_0 \overset{x?(a)}{\mapsto} s_1$*, where we have assumed the name* $a = min(\mathcal{N} - \{x\})$.

5. From π-logic to ACTL

In this section we define the translation function that associates an ACTL formula with a formula of π-logic. The translation we propose is defined by having in mind a precise soundness result: we want a π-logic formula to be satisfied by a π-calculus agent P if and only if the finite state LTS associated with P satisfies the corresponding ACTL formula. The translation of a formula is thus not unique, but depends on the agent P. Specifically, it depends on the set S of the fresh names of the LTS associated with the agent P.

The translation of the (strong or weak) π-logic next modalities indexed by a free output or tau action is immediate: it is given by the (strong or weak) ACTL next operator in which the action label is simply the string [3] corresponding to the action argument of the next modality of π-logic to be translated.

The translation of the (strong or weak) next modalities indexed by input or bound output actions is more complex, as it depends on the given set of fresh names S. When the translation of a π-logic next bound output or input modality is considered, the action label of the corresponding ACTL next modality cannot be simply the action that indexes the next modality of π-logic. In fact, we have to consider that a bound output or input action μ, performed by the π-calculus agent P, is represented in the LTS of P by labels that can be different from μ. Actually, whenever a π-logic next bound output modality has to be translated, we have to map its action, say $x!(y)$, to all the possible transition labels of the LTS of P that can represent $x!(y)$, that is to labels of the form $x!(s)$, where $s \in S$. When a next input modality of π-logic has to be translated, two possibilities have to be taken into account. The first is that an explicit transition labelled by the action of the π-logic next modality, say $x?y$, may exist (this is the case if y is an active name of the agent P), hence we have to map $x?y$ directly to the label $x?y$. The second is that the LTS may not have an explicit transition labelled by $x?y$. This means that we also have to map $x?y$ to all the possible labels that can represent $x?y$ in the LTS, that is to labels of the form $x?(s)$, where $s \in S$.

The translation of the π-logic formula $EF\ \phi$ is given simply by the ACTL formula $EF\ \phi'$, where ϕ' is the ACTL translation of ϕ.

DEFINITION 11. *Let $\theta = \{\alpha/y\}$. We define $\mu\theta$ as being the action μ' obtained from μ by replacing the occurrences of the name y with the name α. Moreover, we define* $\mathtt{true}\theta = \mathtt{true}$, $(\phi_1 \& \phi_2)\theta = \phi_1\theta \& \phi_2\theta$, $(\sim \phi)\theta = \sim \phi\theta$, $(EX\{\mu\}\phi)\theta = EX\{\mu\theta\}\phi\theta$, $(< \mu > \phi)\theta = < \mu\theta > \phi\theta$ *and* $(EF\phi)\theta = EF\phi\theta$.

DEFINITION 12. (Translation function). *Given a π-logic formula ϕ and a set of names S, the* ACTL *translation of ϕ is the* ACTL *formula $\mathcal{T}_S(\phi)$ defined as follows:*

- $\mathcal{T}_S(\mathtt{true}) = \mathtt{true}$
- $\mathcal{T}_S(\phi_1 \& \phi_2) = \mathcal{T}_S(\phi_1) \& \mathcal{T}_S(\phi_2)$
- $\mathcal{T}_S(\sim \phi) = \sim \mathcal{T}_S(\phi)$

[3] Note that in the action argument of a π-logic next modality we can distinguish the type of action (silent, output, bound output, input) and eventually a subject and an object. On the contrary, the action labels of the ACTL next modalities are simply labels, that is strings of characters.

- $\mathcal{T}_S(EX\{\mathtt{tau}\}\phi) = EX\{\mathtt{tau}\}\mathcal{T}_S(\phi)$

- $\mathcal{T}_S(EX\{x!y\}\phi) = EX\{x!y\}\mathcal{T}_S(\phi)$

- $\mathcal{T}_S(EX\{x!(y)\}\phi) = \bigvee_{\alpha\in S} EX\{x!(\alpha)\}\mathcal{T}_S(\phi\theta)$, *where* $\theta = \{\alpha/y\}$

- $\mathcal{T}_S(EX\{x?y\}\phi) = EX\{x?y\}\mathcal{T}_S(\phi) \vee \bigvee_{\alpha\in S} EX\{x?(\alpha)\}\mathcal{T}_S(\phi\theta)$, *where* $\theta = \{\alpha/y\}$

- $\mathcal{T}_S(< \mathtt{tau} > \phi) =< \mathtt{tau} > \mathcal{T}_S(\phi)$

- $\mathcal{T}_S(< x!y > \phi) =< x!y > \mathcal{T}_S(\phi)$

- $\mathcal{T}_S(< x!(y) > \phi) = \bigvee_{\alpha\in S} < x!(\alpha) > \mathcal{T}_S(\phi\theta)$, *where* $\theta = \{\alpha/y\}$

- $\mathcal{T}_S(< x?y > \phi) =< x?y > \mathcal{T}_S(\phi) \vee \bigvee_{\alpha\in S} < x?(\alpha) > \mathcal{T}_S(\phi\theta)$, *where* $\theta = \{\alpha/y\}$

- $\mathcal{T}_S(EF\phi) = EF\mathcal{T}_S(\phi)$

In the above definition we have assumed that when $S = \emptyset$ then $\bigvee_{\alpha\in S}\phi = false$

Note that the complexity of the translation has a worst case complexity which is exponential in the number of names appearing in set S.

EXAMPLE 3. *Consider again agent P in Example 1. Agent P satisfies the π-logic formula $\phi = EX\{x?u\}\mathtt{true}$ for each name u, since $P \xrightarrow{x?u}$* `nil` *for each name u. We want to verify whether the* ACTL *translation of the formula holds in the LTS associated with P, hence we have to consider the* ACTL *translation of the formula with respect to the set of fresh names S used in the LTS of P (see Figure 1), that is $\{a\}$. Hence, the translation of the formula is:*

$$EX\{x?u\}\mathtt{true} \vee EX\{x?(a)\}\mathtt{true}$$

Note that the resulting ACTL *formula holds in the LTS of P.*

THEOREM 2. *Let P be a π-calculus agent, with associated LTS $\mathcal{A}$, and ϕ be a π-logic formula. Then $P \models \phi$ if and only if $\Downarrow P \models \mathcal{T}_S(\phi)$, where S is the set of fresh names of $\mathcal{A}$.*

Outline of the proof: In order to prove the theorem, we use structural induction on ϕ. Since the logic has a negation operator, we only have to show one implication. We will show that whenever $P \models \phi$ then $\Downarrow P \models \mathcal{T}_S(\phi)$.

We will prove the theorem for $\phi = EX\{x?y\}\phi'$, the other cases are easier. We recall that
$\mathcal{T}_S(EX\{x?y\}\phi') = EX\{x?y\}\mathcal{T}_S(\phi') \vee \bigvee_{\alpha \in S} EX\{x?(\alpha)\}\mathcal{T}_S(\phi'\{\alpha, y\})$.
Assume that $P \models EX\{x?y\}\phi'$. Then, there exists P' such that $P \stackrel{x?y}{\longrightarrow} P'$ and $P' \models \phi$. By induction, we get $\Downarrow P' \models \mathcal{T}_S(\phi)$. Now, we have to consider two cases, i.e. $y \in an(P)$ and $y \notin an(P)$.

case $y \in an(P)$
If $y \in an(P)$ then $\Downarrow P \stackrel{x?y}{\mapsto} \Downarrow P'$. Hence, $\Downarrow P \models EX\{x?y\}\mathcal{T}_S(\phi')$, which implies $\Downarrow P \models \mathcal{T}_S(EX\{x?y\}\phi')$.

case $y \notin an(P)$
In this case, the LTS does not contain an explicit input transition labelled by $x?y$ from $\Downarrow P$. Instead, we have that there exists the transition $\Downarrow P \stackrel{x?(\beta)}{\mapsto} \Downarrow P''$, where $\beta = min(\mathcal{N} - an(P))$, $P \stackrel{x?\beta}{\longrightarrow} P''$ and $P'' = P'\{\beta/y\}$. We will show that $\Downarrow P \models EX\{x?(\beta)\}\mathcal{T}_S(\phi'\{\beta/y\})$. In order to show this, we have to show that $\Downarrow P'' \models \mathcal{T}_S(\phi'\{\beta/y\})$. Hence, it is sufficient to prove that $\Downarrow P'\{\beta/y\} \models \mathcal{T}_S(\phi'\{\beta/y\})$. To do this we prove (by using structural induction on ϕ) that $P' \models \phi'$ implies $P'\{\beta/y\} \models \phi'\{\beta/y\}$, and then apply the induction hypothesis.

The model checker of π-logic formulae was developed by implementing a logic translator from π-logic formulae into ACTL formulae, and by using tools that we integrated in the JACK verification environment (Gnesi, 1997; Ferrari et al., 1997). The JACK environment, which combines different specification and verification tools, allows one to generate the LTS associated with a given π-calculus agent (Ferrari et al., 1997), and to reduce its size by using standard automata minimization tools. Then, the model checking of π-logic formulae can be done by using the logic translator from π-logic formulae into ACTL formulae, and the AMC model checker for the ACTL logic, inside JACK.

The complexity of the model checking is due to the construction of the state space of the π-calculus agent to be verified, which is, in the worst case, exponential in the syntactical size of the agent.

6. A verification example

In this section we show an example of the verification of some π-logic properties of a system specified by a π-calculus agent.

The system was inspired by the principles used by a new generation of Web browsers, like the one described in (Gosling and McGilton,

1996). Unlike from other Web browsers, which have a static knowledge of Internet data, protocols and behaviours, the browser we want to specify knows essentially none of them. However, it is able to dynamically increase its capabilities by means of a (transparent) software migration across the network. When a request is raised by a user, the special software applications needed to display the data are automatically installed in the user's system. This allows content developers to feel free to add new features to their programs, without having to provide new browsers and servers with added capabilities. Moreover, the dynamic loading of the protocols used allows the end user to use a unique Web browser, instead of selecting some specialized browser that can access the data. The browser will search the target system for the software code to add to the local system in order to correctly serve the user request.

We will give the system specification by using π-calculus formalism, that can suitably be used to describe the dynamic behaviour of the system. The system receives the user's request (the name of the desired object, its class, and the name of the host on which the object is located) via the channels *lo, lc, lh.* The system then sends the host a request to get the protocol to display the objects of the given class. Once the system has obtained the protocol code from the host, it sends the protocol to the local system (using the channel *lp*) and starts the procedure to get the data. The request for the data object is sent to the host. Once the system has received the data from the host, it sends them to the local system (using the channel *ld*). The local system can now understand the data and display them correctly. Below we report the π-calculus specification of the system. The system specification consists of two modules running in parallel: the *get-data* and the *get-protocol* modules. The *get-protocol* module uses the channel *dp* to communicate to *get-data* the name of the host channel to be used to get the data object. For the sake of simplicity we assume that the channel to be used to get both the protocol and the data is the same.

```
parseterm get-protocol(lc,lh,lp,dp) := (lc?(class). lh?(g).
g!class. g?(protocol). dp!g. lp!protocol. nil) endterm

parseterm get-data(lo,ld,dp) := (lo?(object). dp?(g).
g!object. g?(data). ld!data. nil) endterm

parseterm browser(lc,lh,lo,ld,lp) := ((dp)(
get-data(lo,ld,dp) || get-protocol(lc,lh,lp,dp)))
endterm
```

Some of the formulae that we want to verify on the system to guarantee its correct behaviour are:

1. Whenever a class name c and a host name h are communicated by the user to the system, the system can send a request to the host for the protocol of that class, and receive a protocol p from the host that will be loaded on the local system of the user:
$AG([lc?c][lh?h]EF(< h!c >< h?p >< lp!p > true))$

2. Whenever an object name o and a host name h are communicated by the user to the system, the system can send a request to the host for the data object, receive the data from the host, and send them to the local system:
$AG([lo?o][lh?h]EF(< h!o >< h?d >< ld!d > true))$

3. An object name o and a host name h may be communicated by the user to the system, and so the system may ask h for the data object and h may communicate some data d which will be loaded on the local system:
$EF(< lo?o >< lh?h > EF(< h!o >< h?d >< ld!d > true))$

4. Whenever a host name h and an object name o are communicated by the user to the system, the system can send a request to the host for the data object, receive the data from the host, and send them to the local system:
$AG([lh?h][lo?o]EF(< h!o >< h?d >< ld!d > true))$

5. A host name h and an object name o may be communicated by the user to the system, and so the system may ask h for the data object and h may communicate some data d that will be loaded on the local system:
$EF(< lh?h >< lo?o > EF(< h!o >< h?d >< ld!d > true))$

In order to model check these properties on the system, we first have to generate the LTS associated with the system. To do this we use the tool described in (Ferrari et al., 1997). The LTS was produced in 4.27 seconds (using a Sun Ultra 1), and has 245 states and 463 transitions. However, after a minimization step (with respect to weak bisimulation) we obtained an LTS with 86 states and 180 transitions. Figure 2 reports only a part of the LTS obtained, where the set of the fresh names of the LTS is $\{\sharp 0, \sharp 1, \sharp 2\}$. The unexplored states are labelled u.

Note that the LTS representation of the system allows us to exploit the minimization and the graphical editing facilities offered by tools that were built for LTSs.

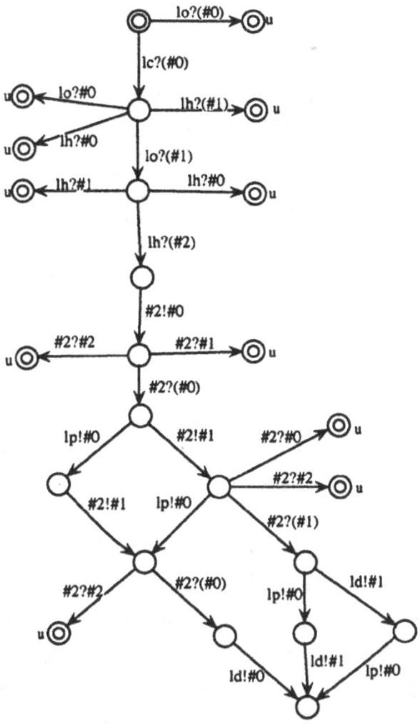

Figure 2. The LTS associated with the system.

The next step is to translate the π-logic formulae into ACTL formulae. To do this we give in input to the automatic formulae translator the π-logic formulae and the LTS on which the translated formulae have to be interpreted. After the translator has captured the information about the fresh names used in the LTS, it translates the given formulae into the corresponding ACTL ones. The resulting ACTL formulae can then be checked on the LTS, by using the ACTL model checker, AMC (Ferro, 1994). All the formulae turned out to be true in the LTS, as we had expected.

```
AG([lc?c][lh?h]EF(<h!c><h?p><lp!p>true)).
The formula is TRUE in state 0 time:
(user: 0.04 sec, sys: 0.00 sec)

AG([lo?o][lh?h]EF(<h!o><h?d><ld!d>true)).
The formula is TRUE in state 0 time:
(user: 0.04 sec, sys: 0.00 sec)

EF(<lo?o><lh?h>EF(<h!o><h?d><ld!d>true)).
The formula is TRUE in state 0 time:
(user: 0.03 sec, sys: 0.00 sec)
```

```
AG([lh?h][lo?o]EF(<h!o><h?d><ld!d>true)).
The formula is TRUE in state 0 time:
(user: 0.04 sec, sys: 0.00 sec)

EF(<lh?h><lo?o>EF(<h!o><h?d><ld!d>true)).
The formula is TRUE in state 0 time:
(user: 0.03 sec, sys: 0.00 sec)
```

7. Conclusions

The proposed model checker for the verification of logical properties of π-calculus agents exploits automata-based approaches to logic verification. Given an automaton and a formula, the model checker determines, using the transition relation among the states and the labelling of the transitions, the set of the states of the automaton that satisfy the formula. Once the finite LTS (an automaton) of a π-calculus agent has been generated (Montanari and Pistore, 1995; Ferrari et al., 1997), efficient model checking (in the style of (Clarke et al., 1986)) of π-logic formulae can be performed.

The complexity of our methodology is due to the construction of the state space of the π-calculus agent to be verified, which is, in the worst case, exponential in the syntactical size of the agent.

Some comparisons have to be made between the logic verification functionalities offered by the proposed model checker and those offered by another existing tool for the verification of logic formulae for the π-calculus: the *Mobility Workbench* (Victor and Moller, 1994). Both the verification methodologies have exponential complexity when one formula has to be checked on an agent. On the other hand, a complexity improvement is obtained with our methodology when a set of formulae are checked on an agent. In fact, in our approach the state space of a π-calculus agent is built once and for all. Then the model checking of a set of formulae can efficiently proceed using the LTS associated with the agent and the AMC model checker. In the MWB, however, the verification of a set of formulae requires the generation of a new, exponential time, proof for each formula in the set.

As a future development, we plan to study classical model checking techniques, based on the same approach we adopted for the verification of π-logic formulae, in order to verify properties expressed by means of a more powerful logic for π-calculus.

Acknowledgements: We wish to acknowledge Gianluigi Ferrari, Giovanni Ferro, Ugo Montanari and Marco Pistore for related works about the topics of the paper.

References

Clarke, E., E. Emerson, and A. Sistla: 1986, 'Automatic Verification of Finite–State Concurrent Systems Using Temporal Logic Specification'. *ACM Transaction on Programming Languages and Systems* **8**(2), 244–263.

Dam, M.: 1993, 'Model checking mobile processes'. In: *Proc. CONCUR'93*, Vol. 715 of *LNCS*.

Emerson, E. A. and J. Halpern: 1986, '"Sometime" and "Not never" revisited: On branching versus linear time temporal logic'. *Journal of ACM* **33**, 151–178.

Emerson, E. A. and C. Lei: 1986, 'Efficient Model Checking in Fragments of the Propositional Mu-Calculus'. In: *Proceedings of Symposium on Logics in Computer Science*. pp. 267–278.

Ferrari, G., G. Ferro, S. Gnesi, U. Montanari, M. Pistore, and G. Ristori: 1997, 'An automata based verification environment for mobile processes'. In: *Third International Workshop on Tools and Algorithms for the Construction and Analysis of Systems, TACAS'97*, Vol. 1217 of *LNCS*.

Ferro, G.: 1994, 'AMC: ACTL Model Checker. Reference Manual'. Technical Report B4-47, IEI.

Gnesi, S.: 1997, 'A formal verification environment for concurrent systems design'. In: *Proc. Automated Formal Methods Workshop, ENTCS*.

Gosling, J. and H. McGilton: 1996, 'The Java Language Environment. A White Paper'. Technical report, Sun Microsystems.

Hennessy, M. and R. Milner: 1985, 'Algebraic laws for nondeterminism and concurrency'. *Journal of ACM* **32**, 137–161.

Kozen, D.: 1983, 'Results on the propositional μ-calculus'. *Theoretical Computer Science* **27**, 333–354.

Milner, R.: 1989, *Communication and Concurrency*. Prentice-Hal.

Milner, R., J. Parrow, and D. Walker: 1992a, 'A calculus of mobile processes (parts I and II)'. *Information and Computation* **100**, 1–77.

Milner, R., J. Parrow, and D. Walker: 1992b, 'Modal logic for mobile processes'. In: *Proc. CONCUR'91*, Vol. 527 of *LNCS*.

Montanari, U. and M. Pistor: 1997, 'History Dependent Automata'. Technical report, Department of Computer Science, University of Pisa.

Montanari, U. and M. Pistore: 1995, 'Checking bisimilarity for finitary π-calculus'. In: *Proc. CONCUR'95*, Vol. 962 of *LNCS*.

Nicola, R. D. and F. W. Vaandrager: 1990, 'Action versus state based logics for transition systems'. In: *Proc. Ecole de Printemps on Semantics of Concurrency*, Vol. 469 of *LNCS*.

Sangiorgi, D.: 1992, 'Expressing mobility in process algebras: first-order and higher-order paradigms'. Ph.D. thesis, University of Edinburgh.

Stirling, C. and D. Walker: 1989, 'Local model checking in the modal μ-calculus.'. *Theoretical Computer Science* pp. 161–177.

Victor, B. and F. Moller: 1994, 'The Mobility Workbench — A tool for the π-calculus'. In: *Proc. CAV'94*, Vol. 818 of *LNCS*.

Walker, D.: 1991, 'π-calculus semantics for object-oriented programming languages'. In: *Proc. TACS'91*.

Interleaving Model and Verification of Distributed Probabilistic Real-time Systems

Tiegeng Luo (tgluo@csit.edu.cn) * and Huowang Chen, Bingshan Wang, Ji Wang, Zhichang Qi, Zhenghu Gong
Department of Computer Science,
Changsha Institute of Technology, Changsha, 410073, P.R. China

Abstract. In this paper, we present a quantitative model checking algorithm for verification of distributed probabilistic real-time systems (DPRS). First, we define a new model of DPRS, called alternating real-time probabilistic process model (ARPPM), which is over continuous time domain. In ARPPM, there are three kinds of states: *probabilistic states* , *nondeterministic states*, and *execution states* . The introduction of *execution states* is to distinguish nondeterminism of system scheduler choosing event from probabilistic distribution of delay time of transition. We describe the properties of DPRS by using TPCTL. The model checking algorithm depends on a suitable equivalence class partition for continuous time. In order to deal with quantified probabilities, the algorithm reduces the computation of the probability that a logic formula holds to a collection of finitely many linear recursive equations.

1. Introduction

The digital systems to control and interact with physical processes have been integrated in human society, such as computer network, vendor, control system of air traffic, rocket, etc.. Many processes work in real-time — the actual time of events is crucial for them, are probabilistic — no one knows which process is being scheduled and the accurate event delay time though they are usually from a probability distribution, and are nondeterministic — sometimes no one knows which process is performing the next instruction . These systems are called *distributed probabilistic real-time systems (DPRS)* .

Because physical processes are likely to fail, there is increasing awareness that the unexpected behavior of interacting processes may cause serious problems. Requirements for DPRS include functional properties as well as performance ones. It is often expected to calculate the probability that a system satisfies some required properties within a specified period of time. Therefore, it is necessary that a formal model should be capable of specifying temporal, probabilistic and nondeter-

* This work has been partially supported by the National 863 High Tech. Development Program of China (863-306-05-04-5), the National Natural Science Foundation of China (Grant NO. 69583002 and 69603010).

H. Barringer et al. (eds.), Advances in Temporal Logic, 359–375.

ministic aspects of DPRS. Consequently, it makes the analysis of DPRS even more challenging.

In the existing work, most of the researches (Alur et al., 1991a; Alur et al., 1991b) is concerned with properties that hold with probability 1 (qualitative method), and only (Hansson, 1991) deals with quantified probabilities when time domain is discrete (quantitative method). The most difficult problems in verification are how to solve state explosion and how to deal with continuous time with quantitative method.

Our aim is to provide a formal framework, including formal model and automatic verification for DPRS over continuous time domain. In the previous work (Luo et al., 1998), we have developed a better algorithm for checking whether a DPRS satisfied a deterministic timed automaton (DTA) specification with qualitative method. Whereas, quantitative method is more important in reality. In this paper, we present a new model for DPRS, called alternating real-time probabilistic process model (ARPPM), over continuous time domain, and develop a quantitative model checking algorithm that checks ARPPM w.r.t. TPCTL (Hansson, 1991) specification. Due to lack of space, we omit the introduction of TPCTL. Please refer to (Hansson, 1991) for details.

The rest of this paper is organized as follows. In section 2, we present ARPPM as a process model for DPRS. Then, model checking algorithm is presented based on equivalence class partition and labeling techniques. Finally, we give the concluding remarks.

2. Alternating Real-time Probabilistic Process Model

In this section, we define alternating real-time probabilistic process model (ARPPM) for DPRS, which is based upon GSMP (W.Whitt, 1980; Alur et al., 1991a; Alur et al., 1991b) and LCMC (Hansson, 1991).

DEFINITION 1. *(ARPPM)*
An ARPPM of DPRS is a 6-tuple $\mathcal{M} = < S, \Sigma, E, F, \Delta, s_0 >$, *where*

- $S = S_N \cup S_P \cup S_E$, *is a non-empty finite set, whose elements are referred to as states.* $S_N \cap S_P = \Phi$, $S_N \cap S_E = \Phi$, $S_E \cap S_P = \Phi$, $S_E \neq \Phi$. $s_0 \in S$ *is the initial state.* $S_N \subset S$ *is the set of non-deterministic states,* $S_P \subset S$ *is the set of probabilistic states,* $S_E \subseteq S$ *is the set of execution states.*
- Σ *is a non-empty finite set,* $a \in \Sigma$ *is referred to as an event.*
- *A mapping* $E : S \longrightarrow 2^{\Sigma}$. *For any* $s \in S$, $E(s)$ *is the set of events that are enabled when the process is in state* s. *If* $s \in S_E$,

then all events in $E(s)$ will be scheduled, not be discarded (disabled).

– *Delay distributions F. The set of events is partitioned into two sets: Σ_e are the events with exponential distribution, and Σ_b are the events with bounded distribution. All the events are independent. The functions $l : \Sigma_b \longrightarrow N$ and $u : \Sigma_b \longrightarrow N$, giving the lower and upper bounds on the times at which bounded events get scheduled. A bounded event x is called a fixed-delay event if $l_x = u_x$, and a variable-delay event if $l_x < u_x$. For each $x \in \Sigma_e$, $l_x = 0, u_x = +\infty$. Associated with each variable-delay event x and $x \in \Sigma_e$ is a probability density function $f_x(t)$ such that $\int_{l_x}^{u_x} f_x(t)\,\mathrm{d}t$ equals 1, and for any $l_1, u_1 : l_x \leq l_1 < u_1 \leq u_x$, $\int_{l_1}^{u_1} f_x(t)\,\mathrm{d}t > 0$. The function $pri : \Sigma \longrightarrow N$ gives the priorities of events, and any two events have different priorities.*

– *Transitions $\Delta = \Delta_N \cup \Delta_P \cup \Delta_E$. $\Delta_N \cap \Delta_P = \Phi$, $\Delta_N \cap \Delta_E = \Phi$, $\Delta_E \cap \Delta_P = \Phi$, $\Delta_E \neq \Phi$. $\Delta_N : S_N \longrightarrow 2^S$, is the non-deterministic transition relation. If $s_2 \in \Delta_N(s_1)$, it can also be represented as $s_1 \longmapsto s_2$. $\Delta_P : S_P \times (0,1] \longrightarrow 2^S$, is the probabilistic transition relation. If $s_2 \in \Delta_P(s_1, p)$, it can also be represented as $s_1 \overset{p}{\longmapsto} s_2$. There exists $\epsilon > 0$ such that $p > \epsilon$ for any $s_1 \overset{p}{\longmapsto} s_2$, and for any state $s_1 \in S_P$, $\sum_{s_2 \in \Delta_P(s_1,p)} p = 1$. $\Delta_E : S_E \times \Sigma \longrightarrow S$, is the execution transition relation. If $s_2 = \Delta_E(s_1, a)$, it can also be represented as $s_1 \overset{a}{\longrightarrow} s_2$. If the current state of $\mathcal{M}$ is $s_1 \in S_E$, then the necessary and sufficient condition of $\mathcal{M}$ performing transition $s_1 \overset{a}{\longrightarrow} s_2$ is: if $x \in old(s_1, a, s_2) = E(s) - \{a\}$, then its remaining delay time in s_2 is $rdt_{s_2}(x) = rdt_{s_1}(x) - rdt_{s_1}(a) \geq 0$, and if $rdt_{s_2}(x) = 0$, then $pri(a) > pri(x)$. If $x \in new(s_1, a, s_2)$, then its remaining delay time in s_2, $rdt_{s_2}(x)$, is assigned a time value, independently of the present and past history of the system, from the probability distribution on $[l_x, u_x]$ in F.*

where, the function $rdt : S \times \Sigma \longrightarrow R^{\geq 0}$ records the remaining delay time of all enabled events when system reaches a state. $rdt_s(x)$ denotes the remaining delay time of x when system just reaches state s. $rdt_{s_0}(x) \in [l_x, u_x]$ when system starts from s_0 at its first time. For each execution transition $s_1 \overset{a}{\longrightarrow} s_2$, the functions $old(s_1, a, s_2)$ and $new(s_1, a, s_2)$ are defined as follows.

– The set $old(s_1, a, s_2) = E(s_2) \cap (E(s_1) - \{a\})$ is the set of events that were enabled previously, and

– The set $new(s_1, a, s_2) = E(s_2) - old(s_1, a, s_2)$ is the set of events that are newly enabled.

ARPPM consists of three kinds of states: *non-deterministic states*, *probabilistic states* , and *execution states*. In *nondeterministic states*, scheduler can non-deterministicly choose an event, then system reaches a new state. In *probabilistic states*, scheduler can choose different event according to their distribution probabilities. In every *execution state*, all enabled events in $E(s)$ would be scheduled, not be discarded (disabled). The introduction of *execution states* is to distinguish nondeterminism of system scheduler choosing event from probabilistic distribution of delay time of transition. Sometimes scheduler choosing event is not constrained by any probabilistic distribution, and suitable to be described by nondeterministic state and nondeterministic transition. Whereas, in execution states, the delay time of an enabled event distributes according to a probability density function, hence, choosing a transition with triggering event and its delay time is constrained by a probabilistic distribution. It is suitable to be expressed by probabilistic states, probabilistic transition, and execution states, execution transition. The execution transition in ARPPM implies a group of probabilistic transition (choosing delay time) and a group of execution transition with exact delay time.

In ARPPM, the process starts in state s_0. The set of initially scheduled events is $E(s_0)$. In every state s, a time value $rdt_s(x)$ is associated with each enabled event x, and it shows delay time remaining till the occurrence of the event when system just enters state s. The initial time value of event x lies in $[l_x, u_x]$. At probabilistic states and nondeterministic states, a state-transition is triggered without passage of time. At execution states s, a state-transition occurs by event x when the elapsed time t_e is equal to the time value $rdt_s(x)$ of event x in state s. With this transition some events get descheduled, whereas some events get newly scheduled. Because all the time value count down at the same rate as the real-time, the remaining delay time of event y that was scheduled previously must sub t_e $(= rdt_s(x))$, and the remaining delay time of event y that was newly scheduled lies in $[l_x, u_x]$.

The ARPPM is a finite state transition system, but we can not directly calculate the probabilities concerning with time on ARPPM because its state doesn't include exact time value. Therefore, it is necessary to define a new concept with exact time value. A *generalized state* of system is described by giving its state and a vector giving the remaining delay time corresponding to all the currently scheduled events. A generalized state is then represented by $< s, v >$, where s is a state of ARPPM, and v is a vector of time of length $|E(s)|$ assigning remaining delay time. If $s \in S_N$, then $< s, v >$ is called a generalized non-deterministic state; if $s \in S_P$, then $< s, v >$ is called a generalized probabilistic state; if $s \in S_E$, then $< s, v >$ is called a generalized execution state.

A particular behavior (experiment sequence) of a process described by ARPPM $\mathcal{M}$ can be represented by an alternating timed word. In an alternating timed word, we use γ to denote the non-deterministic transition, and ζ_p to denote the probabilistic transition.

DEFINITION 2. *(Alternating timed word) Assume that* $\sigma = \sigma_1\sigma_2\sigma_3\cdots$, *is an infinite word over* $\Sigma \cup \{\gamma, \zeta_p\}$ *(ω sequence); $\vartheta = \vartheta_1\vartheta_2\vartheta_3\cdots$, is an infinite sequence of non-negative real numbers. If there exists an infinite sequence of states on ARPPM $\mathcal{M}$: $s_0, s_1, s_2, \cdots$, such that,*

$$\begin{array}{l} s_{i-1} \xrightarrow{\sigma_i} s_i \quad (i \in N, i \geq 1) \\ if \quad \sigma_i = \gamma \quad then \quad \vartheta_i = 0 \\ else\, if \quad \sigma_i = \zeta_p \quad then \quad \vartheta_i = 0 \\ \qquad else \quad \vartheta_i = rdt_{s_{i-1}}(\sigma_i) \end{array}$$

Where, $rdt_{s_{i-1}}(\sigma_i)$ is the remaining delay time of σ_i in s_{i-1}. Then, the ω sequence of 2-tuple $\rho = (\sigma, \vartheta)$ is called an alternating timed word of $\mathcal{M}$.

Let σ be a finite word over $\Sigma \cup \{\gamma, \zeta_p\}$, $\sigma_1\sigma_2\cdots\sigma_n$ $(n \geq 1)$; and ϑ be a finite sequence of non-negative real numbers, $\vartheta_1\vartheta_2\cdots\vartheta_n$ $(n \geq 1)$. If they satisfy above conditions, then, the finite sequence of 2-tuple $\rho = (\sigma, \vartheta)$ is also called an alternating timed word of $\mathcal{M}$.

3. Verification

In this section, we present a quantitative model checking algorithm for checking if a given ARPPM $\mathcal{M}$ satisfies a TPCTL formula f. We extend the model checking algorithm of (Hansson, 1991) over continuous time domain. The key part in our algorithm is to map infinite generalized states of ARPPM to finite super states. On the finite state graph the algorithm is designed so that when it terminates each transition will be labeled with the set of subformulas of f that are true in the transition. It can be concluded that $\mathcal{M}$ satisfies f if the initial transition of the finite state graph is labeled with f.

Let *BIGGEST* be a big enough integer constant (greater than the biggest time constant in TPCTL formula f). Because all the time constraints in TPCTL formula f is of the form "$\leq t$", the boolean value of f with time constraints will not be relevant to the time interval whose elements are greater than *BIGGEST*. Hence, it is unnecessary to check the time interval greater than *BIGGEST*.

3.1. CONSTRUCTING FINITE STATE GRAPH

An ARPPM $\mathcal{M}$ has infinite generalized states because of the continuous time, but the model checking algorithm can only deal with finite state systems directly. Therefore, we must map infinite generalized states of $\mathcal{M}$ to finite super states by uniting equivalence class of generalized states.

It should be specially noticed that all the time values are integer constants in the TPCTL formula f, and f checks the time constraint only when a execution transition is triggered. Therefore, a generalized execution state $gs_1 =< s, v_1 >$ *is equivalent to* another generalized execution state $gs_2 =< s, v_2 >$ iff for gs_1 and gs_2 the following conditions are satisfied:

1. If the execution path reaching gs_1 is the same as the one reaching gs_2, the system time starting from s_0 lies in the same unit time.
2. The delay time in state s before transition being triggered lies in the same unit time.
3. If the execution path reaching gs_1 is the same as the one reaching gs_2, then, after the execution transition, the system time still lies in the same unit time.
4. The probabilities of unit sub-interval of remaining delay time interval of other events can be calculated after the execution transition.

A relation satisfies above conditions is reflective, symmetrical and transitive, therefore it is an equivalence relation.

For $t \in R^+$, let $\lceil t \rceil$ be the smallest integer which is greater than or equal to t. Apparently, when the triggering event is x_0 the condition 2 is satisfied iff $\lceil v_1(x_0) \rceil = \lceil v_2(x_0) \rceil$. But the condition 1, 3 and 4 are difficult to be satisfied. We find that a supplementary flag variable can help us to solve the problem. The flag variable of each event denotes a certain constraint relation between the remaining delay time of this event and the event triggering the execution transition. The constraint relation can ensure that the condition 1, 3 and 4 will be satisfied.

DEFINITION 3. *(Flag variable)*
For a generalized execution state $gs =< s, v >$, we assume that x_0 is the triggering event, $E(s) = \{x_0, x_1, \cdots x_n\} (n \geq 0)$. For $0 \leq i \leq n$, let $\bar{u_i} = \lceil v(x_i) \rceil$. For $1 \leq i \leq n$, the flag variable value $ut(x_i)$ of event x_i is:

$$ut(x_i) = 0 \textit{ iff } v(x_i) > v(x_0) + \bar{u_i} - \bar{u_0}$$
$$ut(x_i) = 1 \textit{ iff } v(x_i) \leq v(x_0) + \bar{u_i} - \bar{u_0}$$

And, we assume that x_s *is the system clock,* $v(x_s)$ *is the system time, let* $\bar{u_s} = \lceil v(x_s) \rceil$, $\bar{l_0} = \bar{u_0} - 1$, *then,*

$$ut(x_0) = 0 \text{ iff } v(x_s) \leq \bar{l_0} + \bar{u_s} - v(x_0)$$
$$ut(x_0) = 1 \text{ iff } v(x_s) > \bar{l_0} + \bar{u_s} - v(x_0)$$

DEFINITION 4. *A generalized execution state* $gs_1 =< s, v_1, ut >$ *is equivalent to* $gs_2 =< s, v_2, ut >$ *iff* $\lceil v_1(x_0) \rceil = \lceil v_2(x_0) \rceil$, *where* x_0 *is the triggering event.*

DEFINITION 5. *(Super state) A super state* $[s, u, ut]$ *denotes the equivalence class to which generalized execution state* $< s, v, ut >$ *belongs. In* $[s, u, ut]$, s *is an execution state of ARPPM,* u *is a vector of time intervals of length* $|E(s)|$, *whose element is assigned the remaining delay time interval of a event.* ut *is a vector of event flag variable,* $ut(x)(x \in E(s))$ *is the flag variable value of event* x.

THEOREM 1. *Any generalized states corresponding to a super state* $[s, u, ut]$ *satisfies conditions 1, 2, 3 and 4.*

Proof: We assume that x_0 is a triggering event, $E(s) = \{x_0, x_1, \cdots x_n\}(n \geq 0)$. For $0 \leq i \leq n$, let $u(x_i) = [\bar{l_i}, \bar{u_i}]$. And assuming that next super state is $[s_2, u_2, ut_2]$ after the execution transition from super state $ss = [s, u, ut]$. The proof is trivial when x_0 is a fixed-delay event.

Condition 2. By Definition 4 and 5, trivially satisfied.

Condition 1, 3. When the system enters super state $[s_0, u_0, ut_0]$ at its first time, the system time is 0, therefore, the system time still lies in the same unit time after the first execution transition.

Assuming that s is the *i-th* execution state in an execution path, system time $v(x_s)$ (of system clock x_s) lies in the same unit time interval $u(x_s) = [\bar{l_s}, \bar{u_s}]$ in super state ss. Meanwhile, we assume that the system time $v_2(x_s)$ lies in $u_2(x_s) = [\bar{l_s^2}, \bar{u_s^2}]$ after the *ith* execution transition from ss, then, for $< s, v, ut >\in ss$,

Case 1: $ut(x_0) = 0$.
Because $v(x_s) \leq \bar{l_0} + \bar{u_s} - v(x_0)$, $v_2(x_s) = v(x_s) + v(x_0) \leq \bar{l_0} + \bar{u_s}$. Therefore, $\bar{u_s^2} = \bar{u_s} + \bar{l_0}$. And because $v_2(x_s) = v(x_s) + v(x_0) \geq \bar{l_s} + \bar{l_0}$, $\bar{l_s^2} = \bar{l_s} + \bar{l_0}$. Therefore, $u_2(x_s) = [\bar{l_s} + \bar{l_0}, \bar{u_s} + \bar{l_0}]$.

Case 2: $ut(x_0) = 1$. Similar to Case 1.
Therefore, $u_2(x_s)$ is an interval of unit time, condition 1 and 3 are proved by induction.

Condition 4. Without loss of generality, let x_1 be one of the other events. For event $x_1 \in E(s)$, we assume that the remaining delay time of x_1 is $v(x_1)$ in super state ss. Let $\bar{u_1^u} = \lceil v(x_1) \rceil$, $\bar{l_1^u} = \bar{u_1^u} - 1$. And assuming that the remaining delay time of x_1 is $v_2(x_1)$ when reaching super state $[s_2, u_2, ut_2]$, then,

Case 1: $ut(x_1) = 0$.
Because $v(x_1) > v(x_0) + \bar{u_1^u} - \bar{u_0}$, $v_2(x_1) = v(x_1) - v(x_0) > \bar{u_1^u} - \bar{u_0} = \bar{l_1^u} - \bar{l_0}$. And because $v_2(x_1) = v(x_1) - v(x_0) \leq \bar{u_1^u} - \bar{l_0}$. $v_2(x_1)$ lies in the unit time interval $[\bar{l_1^u} - \bar{l_0}, \bar{u_1^u} - \bar{l_0}]$.

Case 2: $ut(x_1) = 1$. Similar to Case 1.
We can conclude that

$$\text{If } ut(x_1) = 0, \text{ then } u_2(x_1) = (\bar{l_1} - \bar{l_0}, \bar{u_1} - \bar{l_0}) \ ;$$
$$\text{If } ut(x_1) = 1, \text{ then } u_2(x_1) = (\bar{l_1} - \bar{u_0}, \bar{u_1} - \bar{u_0}).$$

Let $[\bar{l_1^u}, \bar{l_1^u} + 1] \subseteq [\bar{l_1}, \bar{u_1}]$ is an unit time interval, its distribution probability is p_1^u. Then, because $x_0, x_1, \ldots x_n$ are independent events, the distribution probability of unit time interval with flag variable $< [\bar{l_1^u}, \bar{l_1^u} + 1], ut(x_1) >$ is $0.5 \times p_1^u$.

Therefore, after execution transition, the distribution probability of interval $< [\bar{l_1^u} - \bar{l_0}, \bar{l_1^u} + 1 - \bar{l_0}], 0 >$ is equal to the distribution probability of interval $< [\bar{l_1^u}, \bar{l_1^u} + 1], 0 >$, is $(0.5 \times p_1^u)$. Similarly, the distribution probability of interval $< [\bar{l_1^u} - \bar{u_0}, \bar{l_1^u} + 1 - \bar{u_0}], 1 >$ is also equal to $(0.5 \times p_1^u)$.

Therefore, condition 4 is proved.

Remark: the super state defined above is only the equivalence class of generalized execution states. And, a group of generalized non-deterministic states or generalized probabilistic state as $< s, v, ut >$, are equivalent if s is the same one because there is no elapsed time. This kind of equivalence class is also called super state, and represented as $[s, u, ut]$ too.

DEFINITION 6. *(Super ARPPM (SARPPM)) A SARPPM is a 4-tuple* $< S^s, \Sigma, \Delta^s, s_0^s >$, *where*

- $S^s = S_N^s \cup S_P^s \cup S_E^s$, *is a non-empty finite set, whose elements are referred to as super states,* $S_N^s \cap S_P^s = \Phi$, $S_N^s \cap S_E^s = \Phi$, $S_E^s \cap S_P^s = \Phi$, $S_E^s \neq \Phi$. $s_0^s = [s_0, u, ut] \in S^s$ *is the initial super state.* S_N^s *is the set of super non-deterministic states;* S_P^s *is the set of super probabilistic states;* S_E^s *is the set of super execution states.*

- *Super transitions set* $\Delta^s = \Delta_N^s \cup \Delta_P^s \cup \Delta_E^s$. $\Delta_N^s \cap \Delta_P^s = \Phi$, $\Delta_N^s \cap \Delta_E^s = \Phi$, $\Delta_E^s \cap \Delta_P^s = \Phi$, $\Delta_E^s \neq \Phi$. $\Delta_N^s : S_N^s \longrightarrow 2^{S^s}$, *is a super non-deterministic transition relation. If* $s_2^s \in \Delta_N^s(s_1^s)$, *it can also be represented as* $s_1^s \longmapsto s_2^s$. $\Delta_P^s : S_P^s \times (0, 1] \longrightarrow 2^{S^s}$, *is a*

Assigning flag variables.

super probabilistic transition relation. If $s_2^s \in \Delta_P^s(s_1^s, p)$, *it can also be represented as* $s_1^s \overset{p}{\longmapsto} s_2^s$. *If* $tr = s_1^s \overset{p}{\longmapsto} s_2^s$, *then let* $p(tr) = p$. *There exists* $\epsilon > 0$ *so that* $p(tr) > \epsilon$ *for any* $tr \in \Delta_P^s$, *and for any state* $s_1^s \in S_P^s$, $\sum_{s_2^s \in \Delta_P^s(s_1^s, p)} p = 1$. $\Delta_E^s : S_E^s \times \Sigma \longrightarrow S^s$, *is a super execution transition relation. If* $s_2^s = \Delta_E^s(s_1^s, a)$, *it is also represented as* $s_1^s \overset{a}{\longrightarrow} s_2^s$. *The necessary and sufficient condition of* $tr = s_1^s \overset{a}{\longrightarrow} s_2^s \in \Delta_E^s$ *is: assuming that* $s_1^s = [s_1, u_1, ut_1]$, $s_2^s = [s_2, u_2, ut_2]$, $u_1(x) = [\bar{l_x}, \bar{u_x}](x \in E(s_1))$, *for* $x \in E(s_1) - \{a\} \subseteq E(s_2)$,

$$\textit{if } ut(x) = 0, \textit{ then } u_2(x) = (\bar{l_x} - \bar{l_a}, \bar{u_x} - \bar{l_a}) \; ;$$
$$\textit{if } ut(x) = 1, \textit{ then } u_2(x) = (\bar{l_x} - \bar{u_a}, \bar{u_x} - \bar{u_a}).$$

which satisfies condition: $\bar{l_x} - \bar{u_a} \geq 0$; *if* $x \in new(s_1, a, s_2)$, $u_2(x) = [l_x, u_x]$ *is the delay time interval of event* x. *For* tr, *let* $\mathcal{L}(tr) = a$, $ut(tr) = ut_1(a)$, $d(tr) = u_1(a)$.

SARPPM is a finite state graph, the delay time of its every super execution transition is a unit time interval.

It seems that we can construct SARPPM G from ARPPM $\mathcal{M}$. Now the problem is how to assign flag variables. Directly, flag variable denotes two kinds of constraint relation, which can be both 0 and 1. Whereas, for the events in $old(s_1, x, s)$, it is not obvious. We present the following algorithm to assign flag variables before constructing SARPPM. In algorithm 3.1, we, with set EOR and $EOR2$, record the constraint relation between event remaining delay time due to assigning flag variable in the previous execution transition. The element of EOR is 2-tuple $< a, b >$, where $ut(a) = 1$, $ut(b) = 0$. Similarly, $EOR2$ also record the constraint relation between event remaining delay time due to the assigning flag variable in the previous execution transition, of which the element is also 2-tuple $< a, b >$, where, the remaining delay time $v(a), v(b)$ of events a, b satisfy: $v(b) > v(a) + \lceil v(b) \rceil - \lceil v(a) \rceil$, but $ut(a) and ut(b)$ have same value.

GIVEN: For a super execution state $s_1^s = [s_1, u, ut]$ $(s_1 \in S_E)$, we assume that the triggering event is x_0, $E(s_1) = \{x_0, x_1, \cdots x_n\}$ $(n \geq 0)$, the system clock event is x_s, and $u(x_i) = [l_i, u_i](0 \leq i \leq n)$.

STEP 1. Assigning flag variables. Let $TE = \Phi$.

If the time value of event is fixed (the lower and upper bounds on the time is same), the flag variable value of the event is 1.

If unit time interval $[l_0, u_0] \subseteq u(x_i)$, then flag variable value of event x_i on unit time interval $[l_0, u_0]$ must be 0.

If $< x_0, b > \in EOR2$, then $ut(b)$ must be 0.

If $< x_0, b > \in EOR$, then $ut(b)$ must be 1.

If $< a, x_0 > \in EOR2$, then $ut(a)$ must be 1.

If $< a, x_0 > \in EOR$, then $ut(a)$ must be 0.

If $< a, b > \in EOR2, (a, b \neq x_0)$, and $ut(a) = 0$, then $ut(b) = 0$, $TE = TE \cup \{< a, b >\}$.

If $< a, b > \in EOR, (a, b \neq x_0)$, and $ut(a) = 1$, then $ut(b) = 1$, $TE = TE \cup \{< b, a >\}$.

Otherwise, $ut(x) \in \{0, 1\}$.

$ut(x_0) = ut(x_s)$.

STEP 2. Refreshing $EOR, EOR2$.

$$\begin{aligned} EOR \;\; = \;\; & \{< a, b > \mid a, b \in E(s_1), ut(a) = 1, ut(b) = 0\} \cup \\ & \{< a, b > \mid a, b \in E(s_1), a \text{ is clock or event with fixed time value} \\ & \qquad \text{(the lower and upper bounds on the time is same)}\} \end{aligned}$$

$EOR2 = TE - EOR$.

Replace all the x_0 in $EOR, EOR2$ with x_s.

Proof: We only consider the case that $< x_0, b > \in EOR(b \neq x_s)$. Assuming that the triggering event of previous super execution transition is x_t, and in previous super execution state, the remaining delay time of x_t is $v_0(x_t) \in [u_t - 1, u_t]$, the remaining delay time of x_0, and b are $v_0(x_0) \in [u_0 - 1, u_0]$, and $v_0(b) \in [u_b - 1, u_b]$ respectively. Then, from $ut(x_0) = 1, ut(b) = 0$, we have

$$u_0 - 1 \leq v_0(x_0) \leq v_0(x_t) + u_0 - u_t$$

$$u_b \geq v_0(b) > v_0(x_t) + u_b - u_t$$

After transition, the remaining delay time of x_0 and b are,

$$v(x_0) = v_0(x_0) - v_0(x_t) \in [u_0 - u_t - 1, u_0 - u_t]$$

$$v(b) = v_0(b) - v_0(x_t) \in [u_b - u_t, u_b - u_t + 1]$$

Then,

$$\begin{aligned} v(b) - v(x_0) & \leq u_b - v_0(x_t) - (u_0 - 1 - v_0(x_t)) \\ & = u_b - u_0 + 1 \\ & = (u_b - u_t + 1) - (u_0 - u_t) \end{aligned}$$

Therefore, $ut(b) = 1$.

The verification of other cases is similar.

Constructing SARPPM G.

Now, we can construct SARPPM G from ARPPM $\mathcal{M}$. In algorithm 3.2, for every super state $[s, u, ut]$, let p record the distributed probability of unit event delay time. For event x, we assume that $u(x) = [l_x, u_x]$, then p_l^x denotes the distributed probability of unit delay time interval $[l_x + l - 1, l_x + l](1 \leq l \leq u_x - l_x)$ of event x.

STEP 1. The initial super state of G is s_0^s. Let $NS = \{s_0^s\}$, $ECS = \Phi$, $OS = \Phi$, $EOR = \Phi$, $EOR2 = \Phi$. We assume that $s_0^s = [s_0, u, ut]$. For every $x \in E(s_0)$,

1. if $x \in \Sigma_b$, $u(x) = [l_x, u_x]$;

2. if $x \in \Sigma_e$, $u(x) = [0, +\infty]$;

STEP 2. Partitioned into equivalence classes.

Considering a super state $s_1^s = [s_1, u, ut] \in NS$. . *Case 1:* $s_1 \in S_N$.

For any transition $tr = s_1 \longmapsto s_2 \in \Delta_N$, $s_2^s = [s_2, u, ut]$ is a super state of G, $s_1^s \longmapsto s_2^s$ is a super non-deterministic transition of G, and $NS = NS \cup \{s_2^s\}$ if $s_2^s \notin OS$.

$NS = NS - \{s_1^s\}$, $OS = OS \cup \{s_1^s\}$.

Case 2: $s_1 \in S_P$.

For any transition $tr = s_1 \overset{p}{\longmapsto} s_2 \in \Delta_P$, $s_2^s = [s_2, u, ut]$ is a super state of G, $s_1^s \overset{p}{\longmapsto} s_2^s$ is a super probabilistic transition of G, and $NS = NS \cup \{s_2^s\}$ if $s_2^s \notin OS$.

$NS = NS - \{s_1^s\}$, $OS = OS \cup \{s_1^s\}$.

Case 3: $s_1 \in S_E$.

We assume that $E(s_1) = \{x_0, x_1, \cdots x_n\}(n \geq 0)$, $p_{s_1^s}$ denotes the product of distributed probabilities of the remaining delay time intervals of all events in $E(s_1)$. For any transition from s_1 in $\mathcal{M}$: $s_1 \xrightarrow{x_0} s_2$,

If x_0 is a fixed delay event, then, $NS = NS - \{s_1^s\}$, $ECS = ECS \cup \{s_1^s\} - OS$, $OS = OS \cup \{s_1^s\}$. The decomposition of s_1^s terminates for x_0.

Otherwise, assuming that $u(x_i) = [l_{x_i}, u_{x_i}]$ $(0 \leq i \leq n)$, let

$$U_{s_1^s} = min\{BIGGEST, min\{u_x | x \in E(s_1), u(x) = [l_x, u_x]\}\}$$

$[\bar{l_{x_0}}, \bar{u_{x_0}}] = [l_{x_0}, u_{x_0}] \cap [0, U_{s_1^s}]$. Decomposing $[\bar{l_{x_0}}, \bar{u_{x_0}}]$ into unit time intervals:

$$[\bar{l_{x_0}}, \bar{l_{x_0}} + 1], \ \cdots \ [\bar{l_{x_0}} + m - 1, \bar{l_{x_0}} + m(= \bar{u_{x_0}})] \quad (1 \leq m)$$

Hence, we obtain a set $SUB_{s_1^s}$ of sub-state (super state) from s_1^s. For all $ss_1^s \in SUB_{s_1^s}$, $ss_1^s = [s_1, u_s, ut_s]$, where $ut_s(x_i)$ are assigned according to step 1 of algorithm 3.1. $u_s(x_i)$ are defined as follows:

$$\begin{array}{lll} u_s(x_0) = & (\bar{l_{x_0}} + k - 1, \bar{l_{x_0}} + k) & (1 \leq k \leq m) \\ u_s(x_i) = & [\bar{l_{x_i}}, \bar{u_{x_i}}] = (l_{x_i}, u_{x_i}) \cap (\bar{l_{x_0}} + k - 1, u_{x_i}) & (1 \leq i \leq n) \end{array}$$

Calculating $EOR, EOR2$ according to step 2 of algorithm 3.1.

Then, ss_1^s is a super state of G, $s_1^s \overset{p}{\longmapsto} ss_1^s$ is a super probabilistic transition of G, where,

$$p = p_k^{x_0} \times \prod_{i=1}^{n} \left(\frac{1}{2} \times \sum_{j=1}^{u_{x_i} - l_{x_i}} p_j^{x_i}\right) \div p_{s_1^s}$$

$NS = NS - \{s_1^s\}$, $ECS = ECS \cup SUB_{s_1^s} - OS$, $OS = OS \cup \{s_1^s\}$.

STEP 3. Super transitions.

Considering a super state $s_1^s = [s_1, u, ut] \in ECS$.

Case 1: $s_1 \in S_N$.

For any transition $tr = s_1 \longmapsto s_2 \in \Delta_N$, and $s_2^s = [s_2, u, ut]$ is a super state of G, $s_1^s \longmapsto s_2^s$ is a super non-deterministic transition of G, and $ECS = ECS \cup \{s_2^s\}$ if $s_2^s \notin OS$.

$ECS = ECS - \{s_1^s\}$, $OS = OS \cup \{s_1^s\}$.

Case 2: $s_1 \in S_P$.

For any transition $tr = s_1 \overset{p}{\longmapsto} s_2 \in \Delta_P$, and $s_2^s = [s_2, u, ut]$ is a super state of G, $s_1^s \overset{p}{\longmapsto} s_2^s$ is a super probabilistic transition of G, and $ECS = ECS \cup \{s_2^s\}$ if $s_2^s \notin OS$.

$ECS = ECS - \{s_1^s\}$, $OS = OS \cup \{s_1^s\}$.

Case 3: $s_1 \in S_E$.

We assume that $E(s_1) = \{x_0, x_1, \cdots x_n\} (n \geq 0)$, $u(x_i) = [\bar{l_{x_i}}, \bar{u_{x_i}}] (0 \leq i \leq n)$. For any execution transition from s_1: $s_1 \xrightarrow{x_0} s_2$,

Assuming that the next super state of s_1^s is $s_2^s = [s_2, u_2, ut_2]$ after this transition, where u_2 and ut_2 are defined as follows.

If x_0 is a fixed delay event, $u_2(x_i) = [\tilde{l_{x_i}}, \tilde{u_{x_i}}] = [\bar{l_{x_i}} - \bar{l_{x_0}}, \bar{u_{x_i}} - \bar{l_{x_0}}] (1 \leq i \leq n)$. And, for any $x \in new(s_1, x_0, s_2)$, $u_2(x) = [l_x, u_x]$. Then s_2^s is a super state of G, and $s_1^s \xrightarrow{x_0} s_2^s$ is a super execution transition of G. Then, $ECS = ECS - \{s_1^s\}$, $NS = NS \cup \{s_2^s\} - OS$, $OS = OS \cup \{s_1^s\}$. The decomposition of s_1^s terminates for x_0.

Otherwise, for $1 \leq i \leq n$,

$$\text{if } ut(x_i) = 0,\ u_2(x_i) = [\tilde{l_{x_i}}, \tilde{u_{x_i}}] = [\bar{l_{x_i}} - \bar{l_{x_0}}, \bar{u_{x_i}} - \bar{l_{x_0}}];$$
$$\text{if } ut(x_i) = 1,\ u_2(x_i) = [\tilde{l_{x_i}}, \tilde{u_{x_i}}] = [\bar{l_{x_i}} - \bar{u_{x_0}}, \bar{u_{x_i}} - \bar{u_{x_0}}].$$

And, for any $x \in new(s_1, x_0, s_2)$, $u_2(x) = [l_x, u_x]$. The distributed probability, $p_k^{x_0}(1 \leq k \leq u_x - l_x)$, of every unit time interval for x_0 can be calculated according to the probability density function of event x_0.

$$p_k^{x_0} = \int_{\bar{l_{x_0}}+k-1}^{\bar{l_{x_0}}+k} f_{x_0}(t)\,\mathrm{d}t$$

Then s_2^s is a super state of G, and

$$tr = s_1^s \xrightarrow{x_0} s_2^s$$

is a super execution transition of G, $d(tr) = [\bar{l_{x_0}}, \bar{u_{x_0}}]$, $ut(tr) = ut(x_0)$. Then, $ECS = ECS - \{s_1^s\}$, $NS = NS \cup \{s_2^s\} - OS$, $OS = OS \cup \{s_1^s\}$.

STEP 4. If $NS = \Phi$ and $ECS = \Phi$, algorithm terminates. Otherwise, repeating STEP 2 and 3.

3.2. The Model Checking Algorithm

Now, we show how to label TPCTL formulas on SARPPM G.

3.2.1. *Add a New Initial Transition*

In SARPPM, there are possible many transitions from initial state, and the initial transitions must not be unique, and different formulas are labeled in different transitions. In order to judge whether the initial transition is labeled with formula f, there must exist an unique initial transition with zero delay time together with a null event that doesn't occur in SARPPM G.

DEFINITION 7. *(Initial transition hiding)*
For a SARPPM $G =< S^s, \Sigma, \Delta^s, s_0^s >$, *we define the new SARPPM* $\vec{G} = < S^s \cup \{s_{00}^s\}, \Sigma \cup \{\tau(\lambda)\}, \Delta^s \cup \{tr_\lambda\}, s_{00}^s >$, *where* $\tau(\lambda) \notin \Sigma$ *is a distinguish event that otherwise does not occur in G, is a null event,* $s_{00}^s \notin S^s$ *is a super execution state, and* tr_λ *is a super execution transition with delay time 0 defined by:*

$$tr_\lambda = s_{00}^s \xrightarrow{\tau(\lambda)/0} s_0^s$$

3.2.2. *Labeling Transitions with Formulas*

Let *label*(tr) denote the set of all subformulas of f that have been checked to be *true* in transition tr. Initially, *label*$(tr) = \Phi$ for all transitions in $\vec{G}$. The labeling starts from the smallest subformulas of f, i.e., the events. Transition tr is labeled with event a if $a \in \mathcal{L}(tr)$. Composite formulas are labeled based on the labeling of their parts. Labeling transitions with boolean formulas ($\neg f$ and $f_1 \wedge f_2$) is straight forward given the labeling of f_1 and f_2, i.e.. In this section, we deal with the modal operators *Until*.

The algorithm for labeling transitions with formulas of type $f = f_1 AU^{\leq t}_{\geq p} f_2$ and $f = f_1 AU^{\leq t}_{> p} f_2$ relies on the following definition and theorem:

DEFINITION 8. *(Neighbour transition.) For $tr_1, tr_2 \in \Delta^s$, tr_1 and tr_2 are neighbor transitions if tr_1 and tr_2 appear in an execution path, and there is no other transition between tr_1 and tr_2. It is also represented as $nbrt(tr_1, tr_2)$. And if tr_1 occurs before tr_2, $tr_2 \in ntt(tr_1)$.*

THEOREM 2. (Minimal satisfaction). *For a SARPPM $\vec{G}$ and a transition tr in $\vec{G}$, we define $P_u(tr, f_1, f_2, t, 0)$ to be the minimal probability that TCTL formula $f_1 AU^{\leq t} f_2$ is true in tr and all the subsequent execution pathes.*

Assuming that $ntt(tr_0) = \{tr_1, \cdots tr_n\}$, and $var \in \{0, 1\}$. Then, $P_u(tr_0, f_1, f_2, t, var)$ is:

1. $tr_0 \in \Delta^s_E$.

$$\begin{array}{lll} if & f_2 \in label(tr_0) \;\; then \;\; P_u(tr_0, f_1, f_2, t, var) = 1 \\ else & if \;\; f_1 \notin label(tr_0) \; OR \; t < 0 \;\; then \;\; P_u(tr_0, f_1, f_2, t, var) = 0 \\ & else \;\; calculating \; according \; to \; the \; following. \end{array}$$

2.If $tr_1, \cdots tr_n \in \Delta^s_E$, *then* $n = 1$. *If* $\mathcal{L}(tr_1)$ *is a fixed-delay event,* $d(tr_1) = [l, l]$, *then,* $P_u(tr_0, f_1, f_2, t, var) = P_u(tr_1, f_1, f_2, t - l, var)$, *otherwise, let* $d(tr_1) = [l, u]$,

$$P_u(tr_0, f_1, f_2, t, 0) = P_u(tr_1, f_1, f_2, t - u, 1)$$

$P_u(tr_0, f_1, f_2, t, 1) = P_u(tr_1, f_1, f_2, t_1, 1)$, *where,*

$$t_1 = \begin{cases} t - l & ut(\mathcal{L}(tr_1)) = 0 \\ t - u & ut(\mathcal{L}(tr_1)) = 1 \end{cases}$$

3.If $tr_1, \cdots tr_n \in \Delta^s_P$, *then,*

$$P_u(tr_0, f_1, f_2, t, var) = \sum_{i=1}^{n} p(tr_i) \times P_u(tr_i, f_1, f_2, t, var)$$

4.If $tr_1, \cdots tr_n \in \Delta^s_N$, *then,*

$$P_u(tr_0, f_1, f_2, t, var) = min_{i=1,\cdots n} P_u(tr_i, f_1, f_2, t, var)$$

Proof: We only verify the second calculation rule, and the other rules are trivial.

Let x_s be the system clock started from tr, and $[l_i, u_i](i = 0, 1)$ be unit time interval. x_s in tr_0 is in $[l_0, u_0]$, while x_s in tr_1 is in $[l_1, u_1]$. Let t_0 be time value appearing in f in tr, which means that remaining time for the following transitions is t_0 at most. Let t be time value appearing in f in tr_0, and t_1 be time value appearing in f in tr_1.

If $\mathcal{L}(tr_1)$ is a fixed-delay event, $d(tr_1) = [l, l]$, time interval of system clock is $[l_1, u_1] = [l_0 + l, u_0 + l]$, therefore, remaining time for the following transitions is

$$[t_0 - u_1, t_0 - l_1] = [t_0 - u_0 - l, t_0 - l_0 - l]$$

Therefore, $t_1 = t_0 - u_0 - l = t - l$.

Otherwise, let $d(tr_1) = [l, u]$,

Case1: $l_0 = u_0$

The time interval of system clock is $[l_1, u_1] = [l_0 + l, l_0 + u]$, therefore, the remaining time for the following transitions is:

$$[t_0 - u_1, t_0 - l_1] = [t_0 - l_0 - u, t_0 - l_0 - l]$$

Therefore, $t_1 = t_0 - l_0 - u = t - u$.

Case2: $l_0 < u_0$

If $ut(\mathcal{L}(tr_1)) = 0$, the time interval of system clock is $[l_1, u_1] = [l_0 + l, u_0 + l]$, therefore, the remaining time for the following transitions is:

$$[t_0 - u_1, t_0 - l_1] = [t_0 - u_0 - l, t_0 - l_0 - l]$$

Therefore, $t_1 = t_0 - u_0 - l = t - l$.

If $ut(\mathcal{L}(tr_1)) = 1$, the time interval of system clock is $[l_1, u_1] = [l_0 + u, u_0 + u]$, therefore, the remaining time for the following transitions is:

$$[t_0 - u_1, t_0 - l_1] = [t_0 - u_0 - u, t_0 - l_0 - u]$$

Therefore, $t_1 = t_0 - u_0 - u = t - u$.

In theorem, $var = 0$ denotes the case: $l_0 = u_0$, $var = 1$ denotes the case: $l_0 < u_0$.

Criterion: The transition tr can be labeled with $f_1 AU^{\leq t}_{\geq p} f_2$ if $P_u(tr_0, f_1, f_2, t, 0) \geq p$ and with $f_1 AU^{\leq t}_{> p} f_2$ if $P_u(tr_0, f_1, f_2, t, 0) > p$.

The algorithm for labeling transitions with formulas of type $f = f_1 EU^{\leq t}_{\geq p} f_2$ and $f = f_1 EU^{\leq t}_{> p} f_2$ relies on the following theorem:

THEOREM 3. *(Maximal satisfaction) For a SARPPM $\vec{G}$ and a transition tr in $\vec{G}$, we define $Q_u(tr, f_1, f_2, t, 0)$ to be the maximal probability* *that TCTL formula $f_1 AU^{\leq t} f_2$ is true in tr and all the subsequent execution pathes.*

Assuming that $ntt(tr_0) = \{tr_1, \cdots tr_n\}$, and $var \in \{0,1\}$. Then, $Q_u(tr_0, f_1, f_2, t, var)$ is:

1. $tr_0 \in \Delta_E^s$.

$$\begin{array}{lll} if & f_2 \in label(tr_0) \quad then \quad Q_u(tr_0, f_1, f_2, t, var) = 1 \\ else & if \quad f_1 \notin label(tr_0) \; OR \; t < 0 \quad then \quad Q_u(tr_0, f_1, f_2, t, var) = 0 \\ & else \;\; calculating \; according \; to \; the \; follows. \end{array}$$

2. *If $tr_1, \cdots tr_n \in \Delta_E^s$, then $n = 1$. If $\mathcal{L}(tr_1)$ is a fixed-delay event, $d(tr_1) = [l, l]$, then, $Q_u(tr_0, f_1, f_2, t, var) = Q_u(tr_1, f_1, f_2, t-l, var)$, otherwise, let $d(tr_1) = [l, u]$,*

   $Q_u(tr_0, f_1, f_2, t, 0) = Q_u(tr_1, f_1, f_2, t-u, 1)$

   $Q_u(tr_0, f_1, f_2, t, 1) = Q_u(tr_1, f_1, f_2, t_1, 1)$, *where,*

$$t_1 = \begin{cases} t - l & ut(\mathcal{L}(tr_1)) = 0 \\ t - u & ut(\mathcal{L}(tr_1)) = 1 \end{cases}$$

3. *If $tr_1, \cdots tr_n \in \Delta_P^s$, then,*

$$Q_u(tr_0, f_1, f_2, t, var) = \sum_{i=1}^{n} p(tr_i) \times Q_u(tr_i, f_1, f_2, t, var)$$

4. *If $tr_1, \cdots tr_n \in \Delta_N^s$, then,*

$$Q_u(tr_0, f_1, f_2, t, var) = max_{i=1,\cdots n} Q_u(tr_i, f_1, f_2, t, var)$$

Criterion: The transition *tr* can be labeled with $f_1 EU^{\leq t}_{\geq p} f_2$ if $Q_u(tr_0, f_1, f_2, t, 0) \geq p$ and with $f_1 EU^{\leq t}_{> p} f_2$ if $Q_u(tr_0, f_1, f_2, t, 0) > p$.

3.3. Complexity analysis

In algorithm 3.2, when constructing SARPPM G, the time complexity for recording *EOR*, *EOR*2 (that is algorithm 3.1) is $|E(s)|^2$. The complexity of algorithm is linear in the number of vertex in SARPPM G.

Let $\delta(E) = max\{|E(s)| \mid s \in S_E\}$, hence, $\delta(E)$ is the maximal number of enabled event simultaneously. Let $\delta(D) = min\{BIGGEST, max\{u_x \mid x \; E(s), s \in S_E\}\}$.

The maximal number of super state in SARPPM G is:

$$|S| \times \delta(D) \times 2^{\delta(E)}$$

Therefore, the complexity of our quantitative model checking algorithm for DPRS is linear in the size of TPCTL formula f, time value in f, is second order polynomial in $\delta(D)$ and $|S|$, but exponential in $\delta(E)$.

4. Concluding Remark

In this paper, we present a new model ARPPM and develop a quantitative model checking algorithm for DPRS. It is the first time to solve the problem to verify quantitative properties over continuous time model for DPRS.

For simplicity, we add a condition in ARPPM: for any $l_1, u_1 : l_x \leq l_1 < u_1 \leq u_x$, $\int_{l_1}^{u_1} f_x(t)\,\mathrm{d}t > 0$. In fact, it can be easily extended so that the algorithm can deal with the case that delay time includes several sub-intervals.

Acknowledgement We are grateful to Prof. C. Courcoubetis and Dr. H.A. Hansson for providing us their research papers.

References

Alur, R., C. Courcoubetis, and D. Dill: 1991a, 'Model-checking for probabilistic real-time systems'. In: *LNCS 510: Proc. 18th ICALP*. Madrid, pp. 115–136.

Alur, R., C. Courcoubetis, and D. Dill: 1991b, 'Verifying automata specification of probabilistic real-time systems'. In: *LNCS 600: Proc. of the REX Workshop "Real-Time: Theory in Practice"*. pp. 28–44.

Hansson, H.: 1991, *Time and Probabilities in Formal Design of Distributed Systems*. Uppsala, Sweden: Dept. of Computer Systems, Uppsala University. Ph.D. thesis.

Luo, T., H. Chen, B. Wang, J. Wang, Z. Gong, and Z. Qi: 1998, 'Verifying Automata Specification of Distributed Probabilistic Real-time Systems'. *Journal of Computer Science and technology*.

W.Whitt: 1980, 'Continuity of generalized semi-markkov processes'. *Mathematics of Operations Research* **5**(4).

Constructive Interval Temporal Logic in Alf

Simon Thompson (S.J.Thompson@ukc.ac.uk)
Computing Laboratory
University of Kent at Canterbury
Canterbury, Kent, U. K.

Abstract. This paper gives an implementation of an interval temporal logic in a constructive type theory, using the Alf proof system. After explaining the constructive approach, its relevance to interval temporal logic and potential applications of our work, we explain the fundamentals of the Alf system. We then present the implementation of the logic and give a number of examples of its use. We conclude by exploring how the work can be extended in the future.

1. Introduction

The traditional approach to executing temporal logics is to execute the formulas of the logic; this is in accord with the logic programming paradigm. The implementation can be deterministic for particular subclasses of formulas, as in the approach taken by Moszkowski and others (Moszkowski, 1986; Hale, 1989; Duan, 1996). On the other hand, all the formulas of a logic can be executed using a backtracking mechanism; this is shown by Gabbay's normal form result in (Gabbay, 1987) and is implemented in the various MetateM systems (Barringer et al., 1995) amongst others.

There is another paradigm for implementing logics, based on a *constructive* philosophy (Martin-Löf, 1985; Thompson, 1991). Instead of formulas being seen as true or false on the basis of truth tables or model theory, a constructive approach takes proof as the means of exhibiting validity; as a slogan, one might say that constructive logic is 'proof functional' rather than 'truth functional'. A constructive proof of a proposition contains more information than a classical version, so that from a proof we can derive all the evidence for the proposition being valid. For instance, a proof of an existential statement will contain a *witness* which is an object for which the statement holds.

Under the constructive approach, then, we execute not the formulas of the logic but their *proofs*, which we can see from the discussion above contain sufficient witnessing information to be executable. Further details of the basics of constructive logics, and their interpretation as programming languages can be found in Section 2.

The system in which we make our implementation is Alf, which comes from the programming logics group at Chalmers University of

H. Barringer et al. (eds.), Advances in Temporal Logic, 377–392.

Technology in Göteborg, Sweden. We give a short introduction to Alf in Section 3.

The logic we look at here is an interval temporal logic, which describes finite intervals. Because of this, besides containing the familiar temporal operators $\Box$, $\Diamond$ and so on, interval temporal logics also contain predicates which can only apply to finite intervals, such as those which measure length or which compose two propositions in sequence (or 'chop'). Introductions to the logic are to be found in (Moszkowski, 1986; Duan, 1996; Bowman et al., 1997) and we refer readers there for further details. One distinctive aspect of our logic is that it involves atomic *actions* which happen at the instants of an interval.

The approach we examine here can equally well be used to give a constructive account of an infinitary linear (or branching time) logic. Details of an implementation in the Coq system are to be found elsewhere (Thompson, 1997).

We see three strengths of the work reported here.

- We provide a single system in which we can model both specifications and their implementations. Specifications can be related by logical inference, and are shown to be consistent by exhibiting an implementation; inconsistent specifications will simply have no implementation.

- We maintain two levels of abstraction in our system. In specifications we can use operators such as $\Diamond$ and 'chop' which can be realised in many ways; we can think of them as non-deterministic. In our implementations or proofs we have determinism. For example, a constructive proof of a formula $\Diamond P$ will show not only that P holds at some point in the future but also will state at precisely which point in the future P holds. This distinction is entirely appropriate to the modelling applications of interval temporal logics.

- An implementation of a logic such as this forces an implementer to check the coherence of his or her definition of the logic. In our work here we see a distinction between the notions of interval and interval proposition which in an informal account may be elided. We also have to maintain a distinction between an action A, say, and the proposition that 'A happens (now)'.

  In our related work on linear-time temporal logics, (Thompson, 1997), the issue of whether the logic is anchored or not has to be addressed at an early stage in writing the implementation.

I am grateful to Erik Poll both for supplying an implementation of basic logic for modification and for making a number of useful com-

$A \wedge B$	A proof of $A \wedge B$ consists of a proof of A and a proof of B.
$A \vee B$	A proof of $A \vee B$ consists either of a proof of A or of a proof of B.
$A \rightarrow B$	A proof of $A \rightarrow B$ is a method (or function) taking proofs of A into proofs of B.
$(\exists x \in A)B(x)$	A proof of $(\exists x \in A)B(x)$ consists of an element a of A together with a proof of $B(a)$.
$(\forall x \in A)B(x)$	A proof of $(\forall x \in A)B(x)$, which we also write as $(x \in A) \rightarrow B$, consists of a function taking x in A to a proof of $B(x)$.

Figure 1. Proof in constructive logic

ments on drafts of the paper. I would also like to thank Howard Bowman, Helen Cameron and Peter King for their collaboration in the *Mexitl* (Bowman et al., 1997) work. It was this which stimulated the investigation reported here.

2. Constructive logic

What counts as a constructive proof of a formula? An informal explanation is given in Figure 1.

That this gives the logic a distinctive character should not be in question; while truth functionally one would accept $A \vee \neg A$ for any A, it is by no means clear that for an arbitrary formula one can find either a proof of A or a proof that A is contradictory. On the other hand we can see that proofs are much more informative than in the classical case. A proof of a disjunction must be a proof of one of the disjuncts, and a proof of an existential statement must provide a witness which is a point where the statement holds, together with a proof that it does indeed hold at that point.

How does a constructive implementation work? We take the formulas of a logic as specifications of behaviour; it is then the *proofs* of these formulas that are implemented. Underlying this is an important correspondence, attributed to Curry and Howard and illustrated in Figure 2, which identifies a constructive logic and a (functional) programming language.

Under the Curry-Howard correspondence the formulas of a logic are seen as the types of an expressive type system which includes not only

Constructive Logic		Programming Language
Formula		Type
Proof		Value
Conjunction	$\wedge$	Product or record type
Disjunction	$\vee$	Sum or union type
Implication	$\rightarrow$	Function space
Existential quantification	$\exists$	'Dependent' record type
Universal quantification	$\forall$	'Dependent' function type

Figure 2. The Curry-Howard correspondence

record, union and function types but also dependent function types

$$(x \in A) \rightarrow B$$

in which B can depend upon x, so that the type of a function application can depend upon the *value* to which the function is applied. These types correspond to universally quantified formulas, while a dependent record type represents an existentially quantified statement — we shall see this in Section 3.1.

Given this explanation we can now see how our implementation is built. The formulas of our interval logic become the types of functions which implement the specifications that the formulas embody.

3. Introducing Alf

The logic used here is a standard formulation of a constructive logic in Alf. As we explained in Section 2 we can view Alf as a functional programming language with a strong type system. It is for this reason that we chose to use Alf here rather than, say, Coq; in Coq the proof terms are implicit rather than explicit, and we wanted to emphasise these functions in our account.

We explain the basics of the system by means of a sequence of examples

3.1. Basic constructive logic in Alf

Built into the system is the type

$$Set \in Type$$

which is the type of sets or alternatively the type of propositions. Types are defined in Alf by inductive definitions, these are a strengthening of the algebraic data types of standard functional languages such as Haskell (Peterson and Hammond, 1996). We first define a trivially true proposition *True* by giving it a single element, **trivial**. Constructors of types are given in **boldface**; here we see that **trivial** is a constant, that is a 0-ary constructor.

$$\left[\begin{array}{l} True \in Set \\ \quad = \textbf{data}\ \{\textbf{trivial}\ (\,)\} \end{array}\right.$$

Thinking set theoretically, *True* is a one element set. A *False* proposition is a proposition with no proof, or an empty type, which has no constructors:

$$\left[\begin{array}{l} False \in Set \\ \quad = \textbf{data}\ <> \end{array}\right.$$

The angled brackets here indicate that the type has *no* constructors, and so is indeed empty. Next, we have a definition of conjunction:

$$\left[\begin{array}{l} And(P, Q \in Set) \in Set \\ \quad = \textbf{data}\ \left\{\textbf{conj}\ \left(\begin{array}{c} p \in P \\ q \in Q \end{array}\right)\right\} \end{array}\right.$$

This definition of a **data** type states that to construct an element of *And P Q* it is necessary to use the single constructor **conj**. This requires two arguments to construct an element of the conjunction, namely elements p of P and q of Q. In other words, it is necessary to supply proofs of both conjuncts to give a proof of the conjunction. We also have a definition of a constructive disjunction

$$\left[\begin{array}{l} Or(P, Q \in Set) \in Set \\ \quad = \textbf{data}\ \left\{\begin{array}{l} \textbf{inl}\ (p \in P) \\ \textbf{inr}\ (q \in Q) \end{array}\right\} \end{array}\right.$$

To supply an element of *Or P Q* we need either to give an element p of P, making **inl** $p \in (Or\ P\ Q)$, or to give an element q of Q, so that **inr** $q \in (Or\ P\ Q)$. This is evidently a constructive disjunction, since a proof of *Or P Q* is a proof of one of the disjuncts; the first disjunct if it is of the form **inl** p and the second disjunct otherwise. As we implied earlier, this explanation is quite different from a classical interpretation, and so the law of the excluded middle, $(Or\ A\ (Not\ A))$ is not valid in general.

The existential quantifier is also constructive:

$$\left[\begin{array}{l} Exists(A \in Set,\ P \in (x \in A) \to Set) \in Set \\ \quad = \textbf{sig}\ \left\{\begin{array}{l} witness \in A \\ proof \in P\ witness \end{array}\right\} \end{array}\right.$$

We can think of this type as giving a **signature**; the elements of the type are **structures** taking the form

$$\textbf{struct}\left\{\begin{array}{l} \mathit{witness} = \ldots \\ \mathit{proof} = \ldots \end{array}\right\}$$

thus containing a *witness* of the point at which the predicate P holds together with a *proof* that the predicate holds at the *witness*, that is an element of P *witness*. Note that we use a *dependent* type here: the type of the second element: P *witness* depends on the first element, *witness*.

The syntax of Alf allows quantifiers to be written in a more readable form, with

$$\mathit{Exists}\ x \in A\ .\ \ldots.x\ldots.$$

replacing

$$\mathit{Exists}\ A\ (\lambda\ x \rightarrow \ldots.x\ldots.)$$

where we use ....x.... for an expression involving x. We use this form in the remainder of the paper.

3.2. Data types

The natural numbers are given by the declaration

$$\left[\begin{array}{l} \mathit{Nat} \in \mathit{Set} \\ \quad = \textbf{data}\left\{\begin{array}{l} \mathbf{0}\ (\,) \\ \mathbf{S}\ (n \in \mathit{Nat}) \end{array}\right\} \end{array}\right.$$

and the constants *zero*, *one*, *two* and so on have the obvious meaning.

In our implementation of interval temporal logic we represent intervals by non-empty lists of propositions. In order to do this we have to define a type constructor for non-empty lists, and this constructor needs to be of the appropriate *kind*: since it is used to build lists of *Set* it needs to take a *Type* to a *Type*. The constructor is called *list*, and takes a *Type* argument, making it polymorphic:

$$\left[\begin{array}{l} \mathit{list} \in (T \in \mathit{Type}) \rightarrow \mathit{Type} \\ \quad = \lambda\ T \rightarrow \textbf{data}\left\{\begin{array}{l} \textbf{sing}\ (x \in T) \\ \textbf{cons}\left(\begin{array}{l} x \in T \\ xs \in \mathit{list}\ T \end{array}\right) \end{array}\right\} \end{array}\right.$$

Because the lists are non-empty, they all have a *first* and a *last* element. Here we use the **case** construction which gives case analysis (and indeed primitive recursion) over a **data** type, by means of pattern matching.

$$\left[\begin{array}{l} \mathit{first} \in (T \in \mathit{Type},\ b \in \mathit{list}\ T) \rightarrow T \\ \quad = \lambda\ T\ b \rightarrow \textbf{case}\ b\ \textbf{of}\left\{\begin{array}{l} \textbf{sing}\ x \ \rightarrow\ x \\ \textbf{cons}\ x\ xs\ \rightarrow\ x \end{array}\right\} \end{array}\right.$$

An arbitrary element of the type *list T* will either have the form **sing** x or **cons** $x\ xs$; the **case** construct requires us to give the value of *first* in both these cases. We can use the variables in the particular pattern in the corresponding part of the definition.

$$\left[\begin{array}{l} last \in (T \in Type,\ b \in list\ T) \to T \\ \quad = \lambda\, T\, b \to \mathbf{case}\ b\ \mathbf{of} \left\{ \begin{array}{lll} \mathbf{sing}\ x & \to & x \\ \mathbf{cons}\ x\ xs & \to & last^\circ\ xs \end{array}\right\} \end{array}\right.$$

Although the function *last* takes two arguments we suppress the first (type) argument, since it is invariably obvious from the context. The absence of one or more parameters is indicated by the superscript in $last^\circ$.

Before we proceed, note that in this presentation of lists we take the length of a list to be the number of elements it contains *minus one*. In particular therefore a single element list has length zero in this formulation.

The functions *take* and *drop* are used to select portions of a list. The natural number argument supplied gives an indication of the number of elements taken or dropped from the front of the list. Specifically

$$take^\circ\ n\ l$$

gives a list of length n (that is comprising $(n+1)$ elements) from the front of l, whilst

$$drop^\circ\ n\ l$$

removes n elements from the front of l. The effect of this choice is that $take^\circ\ n\ l$ and $drop^\circ\ n\ l$ overlap by one element.

$$\left[\begin{array}{l} take \in (T \in Type,\ b \in Nat,\ c \in list\ T) \to list\ T \\ \quad = \lambda\, T\, b\, c \to \mathbf{case}\ b\ \mathbf{of} \left\{ \begin{array}{l} 0 \to \mathbf{case}\ c\ \mathbf{of} \left\{ \begin{array}{lll} \mathbf{sing}\ x & \to & \mathbf{sing}\ x \\ \mathbf{cons}\ x\ xs & \to & \mathbf{sing}\ x \end{array}\right\} \\ \mathrm{S}n \to \mathbf{case}\ c\ \mathbf{of} \left\{ \begin{array}{lll} \mathbf{sing}\ x & \to & \mathbf{sing}\ x \\ \mathbf{cons}\ x\ xs & \to & \\ \quad \mathbf{cons}\ x\ (take^\circ\ n\ xs) & & \end{array}\right\} \end{array}\right\} \end{array}\right.$$

$$\left[\begin{array}{l} drop \in (T \in Type,\ b \in Nat,\ c \in list\ T) \to list\ T \\ \quad = \lambda\, T\, b\, c \to \mathbf{case}\ b\ \mathbf{of} \left\{ \begin{array}{l} 0 \to c \\ \mathrm{S}n \to \mathbf{case}\ c\ \mathbf{of} \left\{ \begin{array}{lll} \mathbf{sing}\ x & \to & \mathbf{sing}\ x \\ \mathbf{cons}\ x\ xs & \to & drop^\circ\ n\ xs \end{array}\right\} \end{array}\right\} \end{array}\right.$$

The function *index* selects an element of a list, numbering the elements from zero. If the index exceeds the number of elements in the list, the last element is returned.

$$\left[\begin{array}{l} index \in (T \in Type,\ b \in Nat,\ c \in list\ T) \to T \\ \quad = \lambda\, T\, b\, c \to \mathbf{case}\ b\ \mathbf{of} \left\{ \begin{array}{l} 0 \to first^\circ\ c \\ \mathrm{S}n \to \mathbf{case}\ c\ \mathbf{of} \left\{ \begin{array}{lll} \mathbf{sing}\ x & \to & x \\ \mathbf{cons}\ x\ xs & \to & index^\circ\ n\ xs \end{array}\right\} \end{array}\right\} \end{array}\right.$$

3.3. Using Alf

We have used the experimental **Alfa** version of Alf, which is implemented using Haskell and the Fudgets library (Carlsson and Hallgren, 93) by Thomas Hallgren. The system contains an interactive graphical editor which allows a user to build complex definitions by point and click. A particularly valuable feature is a menu of options giving possible constructions which it would be type safe to use at any point in an expression; by means of this one constructs type correct programs.

This concludes our introduction to the aspects of Alf used here; more details can be found at `www.cs.chalmers.se/~hallgren/Alfa/`.

4. Interval Temporal Logic

In this section we begin by giving in Section 4.1 our definition of the fundamentals of the implementation, namely definitions of what it is to be an interval, an action and an interval proposition.

Central to interval logic are various connectives or combinators which allow us to combine interval propositions together to give more complex propositions. Apart from the obvious lifting of the propositional connectives and quantifiers of predicate logic, which we look at in Section 4.8, and the standard temporal operators defined in Section 4.7, we introduce two operators characteristic of an interval logic.

The first is *chop* P Q, in Section 4.3, which holds of an interval when the interval can be split into two halves satisfying P and Q separately. Secondly we introduce *proj* P Q which projects one interval, by means of P, onto another which should meet Q; projection is introduced in Section 4.5.

4.1. Actions and Intervals

We take the type of actions as given; for the sake of exposition here we assume it is a data type of constants (or 0-ary constructors):

$$\left[\begin{array}{l} Act \in Set \\ \quad = \textbf{data} \left\{ \begin{array}{c} \textbf{A}\,(\,) \\ \ldots \end{array} \right\} \end{array}\right.$$

There are various means of representing sets in constructive logic; here we choose to model sets of actions by 'characteristic' functions from *Act* to *Set*:

$$\left[\begin{array}{l} ActSet \in Type \\ \quad = (A \in Act) \to Set \end{array}\right.$$

An interval is a *list* of action sets.

$$\left[\begin{array}{l} \mathit{Interval} \in \mathit{Type} \\ \quad = \mathit{list}\ \mathit{ActSet} \end{array}\right.$$

and an interval proposition or *IntProp* is a function which takes an interval to a proposition (that is a *Set*).

$$\left[\begin{array}{l} \mathit{IntProp} \in \mathit{Type} \\ \quad = (I \in \mathit{Interval}) \to \mathit{Set} \end{array}\right.$$

An interval is said to be empty if it contains a single point.

$$\left[\begin{array}{l} \mathit{Empty} \in \mathit{IntProp} \\ \quad = \lambda\ I \to \textbf{case}\ I\ \textbf{of} \left\{ \begin{array}{lcl} \textbf{sing}\ x & \to & \mathit{True} \\ \textbf{cons}\ x\ xs & \to & \mathit{False} \end{array} \right\} \end{array}\right.$$

Generalising this is a proposition *Length n* expressing that the length of an interval is *n*: *Empty* is then given by *Length 0*.

$$\left[\begin{array}{l} \mathit{Length} \in (a \in \mathit{Nat}) \to \mathit{IntProp} \\ \quad = \lambda\ a\ b \to \textbf{case}\ a\ \textbf{of} \left\{ \begin{array}{l} \textbf{0} \to \textbf{case}\ b\ \textbf{of} \left\{ \begin{array}{lcl} \textbf{sing}\ x & \to & \mathit{True} \\ \textbf{cons}\ x\ xs & \to & \mathit{False} \end{array} \right\} \\ \textbf{S}\ n \to \textbf{case}\ b\ \textbf{of} \left\{ \begin{array}{lcl} \textbf{sing}\ x & \to & \mathit{False} \\ \textbf{cons}\ x\ xs & \to & \mathit{Length}\ n\ xs \end{array} \right\} \end{array} \right\} \end{array}\right.$$

Our final example of an atomic proposition is '*A* happens now', that is at the *first* point of an interval

$$\left[\begin{array}{l} \mathit{happensNow} \in (A \in \mathit{Act}) \to \mathit{IntProp} \\ \quad = \lambda\ A\ I \to \mathit{first}^{\circ}\ I\ A \end{array}\right.$$

The expression $\mathit{first}^{\circ}\ I$ is an action set, and so the proposition that *A* holds is given by applying the action set to the action, giving the proposition $(\mathit{first}^{\circ}\ I\ A)$.

4.2. Specifications

A *specification* of an interval can now be seen as an interval property, that is a member *P* of *IntProp*. An *implementation* of such a specification will be an interval *I* for which we can find a proof

$$p \in P\ I$$

The proof *p* contains information about exactly *how* the interval *I* meets the specification *P*. It will, for instance, state a point in an interval at which a $\Diamond$ property holds, and state which of a pair of disjuncts is valid. Examples are given in Sections 4.4 and 4.6 below.

4.3. Chop

An interval *I* satisfies *chop P Q*, where *P* and *Q* are interval propositions, if it can be split (or 'chopped') into two halves with the first satisfying *P* and the second *Q*.

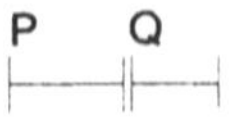

In our logic a proof of a chop formula specifies the point at which the split is made, together with proofs that the resulting halves have the requisite properties themselves. It is formalised thus:

$$\left[\begin{array}{l} chop \in (P,\ Q \in IntProp) \to IntProp \\ \quad = \lambda\ P\ Q\ I \to Exists\ n \in Nat\ .\ And\ (P\ (take^{\circ}\ n\ I))\ (Q\ (drop^{\circ}\ n\ I)) \end{array}\right.$$

Note that because of the way that *take* and *drop* are defined the two halves of the interval, namely ($take^{\circ}$ n I) and ($drop^{\circ}$ n I), share an end point.

4.4. Examples of proof – I

To make some intervals over which to give examples we define the function *makeInt*. When applied to a natural number it makes an interval of that length, at each point of which all actions take place

$$\left[\begin{array}{l} allActs \in ActSet \\ \quad = \lambda\ A \to True \end{array}\right.$$

$$\left[\begin{array}{l} makeInt \in (a \in Nat) \to Interval \\ \quad = \lambda\ a \to \textbf{case}\ a\ \textbf{of} \left\{\begin{array}{lll} \mathbf{0} & \to & \textbf{sing}\ allActs \\ \mathbf{S}\ n & \to & \textbf{cons}\ allActs\ (makeInt\ n) \end{array}\right\} \end{array}\right.$$

Our first example shows that *makeInt two* has length *two*:

$$\left[\begin{array}{l} len2 \in Length\ two\ (makeInt\ two) \\ \quad = \textbf{trivial} \end{array}\right.$$

The proof here is **trivial**. Building on this we can give an example of a proof involving *chop*. Our goal is to prove that

$$chop\ (Length\ two)\ (Length\ one)$$

holds of the interval *makeInt three*. So, we define

$$\left[\begin{array}{l} egChop \in Set \\ \quad = chop\ (Length\ two)\ (Length\ one)\ (makeInt\ three) \end{array}\right.$$

What is a proof of this? Proofs of existential statements are **structs** consisting of two things: a *witness* saying at which point the chop is made and a *proof* that the halves have the appropriate properties. This latter proof will itself be a conjunction of proofs that each half has the requisite property. Our proof is

$$\left[\begin{array}{l} chopPrf \in egChop \\ \quad = \textbf{struct} \left\{\begin{array}{l} witness = two \\ proof = \textbf{conj trivial trivial} \end{array}\right\} \end{array}\right.$$

The chop at position *two* ensures that the first interval has length *two* and the second length *one*. In both cases the proofs of these facts are **trivial**.

4.5. PROJECTION

A more powerful interval operation is that of projection. The proposition *proj P Q* holds for an interval if

- the interval can be split into a sequence of sub-intervals each of which satisfies P and also
- a new interval made up from the end points of the original interval satisfies the interval proposition Q.

In pictures we require that

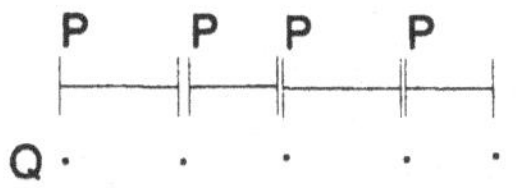

Note that as was the case for *chop*, the end point of the first sub-interval is the first element of the second and so on.

In order to define *proj* we have to define two subsidiary operations, *allSat* and *select* which embody the two cases enumerated above. These in turn take a parameter of type *List Nat* — the type of (empty and non-empty) lists of natural numbers — given by the following declaration. The points in the list indicate the lengths of the sub-sequences of the list which form the projection; we therefore have no case in which the *List* is empty.[1]

$$\left[\begin{array}{l} List(A \in Set) \in Set \\ \quad = \mathbf{data} \left\{ \begin{array}{l} \mathbf{Nil}\ (\,) \\ \mathbf{Cons} \left(\begin{array}{l} head \in A \\ tail \in List\ A \end{array}\right) \end{array}\right\} \end{array}\right.$$

To define *allSat* we perform a primitive recursion over the list of lengths

$$\left[\begin{array}{l} allSat \in (P \in IntProp,\ nums \in List\ Nat,\ I \in Interval) \to Set \\ \quad = \lambda\ P\ nums\ I \to \mathbf{case}\ nums\ \mathbf{of} \left\{ \begin{array}{l} \mathbf{Nil} \to \mathbf{True} \\ \mathbf{Cons}\ head\ tail \to \\ \quad And\,(P\ (take^{\circ}\ head\ I)) \\ \qquad (allSat\ P\ tail\ (drop^{\circ}\ head\ I)) \end{array}\right\} \end{array}\right.$$

In the interesting — **Cons** *head tail* — case *allSat* is a conjunction of two propositions. The first that the sub-interval of length *head* has

[1] Why then use *List* rather than *list* here? The answer is that the former gives a *Set* over which we can existentially quantify, whilst the latter only gives a type, and so we cannot use *Exists* to perform quantification over it.

property P; the second that all the subdivisions specified by *tail* have the property.

The function *select* is defined to pick out the end points specified by the subdivisions in the *List* of natural numbers.

$$\left[\begin{array}{l} select \in (T \in Type,\ b \in list\ T,\ c \in List\ Nat) \rightarrow list\ T \\ \quad = \lambda\ T\ b\ c \rightarrow \textbf{case}\ c\ \textbf{of} \left\{ \begin{array}{l} \textbf{Nil} \rightarrow \textbf{sing}\ (last^{\circ}\ b) \\ \textbf{Cons}\ head\ tail \rightarrow \\ \qquad \textbf{cons}\ (first^{\circ}\ b)\ (select^{\circ}\ (drop^{\circ}\ head\ b)\ tail) \end{array} \right\} \end{array}\right.$$

Now we can put the two components together. To prove *proj P Q I* we need to provide a list of lengths into which I is split. Given this, we then have to prove the two obligations enumerated above.

$$\left[\begin{array}{l} proj \in (P, Q \in IntProp) \rightarrow IntProp \\ \quad = \lambda\ P\ Q\ I \rightarrow Exists\ x \in (List\ Nat)\ .\ And\ (Q\ (select^{\circ}\ I\ x))\ (allSat\ P\ x\ I) \end{array}\right.$$

4.6. Examples of proof – II

Here we give a simple projection proof using the *Length* predicate.

$$\left[\begin{array}{l} egProj \in Set \\ \quad = proj\ (Length\ two)\ (Length\ three)\ (makeInt\ six) \end{array}\right.$$

The proof is

$$\left[\begin{array}{l} projPrf \in egProj \\ \quad = \textbf{struct} \left\{ \begin{array}{l} witness = \textbf{Cons}\ three\ (\textbf{Cons}\ three\ \textbf{Nil}) \\ proof = \textbf{conj trivial}\ (\textbf{conj trivial}\ (\textbf{conj trivial trivial})) \end{array} \right\} \end{array}\right.$$

In this **struct** the *witness* gives the lengths into which the interval is cut, namely two intervals of length three.

In the *proof* part, the outer **conj** separates to give two proofs. The first is

$$\textbf{trivial} \in Length\ two\ (select^{\circ}\ (makeInt\ six)\ (\textbf{Cons}\ three\ (\textbf{Cons}\ three\ \textbf{Nil})))$$

which states that if we split the interval into sub-intervals of length three, the interval made up of the end points will have length two. (Recall that in this exposition the length of an interval is taken to be the number of points in the interval *minus one*.)

The second conjunct is

$$\begin{array}{l} \textbf{conj trivial}\ (\textbf{conj trivial trivial}) \in \\ \quad allSat\ (Length\ three)\ (\textbf{Cons}\ three\ (\textbf{Cons}\ three\ \textbf{Nil}))\ (makeInt\ six) \end{array}$$

which states that the sub-intervals all satisfy (*allSat*) the property that they have length three. These two conjuncts together show that the projection holds.

4.7. Temporal Operators

The traditional temporal operators have straightforward definitions in Alf. Given an interval proposition P, *next* P should be valid over an interval if P holds over the tail, that is the interval starting at the *next* instant.

$$\left[\begin{array}{l} next \in (P \in IntProp) \to IntProp \\ \quad = \lambda\ P\ I \to \textbf{case}\ I\ \textbf{of} \left\{ \begin{array}{lcl} \textbf{sing}\ x & \to & False \\ \textbf{cons}\ x\ xs & \to & P\ xs \end{array} \right\} \end{array}\right.$$

We might, for instance, state that an action takes place at the next instant in time by

$$next\ (happensNow\ \mathbf{A})$$

The proposition *sometime* P should be valid over an interval if P holds over some final segment of the interval (including possibly the whole interval itself). A recursive definition is

$$\left[\begin{array}{l} sometime \in (P \in IntProp) \to IntProp \\ \quad = \lambda\ P\ I \to \textbf{case}\ I\ \textbf{of} \left\{ \begin{array}{lcl} \textbf{sing}\ x & \to & P\ I \\ \textbf{cons}\ x\ xs & \to & Or\ (P\ I)\ (sometime\ P\ xs) \end{array} \right\} \end{array}\right.$$

Because of the constructive nature of our definition of *Or* it is possible to read off from a proof *which* segment it is over which the interval proposition holds. An explicit definition of 'sometime' states

$$\left[\begin{array}{l} sometimeAlt \in (P \in IntProp) \to IntProp \\ \quad = \lambda\ P\ I \to Exists\ n \in Nat\ .\ P(drop^{\circ}\ n\ I) \end{array}\right.$$

Again, a proof of this contains a witness n to the particular segment over which P holds.

The proposition *always* P should be valid over an interval if P holds over every final segment of the interval. A recursive definition is

$$\left[\begin{array}{l} always \in (P \in IntProp) \to IntProp \\ \quad = \lambda\ P\ I \to \textbf{case}\ I\ \textbf{of} \left\{ \begin{array}{lcl} \textbf{sing}\ x & \to & P\ I \\ \textbf{cons}\ x\ xs & \to & And\ (P\ I)\ (always\ P\ xs) \end{array} \right\} \end{array}\right.$$

Since a proof of *And* P Q must contain component proofs of P and Q, in the case of a non-empty interval it is necessary for both $(P\ I)$ and $(always\ P\ xs)$ to be provable, with the consequence that from a proof of $(always\ P\ I)$ it is possible to extract proofs of (P y) for every final segment y. An explicit definition of 'always' states

$$\left[\begin{array}{l} alwaysAlt \in (P \in IntProp) \to IntProp \\ \quad = \lambda\ P\ I \to (n \in Nat) \to P\ (drop^{\circ}\ n\ I) \end{array}\right.$$

It is clear here that the proof of $(alwaysAlt\ P\ I)$ gives a proof for all final segments, since the proof is a function taking a natural number n to a proof of $P\ (drop^{\circ}\ n\ I)$.

4.8. Lifting Propositional connectives

We can lift the standard propositional connectives in a uniform way – for instance to lift binary connectives we can define

$$\left[\begin{array}{l} liftC \in (conn \in (a, b \in Set) \to Set,\ P, Q \in IntProp) \to IntProp \\ \quad = \lambda\ conn\ P\ Q\ I \to conn\ (P\ I)\ (Q\ I) \end{array}\right.$$

and as examples we define the infix functions

$$\left[\begin{array}{l} andT \in (P, Q \in IntProp) \to IntProp \\ \quad = liftC\ And \end{array}\right.$$

$$\left[\begin{array}{l} impliesT \in (P, Q \in IntProp) \to IntProp \\ \quad = liftC\ (\lambda\ a\ b \to (c \in a) \to b) \end{array}\right.$$

We can then lift the theorem *implies* (*And A B*) *A* and supply a proof:

$$\left[\begin{array}{l} liftThm \in (P, Q \in IntProp,\ I \in Interval) \to ((P\ andT\ Q)\ impliesT\ P)\ I \\ \quad = \lambda\ P\ Q\ I\ c \to \textbf{case}\ c\ \textbf{of}\ \{\textbf{conj}\ p\ q \to p\} \end{array}\right.$$

The method we use here is uniform: any result that we have for our standard logic can be lifted to the Interval Logic in a similar way.

4.9. Discussion

The implementation discussed here can be modified in a number of ways. We might choose to represent intervals by finite functions rather than by finite lists; such a representation is more reminiscent of an implementation of a traditional (linear-time) temporal logic by means of functions over *Nat*; an implementation of this in Coq is reported in (Thompson, 1997).

We reiterate that the interpretations of the (temporal) connectives $\vee$, $\diamond$, '*chop*' and '*proj*' are constructive. This means that from proofs of statements using these we can derive witnessing information about how exactly the result is proved; in other words each proof resolves the implicit non-determinism or under-specification in a particular way.

5. Conclusions and Future Work

This paper reports an implementation of an interval temporal logic using the Alf proof system. We have shown how proofs of temporal formulas can be executed, the proofs corresponding to (functional) programs and the formulas to the types of a programming language.

We argued in the Introduction that this separation between specification and implementation is valuable: specifications can be at a higher level of abstraction than their implementations. In particular, specifications can be non-deterministic whilst their implementations are deterministic.

In the future we intend to investigate a variety of topics.

- Type abstraction and (relative) completeness issues: our reasoning here is in terms of the representation of the logic. We seek an abstraction result which allows us to reason directly at the temporal level; from another point of view we look for a (relatively) complete temporal formal system for the primitives of our system.
- Refinement: we can model the refinement of one specification, S_1, by another, S_2, by giving the implication $S_2 \rightarrow S_1$. A proof of this implication in a constructive system provides a transformation from implementations of S_2 to implementations of S_1.
- Efficiency of implementation: we would like to investigate the efficiency of this style of implementation. We also envisage links with tableau methods of proof search, which can automate the construction of implementations in some cases.
- Actions and framing: as a matter of urgency we aim to investigate the constructive perspective on framing.
- Higher-level language: as a part of the *Mexitl* programme we will investigate the possibility of defining a higher-level language for multimedia description; the constructive approach can have a role to play in this definition and its implementation.

References

Barringer, H. et al.: 1995, 'MetateM: An Imperative Approach to Temporal Logic Programming'. *Formal Aspects of Computing* **7**(**E**).

Bowman, H., H. Cameron, P. King, and S. Thompson: 1997, 'Specification and Prototyping of Structured Multimedia Documents using Interval Temporal Logic'. In: *this volume.*

Carlsson, M. and T. Hallgren: 93, 'FUDGETS - A Graphical User Interface in a Lazy Functional Language'. In: *Functional Programming & Computer Architecture.*

Duan, Z. H.: 1996, 'An Extended Interval Temporal Logic and A Framing Technique for Temporal Logic Programming'. Ph.D. thesis, University of Newcastle Upon Tyne.

Gabbay, D.: 1987, 'Declarative Past and Imperative Future: Executable Temporal Logic for Interactive Systems'. In: *Proceedings of the 7th ACM Symposium on the Principles of Programming Languages, Lecture Notes in Computer Science, vol. 398.* pp. 402–450.

Hale, R.: 1989, 'Using Temporal Logic for Prototyping: the Design of a Lift Controller'. In: *Lecture Notes in Computer Science, vol. 379*. pp. 375–408.

Martin-Löf, P.: 1985, 'Constructive Mathematics and Computer programming'. In: C. A. R. Hoare (ed.): *Mathematical Logic and Programming Languages*. Prentice-Hall.

Moszkowski, B.: 1986, *Executing Temporal Logic Programs*. Cambridge University Press.

Peterson, J. and K. Hammond (eds.): 1996, *Report on the Programming Language Haskell, Version 1.3*. `http://haskell.cs.yale.edu/haskell-report/haskell-report.html`.

Thompson, S.: 1997, 'Temporal Logic using the Coq proof assistant'. Technical report, Computing Laboratory, University of Kent at Canterbury, U.K. Available via: `www.cs.ukc.ac.uk/people/staff/sjt/pubs.html`.

Thompson, S. J.: 1991, *Type Theory and Functional Programming*. Addison Wesley.

Two-dimensional Executable Temporal Logic for Bitemporal Databases

Marcelo Finger (mfinger@ime.usp.br) * and Mark Reynolds †
(markr@dcs.kcl.ac.uk)

Abstract. We investigate the application of the executable two-dimensional temporal logic idea to bitemporal databases. It is well-known that (one-dimensional) temporal formulas can be separated and executed, in what constitutes the "imperative future" paradigm. We would like to apply this paradigm to bitemporal temporal databases, i.e. databases that record both the valid- and transaction-time of data.

However, in the two-dimensional temporal logics (2DTL) which serves as a model of bitemporal databases formulas do not separate. We show how this problem can be avoided by separating and executing *vertical 2DTL formulas* only.

These ideas are put in practice in a hotel room booking and room occupancy management system.

1. Introduction

In this paper, we show how the "executable temporal logics" paradigm of (Barringer *et al.*, 1996) can be combined with two-dimensional temporal logics and applied to bitemporal databases.

Bitemporal databases store each fact together with both its validity time and its transaction time (Snodgrass and Ahn, 1985). The expressivity of the database is thus greatly enhanced, for past views of history are kept and can be used to track the "justification" of past actions taken with the database's support.

Bitemporal databases are described using two-dimensional temporal logics (2DTLs), a family of logics that result from joining two one-dimensional temporal logics (Finger and Gabbay, 1992; Finger and Gabbay, 1996).

After being introduced as the the simple "declarative past implies imperative future" approach in (Gabbay, 1989), the idea of direct execution of temporal logic formulas has developed into a popular method of producing programs which are easy to understand and easy to verify. A wide range of applications appear in (Fisher and Owens, 1995) and (Barringer *et al.*, 1996). The basic idea is to read formulas in an imperative way so as to build a structure which makes them true.

There are two hurdles to overcome in putting together executable logics, 2DTLs and bitemporal databases:

* Partly supported by brazilian CNPq, grant 300597/95-9.
† Partly supported by UK EPSRC grant GR/K54946.

H. Barringer et al. (eds.), Advances in Temporal Logic, 393–411.

– The basic 1DTL property of *separation* that give support to the execution of formulas does not hold for 2DTLs.

– The theory is propositional, but any non-toy database application needs (some of) the expressivity of first-order logic.

The first problem is solved by introducing the notion of *vertical separation* that allows a class of 2DTL formulas to be executed. The second problem is solved with the *propositional abstraction* of database facts, queries and rules.

After introducing these concepts in this paper, we describe their application to the case of a hotel room booking and room occupancy management system.

2. Bitemporal Databases

At first sight, it is rather trivial to add a temporal dimension to a database. It suffices to include a new attribute to every relation one wishes to make temporal, and call this attribute a *time-stamp*. Whether a time-stamp represents a single time point, or an interval, or a finite or infinite union of disjoint intervals is considered here a physical-level representation problem; it is assumed only that a time stamp represents an unqualified set of time points in some linear flow of time.

For example, the ubiquitous relation Employee(Name, Salary) is made temporal by adding a time-stamp attribute as shown in Figure 2, where a time stamp is represented as an interval.

Name	*Salary*	*Time-stamp*
Peter	$ 1000	[Feb92–Feb96]

Figure 1. Adding a temporal dimension to the Employee relation

Triviality disappears once it is asked what the meaning of such time-stamp is, with severe consequences for data manipulation and application building. We are faced with two basic choices for the semantics of time-stamps:

– The time-stamp represents *valid-time* if it represents the time that data was true in the modelled world. In this case, Figure 2 represents when Peter's salary was $1000 in the real world.

– *Transaction-time*, on the other hand, records the time the system "believes" the data to hold; in other words, the times between data insertion in and deletion from the database. In Figure 2, Peter's salary of $1000 was inserted in the database on February 1992 and deleted on February 1996, irrespectively of when Peter received such a salary in the real world. The control of transaction time-stamp lies with the database management system, and it cannot be altered (in principle) by a user or programs.

It turns out that those two interpretations are orthogonal, and both types of time-stamp can coexist in a database. Figure 2 illustrates one bitemporal relation, representing the fact that the system believed that Peter's salary was $1000 from February 1992 to February 1996 and that this information was believed to hold by the system from April 1992, when it was inserted, until March 1996, when it was logically deleted but not physically erased from the database.

Name	*Salary*	*Valid Time-stamp*	*Transaction Time-stamp*
Peter	$ 1000	[Feb92–Feb96]	[Apr92–Mar96]

Figure 2. Making Employee a bitemporal relation

A temporal database that allows for and manipulates such doubly time-stamped relations is called a *bitemporal databases* (2TDBs).

For efficiency reasons, until very recently bitemporal databases were just a theoretical object of study, but with the development of double time-indexing mechanisms (Nascimento *et al.*, 1995; Nascimento *et al.*, 1996), it is possible to efficiently access bitemporal information. We therefore believe bitemporal databases will become more common in the near future. It will be possible then to take advantadge of the great representation power of bitemporal databases:

(a) Not only the world history is recorded, but also how the perception of such history evolved.

(b) As a consequence, when actions are performed based on temporal data, both the action and the data that gave support to its execution persist in the database.

(c) Even if that support is logically deleted from the current perception of history, the past perceptions have not been removed, and all past actions have their justifications preserved.

So not only history, but historical changes and their consequences can all be dealt with 2TDBs. It becomes possible to deal with complex situations such as retroactive salary increases, the cancelling of future appointments, changes of strategical plans and, most importantly, error corrections. These are the crucial applications for 2TDBs.

But lest we get confused by the unclear meaning of terms such as "justification", "action support", "historical changes and their consequences" and other expressions we have informally used so far, we need a framework that allows one to precisely define their meanings. For that we introduce next the two-dimensional temporal model and its associated logic.

3. Two-dimensional temporal logic

There are several modal and temporal logic systems in the literature which are called *two-dimensional* (see, for example, (Vlach, 1973; Spaan, 1993; Venema, 1994; Marx and Venema, 1997)); all of them provide some sort of double reference to an underlying modal or temporal structure. More systematically, two-dimensional systems have been studied as the result of combining two one-dimensional logic systems (Finger and Gabbay, 1992; Finger and Gabbay, 1996). In (Finger and Gabbay, 1996) two criteria were presented to classify a logical system as two-dimensional:

- *The connective approach:* a temporal logic system is two-dimensional if it contains two sets of connectives, each set referring to a distinct flow of time.
- *The semantic approach:* a temporal logic system is two-dimensional if the truth value of a formula is evaluated with respect to two time points.

The two criteria are independent and there are examples of systems satisfying each criterion alone, or both. For the purposes of this work, the two-dimensional temporal logic satisfies both criteria, and is thus a *broadly two-dimensional logic*. Both flows of time are assumed to be linear discrete ($\mathbb{N}$ or $\mathbb{Z}$).

So let $\mathcal{L}$ be a countable set of propositional atoms. Besides the boolean connectives, we consider two sets of temporal operators. The *horizontal operators* are the usual "since" ($\mathcal{S}$) and "until" ($\mathcal{U}$) two-place operators, together with all the usual derived operators; the horizontal dimension will be used to represent *valid time* temporal information. We read $\mathcal{U}(\psi, \xi)$ as "ξ until ψ" and similarly for since. The *vertical dimension* is assumed to be a $\mathbb{N}$-like flow and the only operators over such dimensions are the two-place operators "since vertical" ($\bar{\mathcal{S}}$) and the "until vertical" ($\bar{\mathcal{U}}$); in general, we use barred symbols when they refer to the vertical dimension. The vertical dimension will be used to represent *transaction time* information. Two-dimensional formulas are inductively defined as:

- every propositional atom is a two-dimensional formula;
- if φ and ψ are two-dimensional formulas, so are $\neg\varphi$ and $\varphi \wedge \psi$, $\mathcal{S}(\varphi, \psi)$ and $\mathcal{U}(\varphi, \psi), \bar{\mathcal{S}}(\varphi, \psi)$ and $\bar{\mathcal{U}}(\varphi, \psi)$.

On the semantic side, we consider two flows of time: the horizontal one
$(T, <)$ and the vertical one $(\overline{T}, \bar{<})$. Two-dimensional formulas are evaluated with respect to two dimensions, typically a time point $t \in T$ and a time point $\bar{t} \in \overline{T}$, so that a *two dimensional plane model* is a structure based on two flows of time $\mathcal{M} = (T, <, \overline{T}, \bar{<}, \pi)$. The *two-dimensional assignment* π maps every triple $(t, \bar{t}, p)$ into $\{\top, \bot\}$. The model structure can be seen as a two-dimensional plane, where every point is identified by a pair of coordinates, one for each flow of time (there are other, non-standard models of two-dimensional logics which are not planar; see (Finger and Gabbay, 1996)).

The fact that a formula φ is true in the two-dimensional plane model $\mathcal{M}$ at point $(t, \bar{t})$ is represented by $\mathcal{M}, t, \bar{t} \models \varphi$ and is defined inductively by:

$\mathcal{M}, t, \bar{t} \models p$	iff $\pi(t, \bar{t}, p) = \top$.
$\mathcal{M}, t, \bar{t} \models \neg\varphi$	iff it is not the case that $\mathcal{M}, t, \bar{t} \models \varphi$.
$\mathcal{M}, t, \bar{t} \models \varphi \wedge \psi$	iff $\mathcal{M}, t, \bar{t} \models \varphi$ and $\mathcal{M}, t, \bar{t} \models \psi$.
$\mathcal{M}, t, \bar{t} \models \mathcal{S}(\varphi, \psi)$	iff there exists a $t' \in T$ with $t' < t$ and $\mathcal{M}, t', \bar{t} \models \varphi$ and for every $t'' \in T$, whenever $t' < t'' < t$ then $\mathcal{M}, t'', \bar{t} \models \psi$.
$\mathcal{M}, t, \bar{t} \models \mathcal{U}(\varphi, \psi)$	iff there exists an $t' \in T$ with $t < t'$ and $\mathcal{M}, t', \bar{t} \models \varphi$ and for every $t'' \in T$, whenever $t < t'' < t'$ then $\mathcal{M}, t'', \bar{t} \models \psi$.

$\mathcal{M}, t, \bar{t} \models \bar{\mathcal{S}}(\varphi, \psi)$	iff there exists a $\bar{t}' \in \overline{T}$ with $\bar{t}' < \bar{t}$ and $\mathcal{M}, t, \bar{t}' \models \varphi$ and for every $\bar{t}'' \in \overline{T}$, whenever $\bar{t}' < \bar{t}'' < \bar{t}$ then $\mathcal{M}, t, \bar{t}'' \models \psi$.
$\mathcal{M}, t, \bar{t} \models \bar{\mathcal{U}}(\varphi, \psi)$	iff there exists a $\bar{t}' \in \overline{T}$ with $\bar{t} < \bar{t}'$ and $\mathcal{M}, t, \bar{t}' \models \varphi$ and for every $\bar{t}'' \in \overline{T}$, whenever $\bar{t} < \bar{t}'' < \bar{t}'$ then $\mathcal{M}, t, \bar{t}'' \models \psi$.

Note that the semantics of horizontal and vertical operators are totally independent from each other, i.e. the horizontal operators have no effect on the vertical dimension and similarly for the vertical operators. If we consider only the formulas without the vertical operators, we have a one-dimensional horizontal $\mathcal{U}\,\mathcal{S}$-temporal logic: similarly for the vertical temporal logic. Note that our $\mathcal{U}$ is *strict* in the sense that $\mathcal{U}\,(q,p)$ being true says nothing about what is true now. In some presentations of temporal logic, until is defined to be non-strict. We can introduce an abbreviation $\mathcal{U}^+$ for non-strict until: $\mathcal{U}^+(\psi,\xi)$ iff $\psi \vee (\xi \wedge (\mathcal{U}\,(\psi,\xi)))$. As well as the classical abbreviations $\bot$, $\vee$, $\rightarrow$, $\leftrightarrow$ we also have many temporal ones. The only ones that we need in this paper are:

$\bigcirc\varphi$	"φ is true in the next state "	$\mathcal{U}\,(\varphi,\bot)$
$\bullet\varphi$	"there was a last state and φ was true then"	$\mathcal{S}\,(\varphi,\bot)$
$\Diamond\varphi$	"φ will be true in some future state"	$\mathcal{U}\,(\varphi,\top)$
$\Box\varphi$	"φ will be true in all future states"	$\neg\Diamond\neg\varphi$
$\blacklozenge\varphi$	"φ was true in the past"	$\mathcal{S}\,(\varphi,\top)$
$\blacksquare$	"φ has always been true in the past"	$\neg\blacklozenge\,(\neg\varphi)$

Unary temporal predicates can be defined for both dimensions in the usual way, so we get $\Box$, $\blacksquare$, $\Diamond$, $\blacklozenge$, etc, for the horizontal dimension and $\overline{\Box}$, $\overline{\blacksquare}$, $\overline{\Diamond}$, $\overline{\blacklozenge}$, etc, for the vertical one.

3.1. THE TWO-DIMENSIONAL DIAGONAL

The diagonal is a privileged line in the two-dimensional model intended to represent the sequence of time points we call "now", i.e. the time points which an historical observer is expected to traverse. The observer is on the diagonal when he or she poses a query (i.e. evaluates the truth value of a formula) on a two-dimensional model.

So let δ be a special atom that denotes the points of the diagonal, which is characterised by the following property: for every $t \in T$ and every $\bar{t} \in \bar{T}$:

$$\mathcal{M}, t, \bar{t} \models \delta \quad \text{iff} \quad t = \bar{t}.$$

The diagonal is illustrated in Figure 3.1.

The following formulas are true at all points of the two-dimensional plane model:

$$\blacklozenge\,\delta \vee \delta \vee \Diamond\delta \qquad\qquad \blacklozenge\,\delta \leftrightarrow \overline{\Diamond}\delta$$

$$\delta \leftrightarrow (\Box\neg\delta \wedge \blacksquare\neg\delta \wedge \overline{\Box}\neg\delta \wedge \overline{\blacksquare}\neg\delta) \qquad \overline{\blacklozenge}\,\delta \leftrightarrow \Diamond\delta$$

The diagonal divides the two-dimensional plane in two semi-planes. The semi-plane that is to the (horizontal) left of the diagonal is "the

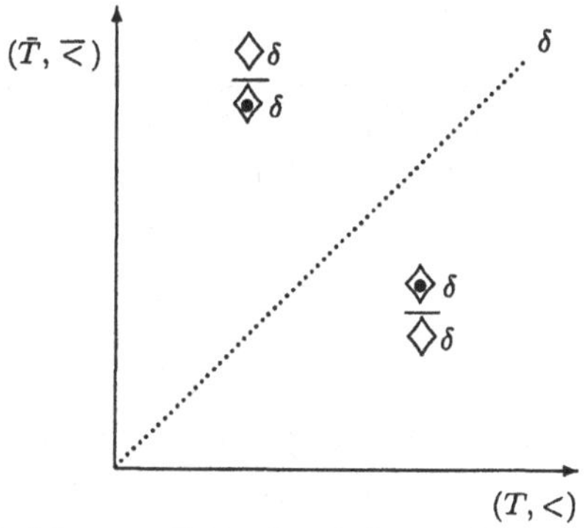

Figure 3. The two-dimensional diagonal

past", and the formulas $\Diamond\delta$ and $\overline{\blacklozenge}\,\delta$ are true at all points of this semi-plane. Similarly, the semi-plane that is to the (horizontal) right of the diagonal is "the future", and the formulas $\blacklozenge\,\delta$ and $\overline{\Diamond}\delta$ are true at all points of this semi-plane. Figure 3.1 puts this fact in evidence.

3.2. Properties of the Logic

Two-dimensional temporal logic is a very expressive logic and hence quite complex to reason with. There are several less expressive two-dimensional fragments that are axiomatizable and decidable over several classes of two-dimensional time. However, the full logic described here is axiomatizable only over some classes of two-dimensional models; in interesting cases such as $\mathbb{Z}$- and $\mathbb{N}$-like two-dimensional grids, it is not, as shown in (Finger and Gabbay, 1996).

The full fragment two-dimensional logic we use here is highly undecidable, see tiling proofs such as in (Harel, 1983) and (Reynolds, 1996). Nevertheless, below we shall see that for controlling database updates we do not need the full power of the logic and so we are able to work effectively.

3.3. From Proposition to First-Order 2DTL

There is a usual "impedance mismatch" between the theory and the practice of systems based on modal and temporal logics. Typically, the theory is propositional and possesses the basic logical properties of complete axiomatisation and decidability, at least in the one-dimensional case. On the other hand, any non-trivial, non-toy application demands the expressivity of predicate logic, but first-order modal/temporal logics are often unaxiomatizable, never decidable (Garson, 1984; Gabbay *et al.*, 1994).

Is it possible to bridge this gap, having the best of both worlds: the simplicity of propositional temporal systems with the desired expressivity? Sure! — as long as those basic logical properties are not needed. Fortunately, in the case of rule execution, the activities to be performed are *querying* and *updating* the database, neither depending on the basic logical properties.

It was shown in (Finger, 1994) that there exists a *propositional abstraction* of a first-order temporal database, provided the database satisfies some conditions. The abstraction is a mapping μ from first order data, queries and updates into propositional data, queries and updates; μ-corresponding queries have μ-corresponding answers, and such property is invariant under μ-corresponding updates. Due to space constraints, we will not reproduce the full *propositional abstraction* construction here — how to get a propositional database, queries, updates and rules from a first-order one, which is therefore covered by the theory developed so far— but we are going to sketch how this can be done, focusing on the main restrictions that should be obeyed.

The main constraint is the *finiteness of the database*, i.e. all relations in the database must contain a finite number of tuples. In the case of temporal relations, this means a finite number of time-stamped tuples (or doubly time-stamped in the case of bitemporal relations). In logic programming terms, we are forcing a finite Herbrand domain. In this case, each time-stamped tuple is *abstracted into* a time-stamped proposition.

The first-order language we use for querying the database contains an enumerable set of constant symbols, no function symbols (as usual in databases) and one predicate symbol for each relation with its associated arity. Special relation symbols with fixed (normally infinite) interpretation may also be considered; in particular, we include equality, $=$, and the useful infinite relation *vtime(x)* which is true at a valid-time t if and only if the interpretation of x coincides with t; *ttime(x)* can analogously be added for transaction-time. A denumerable set of variables is at our disposal, and a *term* is either a variable or a constant. Given a predicate symbol p with arity n, $p(a_1, \ldots, a_n)$ is an atomic formula for $a_1, \ldots, a_n$ terms. Formulas are built in the usual way with boolean connectives, temporal operators and the quantifiers $\forall$ and $\exists$. Semantics is given as usual.

A query $\{x_1, \ldots, x_n | \varphi(x_1, \ldots, x_n)\}$ is a relational calculus-like expression such that φ is a first-order temporal (or bitemporal) formula with free-variables $x_1, \ldots, x_n$. Given the finiteness of the database, the $\forall$ and $\exists$ quantifiers can be abstracted into well-formed, finite conjunctions and disjunctions, respectively. However, database finiteness does not guarantee that queries can be propositionally abstracted. (Finger, 1994) shows that we need queries with the following property:

> *Domain Independence*: Every (bi)temporal query generates as answer a finite (bi)temporal relation containing only the constant symbols that occur in the database and query.

To achieve domain independence we restrict the allowable class of queries to *safe queries.* A query is safe whenever:

- if it contains $\varphi_1 \vee \varphi_2$, then φ_1 and φ_2 have the same set of free variables.
- if it contains a maximal conjunction (i.e. a conjunction that is not a conjunct) $\varphi_1 \wedge \ldots \wedge \varphi_m$, then all the free variables in it must be *limited* in some the φ_j in the following sense. A variable x is limited in φ if:
  - φ is $p(\ldots, x, \ldots)$, with p a database predicate;
  - φ is $x = c$ or $c = x$, with c a constant;
  - φ is $x = y$ or $y = x$ and y is a limited variable;
  - φ is $\mathcal{S}(\psi, \xi)$, $\mathcal{U}(\psi, \xi)$, $\bar{\mathcal{S}}(\psi, \xi)$ or $\overline{\mathcal{U}}(\psi, \xi)$ and x is limited in ψ. The free variables of ξ, when not occurring in ψ, have to be limited in another conjunct.
- it contains a negated formula only in the sense above, that is, in a conjunction where all its free variables are limited in other conjuncts.

PROPOSITION 3.1. *Safe (bi)temporal formulas are domain independent. They can be propositionally abstracted.*

It may seem that the addition of infinite relations such as $=$, $<$, *vtime(x)* and *ttime(x)* renders the the propositional abstraction impossible. This is not the case, however. Since the database is finite, only a finite number of equalities are relevant, namely, those involving the constants in the database, so $=$ is abstracted into a finite set of propositions. As for *vtime(x)*, we use the property that for a database time-stamped with intervals (or finite unions of intervals) over $\mathbb{Z}$, there is a time t_{max} after which all data persists into the future, that is, whatever is true (or false) at t_{max} will be true (or false) at all subsequent points; similarly for t_{min} in the past. We can then, in the abstraction process, introduce a finite number of propositions p_t, each being true at a single valid-time $t \in [t_{min}, t_{max}]$; *vtime(x)* is abstracted into a finite set of time-stamped propositions, just like database relations. Similarly for *ttime(x)*.

To propositionally abstract updates, we need to guarantee that the finiteness of a relation is maintained after the update. This is a consequence of the following:

PROPOSITION 3.2. *Let* R_1 *and* R_2 *be two finite (bi)temporal relations. Then* $R_1 \cup R_2$ *and* $R_1 - R_2$ *are finite.*

It suffices to restrict an update to the specification of a finite relation to be inserted in or deleted from the database. In the case of updates generated by rule execution, we have to guarantee that a rule generates only finite updates; see Section 5.

We have thus maintained a propositional logic as a theoretical basis for our system, while allowing the expressivity of first-order rules to be used with safe queries and safe rules.

4. Using 2DTL to describe bitemporal databases

A bitemporal database is an object which exists over time. We may suppose for convenience that it exists over natural numbers time: there is a moment of its creation and, from then on, it changes in discrete jumps. If the actual physical database ceases to exist we may just suppose that it really goes on indefinitely believing nothing.

If we suppose that the valid-time data we are recording in the database exists over integers time then it is easy to see how the database corresponds to a two dimensional plane model $\mathcal{M} = (T, <, \bar{T}, \bar{<}, \pi)$. We suppose that the the horizontal order $(T, <)$ is $\mathbb{Z}$ under its usual ordering and the vertical order $(\bar{T}, \bar{<})$ is $\mathbb{N}$ under its usual ordering. Thus we set π to map $(t, \bar{t}, p)$ to $\top$ if and only if at (transaction) time $\bar{t}$ the database believes that fact p did (does or will) hold at (valid) time t; p may be thought as the propositional abstraction of a first-order fact.

Formulas of two-dimensional temporal logic may be evaluated at points of this two-dimensional plane model. Below we will see that they can correspond to useful properties of the database. One example of a property is the definition of one proposition in terms of others. For example, in a hotel management system we might have the occupancy of a room `r` by a customer `c` indicated by the truth of a predicate `occ(c,r)` which is defined in terms of more basic predicates to do with the customer arriving or leaving. We require the formula

$$\forall \mathtt{c}.\ \forall \mathtt{r}.\ (\mathtt{occ(\ c,r)} \Leftrightarrow \mathcal{S}(\mathtt{arrives(\ c,r)}, \neg \mathtt{leaves(\ c,r)})$$

to be true at all times.

It is not immediately clear that the effects of updates to the database can be described by such universally true formulas. However, it is best to handle updates by introducing specific predicates whose truth indicates a request for an update. There may be consistency conditions which override the request. We can then say, using 2DTL, that if such a request predicate is true and the consistency checks are passed then an update occurs.

For example, consider the predicate `request_leaves_add(c,r,t)` which will indicate that a desk clerk wants the system to record that customer `c` is leaving room `r` at time `t`. A sensible room booking system will check that the customer actually occupies the room at that time. In that case it may allow the request to translate into the database being updated. The overall model of representing the evolution of the database may then satisfy the following formula at all times:

$$
\begin{array}{lll}
 & \forall \mathtt{c}.\ \forall \mathtt{r}.\ \forall \mathtt{t}. & (\mathtt{request\_leave\_add(c,r,t)} \\
 & \wedge & \blacklozenge\, (\mathtt{occ(c,r)} \wedge \mathtt{vtime(t)}) \\
\rightarrow & \bar{\bigcirc}\, \blacklozenge & (\mathtt{vtime(t)} \wedge \mathtt{leaves(c,r)})).
\end{array}
$$

Note that this only describes requests for leaving data about the past.

5. Executable 2DTL

In this section we will see how 2DTL formulas can be executed i.e. how they can be used to control the evolution of a bitemporal database.

5.1. Two-Dimensional Separation

One-dimensional propositional temporal formulas enjoy a *separation property*: given a one-dimensional formula, there exists an equivalent formula that is a conjunction of formulas of the form

$$\mathit{Past} \wedge \mathit{Present} \rightarrow \mathit{Future},$$

where *Present* contains no temporal operators, *Past* contains no future (i.e. $\mathcal{U}$) operators and *Future* contains no past (i.e. $\mathcal{S}$) operators. See (Gabbay *et al.*, 1994) for a proof of this result, which was originally shown by Gabbay. The separation property is at the heart of Gabbay's *declarative past/imperative future* paradigm (Gabbay, 1987).

The separation result does not hold for the two-dimensional temporal logic above, where *Past* is a formula that contains no future (i.e. $\mathcal{U}$ and $\bar{\mathcal{U}}$) operators, and *Future* contains no past (i.e. $\mathcal{S}$ and $\bar{\mathcal{S}}$) operators. For example, the formula $\Diamond \overline{\blacklozenge}\, p$ cannot be separated.

However, for a very useful class of formulas called *temporalised formulas*, we obtain a restricted notion of separation, called *vertical separation*, which is strong enough for our purposes here.

A *temporalised formula* is a two-dimensional formula in which no vertical temporal operator appears inside the scope of a horizontal temporal operator. Eg. $\overline{\Diamond}\Diamond\varphi$ is a temporalised formula, but $\Diamond\overline{\Diamond}\varphi$ is not. Temporalised formulas can be seen as the result of having the vertical temporal dimension applied *externally* to a one-dimensional temporal logic, as discussed in (Finger and Gabbay, 1992).

A formula is *vertically separable* if it is equivalent to a formula that is a conjunction of formulas of the form

$$v\text{-}Past \wedge v\text{-}Present \rightarrow v\text{-}Future,$$

where *v-Past* contains no vertical future operators (i.e. $\overline{\mathcal{U}}$ and its derived operators), *v-Future* contains no vertical past (i.e. $\overline{\mathcal{S}}$ and its derived operators) and no vertical operator occurs in *v-Present*. The following result was proved in (Finger and Gabbay, 1992).

THEOREM 5.1. *If φ is a temporalised formula then φ is vertically separable.*

Of course, a totally analogous horizontal separation can be obtained, but in our database model it is the transaction (vertical) time that is external to the valid (horizontal) time.

5.2. Executing Separated Formulas

We will use the notation

$$\varphi \Rightarrow \psi$$

where φ and ψ are formulas to be a rule in a database management system (DBMS) which has the following effect. If φ holds at any time point (t, t) on the diagonal in the two-dimensional structure being built by the DBMS, then we require that ψ also holds at the same diagonal point (t, t). Thus, putting this rule in the DBMS, should result in the final structure being a model of

$$\forall \overline{X}.\ (\varphi \wedge \delta \rightarrow \psi)$$

at all time points. Since the formula is vertically separated, φ refers only to the vertical past and present and ψ to the vertical future, so the general principle of the imperative future is preserved.

Allowing the full generality of formulas φ and ψ here would make the running of the DBMS impossible in general. However, the insight of the executable temporal logic idea is that certain quite general forms of formulas have a direct procedural interpretation. Thus, for some tuple $\overline{a}$, if $\varphi(\overline{a})$ only depends on the transaction past and present then we can, in principal determine its truth directly at any point in the construction of the database. Also, if $\psi(\overline{a})$ only depends on the transaction future then we may be able to bring about its truth at the required time by being careful with our choice of truth of database atoms.

Concerning the propositional interpretation of a first-order $\varphi(\overline{x}, \overline{y}) \Rightarrow \psi(\overline{x})$, we constrain rule format in the following way. The two-dimensional query φ has to be safe; this guarantees that there will be only finitely many tuples satisfying it. Moreover, all free variables in ψ must occur freely in φ; so there will be only finitely many formulas (called *actions*) to be enforced at by each rule at each time. There are other restrictions we can impose on ψ to allow it to be directly translated into an action. As in the one-dimensional MetateM approach (Barringer *et al.*, 1996), it is possible to rewrite general formulas in an imperative way but we will not examine this issue here– techniques used include renaming transformations to remove quantifiers from inside temporal contexts and to eliminate certain temporal operators (see (Fisher, 1997) for details). Instead, we may suppose for our purposes that ψ is $\bar{\bigcirc}\alpha$ or is of the form $\bar{\bigcirc} \blacklozenge\ (\mathtt{vtime(t)} \wedge \alpha))$ for some predicate α.

6. Example: a hotel booking and occupancy management system

In this section we see the Executable Bitemporal ideas in action in a very simple example. The example concerns a hotel management system (HMS) for use in managing the occupancy of rooms in a hotel. We will introduce a historical dimension by requiring occupancy data to be recorded. We will also introduce a second temporal dimension by requiring past assumptions about occupancy to be recorded even if those assumptions are not currently held.

The responsibilities of the HMS will include:

- recording data (manually input) about changes in occupants of rooms, expected occupancy of rooms and payment of bills;
- answering queries about rooms and bills in the present, the past and the future;
- correcting incorrectly recorded information;
- and explaining past actions.

A fully comprehensive database for this task needs to be two dimensional so that it can record information about the correction of past mistakes. Such information is needed to explain actions which were taken on the grounds of information subsequently corrected. A simpler, one-dimensional temporal database may be adequate to hold a representation of the most up to date account of the history of the situation in the hotel. Changes to data about the past could be allowed (as mistakes in recording do occur) but we would not necessarily be able to reconstruct the superseded model of the history of the hotel. Such a one-dimensional database is considered in the example of a distributed hospital patient monitoring system in (Finger and Reynolds, 1997).

The users will be primarily interested in the following predicates:

`pays( Customer, Amt, Room)`	`Customer` pays `Amt` towards the bill for `Room`
`occ( Customer, Room)`	`Room` is occupied by `Customer`
`room(Room)`	`Room` is a room

These are the most important predicates which vary in time in the world which the database will try to model. However it is easier to actually record some different predicates in the database. We introduce:

`leaves( Customer, Room )`	`Room` is vacated by `Customer`
`arrives( Customer, Room )`	`Room` becomes occupied by `Customer`

Recall that in the two dimensional approach, the predicates we have introduced take on a slightly expanded semantics: for example, `pays( Customer, Amt, Room)` holding at valid time t and transaction time $\bar{t}$ will mean that at transaction time $\bar{t}$, the database's model of the world contained the information that at valid time t, `Customer` paid `Amt` towards the bill for `Room`.

Note that this notation comfortably handles future room bookings. They are simply occupancy at a future time.

Let us now examine some of the properties of this two-dimensional account and see which need explicit statement.

6.1. Derived Predicates

In general with deductive databases, whether temporal or otherwise, there is a distinction between basic and derived predicates. The database manager must be told which predicates are to have their history recorded. Other predicates play subsidiary roles: either being able to be derived from the recorded ones or appearing temporarily to produce, in combination with executable rules, systematic changes in the database. In our example the database will only record the history of the predicates `pays`, `leaves` and `arrives`.

Thus `occ` is a derived predicate. We have already seen its definition in terms of `arrives` and `leaves`. To render it in an executable format the syntax is

$$\mathcal{S}(\texttt{arrives(c,r)}, \neg\texttt{leaves(c,r)}) \rightarrow \texttt{occ(c,r)}.$$

Note that there is no interesting two-dimensional character to this definition [1] and we do not need to refer to the vertical dimension.

6.2. Manual Record

Information entered by the desk clerk has a slightly indirect effect. Suppose that s/he enters the information that at time t, customer p vacates room r. We use the atom `request_leave_add(c,r,t)` to indicate this. For reasons which we have discussed above, the request is checked against consistency requirements before it is translated into an update. Thus we use

$$\begin{aligned} & \texttt{request\_leave\_add(c,r,t)} \wedge \blacklozenge\, (\texttt{occ(c,r)} \wedge \texttt{vtime(t)}) \\ \Rightarrow \bar{\bigcirc} \quad & \blacklozenge\, (\texttt{vtime(t)} \wedge \texttt{leaves(c,r)}). \end{aligned}$$

Such a request may be refused: we might like to include a rule

$$\texttt{request\_leave\_add(c,r,t)} \wedge \neg\texttt{occ(c,r)} \Rightarrow \bar{\bigcirc} \bigcirc (\texttt{error}).$$

Similarly, we have request predicates such as `request_arrive_add(c,r,t)` and `request_pay_add(c,a,r,t)` as well as corresponding rules.

6.3. Corrections

Corrections to billing, for example, can not be handled by allowing the user to enter in directly the falsity of `pays` for a particular historical time. This is because, as in the case of direct entry of facts, we must check for attempts at silly corrections. Thus we separate the fact of the user requesting a correction from the act of updating in accordance with that correction.

As well as corrections involving removal of recorded facts we also need to be able to add facts about the past. The user's requests are formalized in the predicate

---

[1] In predicate definitions, the defining formula must be safe, with the same free variables as the predicate being defined; such definition can be propositionally abstracted.

$$\texttt{request\_pays\_add(c,a,r,t)}$$

which means that the user requests that the predicate pays(c,a,r) be added for valid time t and similar predicates such as request_pays_remove(c,a,r,t), request_arrive_add(c,r,t) etc.

We use the predicate error whose truth at particular time on the diagonal has the side-effect of notifying the user that she or he has just requested a silly correction. Of course, in a real system the error predicate would also convey some useful information about why the transaction is not allowed.

The rules which define the truth of error look like the following:

$$\begin{aligned} \texttt{request\_pays\_add(c,a,r,t)} \wedge \quad & \blacklozenge(\texttt{vtime(t)} \wedge \texttt{pays(c,a,r)}) \\ & \Rightarrow \bar{\bigcirc}\bigcirc(\texttt{error}) \\ \texttt{request\_pays\_add(c,a,r,t)} \wedge \quad & \blacklozenge(\texttt{vtime(t)} \wedge \neg\texttt{occ(c,r)}) \\ & \Rightarrow \bar{\bigcirc}\bigcirc(\texttt{error}) \end{aligned}$$

When error does not hold, then we use a set of rules to bring about the correction:

$$\begin{aligned} & \texttt{request\_pays\_remove(c,a,r,t)} \\ & \wedge\blacklozenge(\texttt{vtime(t)} \wedge \texttt{pays(c,r,t)}) \\ \Rightarrow \bar{\bigcirc} \quad & \blacklozenge(\texttt{vtime(t)} \wedge \neg\texttt{pays(c,a,r)}) \\ & \texttt{request\_leaves\_add(c,r,t)} \\ & \wedge\blacklozenge(\texttt{vtime(t)} \wedge \neg\texttt{leaves(c,r)}) \\ \Rightarrow \bar{\bigcirc} \quad & \blacklozenge(\texttt{vtime(t)} \wedge \texttt{leaves(c,r)}) \end{aligned}$$

and similarly for arrives.

6.4. A Systematic Change to History

As an example of history being changed by a more complex correction, let us consider the case of the user realising that the database has recorded the wrong room for a customer. Say that Mr Papescu is in room 312 but the system has him recorded as being in 212. We will introduce a new predicate

$$\texttt{req\_corr\_room(c, r1, r2)}$$

which, under certain conditions, corrects the database by moving customer c from room r1 into room r2 from the time of her/his last arrival at room r1. We need to change the record of his arrival and we should also check that any payments he has made towards the bill of 212 now go towards the bill of 312. Thus we require

$$
\begin{array}{lll}
 & & \texttt{req\_corr\_room(c , r1, r2)} \\
 & \wedge & \mathcal{S}(\texttt{arrives(c, r1)} \wedge \texttt{vtime(t)}, \neg\texttt{leaves(c, r1)}) \\
 & \wedge & \blacklozenge\,(\texttt{vtime(t1)} \wedge \texttt{pays(c, amt, r1)} \wedge \blacklozenge\,(\texttt{vtime(t)})) \\
\Rightarrow \bar{\bigcirc}(& & \blacklozenge\,(\texttt{vtime(t1)} \wedge \texttt{pays(c, amt, r2)} \\
 & & \quad \wedge \neg\texttt{pays(c, amt, r1)}) \\
 & \wedge & \blacklozenge\,(\texttt{vtime(t)} \wedge \texttt{arrives(c,r2)} \\
 & & \quad \wedge \neg\texttt{arrives(c,r1)})).
\end{array}
$$

This rule is a very good example of the power of two-dimensional temporal logic. It is a very clear case of a systematic change to history.

6.5. Explanation

Finally we present use a two-dimensional rule to provide a facility which can not be provided in any one-dimensional temporal databases (without a arcane encoding of the history of transactions and the valid time history into one-dimension). This is the ability of the system to provide an explanation for an action which was based on a superceded model of the world. We consider the simple example of `tv_full` which is a derived predicate whose truth is defined as follow. `tv_full( t,v)` is true at any time pair (a, b) if and only if at transaction time `t` the database thought that the hotel would be (is,was) full at valid time `v`. Obviously, this can only be calculated at transaction time $a \geq$ `t`.

In that case the rule we use is

$$
\begin{array}{ll}
 & \overline{\blacklozenge}\,(\texttt{ttime(t)} \\
 & \wedge \blacklozenge\,\Diamond(\texttt{vtime(v)} \wedge (\forall\texttt{r}.\ (\texttt{room(r)} \rightarrow (\exists\texttt{c}.\ \texttt{occ(c,r)})))) \\
\rightarrow & \texttt{tv\_full(t,v)}.
\end{array}
$$

Such a rule then gives the basis for an explanation for refusals to book customers in.

7. Conclusion

In this paper we have shown how the executable temporal logics paradigm can be extended to at least an important and useful class of two-dimensional temporal formulas. We have demonstrated the usefulness of this approach in a small but realistic example.

In future work we would like to draw on the recent insights in (Fisher, 1997) (for one-dimensional first-order temporal logic) and consider the expressiveness of executable bitemporal languages. It will also be important to use recent advances in extending the usefulness of bitemporal databases (such as those in (Nascimento *et al.*, 1995) and (Nascimento *et al.*, 1996)) in combination with our proposed techniques in larger case studies.

References

H. Barringer, M. Fisher, D. Gabbay, R. Owens, and M. Reynolds, editors. *The Imperative Future.* Research Studies Press, Somerset, 1996.

M. Finger and D. M. Gabbay. Adding a Temporal Dimension to a Logic System. *Journal of Logic Language and Information*, 1:203–233, 1992.

M. Finger and D. Gabbay. Combining Temporal Logic Systems. *Notre Dame Journal of Formal Logic*, 37(2):204–232, Spring 1996.

M. Finger and M. Reynolds. Imperative history: two-dimensional executable temporal logic. In H.J. Ohlbach and U. Reyle, editors, *Logic, Language and Reasoning.* Kluwer, 1997. To be published.

M. Finger. *Changing the Past: Database Applications of Two-dimensional Temporal Logics.* PhD thesis, Imperial College, Department of Computing, February 1994.

M. Fisher and R. Owens, editors. *Proceedings of IJCAI Workshop on Executable Modal and Temporal Logics, Chambery, France 1993*, volume 897 of *LNAI.* Springer-Verlag, 1995.

M. Fisher. A normal form for tempral logic and its application in theorem-proving and execution. *Journal of Logic and Computation*, 7(4):?, 1997.

D. M. Gabbay, I. M. Hodkinson, and M. A. Reynolds. *Temporal Logic: Mathematical Foundations and Computational Aspects*, volume 1. Oxford University Press, 1994.

D. M. Gabbay. The Declarative Past and the Imperative Future. In B. Banieqbal et al., editors, *Coloquium on Temporal Logic and Specifications — Lecture Notes in Computer Science 389*, Manchester, April 1987. Springer-Verlag.

D. M. Gabbay. Declarative past and imperative future: Executable temporal lo interactive systems. In B. Banieqbal, H. Barringer, and A. Pnueli, editors, *Proceedings of Colloquium on Temporal Logic in Specifications*, pages 67–89. Springer-Verlag, 1989. Springer Lecture Notes in Computer Science 398.

J. W. Garson. Quantification in Modal Logic. In *Handbook of Philosophical Logic*, volume II, pages 249–307. Reidel Publishing Company, 1984.

D. Harel. Recurring dominoes: Making the highly undecidable highly understandable. In *Conference on Foundations of Computing Theory*, pages 177–194. Springer, Berlin, 1983. Lec. Notes in Comp. Sci. 158.

M. Marx and Y. Venema. *Multi-Dimensional Modal Logic.* Kluwer, 1997.

M.A. Nascimento, M.H. Dunham, and R. Elmasri. Using Incremental Trees for Space Efficient Indexing of Bitemporal Databases. In *Proceedings of the Second International Conference on Application of Databases (ADB'95)*, pages 235–248, Santa Clara, CA, December 1995.

M.A. Nascimento, M.H. Dunham, and R. Elmasri. M-IVTT: An Index for Bitemporal Databases. In *Proceedings of the Seventh International Conference on Databases and Expert Systems Applications (DEXA'96)*, pages 779–790, Zurich, Switzerland, September 1996. Lecture Notes in Computer Science, Vol. 1134.

M. Reynolds. Two-dimensional temporal logic: a survey, 1996. Invited Talk at Logic Colloquium '96, Submitted.

R. Snodgrass and I. Ahn. A Taxonomy of Time in Databases. In *ACM SIGMOD International Conference on Management of Data*, pages 236–246, Austin, Texas, May 1985.

E. Spaan. *Complexity of Modal Logics.* PhD thesis, University of Amsterdam, 1993.

Y. Venema. Completeness through flatness in two-dimensional temporal logic. In D. Gabbay and H.-J. Ohlbach, editors, *Temporal Logic, First International Conference, ICTL '94, Bonn, Germany, July 11-14, 1994, Proceedings*, volume 827 of *Lecture Notes in A.I.*, pages 149–164. Springer-Verlag, 1994.

F. Vlach. *Now and then: A. formal study in the logic of tense anaphora.* PhD thesis, UCLA, 1973.

Execution and Proof in a Horn-Clause Temporal Logic

Clare Dixon (C.Dixon@doc.mmu.ac.uk)
Department of Computing, Manchester Metropolitan University, Manchester M1 5GD, U.K.

Michael Fisher (M.Fisher@doc.mmu.ac.uk)
Department of Computing, Manchester Metropolitan University, Manchester M1 5GD, U.K.

Mark Reynolds (markr@dcs.kcl.ac.uk)
Department of Computer Science, King's College, Strand, London, WC2R 2LS, U.K.

Abstract. Both proof, via clausal resolution, and execution, via the imperative future approach, depend on the use of a normal form for temporal formulae. While the systems developed have centred around the use of an unrestricted normal form, we here consider a Horn clause-like version of the normal form and its effect on both execution and resolution. This refined normal form is as expressive as the original, and represents a natural way to describe systems, yet allows both execution and resolution to be implemented more efficiently in practice.

Keywords: temporal logic, theorem proving, execution, expressiveness and complexity issues, clausal resolution

1. Introduction

The direct execution of temporal logic formulae has become increasingly popular over recent years (Moszkowski, 1986; Fariñas del Cerro and Penttonen, 1992; Fisher and Owens, 1995), with a range of different logics being executed using a variety of programming paradigms. In many of these cases the expressive power of the programming notation is restricted in order to allow efficient implementation. In the METATEM system (Barringer et al., 1995), however, *any* formula within the underlying temporal logic can be executed. This leads to a situation where complex temporal formulae can be defined and where execution of such formulae may be particularly costly.

Whilst execution is the animation of a temporal logic formula or *specification*, *verification* that a temporal logic specification satisfies a particular property requires proof. Thus proof methods for temporal logics have also been developed based on, for example, automata (Vardi and Wolper, 1986), tableau (Wolper, 1985) and resolution (Abadi, 1987; Fisher, 1991). Here we consider the resolution based method given in (Fisher, 1991). This clausal resolution based method for propositional linear-time temporal logics (PTL) applies classical style resolution to

H. Barringer et al. (eds.), Advances in Temporal Logic, 413–433.

formulae occurring at the same moment in time and temporal resolution between formulae that state a condition must occur sometime in the future with formulae that ensure that the condition can never occur.

The basis of both METATEM and the temporal resolution method is a normal form, Separated Normal Form (SNF). Whilst SNF has been defined for both propositional (Fisher, 1991) and first-order (Fisher, 1997) linear-time temporal logics we limit our discussion to the propositional version as the first-order version is beyond the scope of this paper. The translation of any PTL formula into SNF reduces the number of different temporal operators to a core set and requires formulae to be in a particularly simple 'rule' form. If the original formula is satisfiable then so is the resulting set of formulae.

In classical logic it is usual to translate into a normal form, resulting in a set of *clauses*, before applying any resolution rules (Robinson, 1965). If the normal form is restricted such that only clauses with at most one positive literal are allowed (Horn clauses) then efficient strategies such as unit (Wos et al., 1964) or input (Chang, 1970) resolution may be used to guide the proof whilst retaining completeness. Similarly for execution, the programming language Prolog requires Horn clauses. Thus, in this paper we investigate (temporal) Horn clause-like restrictions on SNF, analysing their effect on both execution and resolution. This paper is structured as follows. The syntax and semantics of PTL are outlined in §2 and SNF and its restricted version RSNF are presented. The complexity of RSNF is discussed in §3. In §4 a review of METATEM and temporal resolution is given. The effect of restricting SNF upon such execution and proof is considered in §5 and §6 respectively. Concluding remarks are made in §7.

2. A Linear Temporal Logic

2.1. Syntax and Semantics

The logic used in this paper is Propositional Temporal Logic (PTL), in which we use a linear, discrete model of time with finite past and infinite future (Gabbay et al., 1980). PTL may be viewed as a classical propositional logic augmented with both future-time and past-time temporal operators. Future-time temporal operators include '$\Diamond$' (*sometime in the future*), '$\Box$' (*always in the future*), '$\bigcirc$' (*in the next moment in time*), '$\mathcal{U}$' (*until*), '$\mathcal{W}$' (*unless* or *weak until*), each with a corresponding past-time operator. Since our temporal models assume a finite past, for convenience, we define an operator, **start**, that is only true at the beginning of time.

Models for PTL consist of a sequence of *states*, representing moments in time, i.e.,

$$\sigma = s_0, s_1, s_2, s_3, \ldots$$

Here, each state, s_i, contains those propositions satisfied in the ith moment in time. As formulae in PTL are interpreted at a particular moment, the satisfaction of a formula f is denoted by $(\sigma, i) \models f$, where σ is the model and i is the state index at which the temporal statement is to be interpreted. For any well-formed formula f, model σ and state index i, then either $(\sigma, i) \models f$ or $(\sigma, i) \not\models f$. For example, a proposition symbol, 'p', is satisfied in model σ and at state index i if, and only if, p is one of the propositions in state s_i, i.e.,

$$(\sigma, i) \models p \quad \text{iff} \quad p \in s_i.$$

The semantics of the future-time temporal connectives are defined as follows

$$
\begin{array}{l}
(\sigma, i) \models \textbf{start} \ \text{iff}\ i = 0; \\
(\sigma, i) \models \bigcirc A \ \text{iff}\ (\sigma, i+1) \models A; \\
(\sigma, i) \models \Diamond A \ \text{iff}\ \text{there exists a } j \geq i \text{ s.t. } (\sigma, j) \models A; \\
(\sigma, i) \models \Box A \ \text{iff}\ \text{for all } j \geq i \text{ then } (\sigma, j) \models A; \\
(\sigma, i) \models A \,\mathcal{U}\, B \ \text{iff there exists a } k \geq i \text{ s.t. } (\sigma, k) \models B \\
\qquad\qquad\qquad \text{and for all } i \leq j < k \text{ then } (\sigma, j) \models A; \\
(\sigma, i) \models A \,\mathcal{W}\, B \ \text{iff}\ (\sigma, i) \models A \,\mathcal{U}\, B \text{ or } (\sigma, i) \models \Box A.
\end{array}
$$

2.2. A Normal Form for PTL

Formulae in PTL can be transformed in to Separated Normal Form (SNF), which is used as the basis for both execution, via the METATEM language (Barringer et al., 1995), and proof, via a clausal resolution method (Fisher, 1991). SNF was introduced first in (Fisher, 1991) and has been extended to first-order temporal logic in (Fisher, 1997). Here we use a future-time version of SNF which is equivalent to the original (which contained both future and past-time formulae). While the translation from an arbitrary temporal formula to SNF will not be described here, we note that such a transformation not only preserves satisfiability, but also any model generated from the formula in SNF is a model for the original formula. A formula in (future-time) SNF is of the form:

$$\Box \bigwedge_{i=1}^{n} (\phi_i \Rightarrow \psi_i)$$

where each of the '$\phi_i \Rightarrow \psi_i$' (called *rules*) is one of the following

$$\mathbf{start} \Rightarrow \bigvee_{k=1}^{r} l_k \qquad \text{(an } \textit{initial} \text{ rule)}$$

$$\bigwedge_{j=1}^{q} m_j \Rightarrow \bigcirc \bigvee_{k=1}^{r} l_k \qquad \text{(an } \textit{always} \text{ rule)}$$

$$\bigwedge_{j=1}^{q} m_j \Rightarrow \Diamond l \qquad \text{(a } \textit{sometime} \text{ rule)}$$

and each m_j, l_k, and l are literals. In certain situations we require sets of the always rules to be combined or *merged* where their left hand sides are conjoined and simplified and the right hand sides are conjoined, simplified and rewritten in DNF. Such merged rules are known as SNF_m rules.

2.3. Restricted SNF (RSNF)

In the rest of the paper we further restrict the normal form and analyse the implications of this on both METATEM execution and temporal resolution. Many restrictions mean that the logic becomes uninteresting as either eventualities cannot be constrained or the restricted logic reduces to classical logic. However, the particular restrictions defined below still allow temporal resolution to take place and non-trivial METATEM programs to be defined.

Let the following be the unrestricted form of the always rules

$$\left(\bigwedge_{j=1}^{r} \neg a_j \wedge \bigwedge_{k=1}^{s} b_k\right) \Rightarrow \bigcirc \left(\bigvee_{l=1}^{t} c_l \vee \bigvee_{m=1}^{u} \neg d_m\right)$$

where the a_j, b_k, c_l and d_m are all propositions. Similarly, we consider the unrestricted form of initial rules to be

$$\mathbf{start} \Rightarrow \bigvee_{l=1}^{t} c_l \vee \bigvee_{m=1}^{u} \neg d_m$$

and of sometime rules to be

$$\left(\bigwedge_{j=1}^{r} \neg a_j \wedge \bigwedge_{k=1}^{s} b_k\right) \Rightarrow \Diamond l$$

Now we restrict this general form in the following two ways.

R1: $r = 0$ and $t \leq 1$;

R2: $u \leq 1$ and $s = 0$.

Restriction R1 has always rules with conjunctions of propositions on the left-hand side and Horn clauses enclosed within a next operator on the right-hand side and initial rules with Horn clauses on the right hand side. Restriction R2 has always rules with conjunctions of negated propositions on the left-hand side and clauses with at most one negative literal enclosed within a next operator on the right-hand side. The initial rules similarly have clauses with at most one negative literal on the right-hand side.

In order that the resolution rules may be applied we define $R1^+$ and $R2^+$ restrictions that further restrict the sometime rules.

DEFINITION 1. *A set of SNF rules, R, is said to be $R1^+$ restricted if, and only if, $r \in R$ is of* R1 *format and if r is a sometime rule then the eventuality is a negated proposition.*

Sets of $R2^+$ restricted SNF rules can be similarly defined. Although the results presented throughout the rest of this paper mainly concern the R1 restriction, dual results apply to R2 restricted rules.

2.4. Translation into RSNF

While any arbitrary PTL formula can be translated into a set of SNF rules, the use of RSNF (R1) introduces some restrictions on this. In particular, while the majority of PTL formulae can still be represented, formulae containing (non-conjunctive) binary operators in the scope of an even number of negations, where each of the operands is a positive literal, i.e.. '$a \vee b$', '$a \mathcal{W} b$', or '$a \mathcal{U} b$', where a and b are both propositions, can not. Note that this also includes formulae such as '$(\neg a \wedge \neg b) \Rightarrow \ldots$'.

3. Complexity

In this section we will show that the problem R1-SAT, of deciding the satisfiability of finite sets of R1 restricted formulae is just as hard as deciding satisfiability of general PTL formulae: we will show that it is PSPACE-complete in the total length of the rules.

Recall (from (Sistla and Clarke, 1985)) that satisfiability in the propositional temporal language with just $\bigcirc$ (called **X** by Sistla and Clarke) and $\Diamond$ (called **F**) is PSPACE-complete. It follows that R1-SAT is in PSPACE as any R1 restricted set of rules almost directly represents a formula of almost the same length in Sistla and Clarke's language.

To see that R1-SAT is PSPACE-hard, we will directly code the operation of a PSPACE Turing Machine M by a R1 restricted set of rules. We assume that M is space bounded by some polynomial $S(n)$ in n and $a = a_1 a_2 ... a_n$ is an input to M.

Due to space restrictions we omit full details of the coding but attempt to give an adequate sketch. The idea is to present a set of rules (of size polynomial in n) which is satisfiable, if at all, in a structure which corresponds closely to a successful run of M (if it has one).

This structure will use its first $S(n)$ states to code up the initial contents of M's tape, and the initial state of M. Then there will be a marker state. Then the next length of $S(n)$ states correspond to the contents of M's tape in the second moment of its successful run. This is also followed by a marker state. This pattern of lengths of $S(n)$ states separated by marker states continues indefinitely. The state corresponding to the tape square under the head at each stage will contain a proposition indicating the state in which M is at that stage. To code up this information we need propositions for the symbols in the alphabet and the states in M.

So far, the coding of the Turing machine in a propositional logic should be very familiar–see (Sistla and Clarke, 1985). However, we have an unusually restricted language. To describe the structure using R1 rules is mostly straight forward using a few extra propositions. For example, we need propositions with meaning "the proposition for tape symbol e will be true in k states" for each e and each $k \leq S(n)$. We need rules which count down k. We also need many rules which say that two different tape symbols are not true at the same state. To enforce that there is only one machine state proposition true in each $S(n)$ length we use many rules containing propositions with meaning "the proposition for machine state q will be true in k states".

The rules which say how to change the tape from one step to the next are easy to state. However, due to the restrictions on use of negations it is harder to specify that tape symbols stay the same on other parts of the tape. To do this we introduce propositions δ_z where $-S(n) \leq z \leq S(n)$. The intuitive meaning of δ_z being true at the jth state in a certain length of $S(n)$ states corresponding to the ith step of M, is that at the ith step the head of M is over the $(j - z)$th square on the tape. We also make a new proposition R or L (but not both) true throughout a length of states if the machine moves right or left at that step. To do this, we must check, at a marker state, what machine state proposition lies ahead and which tape symbol lies the same distance ahead. A rule for each entry in the transition table of M will then give the truth of L or R. This can be perpetuated through the length. With the combination of R or L and δ_z it is then easy to tell if a tape symbol will be modified or not.

The rules for setting up the initial state of the machine and tape are straight forward given a. Finally we need a rule which says that the accepting state eventually occurs. For this we need a new proposition which holds at all structure states except where the accepting state proposition holds. It can be shown that the satisfiability of the resulting set of rules is equivalent to the acceptance of a by M. Thus R1-SAT is PSPACE-hard.

4. Applications of SNF

There are two main applications of SNF that have been developed. The first is the direct execution of temporal formulae in SNF, the second is a temporal resolution method that can be applied to formulae in SNF. Both of these make use of the succinct representation provided by the normal form. Before considering the effect of our restrictions on these applications, we will first outline these execution and resolution mechanisms.

4.1. METATEM

The idea behind METATEM (Barringer et al., 1995; Barringer et al., 1996) is to directly use the formula to be executed in order to build a model (in our case, a sequence) for the formula. Rather than providing a detailed description of the METATEM execution mechanism, we will show how SNF can be executed using the basic ideas from METATEM.

If a program, $\mathcal{P}$, is given as a set of SNF rules $\{\Phi_i \Rightarrow \Psi_i \mid 1 \leq i \leq n\}$, then this represents the PTL formula

$$\varphi_{\mathcal{P}} = \Box \bigwedge_{i=1}^{n} (\Phi_i \Rightarrow \Psi_i).$$

The execution of the program, $\mathcal{P}$, is an iterative process of labelling a model structure with the propositions true in each state, which eventually yields a model for the formula, $\varphi_{\mathcal{P}}$ (if the formula is satisfiable). Essentially, the execution follows a forward chaining approach through the SNF rules. Choice during execution comes either from executing disjunctions of sometime rules. In the latter case, we can express this choice by the equivalence:

$$\Diamond a \Leftrightarrow (a \vee \bigcirc \Diamond a)$$

So to make $\Diamond a$ true, the execution mechanism can either make a true immediately, or, make $\Diamond a$ true in the next state. This latter step is done by passing a commitment $\Diamond a$ from the current state to the next state to be conjoined with the consequents of the successful rules. Thus, the execution mechanism has a strategy not only for choosing between disjuncts, but also for choosing when to satisfy formulae of the form $\Diamond a$ (called *eventualities*). In METATEM, the execution mechanism attempts to satisfy as many eventualities as possible at any moment. In the case of conflicting eventualities, e.g., $\Diamond a$ and $\Diamond \neg a$, the oldest outstanding eventuality is satisfied first. Any unsatisfied eventualities are passed on to the next state.

Backtracking

If a contradiction is generated within a state, i.e., execution of the rules has forced us to make both a proposition and its negation true in the current state, then a form of 'backtracking' — undoing previous choices — occurs. This backtracking undoes the model construction and returns the execution to a previous choice point. If there are no more choices left, execution fails, signifying that the program is unsatisfiable.

Loop Checking

An important aspect of METATEM, and one that ensures that the execution mechanism is complete for PTL, is the inclusion of *loop checking*. When the system is executing, it is possible for it to end up in a 'loop'. This is where the same sequence of states is recurring along with the same unsatisfied eventualities. Before constructing a new state, a check for such a situation is made. If this is detected, the execution is forced to backtrack to a previous choice point. Using this loop checking mechanism, it can be shown that the METATEM execution mechanism, when it uses the following strategy for choosing which disjunct to execute, is sound and complete with respect to the semantics of PTL. The choice strategy used is to try to execute disjuncts that will satisfy the longest outstanding eventualities first. Thus, an attempt to satisfy each individual eventuality will occur as the execution proceeds. For a longer description of the correctness of propositional METATEM, see (Barringer et al., 1995).

Applications

Although much of the development of METATEM has been suspended in favour of Concurrent METATEM (Fisher, 1994), the language has applications in system modelling (Finger et al., 1993), databases (Finger et al., 1991) and meta-level representation (Barringer et al., 1991).

4.2. Temporal Resolution

As we also consider the effect of the various restrictions on SNF has on Fisher's temporal resolution method (Fisher, 1991), here we briefly review this method. The clausal temporal resolution method consists of repeated applications of both 'step' and 'temporal' resolution on sets of formulae in SNF, together with various simplification steps.

Step Resolution

'Step' resolution consists of the application of standard classical resolution rule to formulae representing constraints at a particular moment in time, together with simplification rules for transferring contradictions within states to constraints on previous states. Simplification and subsumption rules are also applied. Pairs of initial rules or always rules may be resolved using the following (step resolution) rules.

$$\frac{\begin{array}{lcl}\textbf{start} & \Rightarrow & A \vee r \\ \textbf{start} & \Rightarrow & B \vee \neg r\end{array}}{\begin{array}{lcl}\textbf{start} & \Rightarrow & A \vee B\end{array}} \qquad \frac{\begin{array}{rcl}P & \Rightarrow & \bigcirc(A \vee r) \\ Q & \Rightarrow & \bigcirc(B \vee \neg r)\end{array}}{\begin{array}{rcl}(P \wedge Q) & \Rightarrow & \bigcirc(A \vee B)\end{array}}$$

Once a contradiction within a state is found using step resolution, the following rule can be used to generate extra global constraints.

$$\frac{\begin{array}{rcl}P & \Rightarrow & \bigcirc\textbf{false}\end{array}}{\begin{array}{rcl}\textbf{start} & \Rightarrow & \neg P \\ \textbf{true} & \Rightarrow & \bigcirc\neg P\end{array}}$$

The step resolution process terminates when either no new resolvents are derived, or **false** is derived in the form of the rules **start** $\Rightarrow$ **false** or **true** $\Rightarrow \bigcirc$**false**.

Temporal Resolution

The aim of temporal resolution is to resolve a sometime rule, $Q \Rightarrow \Diamond l$ with a set of rules that together imply $\Box\neg l$, for example a set of rules that together have the effect of $A \Rightarrow \bigcirc\Box\neg l$. However the interaction between the '$\bigcirc$' and '$\Box$' operators in PTL makes the definition of such a rule non-trivial and further the translation from PTL to SNF will have removed all but the outer level of $\Box$-operators. So, resolution will be between a sometime rule and a *set* of rules that together imply a $\Box$-formula contradicting the sometime rule. Thus, given a set of rules in SNF, then for every rule of the form $Q \Rightarrow \Diamond l$ temporal resolution may be applied between this sometime rule and a set of always rules, which taken together force $\neg l$ always to be satisfied. The temporal resolution rule is given by the following

$$\begin{array}{rcl} A_0 & \Rightarrow & \bigcirc F_0 \\ \ldots & & \ldots \\ A_n & \Rightarrow & \bigcirc F_n \\ Q & \Rightarrow & \Diamond l \\ \hline Q & \Rightarrow & (\bigwedge_{i=0}^{n} \neg A_i)\,\mathcal{W}\,l \end{array} \quad \text{with side conditions} \quad \left\{ \begin{array}{l} \text{for all } 0 \leq i \leq n \\ \vdash F_i \Rightarrow \neg l \\ \vdash F_i \Rightarrow \bigvee_{j=0}^{n} A_j \end{array} \right\}$$

where the side conditions ensure that the set of rules $A_i \Rightarrow \bigcirc F_i$ together imply $\square\neg l$. So if any of the A_i are satisfied then $\neg l$ will be *always* be satisfied, i.e.,

$$\bigvee_{i=0}^{n} A_k \Rightarrow \bigcirc \square \neg l.$$

Such a set of rules are known as a *loop* in $\neg l$. We note that the loops described here are like those we must check for in METATEM. However, in temporal resolution we are trying to *detect* loops to allow us to apply the temporal resolution rule and hence derive a contradiction, whilst in the execution of METATEM programs we are trying to *avoid* loops in order to generate consistent execution sequences.

Completeness of the resolution procedure has been shown in (Peim, 1994).

4.2.1. *Loop Search*

Different approaches to detecting loops, i.e. a set of rules that together imply $\square\neg l$, have been investigated in (Dixon, 1996). Here, we describe one of these approaches, namely a Breadth-First Search Algorithm. The graph constructed using this approach is a sequence of nodes that are labelled with formulae in DNF. This represents the left-hand sides of rules used to expand the previous node which have been disjoined and simplified.

4.2.1.1. *Breadth-First Search Algorithm* For each rule of the form $Q \Rightarrow \Diamond l$ carry out the following.

1. Search for all the rules of the form $X_k \Rightarrow \bigcirc\neg l$, for $k = 0$ to b, disjoin the left-hand sides and make the *top node* H_0 equivalent to this, i.e.

$$H_0 \Leftrightarrow \bigvee_{k=0}^{b} X_k.$$

   Simplify H_0. If $\vdash H_0 \Leftrightarrow$ **true** we terminate having found a loop.

2. Given node H_i, build node H_{i+1} for $i = 0, 1, \ldots$ by looking for rules or combinations of rules of the form $A_j \Rightarrow \bigcirc B_j$, for $j = 0$ to m where $\vdash B_j \Rightarrow H_i$ and $\vdash A_j \Rightarrow H_0$. Disjoin the left-hand sides so that

$$H_{i+1} \Leftrightarrow \bigvee_{j=0}^{m} A_j$$

and simplify as previously.

3. Repeat (2) until

   a) $\vdash H_i \Leftrightarrow$ **true**. We terminate having found a Breadth-First loop and return **true**.

   b) $\vdash H_i \Leftrightarrow H_{i+1}$. We terminate having found a Breadth-First loop and return the DNF formula H_i.

   c) The new node is empty. We terminate without having found a loop.

5. The Effect of Restrictions on MetateM

While, in §4, we presented an outline of both temporal execution and resolution, in §5 and §6 respectively, we consider the effect of restricting SNF on these two applications.

We begin with the effect on expressiveness within METATEM.

5.1. EXPRESSIVENESS

In this section we show that for any temporal formula in the full propositional language there is a set of RSNF rules such that running METATEM on the rules will produce a model of the original formula. This shows there is no difference in absolute expressive power between the restricted syntax and full propositional temporal logic in terms of METATEM programs.

Suppose that ϕ is a formula from PTL using the propositions from P. Using the algorithms of Vardi and Wolper (Vardi and Wolper, 1986) or Emerson and Sistla (Emerson and Sistla, 1984) we can build a non-deterministic finite state Büchi 2^P-automaton $A = (S, T, S_0, F)$ which recognizes a P-structure if, and only if, it is a model of ϕ.

We could determinize A using results of McNaughton (McNaughton, 1966) or Safra (Safra and Vardi, 1989) and then translate the deterministic (Rabin) automaton into a METATEM program. Instead, as we are just interested in describing one model of ϕ, we can do this by direct examination of the transition table T of A. We find a loop of states of A which can be reached from some state in S_0 and which contain at least one element from the acceptance set F. Say that an *accepting accessed loop* in A is a pair

$$(((r_0, b_0, r_1), (r_1, b_1, r_2), ..., (r_{m-1}, b_{m-1}, s_0)),\\ ((s_0, a_0, s_1), (s_1, a_1, s_2), ..., (s_{n-1}, a_{n-1}, s_0)))$$

of sequences of triples from T such that:

- $r_0 \in S_0$
- all the $r_0, ..., r_{m-1}, s_0, ..., s_{n-1}$ are distinct
- some s_i is in F

It is clear that there is an algorithm to find any accepting accessed loop in A if there is one and report not if there is not. It is also clear that ϕ has no models if A has no accepting accessed loop. If A has such a loop as above then we can use it to construct a model of ϕ and an RSNF program which constructs the same model. Simply define $l : \mathbb{N} \longrightarrow 2^P$ by $l(0) = b_0, l(1) = b_1, ..., l(m-1) = b_{m-1}, l(m) = a_0, l(m+1) = a_1, ..., l(m+n-1) = a_{n-1}, l(m+n) = a_0, l(m+n+1) = a_1, ..., l(m+2n) = a_0$, etc.

The chosen model of ϕ is that in which $p \in P$ is true at time i if, and only if, $p \in l(i)$. To construct this model we can use the following RSNF METATEM program which uses propositions from P as well as some new ones $B_0, B_1, ..., B_{m-1}, A_0, A_1, ..., A_{n-1}$.

The program contains rules:

- **start** $\Rightarrow B_0$
- $B_i \Rightarrow \bigcirc B_{i+1}$ for $i = 0, ..., m-2$
- $B_{m-1} \Rightarrow \bigcirc A_0$
- $A_i \Rightarrow \bigcirc A_{i+1}$ for $i = 0, ..., n-2$
- $A_{n-1} \Rightarrow \bigcirc A_0$
- $A_i \Rightarrow \bigcirc p$ for all $p \in a_{i+1}$, for all $i = 0, ..., n-2$,
- $A_{n-1} \Rightarrow \bigcirc p$ for all $p \in a_0$
- **start** $\Rightarrow p$ for all $p \in b_0$
- $B_i \Rightarrow \bigcirc p$ for all $p \in b_{i+1}$, for $i = 0, ..., m-2$

– $B_{n-1} \Rightarrow \bigcirc p$ for all $p \in a_0$

Provided that this program is run with a version of METATEM where atoms default to false unless otherwise constrained, this program will construct the chosen model of ϕ.

5.2. Eliminating Backtracking

Using similar automata-theoretic techniques to §5.1 we can show how to eliminate nondeterminism from RSNF programs. Here, we show how, given a set R of RSNF rules, we can construct a set R' of RSNF rules such that R' is a deterministic program and yet it constructs a model of R. In fact, as R directly represents a temporal logic formula we could use the techniques of the previous subsection to accomplish this task. However, we present an alternative algorithm (also exponentially complex), because it proceeds via the interesting step of demonstrating how to represent the METATEM program R as a nondeterministic finite state Büchi automaton with a trivial acceptance condition —an object commonly known as a transducer. If R uses propositional atoms from P, the transducer's language will be 2^P.

First choose new atoms e_c for each proposition c which appears in the head of a sometime rule in R. Let E be the set of such atoms. The set S of states of the transducer will be those $s \in 2^{P \cup E}$ such that:

– if $(\bigwedge_{i \in I} p_i \Rightarrow \Diamond c) \in R$ and each $p_i \in s$ then either $c \in s$ or $e_c \in s$.

We define $A = (S, T, S_0, F)$, with $F = \mathit{true}$ as follows. Let S_0 contain all the subsets s of $P \cup E$ which themselves contain at least one proposition from the head of each initial always rule in R. The transition table $T \subseteq S \times 2^P \times S$ is given by $(s, a, r) \in T$ if, and only if,:

– $a = P \cap s$

– if $(\bigwedge_{i \in I} p_i \Rightarrow \bigcirc \bigvee_{j \in J} \neg q_j) \in R$ and each $p_i \in s$ then there is some $j \in J$ such that $q_j \notin t$;

– if $(\bigwedge_{i \in I} p_i \Rightarrow \bigcirc(\bigvee_{j \in J} \neg q_j \vee q)) \in R$ each $p_i \in s$ and each $q_j \in t$ then $q \in t$;

– if $e_c \in s$ and $c \notin s$ then $e_c \in t$.

It should be clear that an accepting run σ of A corresponds exactly to a possible model of R provided that whenever e_c is an element of a state σ_i then there is some $j \geq i$ such that $c \in \sigma_j$. Using the same reasoning as in the last subsection we deduce that R has a model if, and only if, there is a pair

$$(((s_0, b_0, s_1), (s_1, b_1, s_2), ..., (s_{m-1}, b_{m-1}, r_0)),$$
$$((r_0, a_0, r_1), (r_1, a_1, r_2), ..., (r_{n-1}, a_{n-1}, r_0)))$$

of sequences of triples from T such that:

- $s_0 \in S_0$
- all the $s_0, ..., s_{m-1}, r_0, ..., r_{n-1}$ are distinct
- if $e_c \in r_0 \cup r_1 \cup ... \cup r_{n-1}$ then $c \in r_0 \cup r_1 \cup ... \cup r_{n-1}$.

We can also turn this sequence of states into a new deterministic RSNF program R' by using new propositions to represent the states which appear. This program will produce a model of R. The details are similar to those in the last subsection and we omit them.

5.3. Efficiency

As a typical METATEM program is designed not to terminate, usual measures of computational complexity are irrelevant. However, for the task of comparing different versions of METATEM, such as the restricted syntax version considered here, it is interesting to examine some measure of how long it takes for the execution process to complete one cycle from deciding on the truths of propositions at one state to those at the next state in the structure it is constructing. Note that any such measure ignores complications due to backtracking. In what follows we shall also ignore the process of loop checking which, for the sake of completeness of the process, should be occurring at each step but which, as we have seen earlier, may be a time consuming process.

So what we consider here is how many computation steps does it take to get from the state of truth of propositions at time k to the state of truth of propositions at time $k+1$ when executing an RNSF program, R. The input to this one-cycle processor will be the program R, the truth values for all the propositions and truth values for any of the eventualities mentioned in the program. This latter set of values will be needed to keep track of which eventualities must be satisfied at some future stage. The output of the process may be **false** if the structure can not be extended by one step, or a list of new truth values for propositions and eventualities.

Let us suppose that the size of R is n. This also bounds the number of propositions and the number of eventualities. Here we show that the one-cycle problem is PTIME computable — in fact cubic in n.

This contrasts with general METATEM programs where the problem is NP-complete.

Checking which program rules are "fired" is at most quadratic — we only need check the truth of a conjunction of propositions. Next, we only have at most n non-temporal Horn clauses to solve. This is also quadratic as we may have n rounds of checking the clauses to reach the fixpoint solution. If there is no solution then the structure can not be extended and we are finished. Next we check which eventualities can be satisfied. This is best done by choosing the oldest unsatisfied eventuality (age can be indicated by the order of listing of eventualities in the input), and seeing whether its proposition can be made true consistently with the propositions already determined as being true. Again this involves n rounds of checking clauses. Whether or not this eventuality can be satisfied at this stage, it is also best to move on to check the second oldest eventuality and so on. This is where the process becomes cubic. Finally we need check which eventualities have to be passed on to the $k+1$st step. This is just the union of those left unsatisfied from the last step, with those whose rules are "fired" at this step. Overall, this is a process of cubic complexity.

5.4. "PROGRAMMABILITY"

Note, while the transformations given in §5.1 and §5.2 showed that RSNF has the same expressive power as SNF, each of these required an exponential increase in the number of rules. Here we (briefly) consider how useful RSNF is in its own right, i.e. what types of problems can be naturally represented using sets of RSNF rules. A typical R1 rule is of the form

$$(a_1 \wedge a_2 \wedge \ldots \wedge a_n) \Rightarrow \bigcirc((b_1 \wedge b_2 \wedge \ldots \wedge b_m) \Rightarrow c)$$

where all of a_i, b_j and c are propositions. In order to fire such a rule, the a_i conjunction of propositions must occur. Once fired, this provides a Horn clause to be computed in the next state. Thus, sets of such RSNF rules can be seen as defining which sets of Horn clauses will be used in each state. In this sense, the execution of RSNF rules can be seen as an intuitive generalisation of standard Horn clause execution.

With respect to the types of program that can be coded in RSNF, the rule form presented above covers many programming clichés seen in METATEM. In particular, the forms that are disallowed, i.e. basing current action on a conjunction of negative literals in the previous state, and describing choices such as $a \vee b$ or $a \mathcal{U} b$ are rare. However, it is likely that a practical implementation of RSNF execution will allow both R1 and R2 restricted rules (see §7).

6. The Effect of Restrictions on Resolution

First we show that by restricting the sometime rules all the resolvents produced can be rewritten as RSNF rules.

THEOREM 1. *Given a set of rules, R, that are R1$^+$ restricted then the application of any resolution rule produces resolvents that are R1$^+$ restricted.*

Proof. We must consider the application of both step and temporal resolution rules. First we consider step resolution. The left hand sides of rules remain in the correct format as we either have two initial rules with **start** on their left hand sides or we have two always rules both with conjunctions of propositions (or **true**) on their left hand sides. Applying the step resolution rule to either two initial rules or two always rules, the resolution of two Horn clauses on the right hand side produces a Horn clause, **true** or **false**. Rules with **true** on the right hand side are removed through simplification as they always hold. The generation of $\textbf{start} \Rightarrow \textbf{false}$ or $\textbf{true} \Rightarrow \bigcirc\textbf{false}$ means that we have detected a contradiction. The generation of $P \Rightarrow \bigcirc\textbf{false}$ from the resolution of two always rules means that the rule must be rewritten as $\textbf{start} \Rightarrow \neg P$ and $\textbf{true} \Rightarrow \bigcirc\neg P$. Again these fit the rule structure as P was a conjunction of propositions (if P was **true** we would have terminated at the previous step) so $\neg P$ is a disjunction of negated propositions and is therefore a Horn clause.

When applying the temporal resolution rule we search for a set of rules $A_i \Rightarrow \bigcirc B_i$ for $i = 0 \ldots n$ that satisfy the conditions for being a loop given in §4.2. Each of these rules is made from combining several always rules where each left hand side is a conjunction of propositions. Assume the sometime rule is $Q \Rightarrow \Diamond\neg l$ where from the R1$^+$ restriction Q is a conjunction of propositions and l is a proposition. The resolvent in this case will be

$$Q \Rightarrow (\bigwedge_{i=0}^{n} \neg A_i)\, \mathcal{W} \neg l.$$

When rewriting to the normal form we obtain the rules

$$\begin{array}{rclcrcl} \textbf{start} & \Rightarrow & \neg Q \vee \neg l \vee t & \quad & \textbf{start} & \Rightarrow & \neg Q \vee \neg l \vee \neg A_i \\ \textbf{true} & \Rightarrow & \bigcirc(\neg Q \vee \neg l \vee t) & & \textbf{true} & \Rightarrow & \bigcirc(\neg Q \vee \neg l \vee \neg A_i) \\ t & \Rightarrow & \bigcirc(\neg l \vee t) & & t & \Rightarrow & \bigcirc(\neg l \vee \neg A_i) \end{array}$$

for each A_i where $i = 0 \ldots n$ and where t is a fresh proposition. So as each A_i and Q are conjunctions of propositions the rules are still in the R1$^+$ rule format.

Our first observation is related to loops that can be detected in the R1$^+$ restriction.

LEMMA 1. *For any loop, $A_i \Rightarrow \bigcirc B_i$ for $i = 0 \ldots n$, in the proposition l (where B_i cannot be equivalent to* **false***) formed from merging rules in the R1 format we can find a loop $A'_i \Rightarrow \bigcirc B'_i$ such that $A_i \Rightarrow A'_i$ for each i and B'_i is a conjunction of propositions.*

Proof. From the definition of a loop in l, $A_i \Rightarrow \bigcirc B_i, \vdash B_i \Rightarrow \bigvee_{j=0}^{n} A_j$ and $\vdash B_i \Rightarrow l$ for each i. From the R1 restriction all the left hand sides of the always rules are conjunctions of propositions therefore $\bigvee_{i=0}^{n} A_j$ is a disjunction of conjunctions of propositions. Thus B_i must be a conjunction of propositions or we can delete one or more of the rules that has been combined to make $A_i \Rightarrow \bigcirc B_i$ for a particular i and still have a loop. As we have only deleted rules from the rules that make up each merged rule then each $A_i \Rightarrow A'_i$ and B'_i is a conjunction of propositions as required.

LEMMA 2. *For any loop in l, $A_i \Rightarrow \bigcirc B_i$, a loop in l, $A'_i \Rightarrow \bigcirc B'_i$, can be constructed such that $A_i \Rightarrow A'_i$ by merging only rules of the form $X \Rightarrow \bigcirc p$ where X is a conjunction of propositions and p is a proposition.*

Proof. From Lemma 1 given any loop $A_i \Rightarrow \bigcirc B_i$ we can construct a (possibly more general) loop, $A'_i \Rightarrow \bigcirc B'_i$, such that $A_i \Rightarrow A'_i$ for each i and B'_i is a conjunction of propositions. Either each B'_i has been made up from combining rules with a single proposition on its right hand side and we are done or a set or rules, R, have been merged that contain a rule r with two or more disjuncts on the right hand side. As B'_i is a conjunction of propositions it means that by performing step resolution (and subsumption) we may resolve all such rules with other rules in R until R contains only rules of the required form.

Step resolution is now more efficient because we have Horn clauses on the right hand side of rules so strategies such as unit resolution may be applied. Unit resolution is complete for classical propositional Horn formulae (Wos et al., 1964), and algorithms have been developed that test for satisfiability of classical propositional Horn formulae based on unit resolution in polynomial time (Jones and Laaser, 1977) . Thus although we must repeatedly perform classical style resolution proofs on the right hand side of rules (because when $\bigcirc$**false** is detected the rule is rewritten) the resolution steps to derive each $\bigcirc$**false** on the right hand sides of rules will be more efficient as less intermediate steps need to be carried out. So step resolution will be more efficient.

Loop search is also more efficient as loops consist of sets of merged rules with conjunctions of propositions on the left and right hand sides. Carrying out Breadth-First Search will be more efficient for the following reasons.

1. From Lemmas 1 and 2 we know that for any loop in l we can find a more general loop that is constructed from merging rules with single propositions on the right hand side. Assuming all unit resolution possible has been carried out, in most cases the number of rules that loop search is applied to is reduced as we can ignore any always rules that do not fit into this class. To apply the Breadth-First Search algorithm we must find merged rules that satisfy certain criteria. In the worst case we must check all possible combinations of (always) rules, i.e. for m rules there are 2^m merged rules. Thus we reduce m. In particular, given n propositions there are 2^n possible left hand sides of rules in the required format. Similarly there are $2^n - 1$ different right hand sides that are disjunctions of negated propositions and $n \times 2^{n-1}$ with exactly one proposition and zero or more negated propositions. Therefore given n propositions we may have of the order of $n2^{2n-1}$ rules. As we have the R1 restriction we need only consider rules that have a single proposition on the right hand side, i.e. $n \times 2^n$ rules.

2. Given a Breadth-First Search node H_j to expand using a merged rule $A_i \Rightarrow \bigcirc B_i$ we must ensure that $\vdash B_i \Rightarrow H_j$. In the unrestricted form both these formulae are in DNF. So for n propositions we would need to check 2^n lines in a truth table to ensure validity. Using the R1 restriction B_i is a conjunction of propositions so to check the implication hold we only need to build the lines in the truth table where all the literals in B_i are set to true. Thus in the worst case where B_i is a single literal we only need to check half as many lines, i.e. 2^{n-1} lines.

3. The final governing factor in Breadth-First Search is the number of nodes that must be constructed. For unrestricted Breadth-First search the maximum will be 2^{n-1}. This can be shown by construction a (*behaviour*) graph whose nodes are valuations of the set of propositions. Paths through the graph represent all possible models of the set of always rules. The set of nodes containing the required literal, l is identified and nodes with any edges leading out of this set deleted. Nodes that lead out of the new set are again deleted until the set is empty or no more nodes can be deleted. The latter represents a loop in l and each cycle of deletions corresponds to the construction of a node in the Breadth-First algorithm. Thus the worst case is when the behaviour graph constructed is a sequence of nodes and we need 2^{n-1} deletion cycles for n propositions. For the R1 restriction this must be reduced as it is not possible to construct such a graph.

Other loop search algorithms have been given in (Dixon, 1996). One of these, Depth-First Search becomes feasible in this setting. The Depth-First Search algorithm is similar to Breadth-First Search but instead of using all the rules to expand the graph we use them one at a time. The problem is that each rule we use with two or more disjuncts on the right hand side represents branching in the graph constructed and we must check that each branch leads back into the set of nodes already constructed. As we know loops are constructed from merged rules with conjunctions of propositions on the left and right hand sides we avoid the extra search required to check each branch.

7. Conclusions

In this paper we have considered particular restrictions of the SNF normal form related to Horn clauses (and their duals). While the overall complexity of the decision problem for the logic remains the same (see §3), we have shown that both execution and resolution can be implemented more efficiently in this restricted version. In particular, the complexity of executing one step of METATEM reduces from NP-complete to PTIME, as does the complexity of deriving $P \Rightarrow \bigcirc \mathbf{false}$ in step resolution, while the Breadth-First Search algorithm used in temporal resolution produces improved efficiency at each stage. Thus, our future work concerns the implementation and testing of both execution and resolution, together with further analysis of similar restrictions. Finally, in practical systems, it is likely to be the case that both R1 and R2 restricted rules would be utilised, but that separate mechanisms would be employed on each rule type.

Acknowledgements

This work was partially supported by EPSRC under Research Grant GR/K57282 and GR/K54946.

References

M. Abadi. *Temporal-Logic Theorem Proving.* PhD thesis, Department of Computer Science, Stanford University, March 1987.

H. Barringer, M. Fisher, D. Gabbay, and A. Hunter. Meta-Reasoning in Executable Temporal Logic. In *Proceedings of the International Conference on Principles of Knowledge Representation and Reasoning (KR)*, April 1991.

H. Barringer, M. Fisher, D. Gabbay, G. Gough, and R. Owens. METATEM: An Introduction. *Formal Aspects of Computing*, 7(5):533–549, 1995.

H. Barringer, M. Fisher, D. Gabbay, R. Owens, and M. Reynolds, editors. *The Imperative Future: Principles of Executable Temporal Logics*. Research Studies Press, Chichester, United Kingdom, 1996.

C. L. Chang. The Unit Proof and the Input Proof in Theorem Proving. *ACM Journal*, 17:698–707, 1970.

C. Dixon. Search Strategies for Resolution in Temporal Logics. In *Proceedings of the Thirteenth International Conference on Automated Deduction (CADE)*, volume 1104 of *Lecture Notes in Computer Science*, New Brunswick, New Jersey, July/August 1996.

E. A. Emerson and A. P. Sistla. Deciding full branching time logic. *Information and Control*, 61:175–201, 1984.

L. Fariñas del Cerro and M. Penttonen, editors. *Intensional Logics for Programming*. Oxford University Press, 1992.

M. Finger, P. McBrien, and R. Owens. Databases and Executable Temporal Logic. In *Proceedings of the ESPRIT Conference*, November 1991.

M. Finger, M. Fisher, and R. Owens. METATEM at Work: Modelling Reactive Systems Using Executable Temporal Logic. In *Sixth International Conference on Industrial and Engineering Applications of Artificial Intelligence and Expert Systems (IEA/AIE)*, Edinburgh, U.K., June 1993. Gordon and Breach Publishers.

M. Fisher. A Resolution Method for Temporal Logic. In *Proceedings of the Twelfth International Joint Conference on Artificial Intelligence*, Sydney, Australia, 1991.

M. Fisher. A Survey of Concurrent METATEM — The Language and its Applications. In *First International Conference on Temporal Logic (ICTL)*, Bonn, Germany, July 1994.

M. Fisher. A Normal Form for Temporal Logic and its Application in Theorem-Proving and Execution. *Journal of Logic and Computation*, 7(4), August 1997.

M. Fisher and R. Owens, editors. *Executable Modal and Temporal Logics*, volume 897 of *Lecture Notes in Artificial Intelligence*. Springer-Verlag, February 1995.

D. Gabbay, A. Pnueli, S. Shelah, and J. Stavi. The Temporal Analysis of Fairness. In *Proceedings of the Seventh Symposium on the Principles of Programming Languages*, 1980.

N. D. Jones and W. T. Laaser. Complete problems for deterministic polynomial time. *Theoretical Computer Science*, 3:107–117, 1977.

R. McNaughton. Testing and Generating Infinite Sequences by a Finite Automaton. *Information and Control*, 9:521–530, 1966.

B. Moszkowski. *Executing Temporal Logic Programs*. Cambridge University Press, Cambridge, U.K., 1986.

M. Peim. Propositional Temporal Resolution Over Labelled Transition Systems. Unpublished Technical Note, Department of Computer Science, University of Manchester, 1994.

J. A. Robinson. A Machine–Oriented Logic Based on the Resolution Principle. *ACM Journal*, 12(1):23–41, January 1965.

S. Safra and M. Y. Vardi. On ω-Automata and Temporal Logic. In *STOC*, pages 127–137, Seattle, Washington, May 1989. ACM.

A. P. Sistla and E. M. Clarke. Complexity of propositional linear temporal logics. *ACM Journal*, 32(3):733–749, July 1985.

M. Y. Vardi and P. Wolper. Automata-theoretic Techniques for Modal Logics of Programs. *Journal of Computer and System Sciences*, 32(2):183–219, April 1986.

P. Wolper. The Tableau Method for Temporal Logic: An overview. *Logique et Analyse*, 110–111:119–136, June-Sept 1985.

L. Wos, D. Carson, and G. Robinson. The Unit Preference Strategy in Theorem Proving. In *Proceedings of AFIPS Fall Joint Computer Conference.* Thompson Book Co., 1964.

Specification and Prototyping of Structured Multimedia Documents using Interval Temporal Logic *

Howard Bowman (H.Bowman@ukc.ac.uk)
Computing Laboratory, University of Kent at Canterbury, Canterbury, Kent, CT2 7NF, United Kingdom

Helen Cameron (hacamero@cs.umanitoba.ca) and Peter King (prking@cs.umanitoba.ca)
Department of Computer Science, University of Manitoba, Winnipeg, Manitoba, R3T 2N2, Canada

Simon Thompson (S.J.Thompson@ukc.ac.uk)
Computing Laboratory, University of Kent at Canterbury, Canterbury, Kent, CT2 7NF, United Kingdom

Abstract. This paper explores a formalism for describing a wide class of multimedia document constraints. We outline the requirements on temporal logic specification that arise from the multimedia documents application area. In particular, we highlight a canonical document example. Then we present the temporal logic formalism that we use. This formalism extends existing interval temporal logic with a number of new features: actions, framing of actions, past operators, a projection-like operator called filter and a new handling of interval length. A model theory and satisfaction relation are defined for the logic and a specification of the canonical example is presented.

1. Introduction

This paper explores a formalism for describing a wide class of multimedia document constraints. The term *multimedia* indicates that a document may contain continuous or time-dependent entities (ISO 19744, 1992) known as *media items* (Erfle, 1993). Part of the task facing the author of such a document, therefore, is to describe the dynamic temporal relationships that are to hold between media items. We are interested in documents with rich sets of such relationships, and as a consequence are keenly interested in issues of consistency verification, modelling, proto-typing, and specification refinement. While a number of authoring systems for multimedia documents are extant (Hardman et al., 1993; Bulterman and Hardman, 1995), little investigation of suitable formalisms for such temporal constraints has been done.

* Travel grants to support the research presented here have been provided by the British Council. The second and third authors are supported by individual research grants from the Natural Sciences and Engineering Research Council of Canada.

H. Barringer et al. (eds.), Advances in Temporal Logic, 435–453.

The formalism that we introduce is an Interval Temporal Logic (ITL) specification notation called *Mexitl* (*Multimedia in Executable Interval Temporal Logic*). While the anticipated application area for this notation is multimedia, it is also relevant to other areas of real-time specification. King (King PODP, 1996; King EP, 1996) proposed the use of ITL to specify multimedia documents and the theory presented here stems from that earlier work.

A major difference between ITL and standard linear time temporal logic (Manna and Pnueli, 1992) is that it is interpreted over finite state sequences, called intervals, rather than over infinite models. A number of authors have investigated ITLs, e.g. (Moszkowski, 1986) (Hale, 1989) (Kono, 1993) (Duan, 1996). The restriction to finite intervals prompts consideration of a number of temporal operators not typically found in non-interval temporal logics, e.g. *chop* and *projection*. These turn out to be useful in the multimedia documents application domain and will be discussed in Section 3.1.

We anticipate that complete specifications of multimedia documents will have a number of elements. An *abstract data typing* notation will be used to describe the primitive *operations/actions* of a specification, such as *displayCaption* or *playVideo*. We will not consider this notation here; the specification language that we present takes the primitive actions as given. Mechanisms to define composite actions out of primitive actions can also be added.

We introduce a methodology for developing multimedia artifacts using *Mexitl*. Specifications are written in the logic, and refined according to the rules of the language. Implementations can be developed as either deterministic refinements or as proofs of *Mexitl* formulas interpreted in a constructive logic.

Structure of the Paper. The paper is structured as follows. Section 2 reviews the requirements associated with multimedia documents, and introduces a typical problem from the field. Section 3 presents the specification notation that we advocate. The operators of the language are presented, the model theory is highlighted, and the satisfaction relation is defined. Section 4 applies the defined notation to the requirements of Section 2. Related work is discussed in Section 5 and concluding remarks are presented in Section 6.

2. Multimedia Documents Requirements

Erfle (Erfle, 1993) presents a set of eighteen issues, or functional requirements, which are regarded as being sufficient to describe multimedia documents. This set was obtained by a study of what is provided in many existing authoring systems and standards. King (King PODP, 1996) presents an equivalent set of eight requirements. For the sake of completeness we will provide a summary of these requirements. We divide our summary into two, a set of general requirements, dictated by the authoring aspect of this application, followed by eight individual functional requirements.

2.0.1. *General Requirements*

We first need to represent the *display* of a media item, both *standard* display, where an item is displayed in its normal fashion at its normal rate, and *variations*, such as displaying at half speed, rewind, fast-forward. We also require facilities for both serial and parallel *composition* of sets of constraints. Parallel composition also permits independent development of *channels* (Hardman et al., 1993; Bulterman and Hardman, 1995), which may then be combined so that they occur in the same multimedia presentation. Our use of the term channel generalises its multimedia usage to *independent authorship* and is akin to the term *thread*.

2.0.2. *Functional Requirements*

The following eight individual items are required:

1. Temporal placement of a media item at an absolute (time) point;

2. Specification of the duration of a media item;

3. Determination of the start and finish points of a media item;

4. Relative placement of two or more media items;

5. Repetitive display of a media item;

6. Conditional display of a media item;

7. Scripting, that is, using events or conditions occurring in one media item to control the display of a second; and

8. Exception handling, that is, controlling error situations which may occur during the display of a multimedia document.

We illustrate these requirements by an informal specification of a fairly elaborate multimedia document, to be known as the *Beethoven Problem.* It requires the development of a multimedia document containing an audio of Beethoven's Fifth Symphony, opus 67 in c minor, together with various other media items to illustrate the music.

Beethoven's Fifth Symphony comprises four movements.

1. Play the four movements of the symphony in sequence with a gap of 20 seconds between each movement.

2. Before the symphony, play an audio which announces the name of the symphony, the composer, and the orchestra. Two seconds after this audio starts, display a video still of Beethoven. Stop the video still display as the first movement starts. After the last movement, wait 3 seconds, display a video of Ludwig van Beethoven, and then, after a further 5 seconds, display for 30 seconds information about how to order this presentation.

3. At the start of each movement display a video still of a title for 5 seconds; repeat this 5 second display every 3 minutes during the corresponding movement.

4. The audio introduced in 2 is actually in three parts, corresponding to the name of the symphony, the composer, and the orchestra. During this audio, display, in sequence, three video stills containing the same information for an appropriate time.

5. In the twenty second gap between the second and third movements, show a video/audio display describing the third movement. If this display takes longer than twenty seconds, truncate it.

6. During each crescendo passage of the first movement, display a looped video tape of a bug climbing an inclined plane.

7. Count the number of staccato notes in the first movement.

Figure 1. The Beethoven Problem.

King (King EP, 1996) presents an earlier version of a portion of this example. The seven parts of the Beethoven Problem appear in Figure 1.

3. Introduction to *Mexitl*

We present a *core language* for *Mexitl*, which contains the primitive constructs of the notation. Then we describe the model theory underlying the language; this theory is based upon finite sequences of states (called *intervals*) and we define the satisfaction relation.

3.1. The Core Language

Expressions have the following form:

$$E ::= c \mid v \mid V \mid f(E) \mid \textbf{mylen}$$

where $c \in \mathcal{N}$, $v \in \mathbf{Var}_{static}$, the set of static variables, $V \in \mathbf{Var}_{state}$, the set of state variables, and f is in a set of assumed functions. In addition, **mylen** is a distinguished variable which denotes the length of the current interval.

$P \in \mathcal{P}$ (the domain of logical propositions) is constructed as follows:

$$\begin{array}{rcl} P & ::= & a_X \mid p(E_1, ..., E_n) \mid E = E \mid \textbf{False} \mid P \Rightarrow P \mid P \; ; \; P \mid \\ & & P \;\; \textbf{proj} \;\; P \mid P \;\tilde{;}\; P \mid (\exists x \leq E)P \mid P \;\; \textbf{filter} \;\; P \end{array}$$

where $a \in$ **Act** and $X \subseteq$ **Act** is a 'framing' set; p is in a set of given predicates and E is an expression. Much of this logic will be well known to a reader familiar with interval temporal logic (Moszkowski, 1986); for instance,

- ; is the sequencing operator, *chop*, familiar from (Moszkowski, 1986). An interval satisfies $P \; ; \; Q$ if the interval can be divided into two contiguous sub-intervals, such that P holds over the first subinterval and Q holds over the second.

- **proj** is the projection operator, also described in (Moszkowski, 1986). An interval satisfies P **proj** Q if it can be sub-divided into a series of sub-intervals each of which satisfies P - we call P *the projection formula* - and a new interval formed from the end points of the sub-intervals satisfies Q, which we call *the projected formula*.

The remainder of our operators are not standard, and require more explanation.

Actions. Actions in *Mexitl* are atomic, in the sense that they cannot be analysed into simpler components. Time is discrete, and an action is thought of as taking place in a single state.

In (Bowman et al., 1997) we consider actions with data attributes; here we confine ourselves to basic actions, written a. We assume that actions are given; their definition is not part of the language. From a logical point of view we can think of an action a as an atomic proposition.

An action can appear a number of times in an interval, however, each of these represents a different instance of the action. At the level of interval states, actions do not have duration. However, durational behaviour can be obtained by defining composite actions, which are a shorthand for the occurrence of multiple primitive actions. In particular, primitive actions may correspond to indexing into a composite action. For example, an action *video*[500] (the 500th frame of the video) might be a constituent of the composite action *video*.

Although actions do not have duration, sets of (distinct) actions can occur at the same state. Such sets reflect simultaneous lock-step occurrence of the actions. In this sense, the model employs synchronous parallelism.

Framing of actions. One aspect which distinguishes our usual perception of logical propositions and actions is the idea of framing. An assertion of a, where a is a particular action, is often interpreted as '*a and no other action* happens' whereas a logical interpretation simply reads this as a happening. The former interpretation, in which the action a is "framed", would lead to a non-monotonic logic were we to adopt it.

Instead of this, in our system we subscript the actions with sets X of actions. a_X is interpreted as 'a happens and none of the other actions in X happens'. The set X thus provides an explicit frame within which the action a takes place. Logically the interpretation of a_X is the conjunction of a and $\neg b$ for all b in $X - \{a\}$. We add a distinguished action – **null** – to the set of actions. This action has null effect, but can be used for framing purposes thus: $\mathbf{null}_X$.

Length and Next. In contrast to the standard approach to ITL we have not included the next operator, $\bigcirc$, directly in *Mexitl*. However, standard length operators and $\bigcirc$ can be derived from the expression **mylen**, as follows:

$$\mathbf{len}(E) \;\equiv\; \mathbf{mylen} = E \qquad \bigcirc P \;\equiv\; (\mathbf{mylen} = 1)\ ;\ P$$

We have included **mylen** as primitive for three reasons. It is useful to be able to refer directly to the length of an interval when defining various properties of component parts, such as the two halves into which it is cut by the 'chop' operator. Moreover, we use **mylen** as a bound for existential quantifications in derivations of many temporal operators from the core language. Finally, including **mylen** allows us to avoid a 'next' operator over expressions, thus avoiding expressions which might be undefined.

Quantification. In our core language we have included a limited form of existential quantification, namely quantification in which the value of the variable is *bounded* by (the value of) an expression: $(\exists x \leq E)P$. The effect of an existential quantification is to introduce a new variable in the scope of the quantifier. A bounded quantification has the property that its satisfaction over a given interval remains decidable; this is not the case for unbounded quantification over the natural numbers.

Past Operators. We include a single past operator in *Mexitl*, chop in the past, denoted $\tilde{;}$. An implication of the inclusion of past operators is that the past history of a computation must be recorded; we will show how this is done in Section 3.2. Using such a more sophisticated model theory, $P \mathbin{\tilde{;}} Q$ is satisfied by an interval such that,

1. P holds over the larger interval resulting from moving the start of the interval into the past, and
2. Q holds over the original interval according to a past history that is truncated at the start of the interval over which P holds.

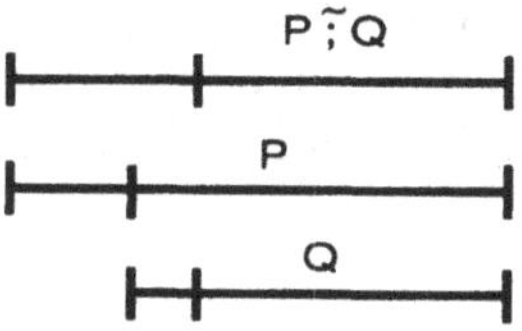

Intervals are depicted as line segments with three reference points. The leftmost point is the start of time, then moving to the right, the next point is the start of the current interval and the final point is the end of the current interval. Amongst other things, $\tilde{;}$ has the effect of chopping up the past, in a dual manner to which ; divides up the future.

The more familiar past operators, since and previous, can be derived from chop in the past as follows. $P \ \tilde{\mathcal{S}} \ Q$ is given by

$$x = \mathbf{mylen} \Rightarrow (((Q \wedge \mathbf{mylen} > x) \wedge \bigcirc\Box(\mathbf{mylen} > x \Rightarrow P)) \mathbin{\tilde{;}} \mathbf{True})$$

where $\Box$ is defined in Figure 3 and $\ominus P \ \equiv \ \mathbf{False} \ \tilde{\mathcal{S}} \ P$. In fact, $\tilde{\mathcal{S}}$ is a strong since operator. The definition of strong since enables previously to be derived; weak since can also be derived in standard fashion. The incorporation of past operators is a significant departure from standard interval temporal logic. The motivation for their inclusion is that the specification of a number of examples is made substantially easier.

Filter. In filtering we have a variant of projection (**proj**). P **filter** Q has the effect of selecting all points which begin an interval over which the property P holds; the formula holds if over this new interval the property Q is true. In the following illustrative figure, propositions are associated with the initial point of the interval over which they hold.

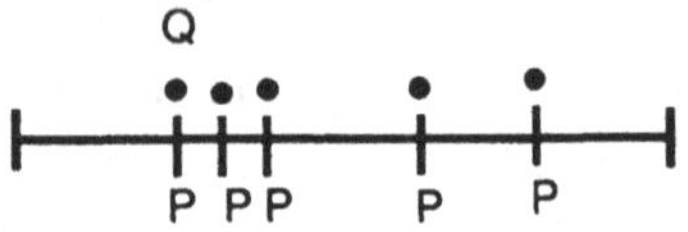

filter generalises **when** as defined in (Hale, 1989); in Hale's definition (based on **proj**) the property P needs to be a local property, that is, one without temporal operators. We see the generalisation as valuable – it allows us to express properties of the form 'whenever a musical performance is within 5 minutes of the end of a movement, show some more of a video sequence' (see for example Part 6 of the Beethoven example presented in Section 2) – but we see no way of deriving it from the other *Mexitl* operators.

Types and definitions. We use a simple mechanism for naming types and values in the remainder of the paper. Note that '::' is used to mean 'is of type', and that definitions of types and values are not allowed to be (mutually) recursive.

3.2. Satisfaction

The order of presentation of this section is as follows. First we describe *computation structures*, which enhance intervals in order that past operators can be handled. Then we consider the interpretation of expressions and finally, we describe the satisfaction relation for the main temporal operators.

The major difference between our interpretation of *Mexitl* and the standard interpretations of interval temporal logics, to be found in (Moszkowski, 1986) (Hale, 1989), for example, is that our satisfaction relation embraces past operators. Consequently, the sequence of states that have already been passed through must be recorded. This record is obtained by defining our satisfaction relation over what we call *computation structures*. These computation structures are pairs, (σ, i) where, in the normal way, σ is a sequence of states, with length $|\sigma|$, i.e., $\sigma = \sigma_0, \sigma_1, ..., \sigma_{|\sigma|}$ (we refer to this as the interval) and i is the *point of reference*. The point of reference identifies the past states of the computation (the *history*) and the interval under current consideration (the *current interval*). We introduce notation for prefix and suffix of intervals,

$$[\sigma]^i = \sigma_0, ..., \sigma_i \qquad (\sigma)^i = \sigma_i, ..., \sigma_{|\sigma|}$$

Since we have actions, our states are slightly more sophisticated than those of standard interval temporal logic. Specifically, each state σ_i is a pair; the first component of the pair A_i is a set of actions and the second component D_i records the current data state; it is a finite function from variables to data values (which come from $\mathcal{N}$, the natural numbers).

Expressions are interpreted in a standard manner apart from the operator **mylen**, which is interpreted as follows:

$$[\![\mathbf{mylen}]\!]_{(\sigma,i)} = |\sigma| - i$$

Our satisfaction relation, $\models$, interprets *Mexitl* propositions over computation structures. The notation $(\sigma, i) \models P$ denotes that the computation structure (σ, i) satisfies the proposition P. The most important clauses of the definition of satisfaction are in Figure 2; others which are standard can be found in (Bowman et al., 1997).

Note that we say that $\sigma' \simeq_{x,E} \sigma$ if σ and σ' are the same length and they have the same value on all actions and variables except (perhaps) x, and that at each point the value of x is less than or equal to the corresponding value of E.

An arbitrary *Mexitl* proposition is interpreted relative to a zero point of reference. Thus, we define that an interval σ *satisfies* a proposition P if and only if $(\sigma, 0) \models P$. In addition, in the usual way, we state that P is *valid* if and only if for all σ in $\mathcal{I}$, $(\sigma, 0) \models P$, where $\mathcal{I}$ denotes the set of all possible intervals. If P is valid we write $\models P$.

(Bowman et al., 1997) defines a full set of ITL derived operators from *Mexitl*. Due to space considerations we can only include the operators that we use in the Beethoven example of Section 4. These operators are presented in Figure 3.

3.3. Reasoning about *Mexitl*

We have developed a range of proof rules for the *Mexitl* logic. Details of these are given in (Bowman et al., 1997). The rules include

- characterisations of the basic operators of the language, such as

$$[P6] \quad ((P \textbf{ proj } R) \vee (Q \textbf{ proj } R)) \Rightarrow ((P \vee Q) \textbf{ proj } R)$$

- derivation of common theorems of temporal logic from the definitions of operators given in (Bowman et al., 1997);

- a characterisation of the framing of actions discussed earlier. From an action a_X we can deduce a and $\neg b$ for all b in $X - \{a\}$; the occurrence of *two* actions in X simultaneously results in a contradiction, and hence an unimplementable specification.

$(\sigma, i) \models a_X$	iff	$a \in A_i$ and $\forall x \in X - \{a\}\ x \notin A_i$
$(\sigma, i) \models P_1; P_2$	iff	$\exists k \in \mathcal{N}\ (i \leq k \leq \|\sigma\|$ and $([\sigma]^k, i) \models P_1$ and $(\sigma, k) \models P_2)$
$(\sigma, i) \models P_1$ **proj** P_2	iff	$\exists m \in \mathcal{N}$ and $\exists \tau_0, \tau_1, ..., \tau_m \in \mathcal{N}$ $(i = \tau_0 < \tau_1 < ... < \tau_m = \|\sigma\|$ and $\forall j < m\ (([\sigma]^{\tau_{j+1}}, \tau_j) \models P_1)$ and $([\sigma]^{i-1}.\sigma_{\tau_0}\sigma_{\tau_1}...\sigma_{\tau_m}, i) \models P_2)$
$(\sigma, i) \models P_1 \ \tilde{;}\ P_2$	iff	$\exists k\ (0 \leq k \leq i$ and $(\sigma, k) \models P_1$ and $((\sigma)^k, i - k) \models P_2)$
$(\sigma, i) \models (\exists x \leq E)P$	iff	$(\sigma', i) \models P$ for some $\sigma' \simeq_{x,E} \sigma$
$(\sigma, i) \models P_1$ **filter** P_2	iff	$\exists \tau_0, \tau_1, ..., \tau_m \in \mathcal{N}$ $i \leq \tau_0 < \tau_1 < ... < \tau_m \leq \|\sigma\|$ and $\forall j < m\ ((\sigma, \tau_j) \models P_1)$ and $\forall k, i \leq k \leq \|\sigma\| \wedge (\sigma, k) \models P_1$ implies $k = \tau_j$ for some j and $([\sigma]^{i-1}.\sigma_{\tau_0}\sigma_{\tau_1}...\sigma_{\tau_m}, i) \models P_2)$ or no points satisfy P_1

Figure 2. Satisfaction for *Mexitl*

The rules have a number of purposes:

- Implication formalises refinement, so that a rule such as $[P6]$ shows that a sound refinement of $(P \vee Q)$ **proj** R is given by a choice between $(P$ **proj** $R)$ and $(Q$ **proj** $R)$.
- We can use the rules to reason about specifications.
- Using the rules we can aim to find a normal form for *Mexitl* formulas (along the lines of Gabbay's work (Gabbay, 1989)) which may form the basis for an implementation of the language.

$$
\begin{array}{rclrcl}
\Diamond_t P & \equiv & \mathbf{mylen} = t \,;\, P & \Diamond P & \equiv & \mathbf{True} \,;\, P \\
\Box P & \equiv & \neg\Diamond\neg P & {\Diamond\mkern-12mu{\scriptscriptstyle i}\mkern6mu}_t P & \equiv & P \,;\, \mathbf{mylen} = t \\
{\Diamond\mkern-12mu{\scriptscriptstyle i}\mkern6mu} P & \equiv & P \,;\, \mathbf{True} & {\Diamond\mkern-12mu{\scriptscriptstyle i}\mkern6mu}_X P & \equiv & P \,;\, \mathbf{null}_X^* \\
{\Diamond\mkern-12mu{\scriptscriptstyle i}\mkern6mu}_{\leq t} P & \equiv & P \,;\, \mathbf{mylen} \leq t & {\Diamond\mkern-12mu{\scriptscriptstyle a}\mkern6mu} P & \equiv & \mathbf{True} \,;\, P \,;\, \mathbf{True} \\
P^* & \equiv & P \;\; \mathbf{proj} \;\; \mathbf{True} & & &
\end{array}
$$

$$
\begin{array}{rcl}
\mathbf{beg}\, P & \equiv & (\mathbf{len}(0) \;\wedge\; P) \,;\, \mathbf{True} \\
\mathbf{halt}\, P & \equiv & \Box(P \Leftrightarrow \mathbf{len}(0)) \\
\mathbf{halt}_X\, P & \equiv & \Box(P \Leftrightarrow \mathbf{len}(0) \;\vee\; \mathbf{beg}(\mathbf{null}_X)) \\
\mathbf{fin}\, P & \equiv & \Box(\mathbf{len}(0) \Rightarrow P)
\end{array}
$$

$$
\begin{array}{l}
\mathbf{for}\; i := 1 \;\mathbf{to}\; E \;\mathbf{do}\; P \;\equiv \\
\quad i := 1 \,;\, ((P \;\wedge\; i \leftarrow i+1 \;\wedge\; i \leq E)^* \;\wedge\; \mathbf{fin}(i > E)) \\
P \;\mathbf{when}\; Q \;\equiv \\
\quad \neg\Diamond Q \;\vee\; (\mathbf{halt}(Q) \,;\, (\bigcirc\mathbf{halt}(Q) \;\; \mathbf{proj} \;\; P) \,;\, \mathbf{keep}\bigcirc(\neg Q))
\end{array}
$$

Figure 3. Derived Operators

We do not claim that the logic we present is complete; we know of no complete axiomatisation of interval temporal logic with projection. Kono claims such an axiomatisation in unpublished work, but we have discovered that one of the rules in that system is unsound.

4. Applying *Mexitl*

We show how *Mexitl* can be employed for multimedia documents. We first describe a development methodology for *Mexitl* and then we show how the individual functional requirements are met by providing a complete formal specification of the Beethoven problem of Figure 1.

4.1. The *Mexitl* Development Methodology

The *Mexitl* interval temporal logic is designed to support the specification, design, prototyping and implementation of multimedia systems. Various aspects of the system are useful in different ways.

- The logical rules mentioned in Section 3.3 and (Bowman et al., 1997) underpin a *refinement* strategy for specification development. A specification S_1 is refined by the specification S_2 if S_2

implies S_1, that is if $S_2 \Rightarrow S_1$ is valid. For instance, if $P' \Rightarrow P$ and $Q' \Rightarrow Q$ then

$$P' \textbf{ proj } Q' \Rightarrow P \textbf{ proj } Q$$

which states that we can refine a projection by refining either of its constituent formulae. We might, for instance, be more specific about the subdivision of an interval in a projection and replace **True** by **len**(2) in **True proj** Q, which is acceptable since **len**(2) $\Rightarrow$ **True**. In a similar way we can refine a specification by conjoining other constraints into it. This meets our formal definition of refinement since $A \wedge B \Rightarrow A$ for any A and B.

– A specification written in *Mexitl* can be prototyped in two ways.

  - A general interval formula can be viewed as a *non-deterministic* description of an interval, since in general it describes a collection of intervals rather than a single interval. A refinement of a specification will be more deterministic, in that it describes fewer intervals, and we can thus see the process of refinement as moving towards a deterministic specification implementable in standard ways.
  - Alternatively, taking a *constructive* view of logic as expounded in (Thompson, 1991), we can view formulas of temporal logic as specifications with *proofs* being implementations. Preliminary work in this direction, which implements interval temporal logic in the Alf system, is reported elsewhere (Thompson, 1997).

– As our first point suggested, we can build refinements of specifications by means of conjunction: we can think of this as a *parallel composition*, in contrast to the sequential composition given by chop.

  If our actions are framed by giving an explicit framing set, as in a_X, then actions provide a means of *synchronising* between different conjuncts. For example in,

$$(a_X \ ; \ b_X \ ; \ c_X) \ \wedge \ (\textbf{True} \ ; \ b_X \ ; \ \textbf{True})$$

  where the framing set X is $\{a, b, c\}$, the only interval satisfying this formula must be one in which the b action in the second conjunct happens at the same point as the b action in the first conjunct. We achieve this synchronisation without adding any specific machinery for that purpose. We are also actively exploring the role that framing plays in the separate development of aspects of multimedia documents.

- We also conclude from our exploration of the sequence of examples in Section 4.2 that the *Mexitl* language is a low-level mechanism for representing such presentations; we aim to investigate higher-level approaches for which *Mexitl* as presented here might play the role of an intermediate language.

We introduce a methodology for developing multimedia artefacts using *Mexitl*. Specifications are written in the logic, and refined according to the rules of the language. Implementations can be developed as either deterministic refinements or as proofs of *Mexitl* formulas interpreted in a constructive logic.

4.2. Specification of Beethoven Example

We begin with a type definition of a symphony with four movements.

type symphony4 = (movement, movement, movement, movement)
type movement = interval – a finite interval of actions
m_1, m_2, m_3, m_4 :: movement

These definitions allow us to use m_1, m_2, etc. both as intervals in *Mexitl* formulae, and also as sequences of primitive actions into which we can index. We distinguish two elements of these sequences, $m_i[1]$ and $m_i[m_i.\text{last}]$, where last is a predefined constant stating how many actions there are in such a sequence. The constants m_1, m_2, m_3 and m_4 are assumed to be set to the contents of the corresponding movements. In addition, the association of a framing set with a composite action, e.g. $m_{2\,X}$, implicitly frames all the primitive actions involved in the composite action, e.g. $m_2[5]_X$.

We now give a *Mexitl* specification of each part of the Beethoven problem as given in Figure 1.

1. In this specification, we frame each movement against any (other) actions from the set of actions comprising the four movements. The bounded temporal operator $\Diamond_{20''}$ is also framed to prevent any of the actions M from any of the movements from playing during the 20 second intervals.

$$S_1 \equiv m_{1\,M} \; ; \; (\textbf{for } i := 2 \textbf{ to } 4 \textbf{ do } \Diamond_{20''\,M} \, m_{i\,M})$$

   where M is the set of all all audio actions forming m_1, m_2, m_3 and m_4.

2. We compose the items at the beginning and end of the presentation in a serial fashion and express the first of these as a parallel

composition of two sub-channels. The video still itself is specified as the repetition (*) of an elementary one-state action displaying a picture of Beethoven.

type still = action ; audio, video = interval
a_1 :: audio ; Ludwig, orderinginfo :: still ; lvb :: video

$$\begin{aligned} S_2 \quad \equiv \quad & (a_1\,{}_A \;\wedge\; (\Diamond_{2''}\,{}_S\text{Ludwig}^*_S)) \;\;;\;\; S_1 \;\;; \\ & (\Diamond_{3''}\,{}_V\text{lvb}_V) \;;\; (\Diamond_{5''}\,{}_S(\text{OrderingInfo}^*_S \;\wedge\; \mathbf{len}(30''))) \end{aligned}$$

where A is the set of audio actions, S is the set of still image actions, and V is the set of video image actions.

3. For each of the four movements we synchronise with the start and finish of each movement m_i using $\mathbf{halt}(m_i[1])$ and $\mathbf{halt}(m_i[m_i.\text{last}])$. Actions are framed against the corresponding class of actions so as to avoid unwanted behaviour. Note that the earlier framing of actions in S_1 ensures, for example, that the occurrence of the first action in m_1 is uniquely defined.

   t_1, t_2, t_3, t_4 :: still – the four titles

$$\begin{aligned} R_{=5''} \quad & \equiv \quad (t^*_{i\,T} \;\wedge\; \mathbf{len}(5'')) & \text{– repeat title for exactly 5 sec} \\ R_{<5''} \quad & \equiv \quad (t^*_{i\,T} \;\wedge\; \mathbf{less}(5'')) & \text{– repeat title for less than 5 sec} \\ S_3 \quad & \equiv \quad S_2 \;\wedge\; \Diamond_T\ \mathbf{for}\ i := 1\ \mathbf{to}\ 4\ \mathbf{do} & \\ & \qquad (\ \mathbf{halt}_T(m_i[1]) \;;\; (\mathbf{halt}(m_i[m_i.\text{last}]) & \\ & \qquad \wedge\; ((\Diamond_{2'55''}R_{=5''})^* \;;\; (R_{<5''} \;\vee\; (\Diamond_{\leq 2'55''}R_{=5''})))\ )\ ) & \end{aligned}$$

   where T is the set $\{t_1, t_2, t_3, t_4\}$.

4. The display of each still is synchronised with the end of the corresponding audio display. The framing here is more subtle. We cannot compose with the original S_2 in Example 2 above, since we deliberately framed S_2 against any other video stills occurring in that prefix interval. Hence, we have provided an alternative S_2' without framing on the item Ludwig.

   $a_1,\ a_2,\ a_3$ :: audio ; $f_1,\ f_2,\ f_3$ :: still

$$\begin{aligned} S_2' \quad & \equiv \quad ((\mathbf{for}\ i := 1\ \mathbf{to}\ 3\ \mathbf{do}\ a_{i\,A}) \;\wedge\; \Diamond_{2''}\text{Ludwig}^*) \;;\; \cdots \\ S_4 \quad & \equiv \quad S_3 \;\wedge\; \Diamond\ \mathbf{for}\ i := 1\ \mathbf{to}\ 3\ \mathbf{do}\ (f_i^* \;\wedge\; \mathbf{halt}(a_i[a_i.\text{last}])) \end{aligned}$$

5. We synchronise the display of the audio-video with an interval where the state previous to the first contains the last action of movement 2, and end the display when the next state contains the

first action of the third movement. In the 20 second interval we either play the entire audio-video, which, if it is less than 20 seconds, will require a (strictly) initial interval, or play as much of it as we can, using a for loop with a bounded existential quantifier.

type videoaudio = interval
va :: videoaudio

$$S_5 \equiv S_4 \wedge \diamondsuit_{\{\text{va[i]}\}}(\mathbf{beg}(\ominus\, m_2[m_2.\text{last}]) \wedge \mathbf{halt}(\bigcirc m_3[1]) \wedge ((\diamondsuit\, \text{va}_{\text{VA}}) \vee (\exists\, t \leq \text{va.last})\ \mathbf{for}\ i\ :=\ 1\ \mathbf{to}\ t\ \mathbf{do}\ \text{va}[i]_{\text{VA}}))$$

where {va[i]} is the set of all actions corresponding to va and VA is the set of all video/audio actions.

In Examples 6 and 7, we have chosen to omit all considerations of framing as they would confuse matters unduly. We assume that the condition staccato corresponds to a state variable whose value is supplied by the application. We assume that each note in the music occupies exactly one state in the interval.

6. C tests if we are in a crescendo by comparing the current volume with the volume in the previous and next states. The filter operation allows us to concatenate the crescendo passages as a single interval, over which we then play the video (of the climbing bug) as often and for as long as we can.

$$C \equiv (\exists x \leq \text{maxVol})((\text{vol} = x) \wedge (\ominus(\text{vol} < x) \vee \bigcirc(\text{vol} > x)))$$
$$S_6 \equiv S_5 \wedge \diamondsuit\ (\mathbf{beg}(\ominus\, a_3[a_3.\text{last}]) \wedge \mathbf{halt}(\bigcirc\, m_1[m_1.\text{last}]) \wedge (C\ \mathbf{filter}\ (\text{video}^*\,;(\exists\, t \leq \text{video.last})\ \mathbf{for}\, i\ :=\ 1\ \mathbf{to}\, t\, \mathbf{do}\ \text{video}[i])))$$

7. Point scripting conditions are specified with **when**. The definition of **gets** can be found in (Bowman et al., 1997).

$$S_7 \equiv S_6 \wedge \diamondsuit\ (\mathbf{beg}(m_1[1]) \wedge \mathbf{halt}(m_1[m_1.\text{last}]) \wedge (\exists\, x \leq (\mathbf{mylen} + 1)((x = 1) \wedge (x\ \mathbf{gets}\ x + 1))\ \mathbf{when}\ \text{staccato}\ )$$

4.3. Serial Composition and Past Operators

In the foregoing we have commented on the issues associated with framing and parallel composition. These issues do not arise in the case of serial composition using the ; operator. Assuming P_A does not contain past operators, P_A implies nothing about the framing of Q in the formula P_A ; Q. Further, the past operators in the formalism allow a coauthor to create piece a of a document a ; b without providing information required by the author of piece b. For example, suppose that a definition P of "pianissimo" should be displayed exactly once. Instead of a having to signal b in some way whether or not a presents P, the author of b could specify **if** $\neg \diamond P$ **then** P.

As a further example, the author of b could use $\mathcal{S}$ (since) to specify ($\neg$ off) $\mathcal{S}$ on to determine whether or not a map display was left on by a.

$\ominus$ is useful for finding a transition point, that is, a point at which some property P becomes true: $\ominus\neg P \wedge P$. Example 6 above provides a specific example.

5. Related Work

There has been little previous work on applying temporal logic to the field of electronic publishing. King (King PODP, 1996; King EP, 1996) justify the use of ITL in this area and present examples. Furuta and Stotts (Stotts et al., 1997; Stotts et al., 1992) make use of a temporal logic for specifying and tracing dynamic link paths in hypertext documents. Their logic is somewhat different from ours, and does not require, for example, any notion of projection, and their application is also rather different.

In contrast, interval temporal logic has a relatively extensive history, e.g. (Moszkowski, 1986; Hale, 1989; Kono, 1993). Our work builds upon this body of literature: *Mexitl* is derived from the ITL notation defined in (Moszkowski, 1986); we have used a number of the derived ITL operators defined in (Hale, 1989) and our proof theory is related to that of (Kono, 1993).

However, in terms of obtaining executability we have taken a rather different approach to other researchers. Ostensibly there are two approaches to obtaining executability. Firstly, a restricted logic (without operators such as eventually) can be used in order that specifications are deterministic and characterise just a single model. This is the approach employed by (Moszkowski, 1986). Alternatively, a richer set of logical operators can be allowed resulting in non-deterministic specifications having to be accommodated. This is the approach employed by (Kono, 1993) and, to take a non interval temporal logic example, by (Barringer et al., 1989). However, the consequence of such an approach is that more complicated execution strategies must be considered, in particular, backtracking must be employed. In contrast to these two alternatives, we are considering multi-levelled development strategies in which abstract *Mexitl* specifications are refined into concrete specifications. In general terms, the abstract specifications will be expressed in the full logic, while refinement entails evolving the specification towards a deterministic executable form. We are also exploring a constructive approach to this issue (Thompson, 1997).

Duan's (Duan, 1996) approach to interval temporal logic is the closest to ours. Duan's work extends Moskowski's interval temporal logic in a number of respects: (1) past operators are added, (2) a new projection operator is defined, (3) framing of variables is investigated, (4) infinite models are incorporated and (5) concurrency and communication primitives are considered. The last two of these are beyond the scope of this paper, however, the other extensions are of interest; we consider each in turn.

1. *Past Operators.* Firstly, we have incorporated past operators for somewhat different reasons to Duan. He is motivated by the desire to give an operational definition of assignment in the presence of framed variables. In contrast, our interest in past operators has arisen because they simplify certain of our example specifications. Importantly though, all the past operators that Duan uses can be defined from our chop in the past operator which, it should be pointed out, is different from the operator *Chopp* that Duan uses. These derivations are included in (Bowman et al., 1997).

   In addition, Duan's chop (in the future) behaves differently to our chop in the presence of past operators in its second argument. Our reason for defining chop in the way we have is to fit with our motivatory examples and to allow $\Diamond$ and $\Box$ to be defined from it in the usual manner. Duan loses this interderivability. In addition, Duan's chop is derivable from ours as demonstrated in (Bowman et al., 1997).

2. *Projection.* Duan defines a new projection operator, denoted *prj*. His motivation for defining this new operator is in order to model concurrency and interaction. However we have remained faithful to the original projection operator of Moskowski (Moszkowski, 1986) which seems much more applicable to our application area. Furthermore, we have discovered how to derive Duan's *prj* operator from the standard projection. This derivation is also included in (Bowman et al., 1997).

3. *Framing of Variables.* In the context in which we are working this has not been important; we have however introduced a notion of framing for actions, without using a non-monotonic logic.

6. Concluding Remarks

Our paper represents the first step in a programme of work looking towards formally based specification, refinement, prototyping and implementation of multimedia systems using interval temporal logic. We have shown how the *Mexitl* formalism is given a semantics; we have highlighted the fundamentals of a proof theory and we have shown how to use the formalism in the development of a substantial example. This is built in a series of steps using the operations of the logic to combine parts of the specification into a coherent whole.

For the future, we intend to investigate in a number of directions. We aim to develop the logic of *Mexitl*, working towards a system which is complete (relative to the theory of Peano arithmetic). We will also look further at a constructive implementation of *Mexitl*; the beginnings of which are reported in (Thompson, 1997).

The methodology supported by *Mexitl* is also a topic of active interest for us: we aim to look further at the interaction between conjunction, framing and the separate development of channels in presentations. This will come from the investigation of further case studies as well as from looking at the theoretical issues involved. In particular, we aim to study higher-level languages for multimedia specification and their translation into the *Mexitl* logic.

References

H. Barringer, M. Fisher, D. Gabbay, G. Gough, and R. Owens. METATEM : A framework for programming in temporal logic. In *Lecture Notes in Artificial Intelligence, vol. 430*. Springer–Verlag, 1989.

H. Bowman, H. Cameron, P. King, and S. Thompson. *Mexitl*: Multimedia in Executable Interval Temporal Logic. Technical Report 3-97 (Kent), Computing Laboratory, University of Kent, 1997.

D.C.A. Bulterman and L. Hardman. Multimedia authoring tools: State of the art and research challenges. In *Computer Science Today: Recent Trends and Developments*, LNCS, No. 1000, pages 575–591. Springer-Verlag, 1995.

Z. H. Duan. *An Extended Interval Temporal Logic and A Framing Technique for Temporal Logic Programming*. PhD thesis, Univ. of Newcastle Upon Tyne, 1996.

R. Erfle. Specification of temporal constraints in multimedia documents using hytyime. *Electronic Publishing*, 6(4):397–411, December 1993.

D. Gabbay. The declarative past and imperative future. In *Temporal Logic in Specification*. LNCS 389, Springer-Verlag, 1989.

R. Hale. Using temporal logic for prototyping: the design of a lift controller. In *Lecture Notes in Computer Science, vol. 379*, pages 375–408. Springer–Verlag, 1989.

L. Hardman, G. van Rossum, and D.C.A. Bulterman. Structured multimedia authoring. *ACM Multimedia*, pages 283–289, 1993.

ISO 19744. *Information Technology – Hypermedia/Time-based Structuring Language (HyTime)*, 1992.

P.R. King. A logic based formalism for temporal constraints in multimedia documents. In *PODP 96*, September 1996. Revised version to appear in LNCS, Springer Verlag.

P.R. King. Modelling multimedia documents. *Elec. Pub.*, 8(2, 3):95–110, 1996.

S. Kono. A combination of clausal and non-clausal temporal logic programs. In *Lecture Notes in AI, vol. 897*, pages 40–57. Springer–Verlag, 1993.

Z. Manna and A. Pnueli. *The Temporal Logic of Reactive and Concurrent Systems.* Springer-Verlag, 1992.

B. Moszkowski. *Executing Temporal Logic.* Cambridge University Press, 1986.

P.D. Stotts, R. Furuta, and J.C. Ruiz. Hyperdocuments as automata: Trace-based browsing property verification. In *Proc. ACM Conf. on Hypertext*, pages 272–281. ACM Press, 1992.

P.D. Stotts, R. Furuta, and J.C. Ruiz. Hyperdocuments as automata: Verification of trace-based browsing properties by model checking. *ACM Transactions on Information Systems*, 1997. To appear.

S. Thompson. *Type Theory and Functional Programming.* Addison-Wesley, 1991.

S. Thompson. Constructive interval temporal logic in Alf. In *this volume*, 1997.